H. Trevor Clifford
Peter D. Bostock
Etymological Dictionary of Grasses

H. Trevor Clifford
Peter D. Bostock

Etymological Dictionary of Grasses

Springer

Authors

PROF. DR. H. TREVOR CLIFFORD

Queensland Herbarium
Brisbane Botanic Gardens Mt Coot-tha
Mt Coot-tha Rd
Toowong, Queensland 4066
Australia

PETER D. BOSTOCK

– Principal Botanist –
Queensland Herbarium
Environmental Protection Agency
Mt Coot-tha Rd
Toowong, Queensland 4066
Australia

Cover photo: *Pennisetum villosum* R.Br. ex Fresen., a species widely distributed in Australia. The photo was kindly provided by Bryan Simon, Queensland Herbarium.

ISBN 978-3-642-07245-1 e-ISBN 978-3-540-38434-2

Springer is a part of Springer Science+Business Media
springer.com

Editor: Dr. Dieter Czeschlik, Heidelberg, Germany
Desk Editor: Dr. Jutta Lindenborn, Heidelberg, Germany
Cover design: WMXDesign GmhH, Heidelberg, Germany

31/3100/YL – 5 4 3 2 1 0 – Printed on acid-free paper

For Gill and Pat

Preface

As employed here the term *grass* applies only to species included in the Poaceae, one of the largest families of flowering plants. However, the word is often applied to any herbaceous plant with long, narrow leaves. A similar view was adopted by the Ancients. The Greeks applied the words *poa*, *poe* and *agrostis* to herbaceous plants in general and the Romans employed the words *gramen* and *herba* in a similar sense. In both cultures, unique names were applied to species of economic or special significance.

As a major source of cereals, pasture plants and even timber, the Poaceae are one of the most important economic plant families. Many have acquired vernacular names but these vary from place to place and so are of limited value for technical purposes. For ease of professional communication vernacular names are replaced with binomials whose use is controlled by an International body.

Initially the binomials were derived mainly from words of Classical Greek or Latin, but the practice was never strictly enforced. Today taxonomists often employ words from their own language or resort to naming grasses after places, people, ships, uses, acronyms to name but a few sources. In the process the names are often Latinized making it difficult for readers, especially those whose language has not been influenced by the European Classics, to recognize their sources and to determine their meanings.

Because it is usually easier to remember a technical name once its meaning is known, the authors hope this work will be of value to ecologists, agronomists and others not primarily interested in grass taxonomy. Those who are concerned can always consult the scientific literature. Even so, without the resources of a large library, determining the origins and meanings of many binomials is impossible. This situation is changing rapidly with the advent of the Internet and the ever increasing amount of information that is available in the public domain. None-the-less, searching the Internet is time consuming and a single reference such as that presented here may be helpful to professional biologists and others interested in the origins of names.

The entries herein include most of the names published during the past 250 years but the seemingly endless torrent of new names being proposed means that a few of the more recent have been overlooked. Some of the older names for which no interpretation was given with the original description have been omitted and the majority of misspellings have been ignored.

The work presented below is based on a previously published dictionary[1] but has been completely revised and expanded. There are about 12 500 entries and the authors apologize to any reader who searches in vain for a name not included.

Trevor Clifford and Peter Bostock
Queensland Herbarium, April 2006

[1] Clifford HT (1996) *Etymological dictionary of grasses* (World Biodiversity Database CD-ROM Series). Joint publication of ETI Expert Center for Taxonomic Identification, Amsterdam, The Netherlands and Springer Berlin Heidelberg New York

Acknowledgements

The authors are particularly grateful to Halina Winters, Librarian, Queensland Herbarium, for her unstinting and cheerful assistance in locating literature, for providing valuable insights into Slavic languages particularly Polish, and for searching the Internet for biographical and geographical information. Meg Lloyd, Librarian, Queensland Museum has also been helpful in searching out obscure references especially to fossil grasses. Our colleagues Bryan Simon and Daniel Healy have taken an active interest in the project throughout, drawing attention to omissions and errors in the text as it evolved. Bryan also provided the photograph of *Pennisetum villosum* which appears on the cover and which was recommended to us by Will Smith, Botanical Illustrator at Queensland Herbarium. Finally it is our pleasure to acknowledge our indebtedness to Jutta Lindenborn for her wise counsel and ongoing support.

Contents

Introduction

Purpose of Nomenclature

For most societies plants are major sources of food, medicine and other essential products and so over time each species has acquired a name which may differ from place to place making it difficult for potential users to share their knowledge. Hence it is not surprising that one of the earliest botanical records is a list of medicinal plants from the Euphrates Valley along with their equivalent names in the Nile Valley. The need for such lists arises whenever the same species is known by different names in the same or different places.

Likewise, it is important to be aware that when the same name is applied to different species failure to distinguish between the two may have disastrous consequences. For example, in Australia, *Solanum nigrum* is widely known as Deadly Nightshade, notwithstanding that its ripe berries are not poisonous (Everist 1979) and often eaten. In contrast eating the similar looking berries of the English species known as Deadly Nightshade (*Atropa belladonna*) could be fatal.

To avoid the confusion that can result from a species having more than one name, or different species having the same name, an international system of nomenclature has been devised. Because the originators of the system were Europeans, for whom the language of scholarship was Latin, the Swedish author of the pioneer text on the subject (Linnaeus 1753) wrote in that language. The Latin he employed, specially for describing plants, differed considerably from that of Classical Times and like Ecclesiastical Latin is always evolving with an expanding vocabulary to account for new structures and ideas. This subject is admirably dealt with by Stearn (1992) who in his "Botanical Latin" discussed many of the problems associated with the formation of binomial names and provided a synopsis of the views propounded by Linnaeus (1753) on the subject.

In "Species Plantarum" Linnaeus assigned every species to a genus and each was briefly described. In the margin beside each description was a single italicized word which usually referred to some salient feature of the species. This word together with the generic name became known as a binomial. Today the binomial is the basis of the nomenclature by which all species are known internationally.

The application of these names is controlled by an "International Code of Botanical Nomenclature" (Greuter 2000) which is subject to periodic revision. The objective of the Code, which encompasses all taxonomic ranks up to and including Family, is to stabilize nomenclature so that each plant has only one name, thereby making it easier to search the literature for information concerning taxa and especially species.

Since the generic and specific names derive from many sources their meanings are often difficult to determine, unless the reader has access to a large botanical library.

Language of Nomenclature

Though not the first botanist to employ the concept of binomial names to plants, Linnaeus produced the first world flora wherein he gave generic and specific names to all the flowering plants of which he was aware. In the binomial, the generic name precedes the specific and is always written with an initial capital letter. The initial letter of the specific name is nowadays written with a lower case initial but in former times it was customary to capitalize the initial letter if the species was named after a proper noun such as a person or a genus.

Binomials must be written using the letters of the Latin alphabet and are treated as a shorthand version of a sentence in that language. Accordingly, the grammatical rules of Latin are followed, treating the genus as a noun and the species as either an adjective or a noun. If the specific name is an adjective it will agree in gender with that of the genus.

Because the Code was formalized relatively late in the history of taxonomic botany, strict application of its recommendations may result in minor changes to the spelling of older names. For example, prior to the acceptance of early versions of the Code it was not uncommon for botanists to emend spellings on quite arbitrary grounds, which usually reflected how the writer chose to transliterate the spelling, into the Latin alphabet, of words from other languages.

For example, Jacquemont in 1809 coined the generic name *Dinebra* basing it on the Arabic vernacular name of the type species; a few years later in 1830 Presl emended the spelling to *Dineba* claiming his transliteration of the Arabic to be better than that of Jacquemont. Whether or not he was correct, the Rule of Priority established by the Code demands that the original spelling of the name must be accepted, unless a compelling case can be made to the contrary.

With transliteration from Greek to Latin the problem is complicated by a lack of consistency amongst scholars from different countries. When Loureiro proposed the generic name *Rhaphis* (1790) he transliterated the Greek *rho* (ρ) as 'rh' whereas twenty-two years later Palisot de Beauvois (1802) transliterated the same letter as 'r' when he coined the name *Rabdochloa*.

In Classical Greek times it was the custom when compounding two words to double *rho* when it was the initial letter of the second word. A common method of writing such compound words, in Botanical Latin, was to treat the first *rho* as the letter 'r' and the second as 'rh', a practice accepted in modern English for words such as diarrhoea (alternative spelling diarrhea) which derives from the Greek *diarhrhoia* through the Latin *diarrhoea.*

Grass genera that follow this rule are *Tetrarrhena*, *Triarrhena* and *Diarrhena.* The eccentric biologist Rafinesque later spelt *Diarrhena* as *Diarina* – although the earlier name has priority under the Code, the alternate transliteration of the Greek would be acceptable under other circumstances.

The convention adopted for the transliteration of *rho* has varied over time and the original spellings of names have sometimes been revised to suit the fashion of the day. Thus *Haloragis*, a dicotyledon, was spelled so by the authors of the genus in 1775 but during the 19th century the name was often changed to *Halorrhagis*, a practice now abandoned in favor of the original spelling.

Although generic names always assume a Latin form, their spelling, especially if transliteration is involved, is not independent of the nationality of the describing author. For example, *Moorochloa* was described by a Dutch national who based the name upon the Greek word *moros* (μωροσ) transliterating the omega (ω) as 'oo' and the omicron (o) as 'o'. Such a practice would not be followed by an English author who would translate as 'o' both omicron and omega, as in the word 'moron', which is derived directly from the Greek.

Because some phonemes of the Scandinavian languages are not represented by letters present in the Latin alphabet, their transliteration has sometimes led to variant spellings of the same name. Thus, the distinguished Swedish botanist Pehr Forsskål (1732–1763) has been honored by the following species epithets: *Avena forskålei* Vahl, *Aristida forskohlei* Tausch, *Chaetaria forskholii* Nees and *Danthonia forskalii* Trinius. In modern binomials, only the dieresis, denoting separate pronunciation of adjacent vowels (as seen on the '*e*' in *Chloë*) is acceptable. All other diacritic marks and non-Roman letters are to be substituted by Roman letters; thus, the umlauts ä, ö and ü are represented by 'ae', 'oe' and 'ue' respectively, while the Scandinavian *å* becomes 'ao'.

With Russian and other languages that use the Cyrillic alphabet, problems of transliteration can become acute and even in Russian texts the same grass may be known under quite different spellings of the same name. Thus *Agropyrum tschimganicum* was described in 1923 by Drobow who two years later referred to it as *A. czimganicum.*

When a language has no alphabet, as with Japanese, names prior to their latinization must be transliterated into one that does. Thus *Tschonoskia* is based on a German version of Chonosuke, the forename of a Japanese botanist.

Structure of Grass Spikelets

Many specific grass names are based on the spikelet structure, which in the past has been the subject of much debate. For example, it has been interpreted both as a flower and part of the inflorescence. As a consequence, specific epithets based on descriptive terms do not necessarily have equivalent meanings. Presently there is little disagreement as to the structure of the spikelet (Clifford 1987).

In its most generalized form the spikelet consists of several alternating bracts, all but the lower two of which bear short shoots, each of which has a prophyll (palea) beyond which are 2 or 3 scales (lodicules) and then the anthers and pistil. The bracts, also known as glumes, are then divided into sterile or fertile depending upon whether or not they support short shoots. Furthermore, the sterile glumes are referred to as subtending glumes, if they are at the base of the spikelet and the fertile glumes as lemmas. A lemma together with its attendant palea, lodicules and reproductive structures is known as a floret. The flower is generally taken to be the reproductive structures along with the lodicules.

In earlier times the spikelet was sometimes regarded as a flower, as attested to by the name *Monanthochloë* whose inflorescences consist of a single spikelet. However each of the spikelets has several florets and accordingly several flowers and so the name is misleading unless interpreted in an historical context.

As their role changed from words in every day use to technical terms many Latin and Greek words also changed their meanings and over time many of these have become quite different from the originals. A few of the changes especially relevant to grass morphology are given in Table 1.1.

Table 1.1. Changes of Latin and Greek terms especially relevant to grass morphology

Word	Language	Original meaning	Derived meaning
ather	Greek	Spine or prickle, or barb of a spear	Awn, or less commonly spikelet or inflorescence
anthera	Greek	Relating to flowers	Anther
culmus	Latin	Stalk or stem especially of grasses	As for original meaning
lemma	Greek	Husk or scale	Glume or bract subtending palea and flower
lobos	Greek	An ear	Ear-like extension of any structure
lodicula	Latin	Small coverlet	Fleshy or membranous scales subtending a grass flower
palea	Latin	Chaff	Prophyll of floret
panicula	Latin	Inflorescence of millet	Freely branched inflorescence
pedunculus	Latin	Little foot	Axis bearing an inflorescence
racemus	Latin	Stalk of a cluster of grapes	Axis bearing stalked spikelets
spica	Latin	A point; hence, in particular, an ear or spike of grain	Axis bearing sessile florets or groups of florets
stigma	Greek	Mark or brand	Pollen receptive surface of pistil
stachys	Greek	An ear or spike of grain	Spike bearing sessile florets or groups of florets
valva	Latin	Leaf of a folding door	Scale embracing another structure

Changes, such as those given above, in the meaning of Classical words on their adoption as descriptive botanical terms, should not be confused with the transfer of meaning that resulted from the application to plants of terminology originally applicable to animals.

Earlier plant anatomical studies were undertaken mainly by human anatomists who gave terms such as ovary to seed bearing structures and cotyledon to the first pair of leaves on the embryonic seedling. Such terminology implied, intentionally or otherwise, a parallelism in function of the structures in plants and animals.

Origins of Generic and Specific Names

The sources of names are not always indicated in the text accompanying the original description of the taxon. In these circumstances the origin of the name must be inferred from the name itself or extraneous information such as the place and date of collection of the taxon, its preferred habitat, the identity of the collector and economic value of the taxon to name a few possibilities.

The origins of plant names has long been of interest. For example, Rabelais (1546, Volume 3, Essay 50) writing about Pantagruelion (*Cannabis sativa*), named in honor of the giant Pantagruel, observed that "all plants come by their names in a variety of ways. First, from the discoverer; second, from the original source; third, in ironic contradiction; fourth, from their effect; fifth, according to their peculiarities; sixth, by remembrance of their metamorphoses, seventh, by similarity; and eight, morphologically". In reaching these conclusions, Rabelais acknowledges his debt to Pliny, an earlier writer on the subject.

The number of origins provided below exceeds the eight recognized by Rabelais, largely as a result of subdividing some of his categories. They are as follows:

Descriptive

The most valuable names from the viewpoint of information content are those in which both the generic and specific names describe the habit of the plant or one or more of its structures. For example, the name *Anthoxanthum odoratum* implies that the species is scented with yellow flowers; *Neuropoa fax* is a grass whose inflorescence resembles a torch with ascending flames; and the leaves of *Leptochloa ligulata* have a conspicuous ligule.

However, not all descriptive names are helpful, for many were applied before the full morphological variation in the genus or species was known. Accordingly, many genera have species epithets such as *altissimus*, which do not apply to the presently known tallest species in the genus but to the tallest known at the time the name was first applied. Furthermore, subdivision of a genus may lead to monospecific taxa with now inappropriate specific epithets. Thus *Mibora minima* comprises a single species so the contrast that existed between this and other species when the taxon was included in *Agrostis* has been lost. When Trinius described *Panicum uniglumis*, relatively few panicoid species were known. With the passage of time, some of the species with laterally compressed spikelets were segregated into their own genera, one of which was *Tricholaena* in which the lower glume of the spikelet is readily overlooked. With its transfer to *Tricholaena*, the significance of the single glume in *T. uniglumis* is lost for the character is shared with all other members of the genus.

There are many species with names which indicate they resemble other taxa in some respect. Nearly always the significance of such epithets should be interpreted against an historical background. Thus when Trinius described *Arundo triodioides* in 1836 he accepted a concept of *Triodia* which is quite different from that held today. Accordingly, when seeking a descriptive interpretation of *Poa triodioides* (Trinius) Zotov, only scant attention should be given to the present day circumscription of *Triodia*.

Commemorative

Many generic and specific names honor people. The majority of these so honored collected the type species and of the remainder, most honor people with particular claims to distinction. These claims include being other botanists (*Danthonia linkii*); leaders of expeditions (*Triodia mitchellii*); statesmen (*Digitaria smutsii*); politicians (*Bambusa moreheadiana*); poets (*Vossia*); wives (*Axonopus jeanyae*); scientific colleagues (*Stipa macalpinei*); or the wives of colleagues (*Agrostis mackliniae*).

Habitat

Habitat has provided a basis for many generic and specific names. Thus all *Ammophila* taxa grow on sandy seashores and those of *Potamophila* on stream banks. Amongst species names those descriptive of habitats abound. Included here are the epithets *calcarea*, *desertorum*, *nivicola* and *maritima*.

However, it cannot always be assumed that a name correctly identifies the habitat normally favored by the species. For example, the describing author of *Phalaris aquatica* was of the opinion the species was associated with wet habitats whereas it commonly occurs in dry-land pastures.

Geographical Location

Localities provide a basis for many generic and specific names, often referring to where they were first collected. The precision with which the localities are cited varies according to how the author perceives the name to apply. Thus Linnaeus several times employed the epithet *aethiopica* for South African species, presumably because Ethiopia in Classical times referred to African countries south of Libya and Egypt. However, his reasons for describing a grass collected by Osbeck in India as *Poa chinensis* are unclear (Linnaeus 1753).

National boundaries are subject to change and so that species with the epithet *palaestina* may not necessarily come from localities included within the boundary of Palestine as presently recognized.

A somewhat similar situation may arise as when a place retains its name but changes countries. Included here is California, most of which was not incorporated into the United States of America until 1850, nine years after *Poa californica* had been described.

For historical reasons place names may change and so obscure the reason for the choice of species name. For example, the specific epithets *zeylanica* and *ceylanica* suggests the species are natives, as they are, of Ceylon, but that country is now known as Sri Lanka. The names Ceylon or its variant spelling Zeylon were applied to the country in Classical Times and used by Europeans from the 16th century onwards.

Until the name Australia was coined by Flinders (1814), the continent was widely known as New Holland, a name that became the basis of the widely used species epithet *novae-hollandiae*.

Decolonization, especially of Africa, in the mid 20th century, led to many countries changing name but due to the Code the names of species described there-from were retained. One example will suffice to illustrate this situation – species named *nyassae* and *nyassana* came from Nyassaland, now known as Malawi.

Nationality of Taxonomist

The names of geographic features often differ depending upon the nationality of the taxonomist. Thus the majestic peak dominating central Taiwan is referred to by English writers as Mount Morrison, is known to the Chinese as Yu Shan and the Japanese as Niityakayama. Each of these names has entered into grass nomenclature; *Yushania* as a genus and *morrisonensis* together with *niitakayamensis* as specific names.

Classical Geographical Names

Because early taxonomists were familiar with Classical literature they often used Ancient Greek and Latin rather than contemporary names for localities. Such a practice overcomes the problem of locality names changing through the centuries but fails to allow for the redefining of national boundaries with the passage time. Thus, although Cadomum of the Romans corresponds closely with the modern Caen, their Gallia corresponds only roughly with modern France.

These Classical names should not be confused with Latinized versions of modern names such as *novae-hollandiae* for New Holland, that is Australia, a country not known to the Romans. Another example is the epithet *capitis-york* applied to plants from Cape York, thereby generating a hybrid between Latin and English words.

Homonymy

The similarity in spelling of geographical names does not necessarily reflect a common origin and may be fortuitous.

Thus the specific epithet *columbiana* may refer to taxa from Colombia, a State in South America, the District of Columbia in the United States of America or British Columbia one of the Canadian States. The names of all three of these localities derive from the navigator of the same name and discoverer of the New World. Likewise, grasses with the specific *georgiana* may come from Georgia, one of the United States of America, or from Georgia a Republic bordering the Black Sea. In both instances the name derives from George, the Christian Saint of that name.

However, similarity of name does not necessarily indicate a common source, as the epithet *gangetica* derives from the Ganges Valley in India or from the village of Ganges in southern France.

The spelling of place names may differ according to the nationality of the taxonomist as with *kamerunense* and *cameroonensis*, the former being the German and the latter the English spelling of two species names for separate taxa collected in the Cameroon Mountains of West Africa.

Seasonality

All four seasonal names have been employed as species epithets: *Agrostis hiemalis* flowers in winter, *Poa aestivalis* in the summer, *Eragrostis autumnalis* in autumn and *Agrostis vernalis* in spring.

Anagrams

Rearranging the letters of generic names to establish anagrams is a well accepted practice as indicated by the following: *Sartidia* derived from *Aristida*; *Leymus* from *Elymus*; *Tarigidia* from *Digitaria*; *Tuctoria* from *Orcuttia*; *Tosagris* from *Agrostis*; *Miphragtes* from *Phragmites*; *Patis* from *Stipa* and *Relchela* from *Lechlera*.

Rarely a misspelling may inadvertently lead to the formation of an anagram as with *Planotia* which arose from the transposition of the *n* and *t* in *Platonia*.

Acronyms

Few acronyms have been employed as the basis of taxonomic names but more are likely as they abound in modern literature. In recognition of the important role played by the Organisation for the Phyto-Taxonomic Investigation of the Mediterranean Area the epithet *optimae* was coined for a species of *Poa* described from material collected in Turkey.

Allusion

The origins of names derived from allusion are often obscure and they cannot be appreciated without an understanding of the allusion. Thus *Farrago* combines the characters of other genera, and the type species *Odyssea* had previously been placed in several other genera, thereby giving it the reputation of a seasoned traveler. Even more obscure is the origin of the epithet in *Panicum diluta* where doubts as to the reality of the species "dissolved" when further specimens were collected.

In anticlerical France, following the Revolution of 1789–1799, a grass was named *Avena precatoria* because its nodding spikelets suggested to the author the bowed heads of worshippers.

Geography and allusion sometimes combine, as for example in the epithet *ursorum*, of the bears, which has been applied to several Arctic species because they come from the "Land of the Bears".

Finally, unless one knew that Lord Talbot of Malahide was Irish, a people regarded traditionally as lucky, there would be no sense in the species name *fortunae-hibernae*, which was applied to a grass that arrived accidentally in England, having been raised from seed in soil attached to plants sent from Tasmania by Lord Talbot to the Royal Botanic Gardens, Kew.

Hybridity

To indicate that hybrids between species are from different genera, it is customary to coin for them a new generic name based on those of the parents. For example ×*Cynochloris* (sometimes written x*Cynochloris*) is the generic name for interspecific hybrids between *Cynodon* and *Chloris*. The multiplication symbol (×) or the letter x before the name indicates the taxon is of known or presumed hybrid origin.

Occupations

A few names including *metatoris* and *geometra* derive from the occupation of the collectors, in these instances surveying.

Vessels

In a few instances, names have been given to commemorate the vessel in which scientific expeditions have been undertaken. Accordingly, *utowanaeum* is derived from Utowana, the name of a steam yacht made available to the Field Museum of Chicago to transport scientists around the Caribbean, and *nascopieana* commemorates the R.M.S. Nascopie, a Canadian Navy vessel, which sailed regularly in Arctic waters.

Misadventure

The literature abounds in names with have been spelled incorrectly. The reasons for this situation are numerous and include typographical errors, momentary lapses in concentration or ignorance on the part of the writer.

Rarely an incorrect name may be inadvertently substituted for another as when *australis* was inadvertently replaced by *neutralis.*

Conventions Employed in Dictionary

- Throughout, generic names are spelt with a capital initial and, in accordance with the provisions of the Code, species names with a lower case initial.
- The grammar has been simplified. Verbs are given in the first person singular present tense and nouns in the nominative singular, as both are so listed in dictionaries. For adjectives which are inflected, the nominative singular forms are given in alphabetical order and thus do not always follow the order of dictionaries, which list masculine form first and neuter, last.
- The spellings of any place names, when changed from those published, wherever possible follow the usage of the "Times Atlas" or "Times Gazetteer", or have been sourced from Cohen (1998).
- Nationalities of persons commemorated in generic or specific names are wherever possible given in terms of their country of birth. Their subsequent nationalities and countries in which they principally lived or collected are also given where appropriate. For example, José de Acosta (1540–1600) (see *acostae*) is recorded as a Spanish Jesuit and scholar who traveled extensively in Central and South America and for his writings earned the title "Pliny of the New World". In contrast, Ferdinand Jacob Heinrich Mueller (1825–1896) is described herein as a German-born Australian botanist, whilst Friedrich M. Müller (fl. 1853–1855) is recorded only as having collected in Mexico, as his country of origin is uncertain (see *muelleri*).
- Variant spellings of both generic and specific names abound in the literature. No effort has been made to correct orthographic variants, other than those arising from mistakes as to the correct method of converting personal names into a Latin form. This means that many epithets published with a single terminal *-i* following a consonant other than *r* have been omitted from the Dictionary but all appear in the corrected form with a terminal *-ii* (*masculine*) or *-ae* (*feminine*). The neglect of forms with an inappropriate ending should cause no difficulty to the reader for they would usually have been located immediately adjacent to the corrected epithet.
- For homonyms, the derivations given apply only to the usage of the name in the Poaceae.

- Hyphenated epithets have been written as single words, unless this would be contrary to the provisions of the Code.
- It should be noted that in Latin texts, plant names are declined to satisfy the rules of grammar for that language and so may differ from their typical, dictionary form. For example, in the following sentence, *Panico teretifolio* is in the dative as required by the participle *affinis*: "*Species nova Panico teretifolio affinis sed spiculis grandioribus, gluma inferiore breviore et panicula diffusiore differt*". In translation the sentence reads in English as "New species allied to *Panicum teretifolium* but it differs by the larger spikelets, the shorter lower glume and the more diffuse panicle". Here the typical (nominative) form is *Panicum teretifolium*, which is the form under which generic and specific names are recorded in this dictionary.
- The years of birth and death of those honored by generic and specific names have been taken in the main from standard sources. In many instances, those honored have checked their own entries.
- Throughout, the origins of words are given as in non-technical non-specialized dictionaries, thereby enabling most entries to be checked in libraries. However, over the past 250 years Botanical Latin has incorporated many words from Medieval and Late Latin, and so consistency of meaning cannot be expected.
- As noted above, misspellings of specific names resulting from employing a termination that fails to reflect the gender of the generic name, as presently understood, have been corrected. However, whilst the gender of most generic names is unambiguous, disputes regarding gender have arisen from time to time, largely because of differences of opinion as to the proper manner of forming names from Greek roots. Thus, whereas Linnaeus regarded *Andropogon* as neuter, most recent botanists have treated the word as masculine. These differences of opinion are expressed in the specific epithets. Whereas Linnaeus (1753) refers to *Heteropogon contortum*, more recent works which retain the species in *Heteropogon* would refer to *Heteropogon contortus*. Herein, specific epithets have been cited in terms of their presently accepted correct genders.
- A few grass species were originally mistaken for sedges (Cyperaceae). For example, *Pharus brasiliensis* and *Abildgaardia polystachya* are synonyms. Genera and species now accepted as sedges are not defined in the dictionary.

References

In addition to the references specifically cited in the preceding text, we have included below the principal biographical sources consulted. To have included the sources of the many thousands of original descriptions investigated would have been impractical. Searches of the World Wide Web provided much useful information, especially as to geographic localities, biographical details and places of publication. The principal web sites consulted are appended below.

Brown RW (1979) *Composition of scientific words.* Smithsonian Institution Press, Washington

Brummitt RK, Powell CE (eds) (1992) *Authors of plant names.* Royal Botanic Gardens, Kew

Chase A, Niles CD (1962) *Index to grass species.* 3 vols, G. K. Hall & Co., Boston

Clifford HT (1987) Spikelet and floral morphology. In: Soderstrom TS, Hilu KW, Campbell CS, Barkworth ME (eds) *Grass systematics and evolution, an international symposium held at the Smithsonian Institution, Washington, D.C., 27–31 July, 1986.* Smithsonian Institution Press, Washington, Chap. 3, pp 21–30

Cohen SB (ed) (1998) *The Columbia Gazetteer of the World.* 3 vols, Columbia University Press, New York

Desmond R (1977) *Dictionary of British and Irish botanists and horticulturalists incl. plant collectors and botanical artists.* Taylor & Francis Ltd., London

Everist SL (1979) *Poisonous plants of Australia*, 2nd ed. Angus and Robertson, Sydney

Flinders M (1814) *A voyage to Terra Australis: Undertaken for the purpose of completing the discovery of that vast country, and prosecuted in the years 1801, 1802, and 1803.* 2 vols, W. Bulmer, London

Funk VA, Scott AM (1989) A bibliography of plant collectors in Bolivia. *Smithson Contrib Bot* 70:1–20

Greuter W (Chairman) (2000) *International code of botanical nomenclature (St. Louis Code).* International Association of Plant Taxonomy. Regnum Vegetabile 131. Koeltz Scientific Books, Königstein

Hitchcock AS (1950) *Manual of the grasses of the United States.* United States Dept. of Agriculture, Misc. Publication No. 200, edn 2, revised A. Chase

Holmgren PK, Holmgren NH, Barnett LC (eds) (1990) *Index herbariorum, part 1: The herbaria of the world.* 8th edn. New York Botanical Garden, New York

Lanjouw J, Stafleu FA (1954–1988) *Index herbariorum, part II (nos. 1–7). Collectors.* International Bureau for Plant Taxonomy and Nomenclature, Utrecht (nos. 1–3); Bohn, Sheltema & Holkema, Utrecht & Dr. W. Junk, The Hague (nos. 4–7)

Linnaeus C (1753) *Species Plantarum.* A facsimile of the first edition 1753, vol. 1, with an introduction by W. T. Stearn. Ray Society, London

Rabelais F (1546) *The complete works of François Rabelais.* Translated from the French by Donald M. Frame, 1991. University of California Press, Berkeley

Stafleu FA, Cowan RS (1976–1978) *Taxonomic literature.* A selective guide to botanical publications and collections with dates, commentaries and types. 7 vols. Bohn, Scheltema & Holkema, Utrecht

Stafleu FA, Mennega EA (1992–2000) *Taxonomic literature.* Suppl. I–VI. Koeltz Scientific Books, Königstein

Stearn WT (1992) *Botanical Latin*, 4th edn. David & Charles Publishers, Newton Abbot, Devon

Steenis-Kruseman MJ van (1950) Malaysian plant collectors and collections: Alphabetical list of collectors. *Flora Malesiana* 1(5):5–605
Steenis-Kruseman MJ van (1958) Malaysian plant collectors and collections, supplement I: Alphabetical list of collectors, supplement. *Flora Malesiana* 1(5):ccli–cccxlii
Steinberg CH (1977) The collectors and collections in the Herbarium Webb. *Webbia* 32(1):1–49

Web Sites (Accessed from July 2004 to April 2006)

Australian Plant Name Index (undated) *http://www.anbg.gov.au/cpbr/databases/apni*
Biographical notes on plant collectors and illustrators or others relevant to Australian botany, Australian National Botanic Gardens (undated) *http://www.anbg.gov.au/biography/*
Christian Cyclopedia, Lutheran Church, Missouri Synod (2000) *http://www.lcms.org/ca/www/cyclopedia/02/*
Fuzzyg, the Fuzzy Gazetteer (2002–2005) *http://tomcat-dmaweb1.jrc.it/fuzzyg/query/*
Google (undated) http://www.google.com/
Holmgren PK, Holmgren NH (1998 onwards, continuously updated) Index Herbariorum. New York Botanical Garden. *http://sciweb.nybg.org/science2/IndexHerbariorum.asp*
Index of Botanists, Harvard University Herbaria (2005) *http://cms.huh.harvard.edu/databases/botanist_index.html*
The International Plant Names Index (2004) *http://www.ipni.org/*
Lonicera, School of Environmental Studies, Agric. Univ. of Szczecin, Poland (undated) *http://www.lonicera.hg.pl/*
Malahide Historical Society, Co. Dublin, Ireland (2004) *http://www.malahideheritage.com/*
The Perseus Digital Library, Tufts University (undated) *http://www.perseus.tufts.edu/*
W3Tropicos, nomenclatural database of the Missouri Botanical Garden (undated) *http://mobot.mobot.org/W3T/Search/vast.html*
Wikipedia, the Free Encyclopedia (2001) *http://www.wikipedia.org/*

Abbreviations

c.	Near to, with respect to dates
fl.	L. *floruit*, (of a person) he or she flourished or lived at this period, that is, collecting or publishing at the date or dates indicated
Gk	Greek
L.	Latin
?	Year of birth or death not determined
C.E.	Common Era i.e. system of naming years from the birth of Jesus
B.C.E.	Before Common Era

Dictionary

A

abadiana L. *-ana*, indicating connection. From Abadia, Brazil

abakanensis L. *-ensis*, denoting origin. From Abakan, Siberia

abata Gk *abatos*, pure. Grows in dense pure stands or the segregation of the species left related taxa more clearly defined

abbreviata L. *abbrevio*, shorten. Culms short

abchazicum L. *-icum*, belonging to. From Abchaza, Southern Caucasus

aberrans L. *aberro*, wander away. Unlike related species, the awn is scarcely exserted from spikelet

abessinica L. *-ica*, belonging to. From Abessin, now Ethiopia

abietifolia L. *folium*, leaf. The leaf-blades resemble the leaves of *Abies*

abietina L. *-ina*, indicating possession. Growing in *Abies* forests

abietorum L. *-etum*, place of growth. Growing in *Abies* forests

abludens L. *abludo*, be unlike. Differing markedly from related species

abnormis L. *ab-*, away from; *norma*, model. Differing from the expected

Abola Meaning obscure, derivation not given by author

abolinii In honor of Robert Ivanovic Abolin (1886–?) Latvian-born Russian botanist

abortiv-a, -um L. aborted. – (1) The sessile spikelet of each pair is sterile. *Digitaria abortiva* – (2) the apex of rhachis projects as a bristle beyond the uppermost spikelet. *Panicum abortivum*

abrahamii In honor of – (1) A. Abraham, Indian botanist. *Ischaemum abrahamii* – (2) A. A. Abraham (fl. 1919) who collected in Guyana. *Paspalum abrahamii*

abregoens-e, -is L. *-ense*, denoting origin. From Abrego, Colombia

abromeitiana L. *-ana*, indicating connection. In honor of Johannes Abromeit (1857–1946) German botanist

abrumpens L. *abrumpo*, break off. The spikelets break off below the glumes

abscissum L. *abscindo*, divide. The inflorescence comprises several axillary panicles

absimil-e, -is L. unlike. Readily distinguishable from related species

abstrusum L. concealed. Inflorescence partially enclosed in the sheath of subtending leaf

abyssinic-a, -um, -us L. *-ica*, belonging to. From Abyssinia, now Ethiopia

academica Gk *-ica*, belonging to; *Akademia*, a school in Athens in Classical times. Found growing in the vicinity of the University City of Cordoba, Argentina

acadiense L. *-ense*, denoting origin. From Acadia, which in the 1600s was the name given to the area of North America comprising the present Nova Scotia, Prince Edward Island and a section of New Brunswick

acamptoclada Gk *a-*, without; *kampto*, bend; *klados*, branch. Primary panicle branches rigid

Acamptoclados Gk *a-*, without; *kampto*, bend; *klados*, branch. Culms stiff

acamptophylla Gk *a-*, without; *kampto*, bend; *phyllon*, leaf. The leaf-blade is rigid and held erect

acanthoneuron Gk *akantha*, prickle or thorn; *neuron*, nerve. Glume nerves bear conspicuous curved spines
acanthophylla Gk *akantha*, prickle or thorn; *phyllon*, leaf. Leaf-blades sharp pointed
acarifer-a, -um L. *acarina*, mite; *fero*, carry or bear. Spikelets have the appearance of mites
accedens L. *accedo*, resemble. Similar to another species
accrescens L. *accresco*, grow. Spikelets growing larger after anthesis
acerosa L. *acer*, sharp; *-osa*, abundance. Lemmas terminating in a sharp point
Achaeta Gk *a-*, without; *chaete*, bristle. Rhachilla extension feathery rather than bristle-like as in associated genera
achalensis L. *-ensis*, denoting origin. From Sierra Achala, Argentina
Achlaena Gk *a-*, without; *chlaena*, cloak. Glumes missing or reduced to a cupule
achlysophila Gk *achlys*, mist; *phileo*, love. Grows on wet rocks adjacent to waterfalls
achmadii In honor of Achmadi (fl. 1917–1920) Indonesian plant collector
Achnatherum Gk *achne*, scale; *ather*, barb or spine. Lemma awned
Achnella Hybrids between species of *Nassella* and *Achnatherum*
Achneria Origin obscure, but possibly an incomplete anagram of *Eriachne*. Name has been used for two quite distinct grass genera
Achnodon Gk *achne*, scale; *odous*, tooth. The glumes terminate in a mucro or awn
Achnodonton See *Achnodon*
Achrochloa See *Airochloa*
Achroostachys See *Athroostachys*
achtarovii In honor of Boris T. Achtarov (1885–1959) who collected in Bulgaria
Achyrodes Gk *achyron*, chaff; *-odes*, resembling. The sterile spikelets have as many as ten or more lemmas lacking flowers
Aciachne Gk *akis*, pointed object; *achne*, scale. The lemma is drawn out into a point
Acicarpa Gk *akis*, pointed object; *karpos*, fruit. Grain apex acute and surmounted by base of persistent style
acicular-e, -is L. *acus*, needle; *-ulus*, diminutive; *-are*, pertaining to. – (1) Leaf-blades sharp-pointed. *Panicum aciculare* – (2) callus sharp pointed. *Andropogon acicularis*, *Raphis acicularis*
acicularifolium L. *acus*, needle; *-ulus*, diminutive; *-aris*, pertaining; *folium*, leaf. Leaf-blades pungent
acicularis See *aciculare*
aciculatus L. *acus*, needle; *-ulus*, diminutive; *-atus*, possessing. – (1) Possessing spikelets with a needle-like callus. *Andropogon aciculatus*, *Chrysopogon aciculatus* – (2) lemma with a long thin awn. *Campulosus aciculatus*
Acidosasa L. *acidus*, disagreeable; Japanese *sasa*, dwarf bamboo. The first species described did not fit into any of the related genera
acinaciformis L. *akinakes*, short sword; *forma*, appearance. Lemma the shape of a short Persian sword
acinaciphylla Gk *akinakes*, short sword; *phyllon*, leaf. Leaf-blade the shape of a short Persian sword
acinifolius L. *acinaces*, short sword; *folium*, leaf. Leaf-blades inrolled, rigid and usually quite pungent
acinosa L. *acinus*, berry; *-osa*, abundance. Inflorescence a contracted panicle resembling a bunch of grapes
aciphylla Gk *akis*, pointed object; *phyllon*, leaf. Leaf-blades rigid and pungent
Acophorum Gk *akoe*, ear; *phero*, bear. Origin in doubt, not given by author
acostae In honor of José de Acosta (1540–1600) Spanish Jesuit and scholar who travelled extensively in Central and South America and for his writings earned the title "Pliny of the New World"
Acostia In honor of Misael Acosta-Solis (1910–1994) Ecuadoran pharmacologist
acostia A species name derived from the monospecific *Acostia*, to avoid forming a homonym, if the genus is included in *Panicum*
Acrachne Gk *akros*, at the tip; *achne*, scale. Racemes terminate in an aborted spikelet
acraea Gk *akrea*, dwelling on heights. A species of high mountain grasslands

Acratherum Gk *akros*, at the tip; *ather*, barb or spine. The upper glume is awned

Acritochaete Gk *akritos*, disorderly; *chaete*, bristle. Awn of proximal lemma irregularly twisted at maturity

acroanthum Gk *akros*, at the tip; *anthos*, flower. Inflorescence an open panicle with long pedicels bearing few spikelets

Acroceras Gk *akros*, at the tip; *keras*, horn. The upper subtending glume and sterile lemma each contract into a horn-like structure

acrochaeta Gk *akros*, at the tip; *chaete*, bristle. Lemma shortly awned

Acrochaete Gk *akros*, at the tip; *chaete*, bristle. The inflorescence branches end in awn-like bristles which exceed the spikelets in length

acrociliata Gk *akros*, at the tip; L. *cilium*, hair; *-ata*, possessing. Glume apices hairy

Acroelytrum Gk *akros*, at the tip; *elytron*, cover. The spikelets are one-flowered with apical tufts of empty lemmas

acroleuca Gk *akros*, at the tip; *leukos*, white. Lemma tips bear white hairs

acromelaen-a, -um Gk *akros*, at the tip; *melaeno*, make black. Apex of anthoecium darkly pigmented

Acrospelion Gk *akros*, at the tip; *spelaion*, pit. The lemma apex is bifid

acrotrich-a, -um, -us Gk *akros*, at the tip; *thrix*, hair. – (1) Sterile lemma with a well developed mucro. *Eriochloa acrotricha*, *Helopus acrotrichus* – (2) sterile lemma bearing stiff hairs. *Panicum acrotrichum*

acroxantha Gk *akros*, at the tip; *xanthos*, yellow. Lemma green with yellow apex

Acroxis Gk *akros*, at the tip; *oxys*, pointed. Glumes cuspidate

actae L. *acta*, sea-shore. Growing on sea-shores

Actinochloa Gk *aktinos*, ray; *chloa*, grass. The inflorescence comprises spicate branches

Actinochloris Gk *aktinos*, ray. The racemes of inflorescence arising collectively from the tip of the peduncle as with *Chloris*

actinoclad-a, -us Gk *aktinos*, ray; *klados*, branch. Inflorescence branches verticillate

Actinocladum Gk *aktinos*, ray; *klados*, branch. Multiple axillary buds give rise to bunches of shoots at the lower nodes

actinocladus See *actinoclada*

actinostachys Gk *aktinos*, ray; *stachys*, spike as of an ear of wheat. Racemes sessile and densely fasciculate

actinotrich-a, -us Gk *aktinos*, ray; *thrix*, hair. With hairs radiating from the orifice of the leaf-sheath

aculeat-a, -um, -us L. *acus*, needle; *-ulus*, diminutive; *-ata*, possessing. – (1) Stems spiny. *Bambusa aculeata*, *Guadua aculeata* – (2) subtending bracts of inflorescence sharp pointed. *Crypsis aculeata*, *Schoenus aculeatus* – (3) sterile lemma abruptly acute. *Panicum aculeatum*

aculeolata L. *acus*, needle; *-ola*, diminutive; *-ata*, possessing. Leaf-blades involute and somewhat rigid

acuminat-a, -um, -us L. *acumen*, sharp point; *-ata*, possessing. – (1) Lemmas or glumes acute. *Anthephora acuminata*, *Aristida acuminata*, *Arundinaria acuminata*, *Chloris acuminata*, *Crypsis acuminata*, *Dichanthelium acuminatum*, *Eragrostis acuminata*, *Glyphochloa acuminata*, *Heteropogon acuminatus*, *Manisuris acuminata*, *Melica acuminata*, *Oplismenus acuminatus* – (2) leaf-blades acute. *Deyeuxia acuminata*, *Festuca acuminata*, *Lasiacis acuminata*, *Loudetia acuminata*, *Puelia acuminata*, *Streptostachys acuminata*, *Trichopteryx acuminata*, *Vilfa acuminata*

acuminatissim-a, -um L. *acumen*, sharp point; *-ata*, possessing; *-issima*, most. Spikelets acuminate in outline

acuminat-um, -us See *acuminata*

acut-a, -um L. *acuo*, sharpen. – (1) Spikelets acute to acuminate. *Agropyrum acutum*, *Eragrostis acuta*, *Panicum acutum*, *Paspalum acutum*, *Reimaria acuta*, *Reimarochloa acuta*, *Setaria acuta* – (2) culm-buds acute. *Phyllostachys acuta* – (3) leaf-blades sharp-pointed. *Festuca acuta* – (4) callus sharp-pointed. *Aristida acuta*, *Stipa acuta*

acuticontracta L. *acutus*, acute; *contraho*, draw together

acutiflor-a, -um L. *acuo*, sharpen; *flos*, flower. Paleas, lemmas or glumes with sharp apices
acutifolium L. *acuo*, sharpen; *folium*, leaf. Leaf-blades sharply tapering
acutiforme L. *acuo*, sharpen; *forma*, appearance. Leaf-blades rigid
acutiglum-a, -is, -um L. *acuo*, sharpen; *gluma*, husk. Glumes acuminate
acutipes L. *acuo*, sharpen; *pes*, foot. Callus at spikelet base long and sharp-pointed
acutispathaceus L. *acuo*, sharpen; *spatha*, sheathing base and false petiole of a palm leaf; *-aceus*, indicating resemblance. Apex of spathe acute
acutispicula L. *acutus*, sharp; *spica*, a point; hence, in particular, an ear or spike of grain. Spikelets terete with long pungent calluses
acutissima L. *acuo*, sharpen; *-issima*, most. Glumes and/or lemmas long tapering
acutiuscula L. *acutius*, more acute; *-ula*, tending to. Lemma apex more acute than in related species
acutivagina L. *acuo*, sharpen; *vagina*, sheath. Culm leaf-sheaths narrowly-acuminate at the apex
acutum See *acuta*
adamaouensis L. *-ensis*, denoting origin. From Adamaou, Republic of Cameroon
adamovicii In honor of Lulji Adamovic (1864–1935) Dalmatian botanist
adamsii In honor of – (1) Laurence George Adams (1929–) Australian botanist. *Micraira adamsii* – (2) John Adams (1872–1950) Irish-born Canadian botanist. ×*Agroelymus adamsii*
adamsonii In honor of Frederick M. Adamson (1836–1858) Australian farmer
adanensis L. *-ensis*, denoting origin. From Adana, Anatolia, Turkey
addisonii In honor of Addison Brown (1830–1913) United States botanist
adelogaeum Gk *adelos*, secret; *ge*, world; *-eum*, belonging to. From Japan, a country until the mid-nineteenth century closed to Europeans
adenense L. *-ense*, denoting origin. From *Aden*, Yemen
adenocoleos Gk *aden*, gland; *koleos*, sheath. Leaf-sheath bearing glands
adenophorum Gk *aden*, gland; *phero*, bear. The leaf-blades bear gland-tipped hairs
adenophyllum Gk *aden*, gland; *phyllon*, leaf. Hairs on the leaf-margin gland-tipped
adenorhachis Gk *aden*, gland; *rhachis*, backbone. The leaf-blades and panicles bear glands
adhaerens L. *adhaero*, cling. Culms scrambling
adjaricus L. *-icus*, belonging to. From Adjaria, an autonomous region within the Republic of Georgia
admirabilis L. wonderful. Attractive in appearance
adoens-e, -is L. *-ense*, denoting origin. From Ado, Ethiopia
adoperiens L. *adoperio*, cover. Plants forming a dense sward
adpress-a, -um L. *ad-*, towards; *presso*, press. – (1) Culm-sheaths closely adpressed to culms. *Fargesia adpressa* – (2) hairs of glumes and sterile lemma closely adpressed. *Panicum adpressum* – (3) panicle branches adpressed to main axis. *Puccinellia adpressa*
adpressiramea, adpressi-ramea L. *ad-*, towards; *presso*, press; *ramus*, branch. Panicle branches held erect
adpressum See *adpressa*
adscendens L. *ascendo*, ascend. Culms erect
adscensionis From Ascension Island in the Atlantic Ocean
adspers-a, -um L. *ad-*, towards; *spargo*, sprinkle. Upper glume and fertile lemma sparsely hairy towards their apices
adstricta L. *ad-*, towards; *stricta*, erect. Culms arising in fascicles
adtenuatum L. drawn out or narrowed. Panicles narrow
adust-a, -um, -us L. blackened. Spikelets dark-colored
advena L. a stranger. Adventive species described from individuals found growing in countries in which they are not native
adzharica L. *-ica*, belonging to. From Adzhar, Republic of Georgia

Aechmophora Gk *aichme*, spear; *phero*, bear. Spikelets the shape of a spear-head

Aegialina Gk *aigialos*, sea-shore; L. *-ina*, indicating possession. Growing near the sea-shore

Aegialitis Gk *aigialos*, sea-shore; *-itis*, closely connected. Sea-shore or salt-marsh species

aegiceras Gk *aix*, goat; *keras*, horn. The lemma terminates in a long awn

Aegicon An alternate name for *Agrostis*

Aegilemma Gk *lemma*, husk. Lemma like that of *Aegilops* but adheres to grain

Aegilonearum Gk *nearon*, youthful. The spicate inflorescence resembles that of an immature *Aegilops* inflorescence

Aegilopodes Gk *pous*, foot. Resembling *Aegilops* but differing in some respect not given by the author

aegilopodioides See *aegilopoides*

aegilopoides Gk *-oides*, resembling. Resembling *Aegilops* in the form of the inflorescence

Aegilops Gk *aegiles*, preferred by goats; *ops*, appearance. On account of its presumed similarity to *aegiles*, a plant whose identity is uncertain other than it was a herb liked by goats

aegilopsoides Gk *-oides*, resembling. Inflorescence somewhat similar to *Aegilops*

Aegilosecale Hybrids between species of *Aegilops* and *Secale*

Aegilotricale Hybrids between species of *Aegilops* and *Triticale*

Aegilotrichum See *Aegilotriticum*

Aegilotricum See *Aegilotriticum*

Aegilotriticum Hybrids between species of *Aegilops* and *Triticum*

Aegopogon Gk *aix*, goat; *pogon*, beard. The spikelets are clustered in triads and so the awned glumes, lemmas and paleas form a beard-like fascicle of hairs

Aegylops See *Aegilops*

aegylopsoides See *aegilopsoides*. The *y* spelling is used by Steudel – possibly a parallel with *sylvan* vs. *silvan*

aegyptiac-a, -um, -us L. Egyptian. From Aegyptus, now Egypt

aegyptica L. *-ica*, belonging to. From Aegyptus, now Egypt

aegypti-um, -us L. *-ium*, characteristic of. From Aegyptus, now Egypt

Aelbroeckia In honor of Jean-Louis van Aelbroeck (1755–1846) Flemish agronomist

Aeluropus Gk *aelouros*, cat; *pous*, foot. The inflorescence bears a fanciful resemblance to a cat's paw

aemul-a, -um, -us L. more or less equalling. Subtending glumes more or less equal

aemulans L. *aemulor*, come near to. Rather similar to another species

aemul-um, -us See *aemula*

aene-a, -um, -us L. *aeneus*, of copper. Spikelets copper-colored

aequabile L. similar. Readily mistaken for another species

aequal-e, -is L. *aequus*, equal; *-alis*, pertaining to. Glumes or lemmas similar in length

aequat-a, -um L. *aequo*, make equal. – (1) Glumes similar. *Agrostis aequata* – (2) both florets of spikelet similar. *Panicum aequatum*

aequatoriensis L. *aequator*, equator; *-ensis*, denoting origin. Growing near the equator in Ecuador

aequatum See *aequata*

aequiglum-e, -is L. *aequus*, equal; *gluma*, husk. Glumes subequal

aequipaleata L. *aequus*, equal; *palea*, scale; *-ata*, possessing. Glumes equal in length

aequiramea L. *aequus*, equal; *ramus*, branch. Arms of three-partite awn equally long

aequiramosum L. *aequus*, equal; *ramus*, branch; *-osum*, abundance. Branch clusters, unlike those of related species, equal in size

aequivaginatum L. *aequus*, equal; *vagina*, sheath; *-ata*, possessing. Leaf-sheaths about equal in length

aequivalvis L. *aequus*, equal; *valvus*, scale. Glumes of similar length

Aera Classical Greek name for darnel or possibly another species of *Lolium*. Name now applied to quite a separate genus

aeria L. *aer*, atmosphere. Aerial roots grow from the culms and stolons

A

Aeropsis Gk *opsis*, resemblance. Similar to *Aera*
aestivalis L. *aestas*, summer; *-alis*, pertaining to. Summer-flowering species
aestivum L. *aestas*, summer; *-ivum*, property of. Grown in the summer by sowing in the spring
aethiopic-a, -um, -us L. *-ica*, belonging to. In classical usage Aethiopia was south of the Sudan and widely used in that sense
Aethonopogon Gk *aethon*, fiery; *pogon*, beard. Spikelet invested in long reddish hairs
aetnens-e, -is L. *-ense*, denoting origin. From Mt Etna, Sicily
aetolic-a, -um L. *-ica*, belonging to. From Aitolia Province, Greece
af-er, -rum L. Africa. From Africa
affghanica See *afghanica*
affin-e, -is L. allied to. Closely allied to another species
afghanic-a, -um L. *-ica*, belonging to. From Afghanistan
Afrachneria Africa combined with *Achneria*
afraurita L. *afer*, African; *aurita*, eared. Pedicels and glumes with a wing-like appendage
african-a, -um, -us L. *-ana*, indicating connection. From Africa
afronardus Scented and so reminiscent of *Nardus* but from Africa
Afrotrichloris Resembling *Trichloris* an American genus but endemic to Somalia in north-east Africa
afrum See *afer*
afzelian-a, -um L. *-ana*, indicating connection. In honor of Adam Afzelius (1750–1837) Swedish botanist
afzelii As for *afzeliana*
agadiriana L. *-ana*, indicating connection. From Agadir, Morocco
agassizensis L. *-ensis*, denoting origin. From lowlands occupying the place of Lake Agassiz, a periglacial Canadian lake
agasthyamalayana L. *-ana*, indicating connection. From Agasthyamalai region of Kerala, Western Ghats, India
Agenium, agenium Gk *ageneios*, boyish, hence unbearded. Lemmas unawned
agglutinans L. *agglutino*, glue to. Plant invested with sticky hairs
Aglycia Gk *a-*, without; *glykus*, sweet. Meaning obscure, not given by author
Agnesia As for *Chasea*
agnesiae As for *Chasea*
agraria L. relating to the land. Cultivated species
Agraulus Gk *agraulos*, rural. Species not cultivated
Agrestis An alternate spelling of *Agrostis*
agrestis L. of the land, uncultivated. Growing in fields, sometimes among crops
agrimonoides Gk *-oides*, resembling. Resembling *Agrimonia*
agriocrithon Gk *agros*, field; *krithe*, barley. A barley raised from seed collected in the wild
Agriopyrum See *Agropyron*
Agrocalamagrostis Hybrids between species of *Agrostis* and *Calamagrostis*
agroelymoides Gk *-oides*, resembling. Similar to *Agroelymus*
Agroelymus Hybrids between species of *Agropyron* and *Elymus*
Agrohordeum Hybrids between species of *Agropyron* and *Hordeum*
Agropogon Hybrids between species of *Agrostis* and *Polypogon*
Agropyrohordeum Hybrids between species of *Agropyrum* and *Hordeum*
agropyroides Gk *-oides*, resembling. Similar to *Agropyron* with respect to the inflorescence
Agropyron, Agropyrum Gk *agros*, field; *pyros*, wheat. Resembling wheat but not cultivated as a cereal
Agropyropsis Gk *opsis*, resemblance. The inflorescences resemble those of *Agropyron*
Agropyrum See *Agropyron*
Agrositanion Hybrids between species of *Agropyron and Sitanion*
Agrosticula L. *-ula*, diminutive. Spikelets small, otherwise resembling those of *Agrostis*
agrostide-a, -um Gk *-idea*, resembling. Similar to *Agrostis* in habit or inflorescence

agrostidiforme Gk *forme*, appearance. Resembling *Agrostis*
agrostiflora L. *flos*, flower. Spikelets *Agrostis*-like
agrostiiformis L. *forma*, appearance. Resembling *Agrostis*
agrostis Resembling *Agrostis* especially with reference to the inflorescence
Agrostis The name for an unidentified Greek fodder plant in Classical times
agrostoidea Gk *-oidea*, resembling. Similar to *Agrostis*
agrostoides Gk *-oides*, resembling. Resembling *Agrostis* in some respect, usually the form of the inflorescence
Agrostomia Gk *agrostis*, unidentified fodder plant; *tome*, the end left after cutting. Lemmas almost awnless in contrast to those of related species
Agrotrigia Hybrids between species of *Agropyron* and *Elytrigia*
Agrotrisecale Hybrids incorporating species of *Agropyron*, *Triticum* and *Secale*
Agrotriticum Hybrids between species of *Agropyron* and *Triticum*
aguana L. *-ana*, indicating connection. From Agua, a volcano in Guatemala
aguascalientensis L. *-ensis*, denoting origin. From Aguascalientes, Mexico
aguilarii In honor of S. Aguilar (fl. 1908) Philippine plant collector
agustinii In honor of Bernardo H. Agustin (fl. 1926–1941)
aikawensis L. *-ensis*, denoting origin. From Aikawa, Sado Island, Japan
Aikinia In honor of Arthur Aikin (1773–1854) English chemist, geologist and naturalist
ailuropodina L. *-ina*, indicating resemblance. From areas inhabited by the Giant Panda *Ailuropoda melanoleuca*
Aira Classical Greek name for darnel or possibly another species of *Lolium*. Name now applied to quite a separate genus
airaeformis L. *forma*, appearance. Resembling *Aira* with respect to the inflorescence
Airella L. *-ella*, diminutive but here used as a name-forming suffix. Resembling *Aira*
Airidium Gk *-idium*, resembling. Similar to *Aira*
Airochloa Gk *chloa*, grass. Resembling *Aira*
airoides Gk *-oides*, resembling. Inflorescences resemble those of *Aira*
Airopsis Gk *opsis*, resemblance. Similar to *Aira*
aitchisonii In honor of James Edward Tierney Aitchison (1836–1898) Indian-born of English parents; physician and plant collector
aizuensis L. *-ensis*, denoting origin. From Aizu, Japan
ajanensis L. *-ensis*, denoting origin. – (1) From Ajan, on the coast of Okhotsk Sea, Russian Far East. *Asperella ajanensis*, *Elymus ajanensis*, *Leymus ajanensis* – (2) from Ajan, Chabarovos Province, Russian Federation. *Calamagrostis ajanensis*
akagiensis L. *-ensis*, denoting origin. From Akagisan, a mountain in Gunma Prefecture, Japan
akasiensis L. *-ensis*, denoting origin. From Akasi, Japan
akhanii In honor of Hossein Akhani (fl. 1995) Iranian botanist
akiensis L. *-ensis*, denoting origin. From Aki Province, now the western part of Hiroshima Prefecture, Japan
akitensis L. *-ensis*, denoting origin. From Kita-Akita, Ugo Province, now the major part of Akita and Yamagata Prefectures, Japan
akmanii In honor of Yildirim Akman (1934–) Turkish plant ecologist
akmolinens-e, -is L. *-ense*, denoting origin. From Akmolinsk, Kazakhstan
akoense L. *-ense*, denoting origin. From Ako, Taiwan
aktauensis L. *-ensis*, denoting origin. From Aktau in the Kyzyl Mountains, Central Asia
alabamens-e, -is L. *-ense*, denoting origin. From Alabama, USA
alaic-a, -um, -us L. *-ica*, belonging to. From the Alaj Valley, Turkestan region of Central Asia
alainii In honor of Enrique E. Alain (1916–) also known as Brother or Hermano Alain or Liogier, Cuban cleric and plant collector

A

alajica See *alaica*

alakaiense L. *-ense*, denoting origin. From Alakai Kauai one of the Hawaiian Islands

alamii In honor of Mohammed Khairul Alam (1952–) Bangladeshi botanist

alamosae From Alamose, Mexico

alamosana L. *-ana*, indicating connection. As for *alamosae*

alany A misspelling of *alang*

alaotrensis L. *-ensis*, denoting origin. From the road to Lake Alaotra, Madagascar

alascana L. *-ana*, indicating connection. From Alaska

alashanica L. *-ica*, belonging to. From Ala Shan Ranges, China

alaskan-a, -um L. *-ana*, indicating connection. From Alaska

alat-a, -um L. *ala*, wing; *-ata*, possessing. – (1) Rhachides broadly winged. *Mesosetum alatum* – (2) glumes broadly winged. *Dimeria alata, Triticum alatum* – (3) lower culm leaves with small blades. *Dinochloa alata* – (4) keel of lower glume winged. *Schizachyrium alatum*

alatavicum L. *-icum*, belonging to. From Alatau, Pamir district, on the border of Kyrgyzstan and Tajikistan

alatum See *alata*

alb-a, -um, -us L. white. Often applied to species some or all of whose parts are covered with long white hairs, but may also indicate that the surfaces of smooth lemmas or fruits are white

albanicum L. *-icum*, belonging to. From Albania

albemarlens-e, -is L. *-ense*, denoting origin. – (1) From Albemarle, now Isabela, one of the Galapagos Islands. *Leptochloa albemarlensis* – (2) from Albemarle County, Virginia, USA. *Panicum albemarlense*

albens L. *albeo*, be white. – (1) Glumes white. *Isachne albens, Panicum albens* – (2) seed white. *Sporobolus albens*

albensis L. *-ensis*, denoting origin. From Alba, central Rumania

albertense L. *-ense*, denoting origin. From Albert, northern Australia

albertii In honor of Albert Regel (1845–1908) Swiss botanist

albertsonii In honor of Frederick William Albertson (1892–1961) United States botanist

albescens L. *albesco*, become white. – (1) Lemmas rendered hoary by hairs. *Eragrostis albescens, Koeleria albescens, Megaloprotachne albescens, Tridens albescens, Triodia albescens* – (2) pedicels invested in long silvery hairs. *Andropogon albescens* – (3) whole plant whitish. *Poa albescens*

albicans L. *albico*, make white. Spikelets greenish-white

albicauda L. *albus*, white; *cauda*, tail. Inflorescence pallid

albicom-a, -um L. *albus*, white; *coma*, the hair of the head. – (1) Spikelets densely hairy. *Brachiaria albicoma, Panicum albicomum* – (2) leaf-sheaths densely hairy. *Digitaria albicoma*

albid-a, -um, -us L. whitish. Plant in whole or in part white usually due to an indumentum of short hairs

albidulum L. *albidus*, whitish; *-ulum*, diminutive. Whole plant glaucescent

albid-um, -us See *albida*

albilanata L. *albus*, white; *lana*, wool; *-ata*, possessing. A prominent band of white, woolly hairs present just below each node

albimontana L. *albus*, white; *mons*, mountain; *-ana*, indicating connection. From Wittebergen (Dutch, white-washed mountains) near Cape Town, South Africa

albinervis L. *albus*, white; *nervum*, vein. Veins of lemma white

albocerea L. *albus*, white; *ceres*, wax. Culms densely farinose

albociliat-a, -um L. *albus*, white; *cilium*, hair; *-ata*, possessing. Invested in part or total with white hairs

alboffii In honor of Nicolas Mikhailowitch Alboff (1866–1897) Russian botanist and traveller

albohispidula L. *albus*, white; *hispidus*, bristly; *-ula*, diminutive. New culms densely invested with short, white hairs

albomaculatum L. *albus*, white; *macula*, spot; *-atum*, possessing. Culm-sheaths purplish and white-spotted

albomarginat-a, -um L. *albus*, white; *margo*, edge; *-ata*, possessing. – (1) Glumes or lemmas white-edged. *Digitaria albomarginata* – (2) leaf-sheath white-edged. *Sasa albomarginata* – (3) leaf-blade white edged. *Isachne albomarginata*, *Panicum albomarginatum*

albosericea L. *albus*, white; *sericeus*, silken. Dense white hairs are abundant immediately above the nodes

albospiculatum L. *albus*, white; *spica*, a point; hence, in particular, an ear or spike of grain; *-ula*, diminutive; *-atum*, possessing. Anthoecia light-yellowish

albovellereum L. *albus*, white; *vellus*, fleece. Origin uncertain. The name has no accompanying description

albovii See *alboffii*

albovillosum L. *albus*, white: *villus*, shaggy hair; *-osum*, abundance. Plant in whole or in part covered with long, white hairs

albowianus L. *-anus*, indicating connection. As for *alboffii*

alb-um, -us See *alba*

alcobense L. *-ense*, denoting origin. From Cerro Alcoba, Guatamela

aldabrens-e, -is L. *-ense*, denoting origin. From Aldabra, an island of the Seychelles Republic in the Indian Ocean

Alectoridia, alectoridia L. *alectorideus*, chicken. Upper glume markedly keeled thereby resembling the comb of a young fowl

alectorocnemum Gk *alectoris*, cock; *kneme*, leg as between knee and ankle. Inflorescence resembling a cock's foot

aleppica L. *-ica*, belonging to. From Aleppo, Syria

aleutensis L. *-ensis*, denoting origin. See *aleutica*

aleutic-a, -us L. *-ica*, belonging to. From the Aleutian Islands, Alaska, USA

alexandrae In honor of Annie Montague Alexander (1867–1950) Hawaiian-born United States collector

alexeenchoana L. *-ana*, indicating connection. As for *alexeenkoi*

alexeenkoi In honor of Th. Alexeenko (fl. 1900) Russian plant collector

alexeji In honor of Aleksey Konstantinovich Skvortsov (1920–) Russian botanist

Alexfloydia In honor of Alexander Geoffrey Floyd (1926–) Australian botanist

alfrediana L. *-ana*, indicating connection. As for *Yvesia*

algeriensis L. *-ensis*, denoting origin. From Algeria

algida L. cold. – (1) From the Arctic. *Catabrosia algida*, *Phippsia algida* – (2) from high mountains such as the Andes. *Poa algida*

algidiformis L. *forma*, appearance. Resembling a related species with the epithet *algida*

alien-a, -us L. different from. – (1) Readily confused with related species. *Festuca aliena*, *Lappago aliena*, *Nazia aliena*, *Roegneria aliena*, *Stipa aliena*, *Tragus alienus* – (2) from another country. *Oxytenanthera aliena*

Allagostachyum Gk *allage*, a change; *stachys*, spike as of an ear of wheat. The lower lemmas are sometimes sterile

allang Vernacular name for the species in Malaya

alleizettei In honor of Charles d'Alleizette (1884–1967). French administrator who collected in Madagascar

Allelotheca Gk *allelon*, one another; *theke*, box. Spikelets alternate in depressions along the raceme

allenii In honor of Timothy Field Allen (1837–1902) United States physician and botanist

allionii In honor of Carlo Allioni (1725–1804) Italian botanist

Alloeochaete Gk *alloios*, of a different kind; *chaete*, bristle. With hair tufts on the lemma

Alloiatheros Gk *alloios*, of a different kind; *ather*, barb or spine. Fertile and sterile lemmas both awned but differing in appearance

Allolepis Gk *allo*, strange; *lepis*, scale. Palea keels winged

Alloteropsis Gk *allotrios*, belonging to another; *opsis*, appearance. The spikelets and inflorescences somewhat resemble those of *Panicum*

Alloterrhopsis See *Alloteropsis*

almadens-e, -is L. *-ense*, denoting origin. From Almade, Brazil

almasovii In honor of Almasov

almaspicata L. *alma*, bountiful; *spica*, a point; hence, in particular, an ear or spike of grain; *-ata*, possessing. Spike-like inflorescence large for genus

almeriensis L. *-ensis*, denoting origin. From Almeria Province, Spain

almum L. nourishing. A nutritious forage grass

alnasteretum L. *-etum*, place of growth. Growing amongst *Alnaster fruticosus* (Alder) woodlands

alonsoi In honor of José Mario Alonso (1926–1991) Argentinian plant geneticist

alopecuroide-a, -um Gk *-oidea*, resembling. Inflorescence like that of *Alopecurus*

alopecuroides Gk *-oides*, resembling. Similar to *Alopecurus*, usually in respect of the inflorescence

alopecuroideum See *alopecuroidea*

Alopecuropsis Gk *opsis*, resemblance. Resembling *Alopecurus*

alopecuros Resembling *Alopecurus*

Alopecurus Gk *alopex*, fox; *oura*, tail. The spicate inflorescence resembles a fox tail

alopecurus Resembling *Alopecurus*

Alophochloa Gk *a-*, without; *lophos*, crest; *chloa*, grass. Lemma unawned

alpestr-e, -is L. *alpes*, high mountain; *-estre*, place of growth. Growing on high mountains

alpicola L. *alpes*, high mountain; *-cola*, dweller. Growing on high mountains

alpigena L. *alpes*, high mountain; *gigno*, beget. High mountain species

alpin-a, -um, -us L. *alpes*, high mountain; *-ina*, indicating possession. Species growing at high altitudes

alsinoides Gk *-oides*, resembling. Prostrate with ovate hairy leaf-blades resembling those of *Alsine*

alsodes Gk *alsodes*, woodland. Woodland species

alsophilum Gk *alsos*, grove; *phileo*, love. Growing in woodlands

alt-a, -um, -us L. tall. Culms tall, relative to those of other members of the genus

altaic-a, -um, -us L. *-ica*, belonging to. From the Altai Mountains bordering Mongolia and China

altera L. one of two. The second described species of a genus

alternans L. *alterno*, change. – (1) Rhachis produced into setaceous lobes opposite all or some of the spikelets. *Plagiochloa alternans*, *Tribolium alternans* – (2) panicle branches arising from alternate nodes. *Andropogon alternans*

alternatum L. *alterno*, change. Origin uncertain, not given by author

alterniflora L. *alternu*, placed alternately; *flos*, flower. The spikelets clearly in two rows on one side of the axis

alticola L. *altus*, lofty; *-cola*, dweller. Grows at high altitudes

altijugum L. *altus*, tall; *jugum*, summit of a mountain. Growing on the tops of high mountains

altiligulata L. *altus*, tall; *ligulus*, small tongue; *-ata*, possessing. Ligule long

alti-or, -us L. taller. Taller than related species

altissim-a, -um, -us L. *altus*, tall; *-issima*, most. Very tall compared with other members of the genus

altius See *altior*

Altoparadisium From Alto Paraíso, Brazil

altopyrenaicum L. *altus*, tall; *-icum*, belonging to. From high peaks in the Pyrenees

altsonii In honor of Ralph Abbey Altson (fl. 1925) who collected in British Guiana, now Guyana

alt-um, -us See *alta*

alveiformis L. *alveus*, little hollow; *forma*, appearance. Leaves and culm internodes with pitted or warty glands

Alvimia In honor of Paulo de Tarso Alvim (fl. 1972–1976) Director, Cacao Research Centre in Itabuna, Brazil

Alycia See *Aglycia*

amabil-e, -is L. lovely. Of attractive appearance

amaena See *amoena*

Amagris A contraction of *Calamagrostis*

amahussana L. *-ana*, indicating connection. From Amahussa, near Amboina, Moluccas, Indonesia

amakusensis L. *-ensis*, denoting origin. From Amakusa Island, Japan

amaliae In honor of Amalia Vissers (1949–) wife of J. F. Veldkamp

amanda L. meriting love. Worthy of recognition

amapaensis L. *-ensis*, denoting origin. See *amapana*

amapana L. *-ana*, indicating connection. From Amapá, Brazil

amaroides Gk *-oides*, resembling. Similar to *Panicum amarum*

amarulum L. *amarus*, unpleasant; *-ulum*, diminutive. The foliage has a slightly bitter taste

amarum L. unpleasant. Foliage bitter to the taste

amaur-a, -us Gk *amauros*, dark. Racemes reddish-brown

Amaxitis Gk *amaxa*, carriage road; *-itis*, indicating a close connection. Commonly growing along roadsides

amazonensis L. *-ensis*, denoting origin. From Amazonas Province, Brazil

amazonic-a, -um L. *-ica*, belonging to. From the Amazon Basin, Brazil

ambalavaoensis L. *-ensis*, denoting origin. From Ambalavao District, Madagascar

ambatoensis L. *-ensis*, denoting origin. From Serra de Ambato, Catamarea Province, Argentina

ambiens L. *ambio*, surround. Intermediate in characters between other species pairs

ambigens L. *ambigo*, be uncertain. Part of a species complex whose members are difficult to delineate

ambigu-a, -um, -us L. uncertain. Species that may be readily confused with others or do not necessarily belong in the genus in which they have been placed or are intermediate in characters between other genera

ambilobensis L. *-ensis*, denoting origin. From Ambilobe, a district in Madagascar

ambitiosum L. *ambitio*, desire honor; *-osum*, abundance. The large effuse inflorescence draws attention to the plant

ambleia Gk *amblys*, blunt. Glumes truncate

Amblichloa Gk *amblys*, blunt. Lemma apices rounded

Amblyachyrum Gk *amblys*, blunt; *achryon*, chaff. Apices of glumes obtuse

amblyantha Gk *amblys*, blunt; *anthos*, flower. Lemma apices rounded

Amblyochloa Gk *amblys*, blunt; *chloa*, grass. The glumes are truncate

amblyodes Gk *amblys*, blunt; *-odes*, resembling. Glume apices obtuse

Amblyopyrum Gk *amblys*, blunt; *pyros*, wheat. The spikelets resemble those of wheat but with blunt glumes

Amblytes Gk *amblys*, blunt. Lower glume neither awned nor aristate

ambohibengensis L. *-ensis*, denoting origin. From Mt Ambohibenga, Madagascar

amboinensis L. *-ensis*, denoting origin. From Amboina, Moluccas, Indonesia

amboinic-a, -us L. *-ica*, belonging to. See *amboinensis*

amboinicea See *amboinensis*

ambongens-e, -is L. *-ense*, denoting origin. From Ambongo, Madagascar

ambositrensis L. *-ensis*, denoting origin. From Ambositra, Madagascar

ambrensis L. *-ensis*, denoting origin. From Massif d'Ambre, Madagascar

ambustum L. *amburo*, injure. Margins of leaf-blades very scabrid and likely to cut if not carefully handled

ameghinoi In honor of Carlos Ameghino (1865–1936) botanical collector in Patagonia and Argentina

american-a, -um, -us L. *-ana*, indicating connection. From North or South America

amethyste-a, -um L. *amethysteus*, violet. Spikelets violet

amethystin-a, -um, -us L. *amethysteus*, violet; *-ina*, indicating resemblance. Spikelets dark purplish-red

amgunens-e, -is L. *-ense*, denoting origin. From Amgun River, Russian Far East

amherstiana L. *-ana*, indicating connection. In honor of William Pitt Amherst (1773–1857) British diplomat and naturalist

amistadensis L. *-ensis*, denoting origin. From the Parque Internacional La Amistad, a World Heritage area in Republic of Costa Rica and Panama

ammobia L. *ammobia*, sand dweller. Growing in damp sand

Ammocalamagrostis Hybrids between species of *Ammophila* and *Calamagrostis*

Ammochloa Gk *ammos*, sand; *chloa*, grass. Growing on sand

ammodes Gk *ammos*, sand. Growing in sandy soils

Ammophila, -a, -um Gk *ammos*, sand; *phileo*, love. Growing in sandy habitats

ammophyla See *Ammophila*

amnigenum L. *amnis*, river; *gigno*, bear. Growing near water

amoen-a, -um L. beautiful. Attractive in appearance

Ampelocalamus Gk *ampelos*, vine; *kalamos*, reed. Culms slender and scandent

Ampelodesma, Ampelodesmos, Ampelodesmus Gk *ampelos*, vine; *desmos*, band. The leaves provide twine for tying up grapevines

Ampelodonax Intermediate between *Ampelodesmos* and *Donax*

amphibium L. living both in water and on land. Growing on swamp margins

amphibolum Gk *amphibolos*, doubtful. Species readily confused with another

Amphibromus Gk *amphi-*, double. The spikelets resemble those of *Bromus*

amphicarpa Gk *amphi-*, both; *karpos*, fruit. Inflorescences either long or short

Amphicarpon See *Amphicarpum*

Amphicarpum Gk *amphi-*, double; *karpos*, fruit. Inflorescences with both aerial and subterranean spikelets

Amphidonax Gk *amphi-*, double; *donax*, reed. Resembling *Donax*

Amphigenes Gk *amphi-*, double; *genos*, descent. Intermediate between *Festuca*, *Poa* and *Molinia*

Amphilophis Gk *amphi-*, double; *lophos*, crest or tail. Both the inflorescence branches and their peduncles are hairy

Amphipogon Gk *amphi-*, double; *pogon*, beard. Both lemmas have several awns arising from incisions at their apices

amphipogonoides Gk *-oides*, resembling. Similar to *Amphipogon*

amphistemon Gk *amphi-*, double; *stemon*, thread. Two of the rhachilla internodes are elongated and thread-like

amphitricha Gk *amphi-*, double; *thrix*, hair. Indumentum different on upper and lower leaf surfaces

Amphochaeta Gk *ampho-*, around; *chaete*, bristle. The spikelets, either solitary or in clusters of two or three, are subtended by an involucre of bristles

amphora L. *amphora*, vase, usually two-handled. Anthoecium inflated-fusiform and lacking handles

amphoralis L. *amphora*, vase, usually two-handled; *-alis*, pertaining to. Anthoecium inflated-fusiform and lacking handles

ampl-a, -um, -us L. large. – (1) Spikelets large. *Festuca ampla* – (2) inflorescences large. *Agrostis ampla*, *Andropogon amplus*, *Bromus amplus*, *Deyeuxia ampla*, *Sorghastrum amplum* – (3) culms large. *Ichnanthus amplus*. – (4) plant robust in all respects. *Sorghum amplum*

amplectens L. *amplecto*, encircle. lower glumes reniform

amplexicaul-e, -is L. *amplexatio*, embrace; *caulis*, stem. The connate leaf-sheath and ligule encircle the stem

amplexifoli-a, -um L. *amplexus*, encircling; *folium*, leaf. Leaf-blades cordate and amplexicaule

amplexum L. encircling. The inflorescence only shortly exserted from the flag-leaf

ampliculmis L. *amplus*, large; *culmus*, stalk. Culms large

ampliflor-a, -um, -us L. *amplus*, large; *flos*, flower. Spikelets large

amplifoli-a, -um L. *amplus*, large; *folium*, leaf. Leaf-blades large

amplissim-a, -um L. *amplus*, large; *-issima*, most. – (1) Culms taller than many other species of the genus. *Arundinaria amplissima*, *Aulonemia amplissimum*, *Festuca amplissima*, *Panicum amplissimum*, *Sasa amplissima* – (2) inflorescence a large panicle. *Aristida amplissima*

amplopaniculata L. *amplus*, large; *panus*, thread; *-ula*, diminutive; *-ata*, possessing. Inflorescence large panicle

ampl-um, -us See *ampla*

ampullacea L. *ampulla*, flask; *-acea*, indicating resemblance. Spikelets flask-shaped

ampullaris L. *ampulla*, flask; *-aris*, pertaining to. Culm-sheaths in upper part contracted into the shape of a bottle neck

amurens-e, -is L. *-ense*, denoting origin. From the Amur District, Russian Far East

amylacea L. *amylum*, starch; *-acea*, indicating resemblance. Grain with abundant starch

amyle-a, -um L. *amylum*, starch. Used for the manufacture of starch

anabaptistum L. rebaptize. Until described, the species had been masquerading under another name

Anachortus Gk *ana*, alike; *chortus*, fodder. Species of related genera useful fodder grasses

Anachyra See *Anachyris*

Anachyris Gk *an-*, without; *achryon*, chaff. The spikelets lack glumes

Anachyrium A misspelling of *Anachyris*

anaclasta Gk *ana*, not; *klastos*, broken in pieces. Awn without an articulation

anacrantha Gk *an-*, without; *akros*, at the end; anthos, flower. Inflorescence branch tips lacking spikelets

anacranthoides Gk *-oides*, resembling. See *anacrantha*

anadabolavensis L. *-ensis*, denoting origin. From Anadabolava, Madagascar

Anadelphia Gk *an-*, without; *adelphos*, brother. Racemes unpaired

anadyrensis L. *-ensis*, denoting origin. From the banks of the Anadyr' River, Russian Far East

anadyrica L. *-ica*, belonging to. See *anadyrensis*

anae In honor of Ana María Crespo de Las Casas (1948–) Spanish botanist

analabensis L. *-ensis*, denoting origin. From Analaba, Madagascar

analamazaotrensis L. *-ensis*, denoting origin. From the forest of Analamazaotra, Madagascar

anamesa Gk *anamesos*, in the heart of a country. Throughout eastern Africa from the Sudan to South Africa

anamitica L. *-ica*, belonging to. From Annam

anantaswamianus In honor of M. Anantaswami, Indian botanist

Anastrophus Gk *an-*, without; *strophe*, a twist. The spikelets are oriented in such a way that their pedicels are not interpreted as twisted, as they are in a related genus

anather-a, -us Gk *an-*, without; *ather*, spike or ear of wheat. Glumes and/or lemmas lacking an awn

Anatherostipa Gk *an*, without; *ather*, barb or spine. Awn of type species reduced to a subuliform mucro

Anatherum Gk *an-*, without; *ather*, barb or spine. The lemma is weakly awned or unawned. The name has been applied to two distinct genera

anatolic-a, -um, -us L. *-ica*, belonging to. From Anatolia, Turkey

anaurita L. *an-*, without; *aurita*, long-eared. The leaf-blades lack auricles

ancachsana L. *-ana*, indicating connection. From Anchachs, Huari Province, Peru

ancashensis L. *-ensis*, denoting origin. From Ancash Region, Peru

anceps L. two-edged. Culms laterally compressed

ancestrale French *ancêtre*, ancestor; L. *-ale*, pertaining to. Regarded as ancestral especially with reference to cultivated species

Ancistrachne Gk *agkistron*, fish-hook; *achne*, chaff. The subtending glumes and sterile lemma bear hooked hairs

Ancistragrostis Gk *agkistron*, fish-hook; *agrostis*, grass. Lemma terminates in a hooked awn

Ancistrochloa Gk *agkistron*, fish-hook; *chloa*, grass. The rhachilla extends beyond the florets and is hairy basally but glabrous terminally

ancoraimensis L. *-ensis*, denoting origin. From Ancoraimis, La Paz Department, Bolivia

ancylocarpum Gk *agkylos*, curved; *karpos*, fruit. Anthoecium gibbous in outline

ancylochaete Gk *agkylos*, curved; *chaete*, bristle. Long bristle of involucre recurved

ancylotrich-a, -um Gk *agkylos*, curved; *thrix*, hair. Lower glume with a few uncinate hairs

andamanic-a, -um L. *-ica*, belonging to. From Andaman Islands, India

andersonii In honor of Edgar Shannon Anderson (1897–1969) United States botanist

anderssonii In honor of Nils Johan Andersson (1821–1880) Swedish botanist

andicola L. *-cola*, dweller. Andean species

andin-a, -um, -us L. *-ina*, indicating possession. – (1) From the Andes. *Axonopus andinus*, *Chaetotropis andina*, *Elymus andinus*, *Hemimunroa andina*, *Hordeum andinum*, *Poa andina*, *Triniochloa andina* – (2) from other high mountains such as those of the north-western United States. *Agropyrum andinum*, *Calamagrostis andina*, *Danthonia andina*, *Muhlenbergia andina*

andongens-e, -is L. *-ense*, denoting origin. From Pungo Andongo District, Angola

andoniensis L. *-ensis*, denoting origin. From Andoni, Namibia

andraei In honor of Karl Justus Andrä (1816–1855) German palaeontologist

andreanszkyi In honor of Gábor Andreánszky (1895–1967) Hungarian botanist

andreanum L. *-anum*, indicating connection. In honor of Edouard-Francois André (1840–1911) who collected extensively in Central and South America

andringitrens-e, -is L. *-ense*, denoting origin. From Mount Andringitra, Madagascar

androgyna Gk *aner*, man; *gyne*, woman. Spikelets three-flowered, the lower functionally male, the upper two functionally female

androphil-a, -us Gk *aner*, man; *phileo*, love. Most spikelets of the racemes are male

andropogoides Gk *-oides*, resembling. Similar to *Andropogon*

Andropogon, Andropogum Gk *aner*, man; *pogon*, a beard. The pedicels of many species are invested with long hairs

andropogonoides Gk *-oides*, resembling. Resembling species of *Andropogon*

Andropogum See *Andropogon*

Andropterum Gk *aner*, man; *pteron*, wing or feather-like. The lower glume of the stalked male spikelets is winged

Androscepia Gk *aner*, man; *skepas*, covering. The sessile hermaphrodite is surrounded by an involucre of four male spikelets

androssovii In honor of N. V. Androssov (fl. 1934)

anelythra Gk *a-*, without; *elytron*, cover. Lower pair of glumes much reduced

anelythroides Gk *-oides*, resembling. Similar to *Chusquea anelytra*

anelytra See *anelythra*

anelytroides Gk *-oides*, resembling. Similar to *Chusquea anelytra*

Anelytrum Gk *an-*, without; *elytron*, cover. The spikelets lack glumes

Anemagrostis Gk *anemos*, wind; *agrostis*, grass. An allusion to *spica-venti*, which see

Anemanthele Gk *anemos*, wind; *anthele*, plume. From its vernacular name of Wind Plume-Grass

anemopaegma Gk *anemos*, wind; *paigma*, play. The long hairy awn on the lemma of the sessile spikelet causes the inflorescence to shake in the slightest breeze

anemotum Gk *anemos* wind. Growing on an island subject to strong breezes during the visit of the collector

Aneurolepidium Gk *a-*, without; *neuron*, nerve; *lepis*, scale. The glumes or lemmas are weakly nerved

anfamensis L. *-ensis*, denoting origin. From Custa de Anfama, Argentina

angarense L. *-ense*, denoting origin. From the Angara River, on the border of the Russian Far East and China

angladei In honor of L. Anglade (fl. 1922–1926) cleric and amateur botanist who collected in India

anglic-a, -um L. *-ica*, belonging to. From Anglia, that is England

angolens-e, -is L. *-ense*, denoting origin. – (1) From Angola. *Aristida angolensis, Eragrostis angolensis, Jardinea angolensis, Panicum angolense, Pennisetum angolense, Rottboellia angolensis* – (2) from Angol, Chile. *Piptochaetium angolense*

angrenic-a, -us L. *-ica*, belonging to. From the Angren Basin, Central Asia

angular-e, -is L. *angulus*, angle; *-are*, pertaining to. Origin uncertain, not given by the author

angulat-a, -um, -us L. *angulus*, angle; *-atus*, possessing. – (1) Culm angular. *Coix angulata, Dactylodes angulatum, Tetragonocalamus angulatus* – (2) rhachis angular. *Elymus angulatus*

angulosum L. *angulus*, angle; *-osum*, abundance. Culms spreading, branching at the nodes

angust-a, -um, -us L. narrow. Narrow, with respect to leaf-blades or spicate panicles

angustata L. *angusta*, narrow; *-ata*, possessing. Panicle very narrow

angustiflor-a, -um L. *angustus*, narrow; *flos*, flower. Spikelets terete

angustifoli-a, -um, -us L. *angustus*, narrow; *folium*, leaf. Leaf-blades narrow

angustiglum-e, -is L. *angustus*, narrow; *gluma*, husk. Glumes narrow

angustispiculatus L. *angustus*, narrow; *spica*, a point; hence, in particular, an ear or spike of grain; *-ulus*, diminutive; *-atus*, possessing. Spikelets long and narrow

angustissim-a, -um, -us L. *angustus*, narrow; *-issima*, most. Leaf-blades very narrow

angust-um, -us See *angusta*

anhispidis See *anhispidus*

anhispidus Gk *an-*, without; L. *hispidus*, rough. Culm leaf-sheaths glabrous at the base

anias Vernacular name of the species in the Philippines

animarum Of the Sierra Ánimas, Brazil

Anisachne Gk *anisos*, unequal; *achne*, scale. The glumes differ in size and shape

Anisantha Gk *anisos*, unequal; *anthos*, flower. Upper florets in the spikelet smaller than the lower and usually sterile

Aniselytron Gk *anisos*, unequal; *elytron*, cover. The glumes are conspicuously unequal

anisochaeta Gk *anisos*, unequal; *chaete*, bristle. Central branch of trifid awn much longer than the two laterals

anisoclada Gk *anisos*, unequal; *klados*, branch. Panicle branches unequal

Anisopogon, anisopogon Gk *anisos*, unequal; *pogon*, beard. The lemma has a long central awn and two shorter equal lateral awns

Anisopyrum Gk *anisos*, unequal; *pyros*, wheat. Glumes of unequal length

anisostachium See *anisostachyum*

anisostachyum Gk *anisos*, unequal; *stachys*, spike as of an ear of wheat. Paired spikelets on pedicels of different length

anisotrichum Gk *anisos*, unequal; *thrix*, hair. Indumentum of two hair types

ankaratrensis L. *-ensis*, denoting origin. From Ankaratra, Madagascar

ankarense L. *-ense*, denoting origin. From Ankarana Province, Madagascar

ankolib Vernacular name of the species in Ethiopia

annableae In honor of Carol Ruth Annable (fl. 1993) who collected in Bolivia

annamens-e, -is L. *-ense*, denoting origin. From Annam now included within Vietnam

annu-a, -um, -us L. *annuus*, lasting a year. Annuals

annuale L. *annuus*, lasting a year; *-ale*, pertaining to. Annuals

annularis L. *annulus*, ring; *-aris*, pertaining to. The subtending bristles form a cup below the spikelet

annulat-a, -um, -us L. *annulus*, ring; *-ata*, possessing. – (1) Furnished with a ring, as with the thickened lower subtending glume forming a fleshy ring at the base of the spikelet. *Eriochloa annulata* – (2) the pedicels with a glandular ring. *Eragrostis annulata* – (3) nodes bearded. *Andropogon annulatus, Dichanthium annulatum, Lipeocercis annulata*

annulifera L. *annulus*, ring; *fero*, carry or bear. Culm nodes thick, black and tyre-like

annulum L. ring. Nodes pubescent

annu-um, -us See *annua*

anomal-a, -um, -us Gk *anomalos*, irregular. Diverging from the normal, often with respect to number of lemmas in the spikelet or otherwise unusual for the genus in some respect

Anomalotis Gk *anomalos*, irregular; *os*, ear. Lemma apex five-awned unlike that of related genera whose lemmas have fewer awns

anomal-um, -us See *anomala*

Anomochloa Gk *anomalos*, irregular; *chloa*, grass. The spikelet structure is unique amongst grasses

anomoplexis Gk *anomos*, irregular; *plexis*, may be applied to anything that strikes. The bristles that subtend the spikelet vary in length and with their retrorse barbs, resemble spear heads

Anoplia Gk *anoplos*, without armour. Lemma unawneds

ansat-a, -um L. *ansa*, handle; *-ata*, possessing. Lower glume bent at the base to form a handle as a cup

antarctic-a, -um, -us L. *anti*, opposite; *arctica*, the Arctic. With southern hemisphere distributions south of about the latitude of the Shetland Islands

Antephora See *Anthephora*

antephoroides Gk *-oides*, resembling. Similar to *Anthephora*

Anthaenantia See *Anthenantia*

Anthaenantiopsis Gk *opsis*, resemblance. Similar to *Anthenantia* in that the palea and lemma gape at maturity revealing the grain

Anthaenantropsis See *Anthaenantiopsis*

Anthenantia Gk *anthos*, flower; *enantios*, contrary. Palisot de Beauvois interpreted the palea as being at right angles to the lemma

Anthephora Gk *anthos*, flower; *phero*, bear. The involucre may be taken to resemble a calyx

Anthersteria See *Anthistiria*

Anthesteria See *Anthistiria*

Anthestiria See *Anthistiria*

Anthipsimus Gk *anthos*, flower; *psimythos*, white lead. Origin uncertain, not given by the author, but may refer to the color of the lemmas

Anthisteria See *Anthistiria*

Anthistiria Gk *anthos*, flower; *steira*, forepart of a ship's keel. The spathes subtending each cluster of spikelets resembles the prow of a ship

anthistirioides Gk *-oides*, resembling. Similar to *Anthistiria* in the form of the inflorescence

Anthochloa Gk *anthos*, flower; *chloa*, grass. The fan-shaped lemmas are white and petal-like

Anthoenantia See *Anthenantia*

Antholithes Gk *anthos*, flower; *lithos*, stone. Fossils resembling grass spikelets

Anthopogon Gk *anthos*, flower; *pogon*, beard. The sterile upper floret grows out into a long awn

Anthosachne Gk *anthos*, flower; *achne*, scale. Upper florets of spikelet sterile

anthosachnoides Gk *-oides*, resembling. Similar to *Anthosachne*

anthoxanthia Gk *anthos*, flower; *xanthos*, yellow. Mature spikelets yellow-green

anthoxanthiformis L. *forma*, appearance. Resembling *Anthoxanthum*

anthoxanthoides Gk *-oides*, resembling. Similar to *Anthoxanthum*

Anthoxanthum, anthoxanthum Gk *anthos*, flower; *xanthos*, yellow. Mature spikelets yellow-green

Anthoxantum See *Anthoxanthum*

Antichloa A misspelling of *Actinochloa*

antidotale L. *antidotum*, antidote; *-ale*, pertaining to. Smoke from burning plants used as a disinfectant against smallpox

antillarum Latinized form of Antilles, an archipelago enclosing the Caribbean Sea

Antinoria In honor of Marchese Vincenzo Orazio Antinori (1811–1882) Italian botanist

antioquensis L. *-ensis*, denoting origin. From Antioquia, Colombia

antipod-a, -um Gk *anti*, opposite; *pous*, foot. Having the feet opposite, that is from the Antipodes

antiquum L. former. Previously regarded as a variety of another species

Antitragus Gk *anti*, like. Similar to *Tragus* in inflorescence form

Antochloa See *Anthochloa*

antofagastensis L. *-ensis*, denoting origin. From Antofagasta, Chile

Antonella L. *-ella*, diminutive but here used as a name-forming suffix. In honor of Ana Mariá Anton (1942–) Argentinian botanist

antoniana L. *-ana*, indicating connection. From San Antonio, Puno Region, Peru

Antoschmidtia See Johann Anton Schmidt under entry for *Schmidtia*

Antoxanthum See *Anthoxanthum*

antsirabens-e, -is L. *-ense*, denoting origin. From Antsirabé, Madagascar

antucensis L. *-ensis*, denoting origin. From Andes de Atuco, Chile

antunesii In honor of José Maria Antunes (1856–1928) Portugese cleric who collected in Angola

aparine Gk *aparine*, bedstraw, also known as cleavers or catchweed, a plant with hooks on the fruits. The spikelets are subtended by barbellate branches

apennina From the Appenines, Italy

Apera Gk *a-*, without; *peros*, mutilated. Spikelets similar to *Calamagrostis* but often with a second floret and thus relatively unreduced, or a euphonous but meaningless name proposed by Adanson for a quite separate genus

apert-a, -um L. open. The glumes of the florets in the central portion of the spikelets tend to be open

apetala Gk *a-*, without; L. *petalum*, petal. Glumes minute

Aphanelytrum Gk *aphanes*, invisible; *elytron*, cover. The subtending glumes are very small in comparison to the length of the spikelet

aphanes Gk invisible. Glume nerves not visible

aphanoneur-a, -um Gk *aphanes*, invisible; *neuron*, nerve. Veins of the glumes inconspicuous

Aphonina Gk *aphona*, broad-leaved plant; *-ina*, indicating resemblance. Leaf-blades broad

aphylla Gk *a-*, without; *phyllon*, leaf. Leaf-blades poorly developed

apiatus L. crisped. Inflorescence invested in curly hairs which turn red at maturity

apiculat-a, -um, -us L. *apiculum*, small point; *-ata*, possessing. Glumes or lemmas sharp-pointed

Aplexia Gk *aplexis*, unplaited. Rhizomes clustered and stouter than the culm base, whereas in related species they are slender and interlaced

Aplocera Gk *aploos*, single garment; *keras*, horn. The upper glume has a long awn in contrast to the lower which is shortly awned or awnless

Apluda L. *apluda*, chaff or bran. The name was used by Pliny for a millet but may be a reference to the chaffy spathes left on the inflorescence after the spikelets have fallen

apludoides Gk *-oides*, resembling. Similar to *Apluda*

Apochaete Gk *apo-*, separate; *chaete*, bristle. Lemma lobes aristulate

Apochiton Gk *apo-*, separate; *chiton*, tunic. The membranous pericarp readily separates from the seed

Apoclada Gk *apo-*, separate; *klados*, branch. The mid-culm branch complements arise independently

Apocopis Gk *apo-*, separate; *kopis*, meat cleaver. The sessile spikelet is truncate

Apocopsis See *Apocopis*

apoensis L. *-ensis*, denoting origin. From Mt Apo, Mindanao, Philippines

Apogonia Gk *a-*, without; *pogon*, beard. Glumes unawned

appendiculat-a, -um, -us L. *appendix*, appendage; *-ulus*, diminutive. *-ata*, possessing. – (1) Lemma of the terminal spikelet drawn out into an appendage. *Anthephora appendiculata, Panicum appendiculatum, Paspalum appendiculatum, Setaria appendiculata* – (2) awn with a pair of basal appendages. *Stipa appendiculata* – (3) stalked male spikelets subtended by a conspicuous appendage. *Andropogon appendiculatus, Leptopogon appendiculatus*

appletonii In honor of Arthur Frederick Appleton (1861–1941) English-born British Army veterinarian

appress-a, -um, -us L. appressed. Lateral branches of panicle held erect

appressifolium L. *appressa*, appressed; *folium*, leaf. Leaf-blades held erect

appress-um, -us See *appressa*

appropinquata L. *appropinquo*, approach. Similar to another species

apric-um, -us L. growing in the sunshine. Savanna or grassland species

apsleyensis L. *-ensis*, denoting origin. From Apsley River, Tasmania, Australia

aptera Gk *a-*, without; *pteron*, wing or feather-like. The palea keels are unwinged

apuanica L. *-ica*, belonging to. From Apuane Alps, Italy

apus A contraction of *pring apus*, the vernacular name of the species in Java, Indonesia

aquarii L. *aquarius*, water carrier. From the Waterhouse Range, Northern Territory, Australia

aquariorum L. *aquarius*, water-man. Swamp species

aquarum L. *aqua*, water. Of waters; habitat regularly flooded

aquatic-a, -um L. *aqua*, water; *-ica*, belonging to. Growing in or close to water

aquehongensis L. *-ensis*, denoting origin. From *Aquehonga*, the Native American name for Staten Island, New York, USA

aquisgranensis L. *-ensis*, denoting origin. From *Aquisgranum*, the Latin name for Aachen, Germany

arabic-a, -um, -us L. *-ica*, belonging to. From Arabia

arabiifelicis From *Arabia Felix*, the name by which Ptolemy designated the northwest portion of the Arabian peninsula

arachnifera L. *arachne*, web; *fero*, carry or bear. With lemmas bearing copious tangled hairs at their base

arachniform-e, -is Gk *arachne*, spider; L. *forma*, appearance. The geniculate branches fan out from the nodes thereby resembling the legs of a spider

arachnoide-a, -um Gk *arachne*, spider; *-oides*, resembling. Densely hairy as of internodes or leaf-sheaths, or of awns

arachnoides Gk *arachne*, web; *-oides*, resembling. Inflorescence a much branched panicle

arachnoideum See *arachnoidea*

arachnopus Gk *arachne*, web; *pous*, foot. Having dense white hairs on the leaf-sheaths

araeanth-a, -um Gk *araios*, slender; *anthos*, flower. Spikelets narrow

aragonense L. *-ense*, denoting origin. From Aragon, Spain

araiostachya Gk *araios*, narrow; *stachys*, spike as of an ear of wheat. Spikelets terete

arakii In honor of Yeiichi Araki (1904–1955) Japanese botanist

arakiyeitiana L. *-ana*, indicating connection. As for *arakii*

aralensis L. *-ensis*, denoting origin. From Aral-Caspian Desert, central Asia

araratic-a, -um L. *-ica*, belonging to. From Mt Ararat on the border of Turkey and Armenia

araucan-a, -um, -us In honor of the Araucana, a Chilean tribe

araxensis L. *-ensis*, denoting origin. From the valley of the Araxes, now Arax River in the Caucasus adjacent to Iran

Arberella L. *-ella*, diminutive but here used as a name-forming suffix. In honor of Agnes Arber (1879–1960) English botanist

arborescens L. *arboresco*, become tree-like. Habit shrub-like

arborum L. *arbor*, tree. Woodland species

arbusculum L. *arbor*, tree; *-ulum*, diminutive. With the habit of a small tree

arcaensis L. *-ensis*, denoting origin. From Cuesta de *Arca*, Tucumán Province, Argentina

arcana L. *arca*, chest; *-ana*, indicating connection. Hidden away as in a chest and so overlooked either because of rarity or confusion with another species

Arcangelina In honor of Giovanni Arcangeli (1840–1921) Italian botanist

archaelymandra Gk *arche*, begin, as of time. With two homogamous pairs at the base of the raceme and therefore more primitive than other *Elymandra* species

archboldii In honor of Richard Archbold (1907–1976) United States explorer and mammologist

Archeoleersia Gk *archaios*, ancient. Fossil grasses resembling *Leersia*

archeri In honor of William Archer (1820–1874) English-born Tasmanian botanist

archiensis See *carchiense*

arcta L. close. Culms densely tufted

Arctagrostis Gk *arktos*, north. Resembling *Agrostis* and growing in the Arctic

arctasianum See *arktasium*

arctatus L. *arcta*, narrow; *-atus*, possessing. Racemes narrow

arctic-a, -um Gk *arktos*, north; *-ica*, belonging to. Occurring in and often extending beyond the Arctic

Arctodupontia Hybrids involving species of *Arctophila* and *Dupontia*

Arctophila Gk *arktos*, north; *phileo*, love. Widely distributed in the Arctic

Arctopoa Gk *arktos*, north; *poa*, grass. An arctic genus

arctostepporum Gk *arktos*, north; L. *steppus*, steppe. Of the northern steppes, that is Arctic Russia

arctum L. close. Distinguished only by careful comparison from a related species

arcuat-a, -um, -us L. *arcus*, curve; *-ata*, possessing. – (1) Spikelets curved. *Panicum arcuatum* – (2) leaf-blades curved. *Stipa arcuata*

arcurameum L. *arcus*, curve; *ramus*, branch. Culms radiate from centre and are geniculately ascending

arduanum L. *arduum*, a steep place; *-anum*, indicating connection. Cliff dweller

arduennensis L. *-ensis*, denoting origin. From Arduenna, now Ardennes, Belgium

arduensis See *arduennensis*

arduini In honor of Pietro Arduino (1728–1805) Italian botanist

arechavaletae In honor of José Arechavaleta y Balpardo (1838–1912) Uruguayan botanist

arechavaletai See *arechavaletae*

arenace-a, -us L. *arena*, sandy place; *-acea*, indicating position. Species of beach dunes or deserts

arenari-a, -um, -us L. *arena*, sandy place; *-aria*, pertaining to. Of sandy habitats

arenicola L. *arena*, sandy place; *-cola*, dweller. Species of sandy habitats

arenicoloides Gk *-oides*, resembling. Similar to *Panicum arenicola*

arenosus L. *arena*, sandy place; *-osa*, abundance. Growing on sandy soils

arfakensis L. *-ensis*, denoting origin. From Arfak Mountains, Papua, Indonesia

argae-a, -um, -us From Montus Argercus, now Erciyas Dagi, Turkey

argentata L. *argentea*, silvery; *-ata*, possessing. Spikelets silver-white terminally, violet towards the base

argente-a, -um, -us L. silvery. – (1) Glumes and/or lemmas with an indumentum of silvery hairs. *Agrostis argentea, Andropogon argenteus, Elionurus argenteus, Elytrigia argentea, Eulalia argentea, Festuca argentea, Koeleria argentea, Melinis argentea, Moorea argentea, Muhlenbergia argentea, Panicum argenteum, Poa argentea, Pollinia argentea, Sesleria argentea* – (2) awns invested with silvery hairs. *Stipa argentea* – (3) panicle branches invested with silvery hairs. *Aristida argentea* – (4) misspelling of *argentata*. *Melica argentea*

argenteopilosus, argenteo-pilosus L. *argenteus*, silvery; *pilum*, hair; *-osus*, abundance. Pedicels of the sterile florets invested in long silvery hairs

argenteostriat-a, -us L. *argenteus*, silvery; *stria*, ridge; *-ata*, possessing. Leaf-blades variegated

argente-um, -us See *argentea*

argentin-a, -us From Argentina

argentinensis L. *-ensis*, denoting origin. From *Argentina*

argentinus See *argentina*

argillacea L. *argillos*, clay; *-acea*, resembling. – (1) Fertile florets dark-colored. *Digitaria argillacea* – (2) spikelets dark-colored. *Sasa argillacea*

Argillochloa L. *argillos*, clay; *chloa*, grass. Grows on shale scree-slopes

argillosa L. *argillos*, clay; *-osa*, abundance. Growing on soils with abundant clay

Argopogon Gk *argos*, shining; *pogon*, beard. Awn glabrous

arguens L. *arguo*, sharp or penetrating but originally meaning the exposure of a flaw in an argument. Callus sharp-pointed

argunensis L. *-ensis*, denoting origin. From the Argun River, on the border of the Russian Far East and China

argut-a, -um, -us L. *arguo*, sharp or penetrating but originally meaning the exposure of a flaw in an argument. – (1) Palea split into two sharply tapering teeth. *Sporobolus argutus*, *Vilfia arguta* – (2) lower glume two-toothed. *Iseilema argutum*

argyre-a, -um, -us Gk *argyreos*, silver. Pedicels invested in long silver hairs

argyrograpt-a, -um Gk *argyreos*, silver; *grapho*, draw. The upper glume and sterile lemma bear bands of silver hairs

argyronema Gk *argyreos*, silver; *nema*, thread. Hairs associated with ligule long and silvery

argyrostachy-a, -um Gk *argyreos*, silver; *stachys*, spike as of an ear of wheat. Glumes and sterile lemmas silver-hairy

argyrotrich-a, -um Gk *argyreos*, silver; *thrix*, hair. The upper glume and sterile lemma are densely covered with whitish-pink hairs

ariani Ariane, in ancient times the eastern provinces of the Persian Empire. From Ariane

arias In honor of Antonio Sandalio de Arias y Costa (1764–1839) Spanish botanist

arid-a, -um, -us L. dry. Growing in arid places

aridicola L. *aridus*, dry; *-cola*, dweller. Growing in areas of low rainfall

arietina L. *aries*, ram; *-ina*, indicating resemblance. Closely related to *Festuca ovina* which derives its name from an association with sheep

ariguensis L. *-ensis*, denoting origin. From Arigue, Chile

arimagunensis L. *-ensis*, denoting origin. From Arimagun Dojohmura, Settsu Province, now part of Hyogo and Osaka Prefectures, Japan

arisanensis L. *-ensis*, denoting origin. From Mt Arisan, Taiwan

arisan-montana L. *mons*, mountain; *-ana*, indicating connection. As for *arisanensis*

Aristaria L. *arista*, bristle; *-aria*, pertaining to. The lemma of the hermaphrodite floret is long awned

aristat-a, -um, -us L. *arista*, bristle; *-ata*, possessing. – (1) The apices of lemmas, paleas or glumes drawn out into a distinct awn. *Agropyron aristatum*, *Anthoxanthum aristatum*, *Andropogon aristatus*, *Apluda aristata*, *Arthrostylidium aristatum*, *Asprella aristata*, *Calamagrostis aristata*, *Chloris aristata*, *Chusquea aristata*, *Dactyloctenium aristatum*, *Dichanthium aristatum*, *Elymus aristatus*, *Eragrostis aristata*, *Hygroryza aristata*, *Ischaemum aristatum*, *Lepturella aristata*, *Melica aristata*, *Meoschium aristatum*, *Nardus aristatus*, *Neurolepis aristata*, *Oropetium aristata*, *Panicum aristatum*, *Psilurus aristatus*, *Stipidium aristatum* – (2) leaf-blade drawn out into a bristle. *Nastus aristatus*

aristat-um, -us See *aristata*

Aristavena L. *arista*, bristle. Lemma awned as in *Avena*

Aristella L. *arista*, bristle; *-ella*, diminutive. Glumes mucronate or shortly awned

aristell-a, -um L. *arista*, bristle; *-ella*, diminutive. – (1) Lemma shortly awned. *Stipa aristella* – (2) glumes shortly awned. *Panicum aristellum*

Aristida Ancient Roman name for an awned Mediterranean grass

aristidea Resembling *Aristida*

aristidis See *Aristida*

Aristidium L. *-ium*, indicating resemblance. Resembling *Aristida*

aristidoides Gk *-oides*, resembling. With spikelets or inflorescences resembling those of *Aristida*

aristiferum L. *arista*, bristle; *fero*, carry or bear. The glumes and lower lemma apices drawn out into a bristle

aristifolia L. *arista*, bristle; *folium*, leaf. The leaf-blades terminate in a fine bristle

aristiglumis L. *arista*, bristle; *gluma*, husk. With awned subtending glumes or lemmas

aristispicula L. *arista*, bristle; *spica*, a point; hence, in particular, an ear or spike of grain; *-ula*, diminutive. Lemma shortly awned

aristoides Gk *arista*, bristle; *-oides*, resembling. Similar to *Aristida*

Aristopsis Gk *opsis*, resemblance. Similar to *Aristida*. Based on an immature spikelet

aristosum L. *arista*, bristle; *-osum*, abundance. Lemma-awn well developed

aristulat-a, -um, -us L. *arista*, bristle; *-ula*, diminutive; *-ata*, possessing. Lemmas and/or glumes shortly awned

arizonic-a, -um, -us L. *-ica*, belonging to. From Arizona, USA

arjinsanensis L. *-ensis*, denoting origin. From Arginsan, Xianjiang Province, China

arkansan-a, -us L. *-ana*, indicating connection. From Arkansas, USA

arktasianum Gk *arktos*, Arctic; L. *-ianum*, characteristic of. From the Asian-Arctic, that is Siberia

armat-a, -um, -us L. armed. – (1) Lower glume has stiff hairs on its margin. *Andropogon armatus, Aristida armata, Dichanthium armatum* – (2) a ring of thorns at the node. *Arundinaria armata*

armen-a, -um, -us From Armenia

armeniac-a, -um L. *-ica*, belonging to. From Armenia

armen-um, -us See *armena*

armitii In honor of William Edington de Margrat Armit (1848–1901) Belgian-born police officer and magistrate in Queensland and Papua New Guinea

armoricana L. *armor*, in Breton meaning "the sea"; *-ic-*, belonging to; *-ana*, indicating connection. From coastal habitats in Brittany

arnacites Gk *arnakis*, sheep fleece; *-ites*, resemblance. Glumes and lower lemma densely woody

arnhemicum L. *-icum*, belonging to. From Arnhem Land, Northern Territory, Australia

arnottian-a, -um, -us L. *-ana*, indicating connection. In honor of George Arnold Walker Arnott (1799–1868) Scots botanist

arnowiae In honor of Lois Goodell Arnow (1921–) United States botanist

aromatic-a, -um L. scented. – (1) Roots aromatic. *Ctenium aromaticum* – (2) foliage aromatic. *Monocera aromatica*

arras Ethiopian *arras* or *adschar*. Vernacular name for the species in Ethiopia

arrect-a, -um, -us L. pointing upwards. Panicle branches held erect

Arrenantherum See *Arrhenatherum*

arrhenatheroides Gk *-oides*, resembling. Similar in habit to *Arrhenatherum*

Arrhenatherum Gk *arrhen*, male; *ather*, barb or spine. The upper floret in each spikelet is male and awned

arrhenobasis Gk *arrhen*, male; *basis*, base. The pair of stalked spikelets at the base of the raceme is male

arriani See *ariani*

Arrozia Spanish *arroz*, rice. In Brazil known as *arroz de mato*

arsenei In honor of Gustav Joseph Brouard Arsène (1867–1938) cleric and botanist who collected in the Americas

Arthragrostis Gk *arthron*, joint. Resembling *Agrostis*, but the panicle disarticulates completely into its component divisions

Arthraterum See *Arthratherum*

Arthratherum Gk *arthron*, joint; *ather*, barb or spine. The column of the awn articulates with the apex of the lemma

Arthraxon Gk *arthron*, joint; *axon*, an axis. At maturity the inflorescence axis break into segments

Arthrochlaena Gk *arthron*, joint; *chlaena*, cloak. Spikelets overlap and so obscure, that is cloak the internodes of the axis on which they are borne

Arthrochloa Gk *arthron*, joint; *chloa*, grass. There are two genera so-called. With one, the glumes fall away with the tip of the pedicel; with the other, the seed is deeply grooved

A

Arthrochortus Gk *arthron*, joint; *chortos*, grass or hay. Following the shedding of its spikelets at maturity, the inflorescence axis resembles a jointed zigzag

Arthrolophis A misspelling of *Athrolophis*

Arthropogon Gk *arthron*, joint; *pogon*, beard. The hairy spikelets fall with their pedicels

Arthrostachya Gk *arthron*, joint; *stachys*, spike as of an ear of wheat. At maturity, the inflorescence disarticulates into separate segments each bearing a single spikelet

Arthrostachys Gk *arthron*, joint; *stachys*, spike as of an ear of wheat. As with *Arthrostachya* the inflorescence disarticulates into segments each of which in this genus bears a pair of spikelets, one of which is hermaphrodite-sessile, the other of which is male or sterile and stalked

Arthrostylidium Gk *arthron*, joint; *stylos*, stalk. The rhachilla readily disarticulates

articulare L. *articulus*, joint. Burrs shortly stalked and readily articulating

articulat-a, -us L. *articulus*, joint; *-ata*, possessing. – (1) Jointed with spikelets readily articulating. *Aira articulata*, *Anachortus articulatus*, *Corynephorus articulatus*, *Elytrophorus articulatus*, *Eragrostis articulata*, *Nardus articulata*, *Pollinia articulata*, *Weingaertneria articulata* – (2) awn jointed along its column. *Aristida articulata*

artvinensis L. *-ensis*, denoting origin. From Artvin, East Anatolia, Turkey

arubensis L. *-ensis*, denoting origin. From Aruba, a Caribbean island

aruensis L. *-ensis*, denoting origin. From Aru, one of the Molucca Islands, Indonesia

Arundarbor L. *arundo*, reed; *arbor*, tree. Culms tall and woody

arundinace-a, -um, -us L. *arundo*, reed; *-acea*, like. Culm tall, thereby resembling a reed

Arundinaria L. *-aria*, pertaining to. Resembling *Arundo* in habit

arundinariae Of *Arundinaria*. With the habit of *Arundinaria*

Arundinella L. *-ella*, diminutive, together with *Arundo*. The plants have the appearance of small reeds

arundinellum Resembling *Arundinella*

arundinifolium L. *folium*, leaf. Leaf-blades like those of *Arundo*

Arundo, arundo Latin name for a reed, stemming from Celtic *aru*, water. Grows in swamps

Arundoclaytonia In honor of William Derek Clayton (1926–) English agrostologist and with reference to its affinity with *Arundo*

arushae From the Arusha District of Tanzania

arvens-e, -is L. *arvum*, arable field; *-ense*, denoting origin. Uncultivated species

arvernensis L. *-ensis*, denoting origin. From Arverna otherwise Alvernia now mostly the Auvergne, France

arzivencoi In honor of Lúcio Arzivenco (fl. 1970) Brazilian botanist

asagishiana L. *-ana*, indicating connection. From Asagishi, Rikuchu Province, now part of Iwate and Akita Prefectures, Japan

asahinae In honor of Y. Asahina (fl. 1929) Japanese botanist

asanoi In honor of Sadao Asano, Japanese botanist

ascendens L. *ascendo*, climb. Scramblers

aschenbornian-a, -um L. *-anum*, indicating connection. In honor of Alwin Aschenborn (1816–1865) German physician who collected in Mexico

aschersoniana L. *-ana*, indicating connection. In honor of Paul Friedrich August Ascherson (1834–1913) German botanist

aschersonii As for *aschersoniana*

ascinodis Gk *askos*, wine-skin; L. *nodus*, knot. Pedicels inflated like puffed-out cheeks

ashei In honor of William Willard Ashe (1872–1932) United States forester and plant collector

asiae-minoris From Turkey in Asia Minor

asiatic-a, -um L. *-ica*, belonging to. From Asia

askelofiana L. *-ana*, indicating connection. In honor of Johan Christopher Askelöf (1787–1848) Swedish botanist

askoldensis L. *-ensis*, denoting origin. From Askold Island, off the coast of the Russian Far East

asoensis L. *-ensis*, denoting origin. From Asosan, a mountain in Kumamoto Prefecture, Japan

asper, -a, -um L. rough. Plants with rough pedicels or leaf-blades

asperat-a, -um, -us L. *asper*, rough; *-atus*, possessing. Leaf-blades and other parts scaberulous

Asperella, asperella L. *asper*, rough; *-ella*, diminutive. The glumes are keeled and shortly ciliate in *Asperella* Schreb. but meaning unclear for *Asperella* Horst

aspericaulis L. *asper*, rough; *caulis*, stem. Culms minutely nodulose

asperiflora L. *asper*, rough; *flos*, flower. Spikelets and rhachides asperous

asperifoli-a, -um, -us L. *asper*, rough; *folium*, leaf. Leaf-blades rough

asperula L. *asper*, rough; *-ula*, diminutive. Awns slightly scabrous

asperum See *asper*

aspidiotes Gk *aspis*, shield; *-otes*, close connection. Upper glume shield-like

aspidistrula L. *-ula*, diminutive. Foliage resembles that of *Aspidistra*

asplundii In honor of Erik Asplund (1888–1974) Swedish botanist

Asprella The Italian name for *Asperella*

Aspris Gk an undetermined species of oak. Growing in open (oak) woodlands

assamensis L. *-ensis*, denoting origin. From Assam State, India

assamic-a, -um L. *-ica*, belonging to. From Assam State, India

assimil-e, -is L. similar. With affinities to another species

assumptionis L. from Assumption Island, Indian Ocean

assurgens L. *assurgo*, rise up into the air. Scandent in habit

Asthenatherum Gk *astheneo*, become weak; *ather*, barb or spine. The awns are relatively small compared with those of *Danthonia* from which the genus was segregated

asthenica Gk *asthenes*, of low specific gravity; *-ica*, belonging to. Growing on very high mountains

Asthenochloa Gk *asthenes*, weak; *chloa*, grass. A decumbent annual

asthenos See *asthenica*

asthenostachys Gk *asthenes*, weak; *stachys*, spike as of an ear of wheat. Inflorescence a slender raceme

astictus Gk *asticos*, of a town. Growing in wasteland about towns

astonii In honor of Bernard Cracroft Aston (1871–1951) English-born New Zealand scientist

astracanicum L. *-icum*, belonging to. Origin not given by author but name may derive from an assumption the seed came from Astrakhan, Russian Federation

Astrebla Gk *a-*, without; *streble*, screw. The awn on the lemma is not twisted

astrepta Gk *-a*, without; *streptos*, collar. The leaf-blades lack the large collar-like glands of related species

astreptoclada Gk *astreptos*, rigid; *klados*, stem. Panicle branches stiff and straight

astroclada Gk *aster*, star; *klados*, stem. Culms with numerous densely fasciculate branches

asymmetric-a, -um, -us Gk *a-*, without; *symmetria*, symmetry; *-ica*, pertaining to. – (1) Leaf-blade width different either side of midrib. *Criciuma asymmetrica* – (2) one side of lemma tuberculate, the other smooth. *Stipidium asymmetricum*

atabapense L. *-ense*, denoting origin. From Depto Atabapo, Venezuela

atacamensis L. *-ensis*, denoting origin. From Provincia de Atacana, Chile

atamiana L. *-ana*, indicating connection. From Atami, Idzu or Izu Province, now part of Shizuoka and Tokyo Prefectures, Japan

Ataxia Gk *a-*, without; *taxis*, order. The spikelet was interpreted as having a lower male floret, neuter middle and terminal bisexual floret

atbassaricum L. *-icum*, belonging to. From Atbassar, Kazakhstan

A

at-er, -a L. dark. – (1) The anthoecium is dark-brown. *Axonopus ater*, *Digitaria atra* – (2) with black hairs on lower culms. *Bambusa atra*, *Lingnania atra* – (3) a variation or misspelling of *atter*. *Gigantochloa ater*

aterrimum L. very black. Sessile floret jet black at maturity

Athenanthia See *Anthenantia*

atherantha, atheranthera Gk *ather*, barb or spine; *anthos*, flower. Lemma awned

atheric-a, -us Gk *ather*, barb or spine; *-ica*, belonging to. Bearing an awn

Athernotus Gk *ather*, barb or spine; *notos*, false. Awn dorsal instead of terminal as in *Triticum*

Atherophora Gk *ather*, barb or spine; *phero*, bear. Both lemma and palea are awned

Atheropogon Gk *ather*, barb or spine; *pogon*, beard. The trifid awns of the subtending glumes give the spikelet a bearded appearance

atherstonei In honor of William Guybon Atherstone (1814–1898) English-born South African medical practitioner, geologist and naturalist

Athrolophis Gk *athroos*, crowded; *lophos*, crest. Inflorescence a delicate much branched dense plumose panicle

athroostachya See *Athroostachys*

Athroostachys Gk *athroos*, crowded; *stachys*, spike as of an ear of wheat. The condensed panicle branching leads to a capitate inflorescence

atjehensis L. *-ensis*, denoting origin. From Mt Atjeh, Sumatra, Indonesia

atlantic-a, -um L. *-ica*, belonging to. – (1) From the Atlantic Coast of North America. *Dactylis atlantica*, *Panicum atlanticum* – (2) from the Atlantic Coast of North Africa. *Avena atlantica*, *Festuca atlantica*, *Stipa atlantica*

atlantigena L. *gigno*, beget. From the Atlantic Coast

atra See *ater*

Atractantha Gk *atraktos*, spindle; *anthos*, flower. The fertile floret is spindle-shaped

Atractocarpa Gk *atraktos*, spindle; *karpos*, a fruit. Achene spindle-shaped

atrat-a, -um, -us L. *ater*, dark; *-ata*, possessing. – (1) Anthoecia dark-colored. *Paspalum atratum*, *Setaria atrata* – (2) glumes dark-brown to black. *Agrostis atrata*, *Elymus atratus*, *Triticum atratum*

atrich-a, -um Gk *a-*, without; *thrix*, hair. – (1) Involucreal bristles mostly lacking long hairs. *Pennisetum atrichum* – (2) plant glabrous. *Panicum atrichum*, *Tristachya atricha*

atriseta L. *ater*, dark; *seta*, bristle. Awn dark-purple

atrisola L. *ater*, dark; *solum*, earth. Growing on black soils

atrocarpum L. *ater*, dark; Gk *karpos*, fruit. Anthoecium dark brown to black

atrocingulare L. *ater*, dark; *cingulum*, girdle; *-are*, pertaining to. Culms with a prominent girdle-like scar on the nodes

atrofusc-a, -um L. *ater*, dark; *fusca*, brown. The fertile floret is dark-brown

atropidiformis L. *forma*, appearance. With the habit of *Atropis convolutae*

atropioides Gk *-oides*, resembling. Resembling *Atropis*

Atropis Gk *a-*, without; *tropis*, keel. Lemma not keeled

atropurpure-a, -um L. *ater*, dark; *purpurea*, purple or dull red. – (1) Culms purple red at their bases. *Aira atropurpurea*, *Arundinaria atropurpurea*, *Deschampsia atropurpurea* – (2) panicles purple-red. *Eragrostis atropurpurea*, *Panicum atropurpureum*, *Poa atropurpurea*

atrorubens L. *ater*, dark; *rubeo*, be red. Plant in whole or in part reddish-colored

atrosanguineum L. *ater*, dark; *sanguineus*, red. Spikelets purplish-brown

atrovaginata L. *ater*, dark; *vagina*, sheath; *-ata*, possession. Culm-sheaths dark-green

atroviolace-a, -um L. *ater*, dark; *violaceum*, violet. – (1) Spikelets dark-purple. *Koeleria atroviolacea*, *Panicum atroviolaceum* – (2) culms purplish. *Gigantochloa atroviolacea*

atrovirens L. *ater*, dark; *virens*, green. – (1) Culms greenish-black. *Bambusa atrovirens* – (2) glumes greenish-black. *Eragrostis atrovirens, Sporobolus atrovirens, Vilfa atrovirens*

atroviridis L. *ater*, dark; *viridis*, green. Glumes greenish-black

attalica L. *-ica*, belonging to. From Atalaya, known to the Romans as Attaleia, Turkey

attenuat-a, -um, -us L. drawn out or narrowed. Spikelets or panicles narrow

attenuatiglum-e, -is L. *attenuatus*, thin; *gluma*, husk. Glumes long tapering

attenuat-um, -us See *attenuata*

atter A contraction of *awi atter* the Sudanese vernacular name of the species

attica From Attica, Greece

aturens-e, -is L. *-ense*, denoting origin. – (1) From "cataractus Aturensis" on the Atabapo River, Venezuela. *Eragrostis aturensis, Homolepis aturensis, Panicum aturense, Poa aturensis* – (2) from Atures Municipality, Venezuela. *Axonopus aturensis*

aubertii In honor of Edgar Aubert de la Rue

auburne Type collected from Auburn, Alabama, USA

aucheri In honor of Pierre Martin René Aucher-Elroy (1792–1838) French botanist

aucklandica L. *-ica*, belonging to. From the Auckland Islands, New Zealand Possessions in the south-western Pacific

aucta L. added to. Two varieties united to form a new species

auctiaurita L. *augeo*, grow; *aurita*, long-eared. Auricles conspicuous

Auena See *Avena*

augeri A misspelling of *aucheri*

augusta L. venerable. Culms robust

Aulacolepis Gk *aulax*, furrow; *lepis*, scale. Palea grooved

aulacosperma Gk *aulax*, furrow; *sperma*, seed. Grain grooved

Aulaxanthus Gk *aulax*, furrow; *anthos*, flower. The glumes are concave with five longitudinal villous furrows

Aulaxia See *Aulaxanthus*

auletic-a, -us Gk *aulos*, flute; *-ica*, belonging to. From Auletus, now Aulet, Spain

Aulonemia Gk *aulos*, flute. Musical instruments are made from the internodes

auquieri In honor of Paul Auquier (1939–1980) Belgian botanist

aurantiac-a, -um L. between yellow and scarlet. Spikelets reddish-yellow

aurasiac-a, -us L. an inhabitant of Aurasius Mons, now Aurès Mountains, Algeria

aurata L. *aureus*, golden-yellow; *-ata*, possessing. Spikelets flecked with gold

aure-a, -um, -us L. golden-yellow. Spikelets, pedicels, bristles or other parts invested with golden-yellow hairs

aurelianum L. *-anum*, indicating connection. From Aurelia the Roman name for Orleans, France but here from New Orleans, USA

aureocephala L. *aureus*, golden-yellow; Gk *kephale*, head. Inflorescence golden-yellow

aureofimbriatum L. *aureus*, golden-yellow; *fimbriae*, fringe; *-atum*, possessing. Oral setae golden-yellow

aureofulv-a, -us L. *aureus*, golden-yellow; *fulvus*, dull-yellow. Spikelets golden-yellow

aureolanata L. *aureus*, golden-yellow; *lana*, wool; *-ata*, possessing. Nodes with a skirt of golden woolly hairs

aureolanta A misspelling of *aureolanata*

aureolatum L. *aureum*, gold; *-ulus*, diminutive; *-atum*, indicating likeness. Ligule reddish-brown, resembling in color a rare form of native gold or Jeweller's gold; that is, possessing a small amount of gold

aureosulcata L. *aureus*, golden-yellow; *sulcus*, furrow; *-ata*, possessing. Culms green and streaked with yellow

aureovagina L. *aureus*, gold; *vagina*, sheath. Leaf-sheath tinged with yellow

aureovillosus L. *aureus*, golden-yellow; *villi*, long weak hairs; *-osa*, abundance. Spikelets, pedicels, bristles or other parts invested with golden-yellow hairs

aure-um, -us See *aurea*

auricom-a, -um L. *aureus*, golden-yellow; *coma*, a head of hair. Leaf-blades golden-yellow

auriculat-a, -um, -us L. *auris*, ear; *-ula*, diminutive; *-ata*, possessing. – (1) Leaf-blades rounded at the base or with auricles. *Hymenachne auriculata* – (2) the apex of the lemma two-lobed. *Danthonia auriculata* – (3) subtending bract of pseudo-spikelet auriculate. *Alvimia auriculata* – (4) leaf-sheaths auricled at the mouth. *Andropogon auriculatus, Sacciolepis auriculata* – (5) callus of upper floret expanded into two membranous wings attached to the base of the lemma. *Panicum auriculatum*

aurigae From Mt Auriga, Papua, Indonesia

aurinuda L. *auris*, ear; *nuda*, bare. Leaves lacking oral setae

aurit-a, -um, -us L. *auritus*, eared. – (1) Leaf-blades auricled. *Elymus auritus, Fargesia aurita, Panicum auritum* – (2) leaf-sheaths auricled. *Triodia aurita* – (3) lower glume asymmetric, half developing terminally into a wing. *Manisuris aurita, Rottboellia aurita*

auronitens L. *aurum*, gold; *niteo*, shine. Spikelets shining golden to olive-brown

aurorae L. *aurore*, redness of dawn. The lemma grades from yellow at the apex to purple at the base, thereby resembling the appearance of a dawn sky

ausserdorferi In honor of Anton Ausserdorfer (1836–1885) German botanist

australasic-a, -us L. *-ica*, belonging to. From Australia

austral-e, -is L. of the south. – (1) From the south in general as from Africa, America, Europe or elsewhere. *Agrostis australis, Aristida australis, Asprella australis, Avena australis, Cenchrus australis, Chionochloa australis, Dactyloctenium australe, Deyeuxia australis, Digitaria australis, Elymus australis, Erythranthera australis, Festuca australis, Gastridium australe, Hierochloe australis, Isachne australis, Ischaemum australe, Lasiagrostis australis, Leersia australis, Monanthochloe australis, Oryza australis, Panicum australe, Phragmites australis, Sasa australis* – (2) from Australia. *Andropogon australis, Ischaemum australe, Panicum australe, Poa australis*

australianus L. *-anus*, indicating connection. From Australia

australiens-e, -is L. *-ense*, denoting origin. From Australia

australindica L. *australis*, of the south; *-ica*, belonging to. From southern India

australis See *australe*

Australopyrum L. *australis*; of the south; Gk *pyros*, wheat. Resembling wheat, and restricted to the Southern Hemisphere

austroaltaica L. *auster*, south; *-ica*, belonging to. From southern Altai Mountains, Kazakhstan

austroasiaticum L. *auster*, south; *-icum*, belonging to. From southern Asia

austrobohemica L. *-ica*, belonging to. From Southern Bohemia, Slovakia

austrocaledonicum L. *auster*, south; *-icum*, belonging to. From southern Caledonia, that is New Caledonia

Austrochloris L. *auster*, south. Resembling *Chloris* and endemic to Australia

Austrodanthonia L. *auster*, the south. A group of largely southern hemisphere species once included in *Danthonia*

austrodensa L. *auster*, south. With southern connections, but here used as a prefix to avoid formation of a homonym

austrodolomitica L. *-ica*, belonging to. From Trentino in the southern Dolomites of Italy

Austrofestuca L. *auster*, south. Resembling *Festuca* and endemic to Australia and New Zealand

austrohercynica L. *auster*, south; *-ica*, belonging to. From the southern part of Hercynia now the south of Germany

austroibericum L. *auster*, south. From southern parts of Spain and Portugal known to the Romans as Iberia

austroitalica L. *auster*, south; *Italia*, Italy; *-ica*, belonging to. From southern Italy

austrokurilensis L. *-ensis*, denoting origin. From southern Kuriles

austromontanum L. *-anum*, indicating connection. From the mountains of southern Tennessee, USA

austroscaberula L. *auster*, south. With southern connections, but here used as a prefix to avoid formation of a homonym
austrosibirica L. *auster*, south; *-ica*, belonging to. From southern Siberia
Austrostipa L. *auster*, south. Genus allied to *Stipa* but restricted to Australia
austrouralensis L. *auster*, south; *-ensis*, denoting origin. From the southern Urals, a mountain range straddling the border between Europe and Asia
autumnal-e, -is L. of the autumn. Flowering in autumn
auyanense L. *-ense*, denoting origin. From Auyana-tepui, Venezuela
Avellinia In honor of Guilio Avellino (fl. 1841) Italian botanist
Avena Latin name for oat possibly an allusion to *aveo*, desire, because it is sought out by cattle
avenace-a, -um, -us L. *-acea*, resembling. With inflorescences and/or spikelets similar to those of *Avena*
avenacellum L. *-ellum*, diminutive. Florets shorter than those of *Piptochaetium avenaceum*
avenace-um, -us See *avenacea*
avenacioides Gk *-oides*, resembling. Similar to *Stipa avenacea*
Avenalla See *Avenella*
Avenaria L. *-aria*, resembling. Similar to *Avena*
Avenastrum Gk *-astrum*, incomplete resemblance. Perennial species of *Avena* if that genus limited to annuals
Avenella, avenella L. *-ella*, diminutive but here used as a name-forming suffix. Spikelets resembling those of *Avena*
Avenochloa Gk *chloa*, grass. A name-forming suffix together with *Avena*
avenoides Gk *-oides*, resembling. Resembling *Avena* in spikelet structure
Avenula Origin uncertain, not given by author but close to *Avena*
aversum L. bent backwards. Spikelets bent backwards from the rhachis
avettae In honor of Carlo Avetta (1861–1941) Italian botanist
axicilium L. *axis*, axis; *cilium*, hair. Axis of inflorescence ciliate
axilis A misspelling of *exile*
axillar-e, -is L. axillary. With panicles arising from the upper leaf axils
Axonopus Gk *axon*, axle; *pous*, foot. The inflorescence branches arising from a common point like the spokes of a wheel
aya From the vernacular name of the species in the Bangli District, Bali, Indonesia
ayacuchensis L. *-ensis*, denoting origin. From Ayacucho, Humanga Province, Peru
ayseniensis L. *-ensis*, denoting origin. From the river Aysén, Chile
azgarica L. *-ica*, belonging to. From Azgar, Caucasus, Russian Federation
azo-cartii In honor of Raphael Azo-Cart or Azocart (fl. 1880) who collected in Chile
azorica L. *-ica*, belonging to. From the Azores, Portuguese islands in the Atlantic
aztecanum L. *-anum*, indicating connection. From Mexico, that is the land of the Aztecs
aztecorum L. of the Aztecs. From Mexico
azuayense L. *-ense*, denoting origin. From Azuay, Ecuador
azucarica L. *-ica*, belonging to. From Cerro Pan de Azucar, Colombia
azutavica L. *-ica*, belonging to. From the Azatau saddle in the Altai Mountains, Kazakhstan

B

babataneyosiana L. *-ana*, indicating connection. In honor of Taneyosi Baba (fl. 1940) Japanese botanist
baccanensis L. *-ensis*, denoting origin. From Bac Thai (Bac Can), Ha Tuyen Province, Vietnam
baccifera L. *baccus*, berry; *fero*, carry or bear. Fruits fleshy
bachmannii In honor of Franz Ewald Bachmann (1856–c. 1916) German-born South African physician and naturalist
bacillata L. *bacillus*, rod; *-ata*, possessing. The rhachilla projects beyond the floret

bacquangensis L. *-ensis*, denoting origin. From Bac Quang, Hu Tuyen Province, Vietnam

bacthaiensis L. *-ensis*, denoting origin. From Bac Thai Province, Vietnam

bactriana L. *-ana*, indicating connection. From Baktroi, a province of the ancient Persian Empire, later Turkestan

baculifera L. *baculum*, rod; *fero*, carry or bear. Culms used for canes and walking sticks

badachschanica L. *-ica*, belonging to. From Badakhshan Province, Afghanistan

badamense L. *-ense*, denoting origin. From Badan Khrebet, a mountain range in Siberia

badamicum L. *-icum*, belonging to. From Badami, Bombay State, India

baddadae From Baddada, Somalia

badensis L. *-ensis*, denoting origin. From Baden, Lower Austria

badi-a, -um L. dull brown. Anthoecium dark-brown

baeoticum L. *-icum*, belonging to. From Baeotia, Greece

baetica From Provincia Baetica, now southern Spain

baffinensis L. *-ensis*, denoting origin. From Baffin Land

bagirmic-a, -us L. *-ica*, belonging to. From Bagirmi District, Republic of Chad

baguirmiensis L. *-ensis*, denoting origin. See *bagirmica*

bahamensis L. *-ensis*, denoting origin. From the Bahamas

bahiana L. *-ana*, indicating connection. From Bahia State, Brazil

bahiens-e, -is L. *-ense*, denoting origin. From Bahia Province, Brazil

baicalensis L. *-ensis*, denoting origin. From the steppes around Lake Baikal, Russian Federation

baileyi In honor of – (1) Frederick Manson Bailey (1827–1915) English-born Australian botanist. *Andropogon baileyi, Digitaria baileyi, Panicum baileyi* – (2) John Frederick Bailey (1866–1932) Australian horticulturalist. *Heterachne baileyi*

baishanzuensis L. *-ensis*, denoting origin. From Baishan, Hainan, China

bajacaliforniana L. *-ana*, indicating connection. From Baja California, Mexico

bajaensis L. *-ensis*, denoting origin. From Baja California, Mexico

bakeri In honor of – (1) Charles Henry Baker (1848–?) United States botanist. *Spartina bakeri* – (2) Charles Fullar Baker (1872–1927) United States botanist. *Agrostis bakeri, Paspalum bakeri* – (3) John Gilbert Baker (1834–1920) English botanist. *Poecilostachys bakeri*

bakuensis L. *-ensis*, denoting origin. From Baku district, Azerbaijan

balanites Gk *balanos*, acorn; *-ites*, close connection. The anthoecium resembles a tiny acorn

balansae In honor of Benedict (Benjamin) Balansa (1825–1891) French botanical explorer

Balansochloa Gk *chloa*, grass. See *balansae*

balbisianum L. *-anum*, indicating connection. As for *balbisii*

balbisii In honor of Giovanni Battista Balbis (1765–1831) Italian botanist

balcanica L. *-ica*, belonging to. From Balcanum, now southern Tyrol, Italy

balcooa Bengali *bhalbua* or *balku*. The vernacular name for the species in Bengal

Baldingera In honor of Ernst Gottfried Baldinger (1738–1804) German physician and naturalist

Baldomiria Meaning uncertain, origin not given by author

baldschuanic-um, -us L. *-icum*, belonging to. From Baldschuan (Baldshuan), Tajikistan

baldshuanicus See *baldschuanicum*

baldwinii In honor of – (1) John Thomas Baldwin (1910–1974) United States botanist. *Loudetia baldwinii, Loudetiopsis baldwinii* – (2) William Baldwin (1779–1819) United States botanist. *Panicum baldwinii, Saccharum baldwinii*

balearica From Balearic Isles

balfouri In honor of – (1) John Hutton Balfour (1808–1884) Scots physician and botanist. *Poa balfouri* – (2) Isaac Bailey Balfour (1833–1922). *Panicum balfouri*

balgooyi In honor of Maximilian Michael Josephus van Balgooy (1932–) Indonesian botanist and long-time resident of the Netherlands

baliana L. *-ana*, indicating connection. From Bali Island, Indonesia

baliensis L. *-ensis*, denoting origin. From Bali Island, Indonesia

ballardii In honor of Francis Ballard (1896–1975) English botanist

balsiensis L. *-ensis*, denoting origin. From Bals, Romania

baltica From one of the countries bordering the Baltic Sea

baltistanicum L. *-icum*, belonging to. From Baltistan, north-east Pakistan

baltodes L. *baltens*, girdle; *-odes*, resembling. Lower glume forms a girdle-like cup around the base of the spikelet

baluchistanicum L. *-icum*, belonging to. From Baluchistan, Pakistan

balui Bornean word for the species, and also the name of a river in Sarawak where the species is cultivated

bamban Vernacular name for the species in Sumatra, Indonesia

bamboa See *Bambusa*

Bambos See *Bambusa*

bambos Resembling *Bambusa*

Bambus See *Bambusa*

Bambusa Latinized version of the Indian bamboo in turn possibly derived from the Malay, *mambu*, which may be a contraction of *rotan semanbu*, malacca cane

bambusaefolium L. *folium*, leaf. Leaf-blades resembling those of *Bambusa*

bambusaeoides Gk *-oides*, resembling. Similar to *Bambusa*

bambusiflor-um, -us L. *flos*, flower. Resembling *Bambusa* with respect to the inflorescence

bambusiform-e, -is L. *forma*, appearance. Similar to *Bambusa* in habit

bambusin-a, -um L. *-ina*, indicating resemblance. Similar to *Bambusa* in habit

bambusioides See *bambusoides*

Bambusites Gk *-ites*, close connection. Fossil leaf-blades resembling those of *Bambusa*

Bambusium L. *-ium*, resembling. See *Bambusites*

bambusiuscul-a, -um L. *-ula*, diminutive. Resembling in habit a dwarf *Bambusa*

bambusoides Gk *-oides*, resembling. Culms wooden resembling those of *Bambusa*

banaoense L. *-ense*, denoting origin. From Sierra de Banao, Cuba

banaticum L. *-icum*, belonging to. From Banat, a former province of Romania, now mostly included in Yugoslavia

bandunduense L. *-ense*, denoting origin. From Bandundu, Zaire

bangweolensis L. *-ensis*, denoting origin. From Bangweolo, now Bangeula, Zambia

banksii In honor of Joseph Banks (1743–1820) English botanist and traveller

bantamensis L. *-ensis*, denoting origin. From Bantam, Java, Indonesia

baojiense L. *-ense*, denoting origin. From Baoji, Shaanxi Province, China

baptarrhenius Gk *baptos*, bright-colored; *arrhen*, male. The persistent anthers are initially yellow, then reddish-brown

Baptorhachis Gk *baptos*, bright-colored; *rhachis*, axis. Inflorescence with a colorful leaf-like rhachis

barbat-a, -um, -us L. *barba*, beard; *-ata*, possessing. – (1) With hairs on the callus, glumes or lemma. *Andropogon barbatus, Anthistiria barbata, Aristaria barbata, Arundo barbata, Avena barbata, Axonopus barbatus, Bouteloua barbata, Briza barbata, Calamagrostis barbata, Chaetochloa barbata, Chloris barbata, Chusquea barbata, Danthoniopsis barbata, Digitaria barbata, Diplachne barbata, Diplocea barbata, Enteropogon barbatus, Festuca barbata, Ischaemum barbatum, Loudetia barbata, Meoschium barbatum, Panicum barbatum, Paspalum barbatum, Phragmites barbata, Schenckochloa barbata, Schismus barbatus, Stipa barbata, Tristachya barbata, Xerochloa barbata, Xerodanthia barbata* – (2) spathe with a long drawn-out tip. *Polytoca barbata* – (3) with bearded nodes. *Poa barbata* – (4) with spikelets subtended by bristles. *Cenchrus barbatus, Chaetochloa barbatum, Setaria barbata* – (5) terminal sterile floret awned. *Diplachne barbata, Gouinia barbata* – (6) glumes awned. *Xystidium barbatum*

B

barbatoides Gk *-oides*, resembling. A suffix designed to distinguish the species from another named *barbata* in a once supposedly closely related genus

barbellatum L. *barba*, beard; *-ella*, diminutive; *-atum*, possessing. Lemma with short hairs at the apex

barberi In honor of Charles Alfred Barber (1860–1933) South African-born English botanist

barbeyana L. *-ana*, indicating connection. In honor of William Barbey (1842–1914) Swiss botanist

barbeyi As for *barbeyana*

barbicall-a, -um L. *barba*, beard; *callus*, callus. Callus bearded

barbicollis L. *barba*, beard; *collum*, neck. Orifice of leaf-sheath bears long hairs

barbiculmis L. *barba*, beard; *culmus*, stem. Culms hairy

barbifult-um, -us L. *barba*, beard; *fulcio*, support. Plants densely invested in long hairs

barbiger-a, -um, -us L. *barba*, beard; *gero*, carry or bear. – (1) Palea barbed. *Agrostis barbigera* – (2) spikelets each subtended by a fine rough bristle. *Axonopus barbigerus, Panicum barbigerum, Setaria barbigera*

barbiglandularis L. *barba*, beard; *glans*, gland; *-ula*, dimimutive; *-aris*, pertaining to. Sterile lemma bearing large gland-tipped barbed hairs

barbiglumis L. *barba*, beard; *gluma*, husk. Glumes with stiff hairs on their margins

barbinod-e, -is L. *barba*, beard; *nodus*, knot. Nodes bearded

barbipedum L. *barba*, beard; *pes*, foot. Pedicel bearing long hairs immediately below the spikelet

barbipulvinat-a, -um L. *barba*, beard; *pulvinus*, cushion; *-atum*, possessing. Axil of all branches pubescent with spreading hairs

barbivaginale L. *barba*, beard; *vagina*, sheath. Leaf-sheath hairy

barbulat-a, -um L. *barba*, beard; *-ula*, diminutive; *-ata*, possessing. Shortly bearded as of nodes or spikelets

barbuligera L. *barba*, beard; *-ula*, diminutive; *gero*, carry or bear. Rhachilla extended into a scaberulous bristle

barceloi In honor of Francese Barcelói Combis (1820–1889) Spanish botanist

barcensis L. *-ensis*, denoting origin. From Bárcaság, Hungary

barcinonensis L. *-ensis*, denoting origin. From Barcinona, also spelt Barcinoa, now Barcelona, Spain

barclayi In honor of George W. Barclay (fl. 1835) Scots-born English gardener and traveller who collected widely in South America

baregense L. *-ense*, denoting origin. From Vallée de Barèges (Hautés-Pyrénées), France

bargusinensis L. *-ensis*, denoting origin. From the Barguzinski Khrebet, a mountain range in Siberia

barnardii In honor of Petres Johannes Barnard (1935–) South African biologist

baronii In honor of Richard Baron (1847–1907) English cleric who collected in Madagascar

baronis See *baronii*

barrancaensis L. *-ensis*, denoting origin. From Barranca, Argentina

barrazae In honor of Osvaldo Barraza Quiroga (1906–1982) Chilean agriculturalist

barrelieri In honor of Jacques Barrelier (1606–1675) French botanist

barretoi In honor of Ismar L. Barreto (?–2000) Argentinian agronomist

barrosiana L. *-ana*, indicating connection. In honor of Manual Barros (1880–1973) Argentinian botanist

barteri In honor of – (1) Charles Barter (?–1859) who collected in Nigeria. *Andropogon barteri, Androscepia barteri, Anthistiria barteri, Hyparrhenia barteri* – (2) Barter (fl. c. 1920) who collected in Nigeria. *Eragrostis barteri, Pennisetum barteri*

bartherei In honor of Louis-Henri Barthère (1822–?) French forester

barthii In honor of Jean-Baptiste Barth (1806–1817) French botanist

bartlettii In honor of Harley Hamis Bartlett (1886–1960) United States botanist

bartowense L. *-ense*, denoting origin. From Bartow, Florida, USA

basalis L. *basis*, base; *-alis*, pertaining to. Lemma awned from near the base

basedowii In honor of Herbert Basedow (1881–1933) South Australian geologist

bashanensis L. *-ensis*, denoting origin. See *Bashania*

Bashania From Ba Shan, Sichuan Province, China

basiaurita L. *basis*, base, *aurita*, eared. Leaf-blades narrow with cordate base

basibarbigera L. *basis*, base; *barba*, beard; *gero*, carry or bear. Bases of leaf-sheaths densely hairy

basicladum L. *basis*, base; Gk *klados*, branch. Culms much branched from lower nodes

basifissa L. *basis*, base; *findo*, cleave. Leaf-blade has two narrow sagittate extensions at the base

basigibbosa L. *basis*, base; *gibbosa*, swollen. Base of leaf-blade swollen

basihirsut-a, -us L. *basis*, base; *hirsutus*, hairy. Leaf-bases hairy

basilepis L. *basis*, base; Gk *lepis*, scale. Lower leaves of culms reduced to scales

basiramea L. *basis*, base; *ramus*, branch. Culms much branched at the base

basiserica L. *basis*, base; *seres*, silk; *-ica*, belonging to. Sheaths of the basal leaves densely hairy

basiset-a, -um L. *basis*, base; *seta*, bristle. Spikelet subtended by a bristle

bassacensis L. *-ensis*, denoting origin. From Bassac in the Mekong Delta, Laos

bastardii In honor of Thomas Bastard (?–1815)

basutorum In honor of the Basuti people in southern Africa

batalinii In honor of Alexander Feodorowicz Batalin (1847–1896) Russian botanist

batavicum L. *-icum*, belonging to. From Batavia, now Jakarta, Java, Indonesia

bathiei In honor of Joseph Marie Henri Alfred Perrier de la Bâthie (1873–1958) French botanist

Bathratherum Gk *bathron*, pedestal; *ather*, barb or spine. The lemma bears a geniculate-awn

batianoffii In honor of George Nicholas Batianoff (1945–) of Russian descent but born in China, Australian botanist

Batratherum See *Bathratherum*

Bauchea In honor of Bauche who, like the author, was French

bauhinii In honor of Kaspar Bauhin (1560–1624) Swiss botanist

baumannii In honor of Ernst Baumann (1868–1933) who collected in Togo

baumgarteniana L. *-ana*, indicating connection. In honor of Johann Christian Gottleb Baumgarten (1765–1843) German botanist

baumgartenii As for *baumgarteniana*

bavicchii In honor of Ferruccio David Ugo Bavicchi (1866–1925) Italian born agriculturalist who worked in the Congo

baviensis L. *-ensis*, denoting origin. From Mount Bavi, Vietnam

bavioensis L. *-ensis*, denoting origin. From Estación Bavio, between La Plata and La Magdalena, Buenos Aires Province, Argentina

bawa Burmese *ba*, father; *wa*, bamboo. Culms large for genus

baytopiana L. *-ana*, indicating connection. In honor of Asuman Baytop (1920–) Turkish pharmacist

bazargiciensis L. *-ensis*, denoting origin. From Bazargia

Bealia In honor of William James Beal (1833–1924) United States botanist and agricultural educator

beamanii In honor of John Homer Beaman (1929–) United States botanist

beccabunga Growing amongst *Veronica beccabunga* (Scrophulariaceae), a low-growing herb

beccarii In honor of Odoardo Beccari (1843–1920) Italian botanist

bechuanense L. *-ense*, denoting origin. From Bechuanaland, now Botswana

bechuanica L. *-ica*, belonging to. From Bechuanaland, now Botswana

Beckera In honor of Johannes Becker (1769–1833) German botanist

beckeri In honor of Alexander Becker (1818–1901) Russian organist and plant collector in the Caucasus Mountains

Beckeria In honor of M. Becker, German cleric and botanist

beckeroides Gk *-oides*, resembling. Similar to *Beckera*

Beckeropsis Gk *opsis*, resemblance. Similar to *Beckera*

beckii In honor of S. G. Beck (fl. 1984)

Beckmannia In honor of Johann Beckmann (1739–1811) German botanist

beckmanniaeforme Gk *forme*, resembling. Similar to *Beckmannia*

beddiei In honor of Andrew Davidson Beddie (1880–1962) New Zealand amateur botanist and stonemason

beddomei In honor of Richard Henry Beddome (1830–1911) English-born Indian forester

bedeliensis L. *-ensis*, denoting origin. From Bedel Pass, Kyrgyzstan

beecheyan-a, -us L. *-ana*, indicating connection. In honor of Frederick William Beechey (1796–1856) English Naval Officer

beecheyi As for *beecheyana*

Beehsa See *Beesha*

Beesha Malabar name for *Ochlandra rheedii* and some species of *Melocanna*

beguinotiana L. *-ana*, indicating connection. In honor of Augusto Beguinot (1875–1940) Italian botanist

behriana L. *-ana*, indicating connection. In honor of Hans Hermann Behr (1818–1904) German-born, Australian botanist

beimushanica L. *-ica*, belonging to. From Beimu Shan, Sichuan Province, China

beisitiku Taiwanese *beisi*, sieve for rice grains; Japanese *tiku*, bamboo. Used for making high quality beisi

belangeri In honor of Charles Paulus Bélanger (1805–1881) French botanist

belensis L. *-ensis*, denoting origin. From Belá, Bohemia

bell-a, -um L. pretty. Attractive in appearance

bellardii See *Bellardiochloa*

Bellardiochloa Gk *chloa*, grass. In honor of Carl Antonio Lodovico Bellardi (1741–1826) Italian botanist

bellariensis L. *-ensis*, denoting origin. From Bellary, Deccan Plateau, India

bellatula L. pretty. Attractive in appearance

bellespicat-a, -um L. *bella*, beautiful; *spica*, a point; hence, in particular, an ear or spike of grain; *-atum*, possessing. Spikelets silky-pilose, pink-purplish or pale yellowish-green

bellula L. pretty. Attractive in appearance

bellum See *bella*

belmonte L. from Belmonte, Bahia, Brazil

belsonii In honor of E. Belson (fl. 1930)

bemarivense L. *-ense*, denoting origin. From Bemarivo, Madagascar

beneckei In honor of Franz Benecke (1857–1903) German botanist

beneckii See *beneckei*

benekenii In honor of Ferdinand Beneken (1800–1859) German apothecary

bengalens-e, -is L. *-ense*, denoting origin. From Bengal, now Bangladesh

bengkalisensis L. *-ensis*, denoting origin. From Bengkalis (Island), Sumatra, Indonesia

benguellens-e, -is L. *-ense*, denoting origin. From Benguella, Mossamedes District, Angola

benjaminii In honor of Ludwig Benjamin (1825–1848) German physician and botanist

benneri In honor of Walter Mackinett Benner (1888–1970) United States botanist

Bennetia A Latinized version of *bennet*, an archaic English word for certain grass stalks and incorporated into the vernacular name Bennet-grass (*Agrostis gigantea* Roth) which has similar florets

bennettense L. *-ense*, denoting origin. Discovered in the park of the Bennett Civil War Memorial near Durham, North Carolina, USA

benoistii In honor of Charles Frappier de Mont Benoist (1813–1885) French botanist who was born on Mauritius and died on La Réunion

benthamian-a, -us L. *-ana*, indicating connection. As for *benthamii*

benthamii In honor of George Bentham (1800–1884) English botanist

bentii In honor of James Theodore Bent (1852–1897) English traveller, archeologist and naturalist

bequaertii In honor of Joseph Charles Corneille Bequa(e)rt (1886–1982) Belgian botanist

berazainae, berazainiae In honor of Rosalina Berazaín Iturralde, Cuban botanist

berchtholdiae As for *Berchtoldia*

Berchtoldia In honor of Friderici de Berchtold (1781–1876) Austrian physician and botanist

berelica L. *-ica*, belonging to. From Berelyukh, far-east of Russian Republic

berezovcanum L. *-anum*, indicating connection. From Berezovca, Siberia

Berghausia In honor of Heinrich C. W. Berghaus (1797–1884) German geographer

bergiana L. *-ana*, indicating connection. In honor of Karl Heinrich Bergius (1790–1818) German naturalist and apothecary who collected at Cape Town, South Africa

bergii In honor of – (1) Frederico Guillermo Carlos Berg (1843–1902) Russian-born Argentinian botanist. *Koeleria bergii, Panicum bergii* – (2) Peter Jonas Bergius (1730–1790) Swedish physician and botanist. *Andropogon bergii*

bergrothii In honor of Ivar Ossian Bergroth (1868–1904) Finnish botanist

beringensis L. *-ensis*, denoting origin. From Bering Island or Straits, Russian Far East

beringiana L. *-ana*, indicating connection. See *beringensis*

berkeleyanum L. *-anum*, indicating connection. An artificial hybrid produced at Berkeley, California, USA

berlandieri In honor of Jean Louis Berlandier (1805–1851) French traveller and plant collector in New World

bernieri In honor of Pierre François Bernier (1779–1803) who collected in Madagascar

berningeri In honor of Otto Berninger (1898–?) who collected in Chile

bernoullianum L. *-anum*, indicating connection. In honor of Carl Gustav Bernoulli (1834–1878) Swiss-born physician and botanist

beroensis L. *-ensis*, denoting origin. From the Bero River, Mossamedes district, Angola

Berriochloa In honor of Edward Willard Berry (1900–1968) United States palaeobotanist

berroi In honor of Mariano B. Berro (1905–1922) Uruguayan botanist

berterian-a, -um As for *berteroniana*

berteroan-a, -um, -us As for *berteroniana*

berteronian-a, -um, -us L. *-ana*, indicating connection. In honor of Carlo Guiseppe Bertero (1789–1831) Italian botanist, physician, pharmacist and traveller

bertlingii In honor of F. Bertling (fl. 1913) who collected in Angola

bertolae In honor of Bertola, Italian botanist who collected in the Italian Piédmont

bertolonii In honor of Antonio Bertoloni (1775–1868) Italian botanist

bertonii In honor of Moisés de Blanquis Bertoni (fl. 1918–1945) who collected in South America

bertrandii In honor of Bertrand, French horticularist who collected in Chile

berythea From Berythea, now Beirut, Lebanon

besczetnoviae In honor of Margaret V. Besczetnov, Kazakstan biologist

Besha See *Beesha*

bessarabic-a, -um L. *-ica*, belonging to. From Bessarabia, now Moldova

besseri In honor of Wilibald Swibert Joseph Gottlieb Besser (1784–1842) Austrian-born Russian botanist

besukiensis L. *-ensis*, denoting origin. From Besukie Province, Java, Indonesia

betafensis See *betafoense*

betafoense L. *-ense*, denoting origin. From Bétafo, Madagascar

betsileensis L. *-ensis*, denoting origin. From Betsiléo, Madagascar

bettyae In honor of Betty Jacobs (1947–) Australian Tertiary Educator

Bewsia In honor of John William Bews (1884–1938) Orkney-born South African botanist

beyeri In honor of M. K. Beyer (fl. 1922) Norwegian traveller in Cuba

B

beyrichian-a, -um, -us L. *-ana*, indicating connection. In honor of Heinrich Carl Beyrich (1796–1834) German-born United States botanist

beyrichii As for *beyrichiana*

Bhidea In honor of R. K. Bhide (fl. 1919) Economic Botanist, Poona, India

bhutanic-a, -us L. *-ica*, belonging to. From the Kingdom of Bhutan

bialata L. *bis*, twice; *ala*, wing; *-ata*, possessing. The keels of both glumes narrowly winged

biannularis L. *bi-*, two; *annulis*, ring; *-aris*, pertaining to. The lemma bears two distinct rows of hairs

biaristat-a, -um, -us L. *bi-*, two; *arista*, bristle; *-atum*, possessing. – (1) Lower glume bifid. *Andropogon biaristatus*, *Microstegium biaristatum* – (2) upper glume and lower lemma awned. *Panicum biaristatum* – (3) upper glume and sterile lemma awned. *Melinis biaristata* – (4) lemmas of both florets in spikelet awned. *Arrhenatherum biaristatum*, *Pogonatherum biaristatum*

Biatherium L. *bis*, twice; *ather*, barb or spine. The sterile floret is strongly two-awned

biaurita L. *bi-*, two; *auris*, ear; *-ita*, associated with. Dispersal unit terminating in two ears

bicicatricatus L. *bis*, twice; *cicatrix*, scar; *-atus*, possessing. The culms bear two conspicuous indentations marking the site where the sheaths fell off

biciliata L. *bis*, twice; *cilium*, hair; *-ata*, possessing. Upper glume densely ciliate on the marginal pair of nerves

bicknellii In honor of Eugene Pintard Bicknell (1859–1925) United States botanist

bicolor L. *bis*, twice; *color*, color. Two-colored, usually with respect to spikelets or florets

bicoloratum L. *bis*, twice; *color*, color; *-atum*, possessing. Spikelets two-colored

biconvexa L. *bis*, twice; *convexa*, arched. Anthoecium conspicuously biconvex

bicorn-e, -is L. *bi-*, two; *cornus*, horn. – (1) spikelets paired. *Digitaria bicornis* – (2) inflorescence of two racemes. *Panicum bicorne* – (3) subtending glumes two-toothed. *Triticum bicorne*

bicorniculat-a, -us L. *bi-*, two; *cornus*, horn; *-ula*, diminutive; *-ata*, possessing. Culm-sheaths with a pair of horn-like auricles at the bases of their reduced laminae

bicornis See *bicorne*

bicornuta L. *bis*, twice; *cornu*, horn; *-ata*, possessing. Lower glume two-horned

bicrinita L. *bis*, twice; *crinis*, hair of the head; *-ata*, possessing. There are lateral tufts of hairs on the lemma

bicrurulum L. *bis*, twice; *crus*, shin bone; *-ulus*, diminutive. Inflorescence a pair of short racemes

bicrurum L. *bis*, twice; *crus*, shin bone. Inflorescence a pair of racemes

bidactyla L. *bis*, twice; Gk *daktylos*, finger. Inflorescence formed of two narrow branches

bidentat-a, -um, -us L. *bis*, twice; *dens*, tooth; *-ata*, possessing. Lemma apex bifid

bidenticulata L. *bis*, twice; *dens*, tooth; *-ula*, diminutive; *-ata*, possessing. Lemma apex shortly two-toothed with a short awn from the sinus

biebersteiniana L. *-ana*, indicating connection. In honor of Friedrich August Marschall von Bieberstein (1768–1826) German-born Russian botanist

biebersteinii As for *biebersteiniana*

bielzii In honor of Eduard Albert Bielz (1827–1898) German teacher

bifalciger-a, -um L. *bis*, twice; *falx*, sickle; *gero*, carry or bear. Inflorescence comprises two falcate-secund racemes

Bifaria, -a, -um L. *bis*, twice; *fasces*, bundle; *-ulus*, diminutive; *-ata*, possessing. Panicle branches in paired fascicles

bifasciculat-a, -um L. *bi-*, two; *fasces*, bundle; *-ulus*, diminutive; *-ata*, possessing. Panicle of two condensed branches

bifid-a, -um, -us L. *bis*, twice; *findo*, divide. With structures such as panicle branches regularly bifurcating

bifidifolium L. *bis*, twice; *findo*, divide; *folium*, leaf. Leaf-apices bifid

bifid-um, -us See *bifida*

biflor-a, -um, -us L. *bis*, twice; *flos*, flower. Florets two per spikelet

biform-e, -is L. *bis*, twice; *forma*, appearance. – (1) Spikelets of two types on the same plant. *Digitaria biformis* – (2) inflorescences of two types on the same plant. *Microstegium biforme, Poa biformis*

bifurcat-a, -um L. *bis*, twice; *furca*, fork; *-ata*, possessing. – (1) Inner ligule bifurcate. *Merostachys bifurcata* – (2) panicle branches bifurcate. *Panicum bifurcatum*

bigelovii In honor of John Milton Bigelow (1804–1878) United States surgeon and botanist

bigeniculata L. *bis*, twice; *geniculata*, kneed. Awn of lemma bent in two places

biglandulare L. *bis*, twice; *glans*, acorn; *-ula*, diminutive; *-are*, pertaining to. With two glands on the sterile lemma

biglandulosa L. *bis*, twice; *glans*, acorn; *-ula*, diminutive; *-osa*, possession. There are two swellings at the pedicel bases

biglume L. *bis*, twice; *gluma*, husk. With two well developed scales comprising the upper glume and sterile lemma

bihariensis L. *-ensis*, denoting origin. From Bihari Mountains, Hungary

bikfayensis L. *-ensis*, denoting origin. From Bikfaya, Lebanon

bilimekii In honor of Dominik Bilimek (1813–1887) Austrian cleric and museum curator in Mexico

bilinguis L. *bis*, twice; *lingua*, tongue. The apex of the lower glume is bifid

billardierei As for *labillardierei*

billbergianum L. *-anum*, indicating connection. In honor of Gustaf Johan Billberg (1772–1844) Swedish biologist

billotii In honor of Paul Constant Billot (1796–1863) French botanist

billyi In honor of F. Billy (fl. 1988) French botanist

biloba L. *bis*, twice; *lobus*, lobe. Lemmas notched

bilykiana L. *-ana*, indicating connection. In honor of Gavriel Ivanovich Bilyk (1904–) who collected in Russia

bimaculata L. *bis*, twice; *macula*, spot; *-ata*, possessing. Spikelets pale-green except for margins of lower glume which may be pigmented

bimucronatum L. *bis*, twice; *mucro*, point; *-atum*, possessing. Lower glume bifid

binat-a, -um, -us L. *bis*, twice; *natus*, born. Racemes borne in digitate pairs

binghamii In honor of Major Bingham, British Army Officer and Conservator of Forests in Tinasserim, Myanmar

binodis L. *bis*, twice; *nodus*, knot. Culms two-noded

bipartita L. *bis*, twice; *partia*, divide. – (1) Inflorescence branches regularly dividing into two. *Aristida bipartita* – (2) lemma apex bifid. *Danthonia bipartita* – (3) inflorescence branches paired. *Chaetaria bipartita*

bipennat-um, -us L. *bis*, twice; *pinna*, feather; *-atum*, possessing. The fertile sessile spikelet is subtended by two feathery pedicels which lack the male spikelets customarily present in related species

bipinnata L. *bis*, twice; *pinna*, feather; *-ata*, possessing. – (1) Inflorescence of sessile spikelets on short lateral shoots arising from a central axis. *Desmostachya bipinnata* – (2) with the spikelets hanging in two rows from the under side of the inflorescence branches. *Uniola bipinnata*

bipollicaris L. *bis*, twice; *pollex*, thumb; *-aris*, pertaining to. Culms to about two pollices (5–6 cm) tall; in older literature the pollex as a measure referred only to the upper joint of the thumb

birandiana L. *-ana*, indicating connection. In honor of Hikmet Birand (1904–1972) Turkish botanist

birmanic-a, -us L. *-ica*, belonging to. From Burma, now Myanmar

biseriata L. *bis*, twice; *series*, row; *-ata*, possessing. Spikelets conspicuously two-rowed

bispiculatum L. *bi-*, two; *spica*, a point; hence, in particular, an ear or spike of grain; *-ula*, diminutive; *-atum*, possessing. Inflorescence of two racemes

bisquamulatus L. *bi-*, two; *squama*, scale; *-ula*, diminutive; *-ata*, possessing. Two scale-like processes growing from the pedicel at the base of the floret

bissei In honor of Johannes Bisse (1935–1984) who collected in Cuba

bissetii In honor of David Andreas Bisset (1892–?) United States Garden's Superintendent

bistipulatum L. *bis*, twice; *stipula*, small stalk; *-atum*, possessing. Ligule with two stipule-like outgrowths

bisulcat-a, -um L. *bis*, twice; *sulcus*, furrow. With two furrows as of glumes and lemmas

bitchuensis L. *-ensis*, denoting origin. From Bitchu Province, now part of Okayama Prefecture, Japan

bitextura L. *bi-*, two; *textus*, tissue. The surfaces of the upper and lower portions of the palea and lemma differ in texture

bitung Vernacular name of the species in Java, Indonesia

biuncial-e, -is L. *bis*, twice; *uncus*, hook; *-alis*, pertaining to. Glumes and/or lemmas terminating in two curved awns

bivestita L. *bis*, twice; *vestio*, clothe. Indumentum a mixture of long and short hairs

bivonae In honor of Antonio Bivona-Bernardi (1774–1837) Sicilian botanist

bjoerkmannii In honor of Sven Oscar Björkmann (1920–1956) Swedish botanist

blackii In honor of – (1) John McConnell Black (1855–1951) Scots-born South Australian botanist. *Stipa blackii* – (2) George A. Black (1910–1957) who collected in Brazil. *Panicum blackii*

bladhii In honor of Per Johann Bladh (1746–1816) Finnish botanist who collected in China and South Africa

blakei In honor of – (1) Stanley Thatcher Blake (1911–1973) Queensland botanist. *Andropogon blakei, Aristida blakei, Digitaria blakei, Ectrosia blakei, Sporobolus blakei, Stipa blakei* – (2) Sidney Fay Blake (1892–1959) United States botanist. *Axonopus blakei, Mesosetum blakei, Panicum blakei*

Blakeochloa Gk *chloa*, grass. As for *blakei*

blanchardiana L. *-ana*, indicating connection. In honor of Ferdinand Blanchard (1851–1892) or William Henry Blanchard (1850–1922) both of whom collected in the USA

blancheanum L. *-anum*, indicating connection. As for *blanchei*

blanchei In honor of Charles Isidore Blanche (1823–1887) who collected in Lebanon

blanchetii In honor of Jacques Samuel Blanchet (1807–1875) Swiss botanist

blancoi In honor of Francisco Manuel Blanco (1778–1845) Spanish cleric, explorer and botanist

blanda L. charming. Habit attractive

blanka From Pic Blanc, in the High Pyrenees

blaringhemii In honor of Louis Blaringhem (1878–1958) French botanist and plant breeder

blasdalei In honor of Walter Charles Blasdale (1871–1960) United States amateur botanist and chemist

blastocaulos Gk *blastos*, shoot; *kaulos*, stem. Culms densely branched from the base

blatteri In honor of Ethelbert Blatter (1877–1934) Swiss-born cleric and Indian botanist

blavii In honor of Otto Blau (1828–1879) German diplomat

bleeseri In honor of F. A. K. Bleeser (fl. 1925)

Blepharidachne Gk *blepharis*, eye-lash; *achne*, scale. The lemma margins are pectinate

blephariphyll-a, -us Gk *blepharis*, eye-lash; *phyllon*, leaf. The leaf margins bear long stiff widely separated hairs

blepharochaeta Gk *blepharon*, eye-lid; *chaete*, bristle. Subtending bristles shortly ciliate

Blepharochloa Gk *blepharon*, eye-lid; *chloa*, grass. The lemma bears a row of hairs along its midrib

blepharodes Gk *blepharon*, eye-lid; *-odes*, resembling. Margins of leaf-blades with long hairs

blepharoglumis Gk *blepharon*, eye-lid; *gluma*, husk. Glumes with stiff hairs along the midrib

blepharogyna Gk *blepharon*, eye-lid; *gyne*, woman. Ovary hairy

blepharolepis Gk *blepharon*, eye-lid; *lepis*, scale. Palea keels with stiff hairs

Blepharoneura See *Blepharoneuron*

Blepharoneuron Gk *blepharon*, eye-lid; *neuron*, nerve. The lemmas bear long hairs on each of the three nerves

blepharophor-a, -um Gk *blepharon*, eye-lid; *phero*, bear. Glumes and/or lemmas invested with long white hairs

blepharophyll-a, -um Gk *blepharon*, eye-lid; *phyllon*, leaf. Leaf-blades with tuberculate hairs on their margins

blepharoporum See *blepharophora*

blepharostachya Gk *blepharon*, eye-lid; *stachys*, spike as of an ear of wheat. Peduncles with abundant white hairs

blodgettii In honor of Charles Osgood Blodgett (1904–1979) United States agronomist

blomii In honor of Carl Blom (1885–1978) Swedish botanist

bloomeri In honor of Hiram G. Bloomer (1821–1874) United States botanist

Bluffia In honor of Matthias Joseph Bluff (1805–1857) German physician and amateur botanist

blumeana L. *-ana*, indicating connection. As for *blumii*

Blumenbachia In honor of Johann Friedrich Blumenbach (1752–1840) German physician and zoologist

blumii In honor of Carl Ludwig Blume (1796–1862) German-born Dutch botanist

Blyttia In honor of Matias Numsen Blytt (1789–1862) Norwegian botanist

bobartii In honor of Jacob Bobart (1641–1719) English gardener

bocquetii In honor of Silbert François Bocquett (1927–1986) of Zürich, Switzerland

boecheri In honor of Tyge Wittrock Böcher (1909–1983) Danish botanist

boehmeri In honor of Georg Rudolf Boehmer (1723–1803) German anatomist and botanist

boehmii In honor of R. Böhm who collected in Tanzania

boelckei In honor of Osvaldo Boelcke (1920–1990) Argentinian botanist

bofillianum L. *-anum*, indicating connection. In honor of Arturo Bofill y Pock (1846–1910)

bofillii See *bofillianum*

bogdanii In honor of – (1) Vassilij S. Bogdan, Russian botanist. *Hordeum bogdanii* – (2) Alexis V. Bogdan (fl. 1949–1953) who collected in East Africa. *Dactylotaenium bogdanii*, *Harpachne bogdanii*, *Sporobolus bogdanii*

boghisensis L. *-ensis*, denoting origin. From Boghis, Romania

bogoriensis L. *-ensis*, denoting origin. From Bogor, Java, Indonesia

bogotensis L. *-ensis*, denoting origin. From Bogotá, Colombia

bogueanum L. *-anum*, indicating connection. In honor of Ernest Everett Bogue (1864–1907) United States forester

bohemic-um, -us L. *-icum*, belonging to. From Bohemia, Czech Republic

boinensis L. *-ensis*, denoting origin. From Boïna, Madagascar

boisii In honor of Désiré Georges Jean Marie Bois (1856–1946) French botanist who collected in Vietnam

Boissiera In honor of Pierre Edmond Boissier (1810–1885) Swiss botanist and traveller

boissieri As for *Boissiera*

Boivinella L. *-ella*, diminutive but here used as a name-forming suffix. In honor of Louis Hyacinthe Boivin (1808–1852) French botanist and traveller

boiviniana L. *-ana*, indicating connection. As for *Boivinella*

boivinii See *Boivinella*

bojieiana L. *-ana*, indicating connection. In honor of Keng Pai-chieh (1917–) Chinese botanist

bolanderi In honor of Henry Nicholas Bolander (1831–1897) United States botanist

bolbodes Gk *bolbos*, onion; *-odes*, resembling. Lower culm nodes swollen

boliana L. *-ana*, indicating connection. In honor of Gerald Bol (1940–1996) United States plant collector and artist

bolivian-a, -us L. *-ana*, indicating connection. From Bolivia

boliviens-e, -is L. *-ense*, denoting origin. From Bolivia

B

bolusii In honor of Harry Bolus (1834–1911) English-born South African business-man and botanist

bomanii In honor of E. Boman (fl. 1903) who collected in Argentina

bombaiens-e, -is L. *-ense*, denoting origin. From Bombay, India

bombycin-um, -us L. *bombyx*, silk; *-inum*, indicating resemblance. Inflorescence or leaves invested with long silky hairs

bomiensis L. *-ensis*, denoting origin. From Pomi, Tibet Autonomous Region, China

bomoensis L. *-ensis*, denoting origin. From Bomo-dez-Tua, Zaire

bonaepartis In honor of Napoleon Bonaparte (1769–1821) Corsican-born French statesman and soldier

bonairense L. *-ense*, denoting origin. From Bonaire, an island of the Netherland Antilles

bonangensis L. *-ensis*, denoting origin. From Bonanga, Niari District, Congo

bonariens-e, -is L. *-ense*, denoting origin. From Provincia de Buenos Aires, Argentina

bonassorum L. *bonas(s)us*, European bison. From the forests of Western Russia, home of the European bison

bongaens-e, -is L. *-ense*, denoting origin. From Bonga, Zaire

bongardii In honor of August Gustav Heinrich Bongard (1786–1839) German-born Russian botanist

Bonia In honor of Henri François Bon (1844–1894) French cleric and amateur botanist who collected in Indo-China, now Cambodia, Laos and Vietnam

boninensis L. *-ensis*, denoting origin. From Bonin Islands, now known as Ogasawara-shoto

boniopsis Gk *opsis*, resemblance. Resembling *Bonia*

bonplandian-um, -us L. *-anum*, indicating connection. In honor of Aimé Jacques Alexandré Bonpland (1773–1858) French-born Brazilian botanist

bonplandii As for *bonplandianum*

bonthainica L. *-ica*, belonging to. From Bonthain Peak, Celebes, Indonesia

boormanii In honor of John Luke Boorman (1864–1938) Australian plant collector

bootanens-e, -is L. *-ense*, denoting origin. From the Kingdom of Bhutan

boraei In honor of Alexandre Boreau (1803–1875) French botanist

borbasii In honor of Vincenz Borbás (1844–1905) Romanian and Hungarian botanist

borbonic-a, -um, -us L. *-ica*, belonging to. From *Insula Borbonia*, now Réunion, one of the Mascarene Islands

borchersii In honor of Augusto Borchers (fl. 1883–1886) who collected in Chile

borderei In honor of Henri Bordère (1825–1889) French teacher and plant collector

boreal-e, -is L. *boreas*, north wind; *-ale*, pertaining to. – (1) Growing in the cold regions of the northern hemisphere. *Agropyron boreale, Agrostis borealis, Deschampsia borealis, Elymus boreale, Enneapogon borealis, Festuca borealis, Glyceria borealis, Hierochloe borealis, Pappophorum boreale, Panicularia borealis, Panicum boreale, Puccinellia borealis, Roegneria borealis* – (2) growing in northern Australia. *Aristida borealis*

boreali-tibetica L. *boreas*, north wind; *-ale*, pertaining to; *-ica*, belonging to. From northern Tibet Autonomous Region, China

boresthenica See *borysthenica*

borhidii In honor of Attila L. Borhidi (1932–) Hungarian-born, Cuban botanist

borian-a, -us As for *Borinda*

borii As for *Borinda*

Borinda In honor of Norman Loftus Bor (1893–1972) Irish-born Indian forest botanist

borisii In honor of Boris (1894–1943) Tsar of Bulgaria

Boriskellera In honor of Boris Aleksandrovich Keller (1874–1945) Russian botanist

borneensis L. *-ensis*, denoting origin. From Borneo

bornmuelleri In honor of Joseph Friedrich Nicolaus Bornmüller (1862–1948) German botanist

borreri In honor of William Borrer (1781–1862) English botanist

borszczowii In honor of Elia Grigorievicz Borszczow (1833–1878) Russian botanist

borumensis L. *-ensis*, denoting origin. From Borum, Mozambique

borussica L. *Borussia*, Prussia; *-ica*, belonging to. An allusion either to the country of collection, Tanzania, then known as German East Africa, or directly to Prussia whose principal city, Berlin, had shortly before become the capital of Germany. The expedition leader was Director of the Berlin Botanical Museum which had connection with both Prussia and Germany

boryan-a, -um L. *-ana*, indicating connection. In honor of Jean Baptiste Geneviève Marcellin Bory de St Vincent (1778–1846) French botanist

borysthenica L. *-ica*, belonging to. From *Borysthenes*, ancient name for the Dnieper River, Ukraine

borzianum L. *-anum*, indicating connection. As for *borzii*

borzii In honor of Antonino Borz (1852–1921) Sicilian botanist

boscian-um, -us L. *-anum*, indicating connection. In honor of Louis Augustin Guillaume Bosc (1759–1828) French botanist

boscii As for *boscianum*

bosniaca L. a Bosnian. From Bosnia

bosseri In honor of Jean M. Bosser (1922–) French botanist

bossii In honor of Georg Boss (?–1972)

Botelua See *Bouteloua*

Bothriochloa Gk *bothrion*, pit; *chloa*, grass. The lower glume of some species has a conspicuous pit

Botriochloa See *Bothriochloa*

botryodes Gk *botrys*, bunch of grapes; *-odes*, resembling. The inflorescence is a congested panicle

botryoides Gk *botrys*, cluster of grapes; *-oides*, resembling. The inflorescence is racemose

botryostachya Gk *botrys*, cluster of grapes; *stachys*, spike as of an ear of wheat. Vegetative and flowering culms more or less discrete

botschantzevii In honor of Victor Petrovic Botschantzev (1910–) Russian botanist

botterii In honor of Mateo Botteri (1808–1877) Italian botanist

bottnica L. *-ica*, belonging to. From Bottnieus, now Bothnia, northern Sweden

boucheanum In honor of Peter Carl Bouché (1783–1856) German-born French horticulturist

bourdillonii In honor of Thomas Fulton Bourdillon (1849–1930) who collected in India

bourgaei In honor of Eugène Bourgeau (1813–1877) French botanist

bournei As for *bourneorum*

bourneorum In honor of Albert Gibbs Bourne (1859–1940) and his wife, collectors in India, Thailand and Myanmar

bourouensis L. *-ensis*, denoting origin. From Bourou, now Buru an island in the Moluccas, Indonesia

Bouteloua, Boutelouae In honor of Claudio Boutelou (1774–1842) Spanish horticulturist

bouteloudoides Gk *-oides*, resembling. Similar to *Bouteloua*

bovonei In honor of Ettore Bovone (1880–1922) Italian veterinarian and plant collector in Zaire

bowdenii In honor of Wray M. Bowden (1914–) Canadian botanist

bowes-lyonii In honor of David Bowes-Lyon (1902–1961) British botanist who collected in Pakistan

boxiana L. *-ana*, indicating connection. In honor of Harold Edmund Box (1898–1973) English born West Indian entomologist

boyacensis L. *-ensis*, denoting origin. From Department of Boyacá, Colombia

Brachatera Gk *brachys*, short; *ather*, ear or spike of wheat. Lemma shortly awned in contrast to Danthonia

Brachiaria L. *brachium*, fore-arm; *-aria*, pertaining to. The inflorescence branches frequently resemble signal arms

brachiariaeformis L. *forma*, appearance. Inflorescence resembling that of *Brachiaria*

B

brachiat-a, -um L. *brachium*, forearm; *-atum*, possessing. Inflorescence a single axis bearing racemes often held in a manner suggesting the arms of a signalling station

Brachyachne Gk *brachys*, short; *achne*, scale. The lemma is shorter than the subtending glumes

brachyanther-a, -um Gk *brachys*, short; *antherix*, ear or spike of wheat. Inflorescence or inflorescence branches short and spicate

brachyanthum Gk *brachys*, short; *anthos*, flower. Panicle depauperate with few branches

Brachyaria See *Brachiaria*

brachyather-a, -um, -us Gk *brachys*, short; *ather*, barb or spine. Lemma awn relatively short

brachychaet-a, -e Gk *brachys*, short; *chaete*, bristle. Awn of lemma shorter than that of related species

brachychaetoides Gk *-oides*, resembling. Spikelets resemble those of *Stipa brachychaeta*, which is in a closely related genus

Brachychloa Gk *brachys*, short; *chloa*, grass. Racemes short in comparison to those of *Leptochloa*

brachyclad-a, -um Gk *brachys*, short; *klados*, stem. Inflorescence branches short

Brachyelytrum, brachyelytrum Gk *brachys*, short; *elytrum*, cover. The subtending glumes are short with respect to the length of the spikelet

brachygloss-a, -us Gk *brachys*, short; *glossa*, tongue. Ligule short

brachylachnum Gk *brachys*, short; *lachnos*, wool. Glumes shortly pubescent

brachylemma Gk *brachys*, short; *lemma*, husk. Lemmas much shorter than glumes

brachylepis Gk *brachys*, short; *lepis*, scale. Glumes shorter than lemmas

brachyloph-a, -um Gk *brachys*, short; *lophos*, crest. Fertile lemma shortly apiculate

brachyphyll-a, -um, -us Gk *brachys*, short; *phyllon*, leaf. Leaf-blades short

brachypod-a, -um, -us Gk *brachys*, short; *pous*, foot. Raceme peduncles very short

brachypodioides Gk *-oides*, resembling. Similar to *Brachypodium* in habit

Brachypodium Gk *brachys*, short; *podion*, little foot. Spikelets borne on very short pedicels

brachypodus See *brachypoda*

brachypogon Gk *brachys*, short; *pogon*, beard. Callus shortly hairy

brachyrhynchus Gk *brachys*, short; *rhynchos*, beak. Spikelets bearing hairs with swollen tips

brachyspermum Gk *brachys*, short; *sperma*, seed. Grains shorter than those of related species

brachystachy-a, -um, -us See *Brachystachyum*

brachystachys See *Brachystachyum*

Brachystachyum Gk *brachys*, short; *stachys*, spike as of an ear of wheat. Inflorescence short and composed of spike-like racemes

brachystachyus See *brachystachya*

brachystephana Gk *brachys*, short; *stephanos*, crown. Base of awn surrounded by a short crown of hairs

Brachystylus, brachystylus Gk *brachys*, short; *stylos*, column. Style short

brachythyrs-a, -um, -us Gk *brachys*, short; *thyrsos*, ornamental wand. Racemes short and congested towards the apex of a long peduncle

brachytrich-a, -um, -us Gk *brachys*, short; *thrix*, hair. With short hairs on the glumes and, or lemmas

brachyur-a, -um Gk *brachys*, short; *oura*, tail. Inflorescence of short racemes

Braconnotia See *Braconotia*

Braconotia In honor of Henry Braconnot (1780–1855) French botanist

bracteat-a, -um, -us L. *bractea*, bract; *-ata*, possessing. Panicle or inflorescence branches subtended by leafy bracts

Bracteola L. a small leaf of gold. The spikelets are light-golden, shining and compressed

bracteolata L. *bractea*, bract; *-ola*, diminutive; *-ata*, possessing. Primary branches of inflorescence subtended by small bracts

bracteosa L. *bractea*, bract; *-osa*, well developed. Panicles with a hyaline bract

bradei In honor of Alexander Curt Brade (1881–1971) German-born Brazilian botanist

bradleyi From Bradley grass cultivated as a turf in South Africa

brainii In honor of Charles Kimberlin Brain (1931–) who collected in Southern Rhodesia

brandegei In honor of Townsend Stith Brandegee (1843–1925) United States civil engineer and botanist

brandisii In honor of Dietrich Brandis (1824–1907) German-born Indian forester

Brandtia In honor of Johann(es) Friedrich Brandt (1802–1879) German-born Russian zoologist

brandzae In honor of Marcel Alex Brândza (1868–1934) Romanian botanist

brasilian-a, -um L. *-ana*, indicating connection. From Brazil

brasiliens-e, -is L. *-ense*, denoting origin. From Brazil

Brasilocalamus Gk *kalamos*, reed. A reed-like genus from Brazil

brassii In honor of Leonard John Brass (1900–1971) Australian explorer and plant collector

braun-blanquetii In honor of Josias Braun-Blanquet (1884–1980) French-Swiss vegetation scientist

braunii In honor of Alexander Carl Heinrich Braun (1805–1877) German botanist

bravum From Valle de Bravo, Mexico

brazzae In honor of Jacques de Brazza (1859–1887) Belgian botanist

brazzavillense L. *-ense*, denoting origin. From Brazzaville, Congo

breazensis L. *-ensis*, denoting origin. From Breaza, Romania

bredoensis L. *-ensis*, denoting origin but here used as a name-forming suffix. In honor of Hans Joseph Anna Eric Richard Brédo (1903–)

breedlovei In honor of Dennis E. Breedlove (1939–) United States botanist

breistrofferi In honor of Maurice André Frantz Breistroffer (1910–1986) French Museum administrator

brennia In honor of the tribe of people known as the Brennii after whom the Brenner Pass between Italy and Austria is named

bresolinii In honor of Antônio Bresolin (1919–) Brazilian botanist

breunia Possibly a misspelling of *brennia*, and thus may commemorate the Brennii, a tribe which inhabited the Alps between Italy and Austria. The protologue also mentions "Brunner" and "Brenner" which further confuses the issue

brev-e, -is L. short. – (1) Culms short. *Avena brevis*, *Brachiaria brevis*, *Bromus brevis*, *Dissanthelium breve*, *Hordeum breve*, *Muhlenbergia brevis*, *Panicum breve*, *Paspalum breve*, *Pennisetum breve*, *Poa brevis*, *Yushania brevis* – (2) lemmas short. *Stipidium breve* – (3) leaf-blades short. *Phalaris brevis*

breviaristat-a, -um, -us L. *brevis*, short; *arista*, bristle; *-ata*, possessing. Glumes or lemmas shortly awned

brevicalyx L. *brevis*, short; Gk *kalyx*, cup. The subtending glumes are much shorter than the lemma

breviculmis L. *brevis*, short; *culmus*, stalk. Culms short

brevicuspidata L. *brevis*, short; *cuspis*, head of a spear; *-ata*, possessing. Lemma shortly three-cusped

brevidentatum L. *brevis*, short; *dens*, tooth; *-atum*, possessing. Lateral lobes of lemma short

brevieri In honor of Brevier

breviflor-a, -um L. *brevis*, short; *flos*, flower. Spikelets short

brevifoli-a, -um, -us L. *brevis*, short; *folium*, leaf. Leaf-blades shorter than those of some other species in the genus or relative to the length of the culm

brevigluma L. *brevis*, short; *gluma*, husk. Upper glume not exceeding the spikelet in length

breviglum-e, -is L. *brevis*, short; *gluma*, husk. One or both glumes short with respect to the length of the spikelet

breviligula L. *brevis*, short; *ligula*, small tongue. Ligule very short

breviligulata L. *brevis*, short; *ligula*, little tongue; *-ata*, possessing. Ligule short

brevinodus L. *brevis*, short; *nodus*, knot. Culms with short internodes

B

brevipaleata L. *brevis*, short; *palea*, scale; *-ata*, possessing. Palea much reduced

brevipaniculata L. *brevis*, short; *paniculus*, panicle; *-ata*, possessing. Panicle short and broad

brevipedicellatum L. *brevis*, short; *pedicellus*, stalk; *-atum*, possessing. Primary branches of inflorescence short

brevipedunculatus L. *brevis*, short; *pedunculatus*, stalked. With short peduncles

brevipes L. *brevis*, short; *pes*, foot. – (1) Pedicels short. *Agrostis brevipes*, *Digitaria brevipes*, *Roegneria brevipes*, *Stipa brevipes* – (2) peduncle short. *Arundinaria brevipes*

brevipil-a, -um, -us L. *brevis*, short; *pilum*, hair. Lemmas or glumes bearing short hairs

brevipilis L. *brevis*, short; *pilum*, hair. See *brevipila*

brevipil-um, -us See *brevipila*

Brevipodium L. *brevis*, short; *pes*, foot. Tussock forming species, whereas those that are related have a rhizomatous habit

breviradiatum L. *brevis*, short; *radius*, spoke of a wheel; *-atum*, possessing. Panicle branches short, whorled

breviramosum L. *brevis*, short; *ramus*, branch; *-osum*, abundance. With many short panicle branches

brevis See *breve*

breviscrobs L. *brevis*, short; *scrobis*, ditch. On drying, a small pit develops at each basal margin of the fertile lemma

breviset-a, -um, -us L. *brevis*, short; *seta*, bristle. – (1) Lemmas short-awned. *Chloris breviseta*, *Danthonia breviseta* – (2) involucral bristles short. *Cenchrus brevisetus* – (3) lower glume shortly awned. *Ortachne breviseta*, *Panicum brevisetum*

brevispica L. *brevis*, short; *spica*, a point; hence, in particular, an ear or spike of grain. Spikelets short

brevispicat-a, -um L. *brevis*, short; *spica*, a point; hence, in particular, an ear or spike of grain; *-ata*, possessing. Inflorescence a single or pair of secund spikes

brevispicula L. *brevis*, short; *spica*, a point; hence, in particular, an ear or spike of grain; *-ula*, diminutive. Spikelets one-floreted

brevissima L. *brevis*, short; *-issima*, most. Leaf-blades very short

brevisubulatum L. *brevis*, short; *subula*, awl; *-ata*, possessing. Glumes or lemma shortly awned

brevivaginata L. *brevis*, short; *vagina*, sheath; *-ata*, possessing. Leaf-sheaths only about half the length of the succeeding internode

breweri In honor of William Henry Brewer (1828–1910) United States botanist

bricchetteana L. *-ana*, indicating connection. In honor of Luigi Robecchi-Bricchette (1855–1926) Italian botanist

brieyi In honor of Comte J. de Briey (fl. 1912) who collected in Zaire

brigalow Grows in the Brigalow (*Acacia harpophylla*) dominated forests of north-eastern Australia

brigantiaca L. *-ica*, belonging to. From Brigantium, now Briançon, France

brigantina L. *-ina*, indicating possession. See *brigantiaca*

brilletii In honor of F. Brillet (fl. 1923) French botanist who collected in Tonkin, now Vietnam

brinkmannii In honor of Friedrich Ludwig Brinkmann (1799–1875) German gardener

brintnellii From the shores of Lake Brintnell, south-western Mackenzie, British Columbia, Canada

briquetii In honor of John Isaac Briquet (1870–1931) Swiss botanist

britannic-a, -us L. *-ica*, belonging to. From Brittania now in part Britain

brittonii In honor of Nathaniel Lord Britton (1859–1934) United States botanist

brittonorum Of the Brittons. In honor of Nathaniel Lord (1859–1934) and Elizabeth Gertrude (1858–1934) Britton, United States botanists

brixhei In honor of Brixhe (fl. 1910) who collected in Zaire

Briza Gk *brizo*, nod. The spikelets are borne on long stalks and so droop. The name was used in Classical Times by Galenos for a species of cereal, probably rye

brizaeform-e, -is L. *forma*, appearance. The spikelets resemble those of *Briza*

brizanth-a, -um Gk *brizo*, nod; *anthos*, flower. The spikelets hang from the horizontal rhachis

briziformis See *brizaeforme*

Brizochloa Resembling *Briza*

brizoides Gk *-oides*, resembling. Inflorescences or spikelets resemble those of *Briza*

Brizopyrum Gk *brizo*, nod; *pyros*, wheat. Inflorescence with drooping branches

brockmanii As for *Drake-Brockmania*

brodiei In honor of David Arthur Brodie (1868–?) Canadian-born United States agronomist

Bromelica Sharing the characters of *Bromus* and *Melica*

bromidioides Gk *-oides*, resembling. Similar to *Bromidium*

Bromidium Gk *-idium*, diminutive but here used as a name-forming suffix. Similar to *Bromus*

Bromofestuca Presumed hybrids between species of *Bromus* and *Festuca*

bromoides Gk *-oides*, resembling. Resembling *Bromus*, usually with respect to the inflorescence

bromoideus See *bromoidea*

Bromopsis Gk *opsis*, resemblance. Similar to *Bromus*

Bromuniola Superficially resembles *Bromus* but in the number of glumes resembles *Uniola*

Bromus Gk *bromo*, food. In Classical times the Greek name for oats

brongniartii In honor of Adolphe Théodore Brongniart (1801–1876) French botanist and palaeontologist

broteri In honor of Félix da Silva Avelar Brotero (1744–1828) Portugese botanist

Brousemichea In honor of M. Brousemiche (fl. 1882) sometime Director of Botanic Garden, Saigon, Vietnam

browneana L. *-ana*, indicating connection. As for *brownii* (1)

brownei See *brownii* (1)

browniana L. *-ana*, indicating connection. As for *brownii* (1)

brownii In honor of – (1) Robert Brown, (1773–1858) Scots-born English botanist. *Amphipogon brownii, Cenchrus brownii, Cinna brownii, Digitaria brownii, Eragrostis brownii, Leptochloa brownii* – (2) Joseph R. Brown, United States rancher. *Aristida brownii*

Bruckmannia See *Beckmannia*

brueggeri In honor of Christian Georg Brügger (1833–1899) Swiss museum director

bruggemannii In honor of Paul F. Bruggemann (1890–1974) German-born Canadian naturalist

bruhnsiana L. *-ana*, indicating connection. In honor of Alexander Bruhns who collected around the Caspian Sea

brunne-a, -um L. dull brown. Spikelets dull-brown

brunnescens L. *brunesco*, become brown. Spikelets pale-brown

brunneum See *brunnea*

brunoana L. *-ana*, indicating connection. In honor of Francesco Bruno (1897–1986) Italian botanist

brunonian-a, -um L. *-ana*, indicating connection. As for *brownii*

brunonis See *brownii*

Brylkinia In honor of A.D. Brylkin (fl. 1859–1863) ethnographer and plant collector in Siberia

bryoides Gk *bryon*, moss; *-oides*, resembling. Habitat moss-like

bryophil-a, -us Gk *bryon*, moss; *phileo*, love. Growing amongst mosses

buar The vernacular name of this species in Sumatra, Indonesia

bucegiensis L. *-ensis*, denoting origin. From Bucegi, Romania

Bucetum L. a cattle pasture. Species a common component of cattle pastures

buchananensis L. *-ensis*, denoting origin. Growing on the shores of Lake Buchanan, Queensland, Australia

buchananii In honor of – (1) John Buchanan (1855–1896) who collected in Malawi. *Digitaria buchananii, Eragrostis buchananii* – (2) John Buchanan (1821–1903) Scotts-born clergyman and amateur botanist. *Andropogon buchananii* – (3) John Buchanan (1819–1898) Scots-born New Zealand artist and botanist. *Poa buchananii* – (4) G. Buchanan, collector of the type. *Setaria buchananii*

bucharica L. *-ica*, belonging to. From Buchara District, Turkestan region of Central Asia

buchingeri In honor of Jean Daniel Buchinger (1803–1888) from whose herbarium the species was described

Buchloe, Buchloë Gk *bukalos*, buffalo; *chloe*, grass. A contraction of the Greek translation of the vernacular name Buffalo Grass

Buchlomimus Gk *mimus*, a mimic. Superficially resembling *Buchloe*

buchneri In honor of Max Buchner (1846–1921) who collected in Angola

buchtarmensis L. *-ensis*, denoting origin. From the Buchtarma River, Kazakhstan

buchtienii In honor of Otto Buchtien (1859–1946) German botanist

buchwaldii In honor of Johannes Buchwald (1869–1927) German botanist

buckleyan-a, -um, -us L. *-ana*, indicating connection. As for *buckleyi*

buckleyi In honor of Samuel Botsford Buckley (1809–1884) United States botanist

buddhistica L. *-ica*, belonging to. Origin uncertain, not given by author but may refer to the species growing in the grounds of Buddhist Temples

budensis L. *-ensis*, denoting origin. From Buda, Hungary

buekeana L. *-ana*, indicating connection. In honor of Bueke

Buergersiochloa In honor of Th. Buergers (1881–?) Dutch physician and educator

buettneri In honor of David Sigmund August Buettner (1724–1768) German botanist

bufensis L. *-ensis*, denoting origin. From the Bufa Mountains, Mexico

bulawayense L. *-ense*, denoting origin. From Bulawayo, Zimbabwe

bulbifer L. *bulbus*, onion; *fero*, carry or bear. Culm bases swollen

Bulbilis L. *bulbus*, onion; *-ilis*, property of. Anthoecium bulb-shaped

bulbillifera L. *bulbus*, bulb; *-illus*, diminutive; *fero*, carry or bear. Lateral shoots short, with swollen culm bases

bulbodes See *bolbodes*

bulbos-a, -um, -us L. *bulbus*, onion; *-osa*, abundance. – (1) Culm-bases swollen. *Alopecurus bulbosus*, *Avena bulbosa*, *Cenchrus bulbosus*, *Digitaria bulbosa*, *Erianthecium bulbosum*, *Glyceria bulbosa*, *Hordeum bulbosum*, *Panicum bulbosum*, *Pappophorum bulbosum*, *Phalaris bulbosa*, *Poa bulbosa* – (2) lower glume inflated. *Sorghum bulbosum*

Bulbulus L. *bulbus*, onion; *-ulus*, diminutive. Culms bulbous at the base

bulgarica L. *-ica*, belonging to. From Bulgaria

bullockii In honor of Arthur Allman Bullock (1906–1980) English botanist

buncei In honor of Daniel Bunce (1813–1872) Australian nurseryman and Garden's Curator

bungean-a, -um L. *-ana*, indicating connection. In honor of Aleksandr Andreevic Bunge (1803–1890) Ukrainian physician and botanist

bungei As for *bungeana*

bunglensis L. *-ensis*, denoting origin. From Bungle Bungle Range, Western Australia

bungoensis L. *-ensis*, denoting origin. From Bungo Province, now Oita Prefecture, Japan

bunicola Gk *bounos*, hill; L. *-cola*, dweller. From the Flinders Ranges, Australia

bunophilum Gk *bounos*, hill; *phileo*, love. Growing on hills

bunyensis L. *-ensis*, denoting origin. From Bunya Mountains, Queensland, Australia

burbidgeae In honor of Nancy Tyson Burbidge (1912–1977) Australian botanist

burchan-buddae Of the Burchan-Buddha Mountains, Tibet Autonomous Region, China

burchellii In honor of William John Burchell (1781–1863) English traveller and plant collector in southern Africa and Brazil

burgu Local name for the species in Niger

burgundiana L. *-ana*, indicating connection. From Burgundy, France

burjatica L. *-ica*, belonging to. From Buryat-Mongol, now Republic of Buryatia, Russian Federation

burkartianum L. *-anum*, indicating connection. As for *burkartii*

burkartii In honor of Arturo Erhardo Burkart (1906–1975) Argentinian botanist

burkei In honor of Joseph Burke (fl. 1830s–1840s) who collected in southern Africa and North America

burkensis L. *-ensis*, denoting origin. From Burke District, Queensland, Australia

burkii (1) In honor of Isaac Burk (1816–1893) United States botanist. *Bouteloua burkii* – (2) in error for *burkei. Aristida burkei*

burkittii In honor of George Burkitt (1830–?) pastoralist who collected in northern Australia

Burmabamba See *Burmabambus*

Burmabambus A woody bamboo from Burma, now Myanmar

burmaensis L. *-ensis*, denoting origin. From Burma, now Myanmar

burmahicum L. *-icum*, belonging to. From Burma, now Myanmar

burmanic-a, -um, -us L. *-ica*, belonging to. From Burma, now Myanmar

burmanii In honor of Alisdair Graham Burman (1942–1992) English-born, Brazilian botanist

burmannii In honor of Nicolaus Lorenz Burmann (1734–1793) Dutch botanist

burmensis L. *-ensis*, denoting origin. From Burma, now Myanmar

burmitis Gk *-itis*, indicating a close connection. From Burma now Myanmar

burnaschewii In honor of Burnaschew

burnatii In honor of Emile Burnat (1828–1920) Swiss engineer, magistrate and amateur botanist

burnoufii In honor of Charles Burnouf (fl. 1850) Corsican educator

burnsiana L. *-ana*, indicating connection. In honor of William Burns (1884–1970) Scots-born Indian botanist

burraensis L. *-ensis*, denoting origin. From the Burra Range, Queensland, Australia

burttdavii In honor of Joseph Burtt-Davy (1870–1940) Scots-born Californian and South African botanist

burttii In honor of Bernard Dearman Burtt (1902–1938) English botanist who collected widely in tropical Africa

buschian-a, -um, -us L. *-ana*, indicating connection. In honor of Elizabeth (Elizaveta) Alexandrovna Busch (1886–1960) or Nicolai Adolfowitsch (Adolfovich) Busch (1869–1941) who jointly collected in the Caucasus

buschirica L. *-ica*, belonging to. From Buschir, Iran

busei In honor of Lodewijk Hendrik Buse (1819–1888) Dutch botanist

bushii In honor of Benjamin Franklin Bush (1858–1937) United States botanist

busseanum L. *-anum*, indicating connection. As for *bussei*

bussei In honor of Walter Carl Otto Busse (1865–1933) German botanist and traveller

Butania Named for the Kingdom of Bhutan

butuluensis L. *-ensis*, denoting origin. From Butulu, Zaire

buxbaumii In honor of Johann Christian Buxbaum (1693–1730) German botanist

buza In honor of Búza

bynoei In honor of Benjamin Bynoe (1804–1865) English Naval Surgeon who collected in Australia

byronis From Byron Bay, Hawaii

byrrangensis L. *-ensis*, denoting origin. From Byrrang, that is Bering Peninsula, Russian Far East

byzantina Gk *-ina*, indicating possession. From Byzantium, either the city now known as Constantinople or in the wider sense of the eastern division of the Roman Empire that corresponds approximately with the present day Near East

C

caaguazuense L. *-ense*, denoting origin. From Caaguazú, Paraguay

caamanoi In honor of José Maria Plácido Caamaño (1838–1901) a former President of Ecuador

C

caatingense L. *-ense*, denoting origin. From the Catinga, Brazil
cabanisii In honor of Jean Louis Cabanis (1816–1906) who collected in south-eastern USA
Cabrera In honor of Antonia Cabrera (1763–1827) Spanish cleric and botanist who collected in South America
cabrerae In honor of Angel Lulio Cabrera (1908–1999) Argentinian botanist
cabrerensis L. *-ensis*, denoting origin. From La Cabrera, Spain
cabreriana L. *-ana*, indicating connection. As for *cabrerae*
cacharensis L. *-ensis*, denoting origin. From the Cachar in the Brahmaputra Valley, India
cachemyriana L. *-ana*, indicating connection. From Emodi Cachemyriana, that is Kashmir
cachimboense L. *-ense*, denoting origin. From Serra do Cachimbo, Brazil
cacuminis (1) L. *cacumen*, extreme point. Leaf-apex markedly acuminate. *Arthrostylidium cacuminis* – (2) L. *cacumen*, summit. From the Tibetan Plateau. *Elymus cacuminis*
caduc-a, -um L. dropping off early. Florets or spikelets shed shortly after anthesis
caduciflora L. *caduca*, dropping off early; *flos*, flower. Spikelets not persistent
caduciseta L. *caduca*, dropping off early; *seta*, bristle. Awn deciduous
caducum See *caduca*
caelachyrium See *Coelachyrum*
caerulans L. *caerulea*, bluish; *-ans*, assuming the appearance of. Spikelets dark-purple
caerule-a, -um, -us L. bluish. Often with bluish-green leaf-blades
caerulescens L. *caerulesco*, become bluish. Foliage glaucous
caeruleus See *caerulea*
caesi-a, -um, -us L. bluish-grey, as of eyes. Plant in whole or in part bluish-grey
caesioglaucum L. *caesius*, bluish-green as of eyes; *glauca*, bluish-green. Leaves bluish-green
caespitans L. *caespes*, grass that has been cut; *-ans*, assuming the appearance of. Forming turf
caespitos-a, -um, -us L. *caespes*, grass that has been cut; *-osa*, abundance. Tufted or forming a turf
caffr-a, -um Pertaining to the Kaffirs of southern Africa
caffrorum See *caffra*. Of the Kaffirs who cultivated the species for grain
caffrum See *caffra*
cagiriensis L. *-ensis*, denoting origin. From Mt Cagiri, Pyrenees, France
cahoonianum Origin obscure, not given by author, but probably referring to Calhoun, a city and county in Georgia, USA, as the name it replaced was *georgianum*, which also refers to the state of Georgia
caianus L. *-anus*, indicating connection. In honor of Lian-bing Cai (fl. 1996) Chinese botanist
cainii In honor of Stanley Adair Cain (1902–1995) United States botanist
cairnesiana L. *-ana*, indicating connection. In honor of Donaldson Delorme Cairnes (1875–1917) Canadian geologist and plant collector
cajamarcae From Cajamarca Province, Peru
cajamarcensis L. *-ensis*, denoting origin. See *cajaramacae*
cajatambensis L. *-ensis*, denoting origin. From Cajatambo Province, Peru
calabrica L. *-ica*, belonging to. From Calabria, Italy
calaccanzense L. *-ense*, denoting origin. From Calaccanz, Luzon Island, Philippines
calamagrostidea Gk *-idea*, resembling. Similar to *Calamagrostis*
calamagrostidiformis L. *forma*, appearance. Inflorescences or spikelets resembling those of *Calamagrostis*
Calamagrostis, calamagrostis Gk *kalamos*, reed; *agrostis*, a type of grass. Many of the species are reed-like
calamari-a, -us L. *calamus*, reed; *-aria*, pertaining to. Habit reed-like
Calamina Gk *kalamos*, reed; *-ina*, indicating resemblance. Habit reed-like

Calammophila Hybrids between species of *Calamagrostis* and *Ammophila*
Calamochloa Gk *kalamos*, reed; *chloa*, grass. Culms cane-like
Calamochloe, Calamochloë See *Calamochloa* but referring to a different genus
Calamogrostis A misspelling of *Calamagrostis*
Calamophila Hybrids between species of *Calamagrostis* and *Ammophila*
Calamovilfa A combination of *Calamagrostis* and *Vilfa*
calantha Gk *kalos*, beauty; *anthos*, flower. Inflorescence of attractive appearance
calarashica L. *-ica*, belonging to. From Kalarash, Moldova
calatajeronensis L. *-ensis*, denoting origin. From Caltajerone, Sicily
calcarata L. *calcar*, spur; *-ata*, denoting possession. Base of upper glume formed into a spur
calcare-a, -us L. *calx*, lime; *-arius*, pertaining to. Growing on limestone soils
calcaria See *calcarea*
calchaquia From the Calchaquia Valley, Argentina
calchaquiensis L. *-ensis*, denoting origin. From Cumbres Calcha-quies, a district of Argentina
calcicola L. *calx*, lime; *-cola*, dweller. Growing on limestone
calciphilus L. *calx*, lime; Gk *phileo*, love. Growing on limestone
calcis L. *calx*, lime. Growing on limestone
caldasii In honor of Mancisco José Caldas (1741–1816) Colombian botanist
calderi In honor of James Alexander Calder (1915–1990) Canadian botanist
calderillensis L. *-ensis*, denoting origin. From Calderillo, Bolivia
Calderonella L. *-ella*, diminutive but here a name-forming suffix; *-ana*, indicating connection. In honor of Cleofé Elsa Calderón (1929–) Argentinian-born United States botanist
calderoniae As for *Calderonella*
calderoniana As for *Calderonella*
caldesii In honor of Lodovico Caldesi (1822–1884) Italian botanist
caledonica L. *-ica*, belonging to. From New Caledonia
caliculatus See *calyculatus*
calicutensis L. *-ensis*, denoting origin. From the Calicut District, Kerala State, India
californic-a, -um, -us L. *-ica*, belonging to. From California, USA
Calliagrostis Gk *kallion*, more beautiful; *agrostis*, an unidentified fodder plant of the Ancients. Regarded by the author as beautiful
Callichloea Gk *kallion*, more beautiful; *chloa*, grass. Attractive in appearance
callichroa Gk *kallion*, more beautiful; *chroia*, color. Spikelets attractively colored
callida L. sly. Somewhat resembling three other species
callieri In honor of Alexis Callier (1850–1925) who collected in the Crimea
calliferum L. *callus*, hard skin of an animal; *fero*, carry or bear. Base of spikelet callus-like
calligera Gk *kallion*, more beautiful; L. *gero*, carry or bear. Spikelets pale-purple
callina A misspelling of *collina*
calliopsis Gk *kallion*, more beautiful; *opsis*, resemblance. Meaning obscure, not given by author
calliphyllum Gk *kallion*, more beautiful; *phyllon*, leaf. Leaf-blades light-green drying yellowish
callitrichus Gk *kallion*, more beautiful; *thrix*, hair. Awns setiform, violet
callopus Gk *kallion*, more beautiful; *pous*, foot. Glumes adnate to the internode forming a brightly colored subglobose stipe
callos-a, -um L. hard-skinned. – (1) Florets in some way thickened. *Arundinaria callosa*, *Avena callosa*, *Melica callosa*, *Panicum callosum*, *Schizachne callosa* – (2) the base of the leaf-lamina is thickened. *Poa callosa*
calochloa Gk *kalos*, beautiful; *chloa*, grass. Attractive in appearance
caloptila Gk *kalos*, beautiful; *ptilon*, feather. Central branch of awn plumose
calostachy-a, -us Gk *kalos*, beautiful; *stachys*, spike as of an ear of wheat. Inflorescence spike-like, attractive

C

Calosteca See *Calotheca*

Calotheca, calotheca Gk *kalos*, beautiful; *theke*, box. The lemma margins extend as lateral wings

Calotheria Gk *kalos*, beautiful; *ather*, barb or spine. Apices of the awns are pigmented

calvescens L. *calvesco*, become bald. Plants in whole or in part glabrous

calviniensis L. *-ensis*, denoting origin. From Calvinia, Cape Province, South Africa

calvum L. bald. Racemes glabrous

calycin-a, -um, -us Gk *kalyx*, cup; *-ina*, indicating possession. The subtending glumes are as long or longer than the lemma thereby resembling a cup

Calycodon Gk *kalyx*, cup; *odous*, tooth. Apices of the glumes conspicuously toothed

calyculatus Gk *kalyx*, cup; L. *-ulus*, diminutive; *-atus*, possessing. Spikelets subtended by a cup-like involucre of bristles

Calyptochloa Gk *kalyptos*, cup; *chloa*, grass. The axillary cleistogamous spikelets are protected by an indurated leaf-sheath

camargoanus In honor of Felisberto C. Camargo (c. 1887) Brazilian agriculturalist

cambessediana L. *-ana*, indicating connection. In honor of Jacques Cambessèdes (1799–1863) French botanist

cambodgiensis See *cambogiense*

cambodiensis L. *-ensis*, denoting origin. From Cambodia

cambogiens-e, -is L. *-ense*, denoting origin. From Cambodia (latinized as *Cambogia*)

cambrica L. *-ica*, belonging to. From Cambria, now Wales

cameronii In honor of Kenneth J. Cameron (fl. 1896–1899) who collected in East Africa

cameroonensis L. *-ensis*, denoting origin. From Cameroon Mountain or Republic of Cameroon, West Africa

camerunensis See *cameroonensis*

campan-a, -um L. *-ana*, indicating connection. From Campania, now a Province of Terra de Lavora, Italy

campbellensis L. *-ensis*, denoting origin. From Campbell Island, a New Zealand possession in the south-eastern Pacific Ocean

Campeiostachys Gk *kampe*, caterpillar; *stachys*, spike as of an ear of wheat. The drooping spikes resemble caterpillars

Campelia See *Campella*

Campella Gk *kampe*, caterpillar; L. *-ella*, diminutive but here used as a name-forming suffix. The awn is hygroscopic and bears a fanciful resemblance to a caterpillar

campestr-e, -is L. *campus*, plain; *-estre*, place of growth. Uncultivated

campicola L. *campus*, plain; *-cola*, dweller. Growing uncultivated

campinarum L. Possessive plural of the Portuguese *campina* treated as a femine noun. Of the campina (grasslands) of Amazonas, Brazil

camporum L. *campus*, plain. Growing on the plains

Campuloa Gk *kampylos*, curve. The racemes of the inflorescence are sickle-shaped

Campulosus Gk *kampylos*, curve; L. *-osus*, abundance. The racemes of the inflorescence are very strongly curved

campyloracheus Gk *kampylos*, curve; *rhachis*, backbone. Racemes flexuose

campylostachy-a, -um Gk *kampylos*, curve; *stachys*, spike as of an ear of wheat. Racemes curved

Camusia In honor of Aimée Antionette Camus (1879–1965) French botanist

camusiana L. *-ana*, indicating connection. As for *Camusia*

Camusiella L. *-ella*, diminutive but here employed as a name-forming suffix. See *Camusia*

can-a, -um L. ash-colored. Densely invested with appressed hairs

canadens-e, -is L. *-ense*, denoting origin. From Canada

canaliculat-a, -um, -us L. *canalis*, channel; *-ula*, diminutive; *-ata*, possessing. – (1) Lemma of the lower floret grooved. *Holcolemma canaliculatum*, *Panicum canaliculatum*, *Paspalum canaliculatum* – (2) leaf-blades deeply channelled. *Agropyrum canaliculatum*, *Andropogon canaliculatus*, *Elymus canaliculatum*, *Roegneria canaliculata*, *Saccharum canaliculatum*

canarae From Canara, a region in Karnataka State, India

canariensis L. *-ensis*, denoting origin. From the Canary Islands

Canastra From Parque National da Serra da Canastra, Brazil

canbyi In honor of William Marriott Canby (1831–1904) United States banker and amateur botanist

candamoana L. *-ana*, indicating connection. In honor of Manuel Candamo (1841–1904) President of Peru

candicans L. *candeo*, shine. Anthoecia glossy-white

candid-a, -um, -us L. glossy white. Spikelets white

candid-um, -us See *candida*

candissimum L. *candida*, glossy white; *-issimum*, most. Glumes white and shining

Candollea As for *Decandolia*

candollei As for *Decandolia*

canescens L. *canesco*, grow white. Leaf-blades or leaf-sheaths densely invested with white or grey hairs

caniflora L. *canus*, greyish-white; *flos*, flower. Spikelets dark-purple and invested with white hairs

canila Spanish *canilla*, small cane or reed. Culms woody

canin-a, -um, -us L. *canus*, greyish-white; *-ina*, indicating resemblance. Foliage or inflorescences grey-green

caninoides Gk *-oides*, resembling. Similar to *Agropyron caninum*

canin-um, -us See *canina*

cannanorensis L. *-ensis*, denoting origin. From the Cannanore District, Kerala, India

cannanorica L. *-ica*, belonging to. See *cannanorensis*

cannavieira The vernacular name of the species in Brazil

canovirens L. *canus*, greyish-white; *virens*, green. Lemma invested with long hairs

canoviridis L. *canus*, greyish-white; *viridis*, green. Culm-sheaths greyish-green

cantabrica L. *-ica*, belonging to. From Cantabria, now northern Spain

canterae In honor of Cornelio B. Cantera (1855–?) Uruguayan horticulturalist

cantonens-e, -is L. *-ense*, denoting origin. From Canton, now Guangzho, China

cantorii In honor of Theodor Edvard Cantor (1809–1860) Danish-born botanist who collected in China and Malaya

canum See *cana*

caobangensis L. *-ensis*, denoting origin. From Cao Bang, Vietnam

caparaoens-e, -is L. *-ense*, denoting origin. From Serra do Caparaó, Brazil

capens-e, -is L. *-ense*, denoting origin. In the vicinity of the Cape of Good Hope, South Africa

caperatum L. *capero*, to be wrinkled. Upper glume and sterile lemma coarsely cross-wrinkled

capillace-a, -um L. *capillis*, a hair; *-acea*, indicating resemblance. – (1) Glume apices drawn out into long threads. *Stipa capillacea* – (2) inflorescence with capillary branches. *Aristida capillacea, Eragrostis capillacea, Panicum capillaceum*

capillar-e, -is L. *capillis*, a hair; *-are*, pertaining to. – (1) Inflorescence with filiform branches. *Achneria capillaris, Agropyron capillare, Agrostis capillaris, Aira capillaris, Anastrophus capillaris, Axonopus capillaris, Muhlenbergia capillaris, Senites capillaris* – (2) leaf-blades filiform. *Sasa capillaris*

capillarifolia L. *capillis*, hair; *-aris*, pertaining to; *folium*, leaf. Leaf-blades hair-like

capillarioides Gk *-oides*, resembling. Similar to *Panicum capillaris*

capillaris See *capillare*

capillata L. *capillis*, hair; *-ata*, possessing. – (1) Leaf-blades thread-like. *Festuca capillata, Raddia capillata, Stipa capillata* – (2) inflorescence branches thread-like. *Cryptochloa capillata* – (3) lower glume thread-like. *Olyra capillata*

capilliflorus L. *capillis*, a hair; *flos*, flower. Pedicels thread-like

capillifoli-a, -um L. *capillis*, a hair; *folium*, leaf. Leaf-blades thread-like

Capillipedium L. *capillis*, a hair; *pes*, foot. Spikelets borne on thread-like pedicels

C

capillipes As for *Capillipedium*
capitat-a, -um, -us L. *caput*, head; *-ata*, possessing. Inflorescence condensed to a sphere-like structure
capitellata L. *caput*, head; *-ella*, diminutive; *-ata*, possessing. Panicle forming a small head
capitipila L. *caput*, head; *pilum*, hair. Glumes and sterile lemmas with exquisitely capitillate hairs
capitis-york From Cape York, Queensland, Australia
capitula L. *caput*, head; *-ula*, diminutive. Inflorescence capitate, that is a small head
capitulifera L. *caput*, head; *-ula*, diminutive; *fero*, carry or bear. Inflorescence densely congested
capituliflora L. *caput*, head; *-ula*, diminutive; *flos*, flower. Spikelets bunched into beads
cappadocic-a, -us L. *-ica*, belonging to. From Cappadocia, a region of Central Turkey
cappattama From the Japanese vernacular *Kappa-shrine*
caprearum Of Capreae, now Capri, an island in the Mediterranean
caprina L. *caper*, goat; *-ina*, indicating resemblance. In contrast to *Festuca ovina* with which the species may be confused
Capriola L. *caper*, goat. In Medieval times the name of the wild goat which fed on the grass in waste rocky places
capuronii In honor of René Paul Raymond Capuron (1921–1971) French botanist
capusii In honor of Jean Guillaume Capus (1857–1931) Luxembourg-born French botanist
caput-medusae L. *caput*, head; *Medusa*, monster with snakes for hair. – (1) Inflorescence a spike-like panicle and the spikelets with long trifid, twisted awns. *Aristida caput-medusae* – (2) spikelets subtended by bristles. *Elymus caput-medusae*
caracarahyensis L. *-ensis*, denoting origin. From Campos Gerais, Municipio de Caracaraí, Brazil
caragana From Tjuk-Caragan Peninsula which projects into the Caspian Sea
carajasensis L. *-ensis*, denoting origin. From Serra dos Carajás, Brazil
carannasense L. *-ense*, denoting origin. From Carannas, Brazil
carautae In honor of Jorge Pedro Pereira Carauta (1930–) Brazilian botanist
carazana L. *-ana*, indicating connection. From Caráz, Peru
carazensis L. *-ensis*, denoting origin. See *carazana*
carchiensis L. *-ensis*, denoting origin. From Carchi Province, Ecuador
cardinalis L. *cardo*, hinge; *-alis*, pertaining to. Senior, in the sense of one on whom decisions depend, but by transfer, red from the color of ceremonial garb adopted by Cardinals, senior Catholic clerics
cardonae As for *cardonum*
cardonum In honor of Félix Cardona Puig (1903–1982) Venezuelan geographer and explorer
cardosoi In honor of João Antonio Cardoso (1857–19375) Portuguese botanist
careyanum L. *-anum*, indicating connection. In honor of William Carey (1761–1834) English-born Indian missionary and botanist
caribaea From one of the Caribbean Islands
carica L. *-ica*, belonging to. Growing amongst *Carex*
caricinus L. *carex*, reed-grass; *-inus*, indicating resemblance. Similar to *Carex*
caricoides Gk *-oides*, resembling. Culms thin and much branched resembling *Carex*
caricos-um, -us L. *carex*, reed-grass; *-osum*, abundance. Densely caespitose with much branched culms
carinat-a, -um, -us L. *carina*, keel; *-atus*, possessing. – (1) Lemmas or glumes keeled. *Bromus carinatus*, *Diplachne carinata*, *Leptopogon carinatus*, *Roegneria carinata* – (2) leaf-blades keeled at the tip. *Deyeuxia carinata*, *Metasasa carinata* – (3) leaf-sheaths keeled. *Leptopogon carinatus*, *Muhlenbergia carinata*, *Panicum carinatum*, *Paspalum carinatum* – (4) fruits keeled. *Metasasa carinata*
carinatovaginatum L. *carina*, keel; *-atum*, possessing; *vagina*, sheath. Sheath markedly keeled

carinat-um, -us See *carinata*
carinifolium L. *carina*, keel; *folium*, leaf. Midrib of leaf-blade prominent on lower surface
cariophyllea See *Caryophyllea*
carmeli From Mt Carmel, Palestine
carmichaelii In honor of Dugald Carmichael (1772–1827) Hebridean-born British soldier and plant collector
carne-a, -um L. *caro*, flesh; *-eus*, resembling. Foliage somewhat succulent in texture
carnei In honor of Walter Mervyn Carne (1885–1952) Australian botanist and plant pathologist
carneovaginatum L. *caro*, flesh; *vagina*, sheath; *-atum*, possessing. Leaf-sheath flesh-colored
carniolic-a L. *-ica*, belonging to. From Carniola, now included in southern Austria and northern Yugoslavia
carnosum L. *caro*, flesh; *-osa*, abundance. Internodes of floating stem spongy
carnuntina L. *-ina*, indicating possession. From Carnuntum, a Roman camp at Petronell, Lower Austria
caroli In honor of Jean Martin François Carolus (1808–1863) Belgian botanist
caroli-henrici In honor of Karl Heinz Rechinger (1906–1998) Austrian botanist
carolinensis L. *-ensis*, denoting origin. From the Caroline Islands, one of the Federated States of Micronesia, Eastern Pacific
carolinian-a, -um, -us L. *-ana*, indicating connection. – (1) From Carolina, USA. *Cenchrus carolinianus, Ctenium carolinianum, Panicum caroliniana, Phalaris caroliniana, Poa caroliniana, Tricuspis caroliniana* – (2) In honor of Roger Charles Carolin (1929–) English-born Australian botanist. *Plectrachne caroliniana*
caroliniensis L. *-ensis*, denoting origin. From Carolina, USA
caroniense L. *-ense*, denoting origin. From Caroni River, Venezuela
carpatic-a, -um From Carpatica Montis, that is the Carpathian Mountains
carphoides Gk *-oides*, resembling. Habit similar to that of *Carpha*
carrenianum, carrenoanum L. *-anum*, indicating connection. In honor of Eduardo Carreño (1816/21–1841) Spanish plant collector
carsei In honor of Henry Carse (1857–1930) English-born New Zealand botanist
cartagana From Cartagena, Colombia
carthaginense L. *-ense*, denoting origin. From Carthago, Costa Rica
carthlicum L. *-icum*, belonging to. A reference to Kazakhstan, according to Nevski, the author of the name
cartilagine-a, -um L. cartilaginous. – (1) Culm-sheaths cartilaginous. *Yushania cartilaginea* – (2) lemmas cartilagenous. *Helictotrichon cartilagineum, Paspalum cartilagineum*
carvalhoi In honor of André Mauricio de Vieira de Carvalho (1951–2002) Brazilian botanist
Caryochloa Gk *karyon*, nut; *chloa*, grass. The grain is free within the indurated palea and lemma of the anthoecium forming a nut-like diaspore
Caryophyllea, caryophyllea Foliage resembling that of *Dianthus caryophyllus*
casapaltensis L. *-ensis*, denoting origin. From Casapalta, Peru
Casiostega Gk *cases*, horse's trappings; *stegos*, roof. The inflorescence is partly protected by a sheathing leaf, which enfolds it as does livery a horse
casiquiarensis L. *-ensis*, denoting origin. From Casiquiare, Amazonas Department, Venezuela
caspia From Caspia, that is the region about the Caspian Sea
caspic-a, -um L. *-ica*, belonging to. See *caspia*
cassa L. empty. Lower lemmas lacking flowers
cassanellii In honor of Gaetano Cassanello (fl. 1895) Italian Naval Officer
cassius From Mount Cassius now Jebel-Okrad, Syria
castane-a, -um L. *castaneum*, chestnut. – (1) Fertile lemma the color of chestnuts. *Eriochloa castanea, Paspalum castaneum* – (2) young shoots bearing chestnut-colored hairs. *Schizostachyum castaneum*

C

castellan-a, -us L. *-ana*, indicating connection. From the Spanish Provinces of New and Old Castille

castellaneana In honor of François Castella (1850–?) Swiss botanist

castellanosii In honor of Alberto Castellanos (1896–1968) Argentinian botanist

castellanus See *castellana*

Castellia In honor of Pietro Castelli (c. 1590–1661) Sicilian physician and amateur botanist

castilloniana L. *-ana*, indicating connection. In honor of Léon Castillon (fl. 1908–1928) Argentinian cleric and botanist

castillonis In honor of Emmanuel Drake del Castillo (1855–1904)

castratus L. *castro*, castrate. Sessile spikelet of pair only one present

castriferrei L. *castrum*, castle; *ferreum*, iron. From Vasvár, a fortified town in western Hungary

catabasis Gk *kata*, below; L. *basis*, base. Lower leaf-blades broad, upper leaf-blades narrow

Catabrosa Gk *katabrosis*, corrosion. The apices of the glumes are uneven

Catabrosella L. *-ella*, diminutive but here used as a name-forming suffix. Some members of the genus resemble those of *Catabrosa*

Catabrosia See *Catabrosa*

catabrosodes Gk *-odes*, resembling. Similar to *Catabrosa*

Catalepis Gk *kata*, below; *lepis*, scale. Lower glume scale-like

catamarcensis L. *-ensis*, denoting origin. From Catamarca, Argentina

catangens-e, -is L. *-ense*, denoting origin. From Katanga Province, Zaire

cataonica L. *-ica*, belonging to. From Cataonia, Classical name for central Turkey

Catapodium Gk *kata*, below; *podion*, little foot. The spikelets have short pedicels

Catatherophora Gk *kata*, below; *ather*, barb or spine; *phero*, bear. Spikelets subtended by a single, often deciduous bristle

catbaensis L. *-ensis*, denoting origin. From Catba Island, Gulf of Tonkin, now Vietnam

Cathariostachys Gk *katharios*, neatly arranged; *stachys*, spike as of an ear of wheat. Inflorescence branches arranged like a fan

cathartic-a, -us L. *catharticus*, purge. If ingested, liable to damage the gut

catherineana L. *-ana*, indicating connection. In honor of Ann Catherine Ryves (1929–) English painter

Cathestecum Gk *kathezomai*, remain seated. Plants prostrate creepers

catulifera L. *catula*, small cup; *fero*, carry or bear. The pedicel tips are hollowed-out by the falling away of the spikelet

catumbens-e, -is L. *-ense*, denoting origin. From Catumba, Angola

caucaiana L. *-ana*, indicating connection. From Caucaia, Brazil

caucasic-a, -um, -us L. *-ica*, belonging to. From the Caucasus, a series of mountain ranges, between the Black and Caspian Seas

cauda-ratti L. *cauda*, tail; *rattus*, rat. Inflorescence resembling a rat-tail

caudat-a, -um L. *cauda*, tail; *-ata*, possessing. – (1) Glumes elongated. *Aegilops caudata*, *Agropyron caudatum*, *Chasmopodium caudatum*, *Chloris caudata*, *Eragrostis caudata*, *Rottboellia caudata*, *Stipa caudata*, *Triticum caudatum* – (2) inflorescence elongated. *Aristida caudata*, *Gymnothrix caudata*, *Imperata caudata*, *Koeleria caudata* – (3) apex of the sterile lemma long, drawn out. *Echinochloa caudata*, *Olyra caudata* – (4) lemma awned. *Anthistiria caudata*, *Sorghum caudata*, *Themeda caudata* – (5) leaf-blades sharply tapering. *Schizostachyum caudatum*

caudicatum L. *caudex*, stem; *-atum*, possessing. Culms stout

caudiceps L. *caudex*, stem; *-ceps*, relating to a head. Leaves retained in clumps on the upper nodes of the culms

caudiculat-a, -um L. *cauda*, tail; *-ula*, diminutive; *-ata*, possessing. Lower glume shortly awned

caudiglume L. *cauda*, tail; *gluma*, husk. Lower glume ovate and apex tapering

caudula L. *cauda*, tail; *-ula*, diminutive. Upper floret finely acuminate

caudulat-a, -um L. *cauda*, tail; *-ula*, diminutive; *-ata*, possessing. Glumes narrow, tail-like

caulescens L. *caulesco*, develop a stem. Culms stout and leafy

Caulinites L. *caulis*, stem; *-ites*, resembling. Fossils resembling grass stems

cava L. *cavus*, hollow. Culms hollow

cavanillesii In honor of Antonio José Cavanilles (1745–1804) Spanish cleric and botanist

cavillieri In honor of François Cavillier (1868–1953) Swiss botanist

caxamarcensis L. *-ensis*, denoting origin. From Caxamarca (Cajamarca), Peru

cayennens-e, -is L. *-ense*, denoting origin. From Cayenne, French Guiana

cayoense L. *-ense*, denoting origin. From El Cayo District, British Honduras

cayouetteorum In honor of Richard Cayouette (1914–1997) and his son, Jacques Cayouette (1944–) Canadian botanists

cazorlensis L. *-ensis*, denoting origin. From Cazorla, now Castula, Spain

cearensis L. *-ensis*, denoting origin. From Ceará Province, Brazil

cebadilla Spanish *cebada*, barley; *-illa*, diminutive

ceinfuegos In honor of Bernard Cienfugos (fl. 16th century) Spanish botanist

celakovskyi In honor of Ladislav Josef Celakovsky (1834–1902) or his son, Ladislav Franz Celakovsky (1864–1916) Bohemian botanists

celebic-a, -um, -us L. *-ica*, belonging to. From Celebes, now Suluwasi, Indonesia

celsa L. lofty. Alpine species

Celtica L. *-ica*, belonging to. Named for the Celts, the ancient people of Western Europe

cenchriformis L. *forma*, appearance. Similar to *Cenchrus*

cenchroides Gk *-oides*, resembling. Resembling *Cenchrus* usually with respect to the inflorescence

Cenchropsis Gk *opsis*, resemblance. Resembling *Cenchrus* in some respect

Cenchrus Gk *kegchros*, a classical Greek name for *Panicum miliaceum* or any plant with small grains

Cencrus See *Cenchrus*

cenisia L. from Mont Cenis, North Italy

cenolepis Gk *kenos*, empty; *lepis*, scale. The proximal lemmas of the spikelets are sterile

Centosteca See *Centotheca*

Centotheca Gk *kenteo*, prick; *theke*, box. The lemmas bear long reflexed bristles

central-e, -is L. *centrum*, centre of a circle; *-ale*, pertaining to. – (1) From Central America. *Axonopus centralis*, *Paspalum centrale* – (2) from Central Australia. *Stipa centralis*

centrasiatic-a, -us L. *-ica*, belonging to. From Central Asia

centrifugus L. *centrum*, centre of a circle; *fugo*, drive away. Plants caespitose but dying away in the centre of the tussock

Centrochloa Gk *kentron*, spur; *chloa*, grass. The spikelets have a narrowly elongate callus

centrolepidoides Gk *-oides*, resembling. Inflorescence resembling that of *Centrolepis*

Centrophorum Gk *kentron*, spur; *phero*, bear. The lemma is awned

Centropodia Gk *kentron*, spur; *pous*, foot. The florets have a short, sharp callus

cepacea L. *cepa*, onion; *-acea*, resembling. Lower internodes of culm swollen

cephalantha Gk *kephale*, head; *anthos*, flower. Inflorescence an ovoid panicle

Cephalochloa Gk *kephale*, head; *chloa*, grass. The inflorescence is capitate

cephalonica L. *-ica*, belonging to. From Cephalonia, a Greek Island

Cephalostachyum Gk *kephale*, head; *stachys*, spike as of an ear of wheat. The spikelets are clustered in heads

cephalotes Gk *kephale*, head; *-otes*, resembling. Inflorescence congested

ceramic-a, -us L. *-ica*, belonging to. From Seram, Indonesia

cerata L. *cera*, wax; *-ata*, possessing. Plant overall or in part glaucous

Ceratochaete Gk *keras*, horn; *chaete*, bristle. The lemmas subtending the pistillate florets are stiff and awned

Ceratochloa Gk *keras*, horn; *chloa*, grass. Awn shorter than the lemma it terminates

C

C

Cerdosurus Gk *kerdo*, name of a fox; *oura*, tail. The inflorescence is a dense cylindrical panicle

cereal-e, -is L. *Ceres*, Roman Goddess of the Harvest; *-ale*, pertaining to. Applying to cultivated grain

Ceresia In honor of Jean Nicolas de Céré (1737–1810) Director of the Botanic Garden at Mauritius

ceresiaeformis See *ceresiiforme*

ceresiiforme L. *Ceres*, Roman Goddess of the Harvest; *forma*, appearance. The racemes somewhat resemble those of wheat

ceriferus L. *cera*, wax; *fero*, carry or bear. Basal nodes waxy

cernu-a, -um, -us L. nodding. Panicle branches pendant

cerosissima L. *cera*, wax; *-issima*, most. Culms and leaf-sheaths densely covered with wax when young

certificandum L. *certus*, definite; *facio*, make. Segregate from another species

cerulescens See *caerulescens*

cervicatum L. *cervix*, neck; *-atum*, pertaining to. Refers to the stiff-necked posture of the spikelets

cespitosa See *caespitosa*

cevallos In honor of Cevallos but origin obscure, not given by the author

ceylanica L. *-ica*, belonging to. From Ceylon, now Sri Lanka

Ceytosis Gk *ketho*, cover up. The capitate inflorescence is usually sheathed by the upper leaves

Chaboissaea In honor of Théodore Chaboisseau (1828–1894) French cleric and amateur botanist

chabouisii In honor of F. Chabouis (fl. 1964) French botanist

chacoens-e, -is L. *-ense*, denoting origin. – (1) From Chaco Province, Argentina. *Bambusa chacoensis*, *Digitaria chacoensis*, *Guadua chacoensis*, *Panicum chacoense*, *Paspalum chacoense* – (2) from Chaco, Bolivia. *Echinochloa chacoensis*

Chaetaria Gk *chaete*, bristle; L. *-aria*, pertaining to. The awn is persistent

Chaetium Gk *chaete*, bristle; *-ium*, resembling. The lower glume has a long slender awn

chaetium Gk *chaete*, bristle; *-ium*, resembling. Glumes and both lemmas awned

Chaetobromus Gk *chaete*, bristle. The spikelets resemble those of *Bromus* but have longer awns on the lemmas

Chaetochloa Gk *chaete*, bristle; *chloa*, grass. The spikelets are subtended by bristles

chaetophor-a, -um Gk *chaete*, bristle; *phero*, bear. Florets borne on slender pedicels

chaetophoron Gk *chaete*, bristle; *phero*, bear. Pedicels beset with long silky hairs

chaetophorum See *chaetophora*

chaetophylla Gk *chaete*, bristle; *phyllon*, leaf. Leaf-blades slender

Chaetopoa Gk *chaete*, bristle; *poa*, grass. Spikelets in clusters, the outer imperfect and forming an involucre around the single fertile floret

Chaetopogon Gk *chaete*, bristle; *pogon*, beard. Lower glume extending into a long slender awn

Chaetostichium Gk *chaete*, bristle; *stichos*, row; *-ium*, resembling. The spikelets are in two rows and the upper glume has a long awn

Chaetotropis, chaetotropis Gk *chaete*, bristle; *tropis*, keel. The lemma bears a dorsal hygroscopic awn

Chaeturus Gk *chaete*, bristle; *oura*, tail. The spicate inflorescence has spikelets with one glume terminating in a long bristle

chaffanjonii In honor of Jean Chaffanjon (1854–1913)

chaixii In honor of Dominique Chaix (1731–1800) French cleric and amateur botanist

chalarantha Gk *chalaros*, slack; *anthos*, flower. Inflorescence branches thin flexuous

chalarothyrsos Gk *chalaros*, slack; *thyrsos*, ornamental wand. Inflorescence an open panicle

chalcantha Gk *chalkos*, copper; *anthos*, flower. Spikelets reddish-brown

Chalcoelytrum Gk *chalkos*, copper; *elytron*, cover. Glumes reddish-brown

chalcophaea Gk *chalkos*, copper; *phaeos*, grey. Lemmas streaked with brown and purple

chalybaea Gk *-ea*, belonging to. From the land of the Chalybes, now Turkey

Chamaecalamus, chamaecalamus Gk *chamai*, low growing; *kalamos*, reed. Resembling a dwarf reed

chamaeclinos Gk *chamai*, low growing; *klino*, couch. Forming a dense short sward

Chamaedactylis Gk *chamai*, low growing. Resembling *Dactylis* but low growing

chamaelonche Gk *chamai*, low growing; *lonche*, spear. Plant shortly tufted with rigid culms

Chamaeraphis, chamaeraphis Gk *chamai*, low growing; *rhaphis*, needle. Creeping or prostrate plants with inflorescences whose central axes terminate in a stout bristle

chamaeraphoides Gk *-oides*, resembling. Similar to *Chamaeraphis*

chamaerhaphis See *Chamaeraphis*

Chamagrostis Gk *chamai*, low growing: *agrostis*, grass. Plants caespitose to only a few cm tall

chambersii In honor of Kenton Lee Chambers (1929–) United States botanist

chambeshii From Chambeshi River, Zambia

chamissonis In honor of Ludolf Adelbert von Chamisso (1781–1838) French-born German poet, explorer, naturalist

champlainensis L. *-ensis*, denoting origin. From Lake Champlain, New York State, USA

Chandrasekharania In honor of Chandrasekharan Nair (1927–) Indian botanist

changduensis L. *-ensis*, denoting origin. From Changdu, China

changii In honor of Chang Heungdo (fl. 1940) who collected in Korea

chapadens-e, -is L. *-ense*, denoting origin. – (1) From Serra da Chapada, Brazil. *Aristida chapadensis*, *Campulosus chapadensis*, *Ctenium chapadense* – (2) from Chapada dos Neadeiros, Brazil. *Altoparadisium chapadense* – (3) Portuguese *chapada*, a plain or clearing in woods. Growing on open plains in Brazil. *Paspalum chapadense*

chaparensis L. *-ensis*, denoting origin. From Chapare Province, Departmento Cochabamba, Bolivia

chapelieri In honor of Louis Armand Chapelier (1779–1800) French botanist who collected in Madagascar

chapmanian-a, -us L. *-ana*, indicating connection. In honor of Alvan Wentworth Chapman (1809–1899) United States botanist

chapmanii (1) As for *chapmaniana*. *Manisurus chapmanii*, *Panicum chapmanii*, *Paspalidium chapmanii*, *Paspalum chapmanii*, *Sieglingia chapmanii*, *Tridens chapmanii*, *Triodia chapmanii* – (2) in honor of Frederick Revans Chapman (1849–1936) New Zealand jurist and naturalist. *Deschampsia chapmanii*

chapulcensis L. *-ensis*, denoting origin. From Chapulco, Mexico

charkeviczii In honor of Sisigmund Semenovich Kharkevich (1921–) Russian botanist

charruana L. *-ana*, indicating connection. In honor of the Charrúas, a group of Indian tribes in Uruguay

chartacea L. *charta*, paper; *-acea*, pertaining to. Used for making paper

chasae See *Chasea*

Chascolytrum Gk *chasko*, gape; *elytron*, cover. At maturity the glumes gape exposing the grain

Chasea, chasea In honor of Mary Agnes Merrill Chase (1869–1963) United States agrostologist

chaseae See *Chasea*

chasean-a, -um, -us L. *-anum*, indicating connection. As for *Chasea*

Chasechloa L. *chloa*, grass. See *Chasea*

chasei (1) In honor of Virginius Heber Chase (1876–1966) United States naturalist. *Bouteloua chasei*, *Calamagrostis chasei* – (2) as for *Chasea*. *Panicum chasei*

chasii In honor of Édouard Chas (fl. 1992–1993) French botanist

Chasmanthium Gk *chasma*, hollow; *anthos*, flower. The glumes gape exposing the grain

Chasmopodium Gk *chasma*, hollow; *podus*, foot. The rhachis disarticulates into segments the base of which is a hemisphere and the tip of which is a hollow

chassanensis L. *-ensis*, denoting origin. From Chassan on the Garamov peninsula, Siberia

chatangensis L. *-ensis*, denoting origin. From the Chatang District, northern Siberia

chathamica L. *-ica*, belonging to. From the Chatham Islands, New Zealand

chaudharyana L. *-ana*, indicating connection. In honor of Shaukat Ali Chaudhary (1931–) Saudi Arabian botanist

Chauvinia In honor of François Joseph Chauvin (1797–1859) French algologist

chauvinii As for *Chauvinia*

cheelii In honor of Edmund Cheel (1872–1951) English-born Australian botanist

cheesemanii In honor of Thomas Frederick Cheeseman (1846–1926) English-born New Zealand botanist

chelariensis L. *-ensis*, denoting origin. From Chelari, Kerala State, India

chelungkiangnica L. *-ica*, belonging to. From Heilongjiang (Heilungkiang) Province, China

cheniae In honor of Shou Liang Chen (1921–) Chinese botanist

chenii In honor of Chen Mou, Chinese plant collector

Chennapyrum Gk *pyros*, wheat. In honor of M. S. Chennaveeraiah (1924–) Indian botanist

chepica L. *-ica*, belonging to. From Chepe, a locality on the Juan Fernández Archipelago, Chile

cheribon From Cheribon, Java, Indonesia

Chevalierella L. *-ella*, diminutive but here a name-forming suffix. In honor of Jean Baptiste Auguste Chevalier (1873–1956) French botanist and collector in tropical Africa, SE Asia and Brazil

chevalieri As for *Chevalierella*

chiangshanensis L. *-ensis*, denoting origin. From Jiand Shan, Zhejiang Province, China

chiapasensis L. *-ensis*, denoting origin. From Chiapas, Mexico

chiapporianus L. *-anus*, indicating connection. In honor of Agostino Chiappori, Italian botanist

chienouensis L. *-ensis*, denoting origin. From Jianou, Fujian Province, China

chigar Nepali name of the species

chihuahuana From Chihuahua, Mexico

chiisanensis L. *-ensis*, denoting origin. From Mt Chiisan, Korea

chikatsuafumiana L. *-ana*, indicating connection. In honor of Chikatsuafumi the name of an otherwise unknown person who gave their name to a wayside shrine in Shegu Prefecture, Japan

Chikusichloa Gk *chloa*, grass. Origin uncertain, not given by author but probably in honor of Chikusi

chilens-e, -is L. *-ense*, denoting origin. From Chile

chilianth-um, -us Gk *chilias*, a thousand; *anthos*, flower. Inflorescence many-flowered

chillagoanum L. *-anum*, indicating connection. As for *chillagoense*

chillagoense L. *-ense*, denoting origin. From Chillago, Queensland, Australia

Chilochloa, chilochloa Gk *chilos*, green fodder; *chloa*, grass. Pasture grasses

chiloense L. *-ense*, denoting origin. From Chiloé Island (also Grand Island of Chiloé), Chile

chimakisasa Vernacular name of the species in Japan

chimanimaniensis L. *-ensis*, denoting origin. From Chimanimani Mountains, Zimbabwe

chimantaensis L. *-ensis*, denoting origin. – (1) See *chimantensis*. *Atractantha chimantaensis* – (2) From Chimantá Massif, Venezuela. *Aulonemia chimantaensis*

chimantensis L. *-ensis*, denoting origin. From Maizo del Chimantá District, Venezuela

chimborazensis L. *-ensis*, denoting origin. From Chimborazo, a volcano in Ecuador

Chimonobambusa Gk *cheima*, winter. Resembling *Bambusa* and the new culms of some species appearing in winter

Chimonocalamus Gk *cheima*, winter. Resembling *Calamus* and the new shoots of some species appearing in winter

chinampoensis L. *-ensis*, denoting origin. From Chinampo, Korea

chinantlae From Chinantla, Mexico

chinens-e, -is L. *-ense*, denoting origin. From China

chingii In honor of Ching Ren-chang (1898–1986) Chinese botanist

chino Japanese, a contraction of *Shinodake*, Japanese vernacular name for a species of small bamboo

chinorossicum L. *-icum*, belonging to. From the border of China and the Russian Federation

Chionachne Gk *chion*, snow; *achne*, scale. The lower glume is hard and white

chionachne Gk *chion*, snow; *achne*, scale. Fertile lemma is white

Chionacne See *Chionachne*

Chionanche See *Chionachne*

chionobia Gk *chion*, snow; *bios*, manner of living. Growing in the tundra near snow pools

Chionochloa Gk *chion*, snow; *chloa*, grass. Inhabits alpine grasslands

chionogeiton Gk *chion*, snow; *geiton*, neighbour. Growing close to the snow line

chippindalliae In honor of Lucy Katherine Armitage Chippindall (1913–1992) South African agrostologist

chiquitaniensis L. *-ensis*, denoting origin. From Chiquitanía, Bolivia

chiribiquetens-e, -is L. *-ense*, denoting origin. From Sierra de Chiribiqueta, Colombia

chiriquiense L. *-ense*, denoting origin. From Chiriquí Province, Panama

chirripoensis L. *-ensis*, denoting origin. From Chirripó Grande, Costa Rica

chishuiensis L. *-ensis*, denoting origin. From Chishui, Guizhan Province, China

chita From Chita, Colombia

chitagana From Chitagá, Colombia

chitosensis L. *-ensis*, denoting origin. From Chitose, Iburi Province, Hokkaido, Japan

chitralensis L. *-ensis*, denoting origin. From the Chitral District, Pakistan

chiyomurensis L. *-ensis*, denoting origin. From Chizymura, Nagano Prefecture, Japan

Chloachne Gk *chloa*, grass; *achne*, scale. The apices of the glumes are more less herbaceous

Chloammia Gk *chloa*, grass; *ammos*, sand. Plants of sandy habitats

Chloamnia See *Chloammia*

Chloothamnus Gk *chloa*, grass; *thamnos*, shrub. Habit shrubby

chloranth-a, -um Gk *chloros*, green; *anthos*, flower. Spikelets green

chloride-a, -us L. *-ea*, indicating resemblance. The inflorescence resembles that of *Chloris*

chloridiantha Gk *anthos*, flower. Spikelets resembling those of *Chloris*

chloridiformis L. *forma*, appearance. With inflorescences resembling those of certain *Chloris* species

Chloridion Gk *-idion*, diminutive. Resembles a small *Chloris*

Chloridiopsis In error for *Chloridopsis*

Chloridopsis Gk *opsis*, resemblance. The inflorescence is similar to that of *Chloris*

Chloris The Greek goddess of flowers

Chlorocalymma Gk *chloros*, green; *kalymma*, head covering. The spikelets are enveloped by leafy wings developed from the rhachis

chlorochloe Gk *chloros*, green; *chloa*, grass. The dried foliage is dark-green

Chloroides Gk *-oides*, resembling. Similar to *Chloris*

Chloropsis Gk *opsis*, resemblance. Resembling *Chloris* with respect to the inflorescence

chlorostachyum Gk *chloris*, green; *stachys*, spike as of an ear of wheat. Plants glabrous and so are green in comparison with related species

Chlorostis Derived from *Chloris* together with *Agrostis*

chloroticum Gk *chlorotes*, pale-green; *-icum*, belonging to. Foliage pale-green

chnoodes Gk *chnoos*, fine down on a peach; *-odes*, resembling. Leaf-blades covered with short hairs

C

chodatiana L. *-ana*, indicating connection. In honor of Robert Hippolyte Chodat (1865–1934) Swiss botanist

chokaiensis L. *-ensis*, denoting origin. From Chokai, Japan

chokensis L. *-ensis*, denoting origin. From Choké Mts, Ethiopia

chondrachne Gk *chondros*, grain; *achne*, scale. The grain is shed along with the glumes and sterile lemma

Chondrachyrum Gk *chondros*, grain; *achyron*, chaff. The lodicules are longer than the grain

Chondrochlaena See *Chondrolaena*

Chondrolaena Gk *chondros*, grain; *klaena*, cloak. The bases of the glumes are gristly and horny and so provide extra protection to the enclosed florets

chondrosioides Gk *-oides*, resembling. Resembling *Chondrosum* in respect of the inflorescence

Chondrosium See *Chondrosum*

Chondrosum Gk *chondros*, grain. The spicate inflorescences and the awned lemmas are reminiscent of *Triticum*

chonotic-a, -us L. *-ica*, belonging to. From Chonos Archipelago, Chile

chordorrhiza Gk *chorde*, string of gut; *rhiza*, root. Rhizome conspicuously knotted

choresmica L. *-ica*, belonging to. From Choresm or Corasmiorum of antiquity, now eastern Iran

chorizanthe Gk *chorizo*, separate; *anthos*, flower. Florets widely separated along the rhachilla

chosenensis L. *-ensis*, denoting origin. From Chosen, Korea

christianii-bernardii In honor of Christian Bernard (fl. 1974) French botanist

christophersenii In honor of Erling Christophersen (1898–1994) Norwegian botanist, geographer and diplomat

chromatostigma Gk *chroma*, color; *stigma*, stigma. Stigmas dark violet

chromostachyum Gk *chroma*, color; *stachys*, spike as of an ear of wheat. Panicle dark-green

chrysanth-a, -um Gk *chrysos*, yellow; *anthos*, flower. – (1) Spikelets golden-bronze. *Calamagrostis chrysantha*, *Colpodium chrysanthum*, *Deyeuxia chrysantha* – (2) spikelets subtended by golden-yellow bristles. *Panicum chrysanthum*

chrysargyre-a, -us L. *chrysos*, yellow; *argyreas*, silver. Raceme pedicels with proximal silver hairs and distal fulvous hairs

chrysather-um, -us Gk *chrysos*, yellow; *ather*, barb or spine. – (1) Joints and pedicels of racemes shining, yellowish. *Ischaemum chrysatherum* – (2) awns yellow. *Andropogon chrysatherus*

chrysites Gk *chrysos*, yellow; *-ites*, closely connected. Pedicels invested with golden-yellow hairs

chrysoblephar-a, -e, -is Gk *chrysos*, yellow; *blepharon*, eye-lash. Pedicels invested with golden-yellow hairs

chrysochaetum Gk *chrysos*, yellow; *chaete*, bristle. Bristles subtending the spikelets yellow

Chrysochloa Gk *chryos*, yellow; *chloa*, grass. Glumes golden shining

chrysocomus Gk *chrysos*, yellow; *coma*, hair-tuft. Hairs of internodes and pedicels yellow

chrysodactylon Gk *chrysos*, yellow; *daktylon*, finger. Pedicels invested with golden-yellow hairs

chrysolepis Gk *chrysos*, yellow; *lepis*, scale. The glumes and lemmas are golden-yellow

chrysophylla Gk *chrysos*, yellow; *phyllon*, leaf. Leaf-blades golden-yellow

Chrysopogon, chrysopogon Gk *chrysos*, yellow; *pogon*, beard. Most species have golden-yellow hairs at the base of the spikelet

chrysopsidifolium Gk *chrysos*, yellow; *opsis*, resemblance; L. *folium*, leaf. Leaf-blades yellow-grey

chrysostachy-a, -um, -us Gk *chrysos*, yellow; *stachys*, spike as of an ear of wheat. Panicle branches invested in golden-yellow hairs

chrysostachys See *chrysostachya*

chrysostachy-um, -us See *chrysostachya*

chrysothrix Gk *chrysos*, yellow; *thrix*, hair. Glumes papillose-hispid with spreading golden hairs

chrysotrichum Gk *chrysos*, yellow; *thrix*, hair. Upper part of leaf-sheath bearing yellow hairs

Chrysurus, chrysurus Gk *chrysos*, yellow; *oura*, tail. Inflorescence a yellow spicate panicle

chubutensis L. *-ensis*, denoting origin. From Chubut Province, Argentina

chudeaui In honor of René Chudeau (1864–1921) who collected in the central Sahara

chumbiensis L. *-ensis*, denoting origin. From Chumbi Valley, Tibet Autonomous Region, China

Chumsriella L. *-ella*, diminutive but here used as a name-forming suffix. In honor of Chumsri Chai-Anan (1930–) Thai botanist

chungii In honor of W. K. Chung, Lingnan University President

churunens-e, -is L. *-ense*, denoting origin. From Churun, Venezuela

chusanensis L. *-ensis*, denoting origin. From Chusan, a mountain in Korea

chusque Growing in the land of the Chusque who live in north-west South America

Chusquea Chibcha, a reed. The vernacular name for members of this reed-like genus throughout Colombia and Equador

chusquea Resembling *Chusquea*

cienfuegos In honor of Bernardo Cienfuegos (fl. 18th century) Spanish botanist

cienkowskii As for *zenkowskii*

cilianensis L. *-ensis*, denoting origin. From Cigliano, Italy

ciliar-e, -is L. *cilium*, eyelid; *-are*, pertaining to. Glumes or lemmas ciliate on nerves or margins

ciliat-a, -um, -us L. *cilium*, eyelid; *-ata*, possessing. Plant hairy overall or in part

ciliatiflor-a, -um L. *cilium*, eyelid; *-atus*, possessing; *flos*, flower. Spikelets with hairy lemmas or glumes

ciliatifoli-a, -um, -us L. *cilium*, eyelid; *-ata*, possessing; *folium*, leaf. Leaf-blades hairy

ciliatissima L. *cilium*, hair; *-ata*, possessing; *-issima*, most. Sterile lemma densely silky pubescent

ciliativertex L. *cilium*, eyelid; *-atus*, possessing; *vertex*, whorl. Callus markedly hairy

ciliatoglume L. *cilium*, eyelid; *-atus*, possessing; *gluma*, husk. Glume keels scabrid-ciliate

ciliat-um, -us See *ciliata*

ciliifer-a, -um L. *cilium*, eyelid; *fero*, carry or bear. Leaf-blades ciliate

ciliocinct-a, -um L. *cilium*, eyelid; *cinctum*, girdle. Lower part of leaf-sheath densely hairy

ciliolat-a, -um, -us L. *cilium*, eyelid; *-ola*, diminutive; *-ata*, possessing. The plant in whole or in part invested with short hairs

cilios-a, -um L. *cilium*, eyelid; *-osum*, abundance. In part or wholly invested in short hairs

cimicin-a, -um, -us L. *cimex*, bug; *-ina*, indicating resemblance. Mature spikelets bear a fanciful resemblance to a small bug

cimmericum L. *-icum*, belonging to. From the region of the Cimmerii, who lived about the river Dnieper, Russian Federation

cincinnat-a, -us L. *cincinnus*, lock of hair; *-ata*, possessing. Leaf-blades spirally twisted when dry

cinct-a, -um, -us L. *cinctus*, girdle. – (1) Lemma with a transverse band of hairs. *Danthonia cincta*, *Merxmuellera cincta* – (2) pedicel with a ring of hairs immediately below the spikelet. *Andropogon cinctus*, *Panicum cinctum* – (3) with a girdle of tissue around the culm immediately above the nodal scar. *Dendrocalamus cinctus*

cinerascens L. *cinerasco*, become ashen. Leaf-blades glaucous

cinere-a, -um, -us L. ashy-grey. Plant grey due to hairs or wax

cinereovestit-a, -um L. *cinereum*, ashy-grey; *vestitum*, clothed. Plants invested with grey hairs

cinereoviride L. *cinereum*, ashy grey; *viride*, green. Plant grey-green

cinere-um, -us See *cinerea*

C

cingularis L. *cingulum*, girdle; *-aris*, pertaining to. Sterile lemma with a transverse fringe of hairs

cingulata L. *cingulum*, girdle; *-ata*, possessing. Culm with a conspicuous scar left by the deciduous leaf-base

Cinna Gk *kinna*. A name given by Dioscorides to an unidentified Cilician grass

Cinnagrostis From *Cinna* plus *Agrostis*

Cinnastrum L. *-astrum*, incomplete resemblance. Resembling *Cinna*

cinnoides Gk *-oides*, resembling. Similar to *Cinna*

cintranum L. *-anum*, indicating connection. Of Serra da Sintra, Spain

cipoense L. *-ense*, denoting origin. From Serra do Cipoó, Minais Gevais, Brazil

circinalis L. *circino*, form into a circle; *-alis*, pertaining to. Leaf-blade forming loose open coils

circinat-a, -us, -a, -us L. *circino*, form into a circle. – (1) Leaf-blades coiled. *Andropogon circinnatus, Cymbopogon circinnatus, Fargesia circinata, Festuca circinata, Muhlenbergia circinata* – (2) branches borne around the culms. *Chusquea circinata* – (3) awn spirally twisted. *Bromus circinnatus*

circulare L. *circulus*, circle; *-are*, pertaining to. Spikelets circular in outline

circumciliata L. *circum*, surrounding; *cilium*, hair; *-ata*, possessing. Spikelets subtended by a ring of long, white hairs

circummediterranea L. *circum*, about. Growing about the Mediterranean

circumpilis L. *circum*, surrounding; *pilis*, a hair. The nodes are hairy for two or three years following the shedding of the culm-sheaths

cirrat-a, -um, -us L. *cirrus*, curl; *-ata*, possessing. – (1) With a twisted hygroscopic awn. *Andropogon cirratus* – (2) with twisted leaf-blades. *Danthonia cirrata*

cirrhosa L. *cirrus*, curl; *-osa*, abundance. Leaf-blades drawn out into a narrow curled tip

cirrhulosa L. *cirrus*, curl; *-ula*, diminutive; *-osa*, abundance. Apices of leaf-blades coiled

cirros-a, -um, -us L. *cirrus*, curl; *-osa*, abundance. – (1) Leaf-blade drawn out into a narrow tip. *Stipa cirrosa* – (2) lemma awn long flexuous. *Ctenium cirrosum, Campulosus cirrosus*

cirrosula L. *cirrus*, curl; *-osa*, abundance; *-ula*, diminutive. The curled leaf-tips probably assist the grass to scramble

cirros-um, -us See *cirrosa*

cirtensis L. *-ensis*, denoting origin. From Cira, now Constantine, Algeria

cita L. swift. The species grows rapidly and colonizes recently denuded ground

citardae In honor of Citarda, Sicilian botanist

citratus L. *citron*, lemon; *-atus*, resembling. Plant lemon-scented

citreus L. pure yellow. Spikelets invested in yellow hairs

cladodes Gk *klados*, stem; *-odes*, resembling. Lower culm nodes swollen resembling cladodes

Cladoraphis Gk *klados*, stem; *rhaphis*, needle. The central axis of the inflorescence ends in a pungent tip

cladostachys Gk *klados*, stem; *stachys*, spike as of an ear of wheat. Inflorescence like that of *Triticum*

claessensii In honor of Claessens (fl. 1909–1933) Belgian botanist who collected in Zaire

clandestin-a, -um, -us L. hidden. The inflorescence is enclosed or partially enclosed in the upper leaf-sheath

clarazii In honor of Georges Claraz (1832–1930) Swiss plant collector, sometime resident of Argentina

clarionis In honor of Jean Clarion (1780–1856) French physician and botanist

clarkeana L. *-ana*, indicating connection. In honor of Charles Baron Clarke (1832–1906) English-born Indian teacher-botanist

clarkei As for *clarkeana*

clarkiae In honor of Lynn Gail Clark (1956–) United States botanist

clarksoniana L. *-ana*, indicating connection. In honor of John Richard Lindsay Clarkson (1950–) Scots-born Australian botanist

clarksonii As for *clarksoniana*

clathrata L. *clatra*, trellis; *-ata*, possessing. Lower glume sculptured in a trellis pattern

clatrata L. *clatra*, trellis; *-ata*, possessing. Hairs on upper glume and lower lemma arranged in overlapping rows

clauda L. defective. Leaf-blades spirally twisted

Claudia In honor of Claude Gay (1800–1873) French botanist

claudopus L. *clauda*, defective; Gk *pous*, foot. The spikelets break off transversely and not obliquely from the pedicel

clausa L. hidden. In habit much smaller than related species and inflorescence sometimes not projecting beyond the foliage

Clausispicula L. *clausum*, enclosed space; *spica*, a point; hence, in particular, an ear or spike of grain; *-ula*, diminutive. Spikelets small

clausonis In honor of Th. Clauson (1817–1860) French-born Algerian botanist

Clausospicula L. *claudo*, shut close; *spica*, a point; hence, in particular, an ear or spike of grain; *-ula*, diminutive. Florets cleistogamous

claussenii In honor of Peter Claussen (c. 1800–1855) Danish botanist who collected in Brazil

clavat-a, -um L. *clava*, club; *ata*, possessing. – (1) Possibly a reference to the obtuse lemma. *Agrostis clavata*, *Trichodium clavatum* – (2) panicle clavate. *Notodanthonia clavata* – (3) L. *clavare*, separate from one's wife. Referring to a closely related species. *Bambusa clavata*

clavatiformis L. *forma*, appearance. Resembling *Agrostis clavata*

clavatum See *clavata*

claviformis L. *clava*, club; *forma*, appearance. Inflorescence club-shaped

claviger-a, -um L. *clava*, club; *gero*, carry or bear. – (1) Inflorescence club-shaped. *Microstegium clavigerum*, *Pollinia clavigera*, *Stenotaphrum clavigerum* – (2) rhachilla club-shaped and projecting beyond the glumes after the florets have been shed. *Elytrostachys clavigera* – (3) rhachilla projecting. *Poa clavigera*

Clavinodum L. *clava*, club; *nodus*, knot. Culm nodes swollen

clavipil-a, -um L. *clava*, club; *pilus*, hair. Hairs swollen at the base

clavitricha L. *clava*, club; Gk *thrix*, hair. Indumentum of minute balloon-shaped hairs

clavuliferum L. *clava*, club; *-ula*, diminutive; *fero*, carry or bear. The glumes and sterile lemma bear capitellate hairs

clavulosa L. *clava*, club; *-ula*, diminutive; *-osa*, abundance. Pedicels club-shaped

claytonii In honor of William Derek Clayton (1926–) English botanist

Cleachne Gk *kleis*, tongue of a clasp; *achne*, scale. The lower glume clasp-shaped

cleefiana L. *-ana*, indicating connection. In honor of Antoine Marie Cleef (1941–) Dutch botanist and plant ecologist

cleefii As for *cleefiana*

Cleistachne Gk *kleistos*, to be locked away; *achne*, scale. The palea is very much reduced

Cleistochloa Gk *kleistos*, to be locked away; *chloa*, grass. Some of the spikelets are hidden in leaf axils

Cleistogenes Gk *kleistos*, to be locked away; *genos*, descent. The plants have cleistogamic spikelets

clelandii In honor of John Burton Cleland (1878–1971) Australian medical pathologist and naturalist

clemensae In honor of Mary Knapp Strong Clemens (1873–1968) United States botanist who collected widely in southeast Asia

clemensiorum In honor of Joseph Clemens (1862–1935) English-born United States missionary in southeast Asia and Mary Knapp Strong Clemens (1873–1968) botanical collector

clementei In honor of Rojas Clemente (1777–1827) Spanish librarian

clementii In honor of Emile Clement (fl. 1895–1910) who collected in Western Australia

clementis In honor of Joseph (1862–1935) or his wife Mary Knapp Strong Clemens (1873–1968). See *clemensae*

Clementsiella L. *-ella*, diminutive, but here employed as a name-forming suffix. In honor of Frederick Edward Clements (1874–1945) United States plant ecologist

Cleomena See *Clomena*

Cliffordiochloa Gk *chloa*, grass. In honor of Harold Trevor Clifford (1927–) Australian botanist

Clineleymus Gk *kline*, grave-niche; *eleymus*, bed. The rhachilla is well developed and the spikelets are partially protected by the bent internodes

clipeata L. *clipeus*, round metal shield; *-ata*, possessing. Spikelet shield-shaped

clival-e, -is L. *clivis*, hill; *-alis*, pertaining to. Growing on hillsides

clivicola L. *clivus*, hill; *-cola*, dweller. Mountain species

clivorum L. *clivus*, hill. Of hills, that is growing on slopes in steppe lands

Clomena, clomena Gk *klao*, break. Lower glume trifid as if broken off, in contrast to the entire upper glume

clomenoides Gk *-oides*, resembling. Similar to *Muhlenbergia clomena*

clusii In honor of Charles de l'Ecluse (1526–1609) Dutch botanist

clutei In honor of Willard Nelson Clute (1869–1950) United States botanist

coahuilana L. *-ana*, indicating connection. From Coahuila, Mexico

coahuilensis L. *-ensis*, denoting origin. From Coahuila, Mexico

coarctat-a, -us L. *coarcto*, compress. – (1) Culms compressed. *Erianthus coarctatus* – (2) panicles contracted. *Gaudinia coarctata*, *Oryza coarctatus*, *Spartina coarctata* – (3) leaf-blades folded along midrib. *Calamagrostis coarctata*

cobonii In honor of James John Cobon (1857–1929) English-born Queensland surveyor, Australia

coccosperm-a, -um Gk *kokkos*, kernel; *sperma*, seed. Anthoecia spherical

cochabambana L. *-ana*, indicating connection. From Cochabamba Department, Bolivia

cochinchinens-e, -is L. *-ense*, denoting origin. From Cochinchina, now part of Vietnam

cochleare L. *cochlea*, spoon; *-are*, pertaining to. Upper leaf-sheath inflated

cochlearispatha L. *cochlea*, spoon; *spatha*, sheathing base and false petiole of a palm leaf. Upper leaf-sheath inflated

cochleata L. *cochlea*, snail; *-ata*, possessing. The glumes bear a fanciful resemblance to a snail's shell

Cockaynea In honor of Leonard Cockayne (1855–1934) English-born New Zealand botanist

cockayniana L. *-ana*, indicating connection. As for *Cockaynea*

cocuyana L. *-ana*, indicating connection. From Sierra Nevada del Cocuy, Colombia

Codonachne Gk *kodon*, crier's bell; *achne*, scale. The rhachilla terminates in a clavate cluster of sterile lemmas, the whole resembling a bell

Coelachne Gk *koilos*, hollow; *achne*, scale. The subtending glumes are firm and basin-shaped

Coelachyrium, coelachyrium See *Coelachyrum*

Coelachyropsis Gk *opsis*, resemblance. Similar to *Coelachyrum*

Coelachyrum Gk *koilos*, hollow; *achyron*, chaff. The grains are concave on the lemma side of the rhachis

Coelarthron Gk *koilos*, hollow; *arthron*, segment. Joints of the rhachis tubular

Coeleochloa Gk *koilos*, hollow; *chloa*, grass. Lemma becoming inflated as the grain expands

coelest-e, -is L. *coeleste*, belonging to heaven. Alpine species

Coelochloa See *Coeleochloa*

Coelorachis, coelorachis Gk *koilos*, hollow; *rhachis*, backbone. The axes of the inflorescences are concave

coenicola L. *caenum*, mud; *-cola*, dweller. Growing in places subject to inundation

coerule-a, -us See *caerulea*

coerulescens L. *coerulesco*, become bluish. Foliage glaucous

coeruleus See *caerulea*

cognat-a, -um L. related. Similar to another species

cognatissimum L. *cognata*, related; *-issimum*, most. Very similar to another species

cognatum See *cognata*

coiron The vernacular name of the species in Argentina

Coix Origin obscure, applied by Theophrastus to an unknown plant but possibly from Greek *koix* since the diaspores resemble the fruits of *Hyparrhenia coriacea*, the doum-palm

cojocnensis L. *-ensis*, denoting origin. From Cojocna, Romania

-cola L. *-cola*, dweller. Although occasionally declined (as in *-cola*, *-colum*, *-colus*) this word should always be used as an indeclinable suffix

Colanthelia Gk *kolos*, shortened; *anthele*, plume or panicle of a reed. Inflorescence of relatively few spikelets

colchaguensis L. *-ensis*, denoting origin. From Colchagua Province, Chile

colchic-a, -um L. *-ica*, belonging to. From Colchis, in Classical times the name for the Region south of the Caucasus and east of the Black Sea

Coleanthus Gk *koleos*, sheath; *anthos*, a flower. Sheaths of upper leaves inflated and enveloping the inflorescence

Coleataenia Gk *koleos*, sheath; *tainia*, band. Spikelets borne on the lower surface of a winged rhachis

colensoi In honor of William Colenso (1811–1899) English-born New Zealand cleric and botanist

coleophorum Gk *koleos*, sheath; *phero*, bear. Leaf-sheath inflated

coleotrich-a, -us Gk *koleos*, sheath; *thrix*, hair. Leaf-sheath with spreading tubercle-based hairs

colinii In honor of H. Colin (1880–1943)

Coliquea Vernacular name of the species of the genus in Argentina and Chile

Colladea See *Colladoa*

Colladoa, colladoa In honor of Louis Collado (fl. 1561) Spanish physician and botanical writer

collar-e, -is L. *collum*, neck; *-are*, pertaining to. The apices of the culm leaf-sheaths contract to a small collar

collettii In honor of Henry Collett (1836–1901) English-born soldier who collected in Myanmar

collicola L. *collis*, hill; *-cola*, dweller. Confined to mountainous areas

colliei In honor of Alexander Collie (1793–1835) Scots born naval surgeon, colonial administrator and amateur botanist

collin-a, -um, -us L. *collis*, a hill; *-ina*, indicating possession. Mountain species

Collinaria L. *collina*, hilly; *-aria*, pertaining to. Growing on mountains

collinita L. *collino*, besmear. Florets sticky and attractively colored

collin-um, -us See *collina*

collocarpa Gk *collis*, loaf of coarse bread; *karpos*, fruit. Grain used for making bread

Colobachne Gk *kolobos*, mutilated; *achne*, scale. The glumes are fused at their bases

colobachnoides Gk *-oides*, resembling. Similar to *Colobachne*, another grass genus

colobantha Gk *kolobos*, mutilated; *anthos*, flower. The upper raceme is reduced to a single male spikelet

Colobanthium, Colobanthus Gk *kolobos*, mutilated; *anthos*, flower. The lower glume is awnless or only shortly awned

colombiana L. *-ana*, indicating connection. From Republic of Colombia

colombiensis L. *-ensis*, denoting origin. From Republic of Colombia

colon-a, -um, -us L. colonist. Cultivated as a cereal

colonarius L. *colonia*, settlement; *-arius*, indicating connection. Habit mat-forming

colon-um, -us See *colona*

coloradensis L. *-ensis*, denoting origin. From Colorado, USA

coloradoense L. *-ense*, denoting origin. From Colorado, USA

colorat-a, -um, -us L. *color*, color; *-ata*, possessing. Colored unusually, especially with reference to lemmas

colpodioides Gk *-oides*, resembling. Resembling *Colpodium*

Colpodium Gk *kolpos*, bay or tidal creek; *-ium*, indicating resemblance. The glume apices are irregularly indented

C

columbian-a, -um L. *-ana*, indicating connection. – (1) From District of Columbia, USA. *Dichanthelium columbianum, Panicum columbianum* – (2) from British Columbia, Canada. *Deyeuxia columbiana, Stipa columbiana* – (3) from Republic of Colombia. *Brachypodium columbianum, Gynerium columbianum*

columbiens-e, -is L. *-ense*, denoting origin. From Colombia

columnar-e, -is L. *column*, column; *-aris*, pertaining to. Spike obconical

colusana L. *-ana*, indicating connection. From Colusa, California, USA

colvillensis L. *-ensis*, denoting origin. From Colville River, Alaska

coma-ardeae L. *coma*, hair of the head; *ardea*, heron. One arm of the tripartite awn is densely hairy resembling the crest of a heron

comans L. *como*, comb the hair and by transfer of meaning the hair itself. Sterile lemma with a terminal tuft of hairs

comasii In honor of Augusto Comas (1949–) Cuban algologist

comat-a, -um, -us L. coma, hair of the head; *-ata*, possessing. – (1) With long hairs imparting a silky appearance to the spikelets. *Axonopus comatus, Brachiaria comata, Isachne comata, Mesosetum comatum, Muhlenbergia comata, Panicum comatum, Paspalum comatum, Stipa comata, Vaseya comata* – (2) with hair tufts in the axils of the inflorescence branches. *Eragrostis comata*

combsii In honor of Robert Combs (1872–1899) United States botanist

comifera L. *coma*, hair of the head; *fero*, carry or bear. Hairs of the lower lemma longer towards the apex

comillensis L. *-ensis*, denoting origin. From Comilla, Bangladesh

Commelinidium Gk *-idium*, diminutive but here used as a name-forming suffix. Foliage resembles that of *Commelina*

commelinifolium L. *folium*, leaf. Leaf-blades resemble those of *Commelina*

commersonii In honor of Philibert Commerson (1727–1773) French botanist and physician

commixtum L. *commiscio*, mix up. Easily confused with another species

commonsianum L. *-anum*, indicating connection. In honor of Albert Commons (1829–1919) United States botanist

commun-e, -is L. ordinary. Commonly encountered species

communissima L. *commune*, common; *-issima*, most. Abundant in certain localities

commutat-a, -um, -us L. *commuto*, change. Of species that in the opinion of the proposer may be confused with another

commutato-japonicus Hybrids between *Bromus commutatus* and *B. japonicus*

comophyllum Gk *kome*, tuft of hair; *phyllon*, leaf. Leaf-blades softly pubescent

Comopyrum Gk *kome*, head of hair; *pyros*, wheat. Glume of uppermost spikelet forming one-three mostly very long awns

comorens-e, -is L. *-ense*, denoting origin. From the Comoros Republic, Indian Ocean

comos-a, -um, -us L. *coma*, hair of the head; *-osa*, abundance. – (1) Upper spikelets of the spike awned thereby forming an apical tuft. *Aegilops comosa, Hordeum comosum, Triticum comosum* – (2) nodes bearded. *Panicum comosum* – (3) lemmas and glumes bearded. *Andropogon comosus, Hyparrhenia comosa, Sesleria comosa* – (4) in error for *convexum. Paspalum comosum*

compact-a, -um, -us L. compact. – (1) Inflorescence a short spike. *Micraira compacta, Triticum compactum* – (2) inflorescence a contracted panicle. *Andropogon compactus, Danthoniastrum compactum, Digitaria compacta, Erianthus compactus, Lasiacis compacta, Panicum compactum, Paspalum compactum, Penicillaria compacta, Triraphis compacta*

compactiflorus L. *compactus*, dense; *flos*, flower. Inflorescences congested

compact-um, -us See *compacta*

complanat-a, -um L. *complano*, level. – (1) Culms flattened. *Aristida complanatum, Gymnothrix complanata, Paspalum complanatum, Pennisetum complanatum, Rottboellia complanata, Stenotaphrum complanatum, Yushania complanata* – (2) pedicles flattened. *Digitaria complanata*

complicatum In error for *complanatus*

composit-a, -um, -us L. *compono*, bring together. Inflorescence with many short branches

compress-a, -um, -us L. *comprimo*, squeeze together. – (1) Culms flattened. *Agrostis compressa, Anastrophus compressus, Axonopus compressus, Digitaria compressa, Eleusine compressa, Gymnothrix compressa, Hemarthria compressa, Milium compressum, Panicum compressum, Paspalum compressum, Pennisetum compressum, Poa compressa, Rottboellia compressa, Schizachyrium compressum, Stipa compressa* – (2) spikelets strongly compressed. *Avena compressa, Avenula compressa, Mesosetum compressum*

compressicaul-e, -is L. *comprimo*, squeeze together; *caulis*, stem. Culms flattened

compressifoli-a, -um L. *comprimo*, squeeze together; *folium*, leaf. Leaf-blades folded along midrib

compress-um, -us See *compressa*

compta L. *comptus*, ornamented. Lemma margin bearing long hairs

comptonii In honor of Robert Harold Compton (1886–1979) South African botanist

concanensis L. *-ensis*, denoting origin. From Concan, India

concava L. concave. Ligule concave

concavum L. concave. Lower lemma concave

concepcionensis L. *-ensis*, denoting origin. From Concepción, Chávez Province, Santa Cruz Department, Bolivia

conchatum L. *conchus*, shell; *-atum*, possessing. Upper glume conchiform

conchifera L. *conchus*, shell; *fero*, carry or bear. Mature lemmas bladder-like resembling a conch-shell

conciliatum L. *concilio*, unite in sentiment. An expression of hope by the author that peace will come to Angola which at the time was suffering from civil strife

concinn-a, -um, -us L. elegant. Panicles or habit attractive

concinnius L. *concinnus*, elegant; *-ius*, characteristic of. Vernal foliage bright-green

concinn-um, -us See *concinna*

condensat-a, -um, -us L. *condenso*, press close together. Spikelets crowded in the inflorescence

condensum L. *condenso*, press together. Panicle branches held erect

conduplicat-um, -us L. *conduplico*, fold. Leaf-sheaths strongly keeled

condylifolia Gk *kondylus*, knuckle; L. *folium*, leaf. Leaf-blade abruptly bent at the junction with sheath

condylotrich-a, -us Gk *kondylos*, knuckle; *thrix*, hair. Lemma awns abruptly bent about the middle

conertii In honor of Hans Joachim Conert (1929–) German botanist

confert-a, -um, -us L. *confercio*, cram together. – (1) Plants forming a dense covering on the forest floor. *Bromus confertus* – (2) inflorescence branches densely crowded. *Deyeuxia conferta, Eragrostis conferta, Imperata conferta, Milium confertum, Paspalum confertum, Poa conferta, Reimaria conferta, Saccharum confertum, Trisetum confertum*

confertiflor-a, -us L. *confercio*, cram together; *flos*, flower. Inflorescence branches densely congested

confert-um, -us See *conferta*

confin-e, -is L. adjoining. – (1) Geographical distribution overlapping that of another species. *Deschampsia confinis, Poa confinis* – (2) growing gregariously. *Andropogon confinis, Arundinella confinis, Calamagrostis confinis, Hyparrhenia confinis, Panicum confine, Piptatherum confine*

confus-a, -um, -us L. confused. – (1) Likely to be mistaken for another species. *Calamagrostis confusa, Danthonia confusa, Elymus confusus, Festuca confusa, Indocalamus confusus, Isachne confusa, Muhlenbergia confusa, Oreochloa confusa, Panicum confusum, Poa confusa, Roegneria confusa, Sasa confusa, Sporobolus confusus, Vilfa confusa* – (2) species variable. *Agropyron confusum*

congdonii In honor of Joseph Wipple Congdon (1834–1910) United States botanist

congest-a, -um L. crowded together. – (1) Spikelets crowded in inflorescence. *Agrostis congesta, Aristida congesta, Eragrostis congesta, Melinis congesta, Phyllostachys congesta* – (2) branches crowded in inflorescence. *Panicum congestum*

congestiflora L. *congesta*, crowded together; *flos*, flower. Panicle branches erect at anthesis

congestum See *congesta*

conglomerat-a, -um L. *conglomero*, entangle. – (1) Culms form a dense intertangled mass. *Panicum conglomeratum* – (2) spikelets closely packed on raceme. *Eleusine conglomerata*

congoens-e, -is L. *-ense*, denoting origin. From the former French or Belgian Congo now Gabon and Zaire

conica L. *conus*, cone; *-ica*, belonging to. Base of lemma cone-shaped

conjugat-a, -um, -us L. *conjugo*, yoke together. Inflorescence branches paired

conjugens L. *conjugo*, yoke together. Intermediate in characters between two other species

conjunctus L. *conjungo*, join together. Panicle branches appressed

connat-a, -us L. fused. Base of pedicel shortly fused to adjacent raceme axis

connectens L. *con(n)ecto*, unite. Spikelets paired and very shortly pedicellate

connivens L. *conniveo*, let pass unnoticed. Overlooked by previous workers

conradeae In honor of Marcelle Conrade (1897–1990) French botanist who studied the flora of Corsica

conradii In honor of Hubert Aloys Conrads (1874–?) who collected in Tanzania

conrathii In honor of Paul Conrath (1861–1931) Czech chemist and naturalist

consanguine-a, -um L. *con*, with; *sanguineus*, blood-red. Spikelets or inflorescence branches reddish

consentanea L. in accordance with something. Coming together in being intermediate between two other species

consimilis L. alike in all respects. Readily confused with another species

consobrina L. cousin. Closely related to another species

conspersum L. *conspergo*, scatter. Glumes in part hairy and in part glabrous

conspicuum L. *conspicuo*, look at attentively. Leaf-blades to one metre long and one cm broad and so attract attention

constantae L. from Constanta, Romania

constrict-a, -um L. *constringo*, bind together. – (1) Lemma constricted at the summit. *Aristida constricta, Stipa constricta* – (2) panicle branches held erect. *Panicum constrictum, Paspalidium constrictum* – (3) internodes of spikes deeply constricted. *Thaumastochloa constricta*

contigu-a, -us L. neighbouring. Closely related to another species

continuata L. *continuus*, uninterrupted; *-ata*, possessing. The rhachilla projects conspicuously beyond the floret

contort-a, -us L. twisted. Awns hygroscopic and so twisted when dry

contract-a, -um, -us L. contracted. – (1) Inflorescence a narrow panicle. *Arundo contracta, Deyeuxia contracta, Fargesia contracta, Festuca contracta, Hyparrhenia contracta, Imperata contracta, Luziola contracta, Macroblepharus contracta, Poa contracta, Saccharum contractum, Sacciolepis contracta, Sporobolus contractus, Stipa contracta* – (2) inflorescences with the spikelets densely crowded. *Paspalum contractum*

contrerasii In honor of Elias Contreras (fl. 1962) who collected in Guatemala

contristata L. *con*, with; *tristis*, dull; *-ata*, possessing. Spikelets dull olive-green

controvers-a, -um, -us L. controversial. Usually applied to species which had been inadvertently misidentified

convallarioides Gk *-oides*, resembling. Leaf-blades similar to those of *Convallaria*

convergens L. *convergo*, approach. Spikelets two-ranked but appearing to be one-ranked

convexum L. convex. Glumes and/or lemmas markedly convex

convolut-a, -um L. rolled up longitudinally. Leaf-blades rolled length-wise

conwentzii In honor of Conwentz

cookei In honor of – (1) William Bridge Cooke (1908–) United States botanist. *Glyceria cookei* – (2) Theodore Cooke (1816–1910) English-born Indian botanist. *Andropogon cookei, Polytoca cookei, Trilobachne cookei* – (3) Charles Montague Cooke (1874–1948). *Panicum cookei*

cookii In honor of James Cook (1728–1779) English navigator

cooperi In honor of Ronald Edgar Cooper (1890–1962) English botanical collector in Sikkim State, India and the Kingdom of Bhutan

copeanus See *copei*

copei In honor of Thomas Arthur Cope (1949–) English botanist

copelandii (1) In honor of Copland, misspelt Copeland, King (1863–1918) Australian cleric, missionary and plant collector in Papua New Guinea. *Schizostachyum copelandii* – (2) origin not given by author but probably in honor of a colleague, Ralph Copeland (1837–1905) whose collections from Trinidad are in the Royal Botanic Gardens, Kew. *Bambusa copelandii*

copiapinus L. *-inus*, indicating possession. From Copiapó, Chile

coquimbensis L. *-ensis*, denoting origin. From Coquimbo, Chile

coracan-a, -us L. *-ana*, indicating connection; Sinhalese *kurakkan*, porridge. The grains are ground and used for flour

corcovadensis L. *-ensis*, denoting origin. From Mt Corcovado, Brazil

cordaense L. *-ense*, denoting origin. From Barra do Corda, Brazil

cordat-a, -um, -us L. *cordus*, heart; *-ata*, possessing. Leaf-blades cordate

cordatifolius L. *cordus*, heart; *-ata*, possessing; *folium*, leaf. Leaf-blades cordate

cordechii In honor of O. M. R. Cordech (fl. 1987) Bolivian botanist

cordifoli-a, -um, -us L. *cordus*, heart; *folium*, leaf. Leaf-blades cordate

cordilleran-a, -us Spanish *cordillera*, mountain range; L. *-anus*, indicating connection. Species of high mountains

cordobensis L. *-ensis*, denoting origin. From Córdoba, Argentina

cordofan-a, -um, -us L. *-anum*, indicating connection. From Cordofan Province, Sudan

cordovense L. *-ense*, denoting origin. From Córdoba, Mexico

cordubensis L. *-ensis*, denoting origin. From Corduba now Córdoba, Spain

corean-a, -um L. *-ana*, indicating connection. From Korea

coreensis L. *-ensis*, denoting origin. From Korea

Corethrum Gk *korethron*, broom. Glumes many awned

coriace-a, -um L. *corium*, leather; *-acea*, indicating resemblance. Lemmas or glumes leathery in texture

Coridochloa Gk *koris*, a crab; *chloa*, grass. The racemes arise close together from the peduncle

corinum L. *corus*, north-west wind; *-inum*, indicating possession. Growing on sites exposed to the north-west wind

coriophorum Gk *koris*, crab; *phero*, bear. Lower inflorescence branches sub-verticillate

corniae In honor of Carolyn Corn (1939–) United States botanist

corniculata L. *cornu*, horn; *-ulus*, diminutive; *-ata*, resembling. Lemma apex surmounted by a short cone

cornigera L. *cornu*, horn; *gigno*, bear. The young curved shoots resemble the horn of a buffalo (*Bos bubalus*)

Cornucopiae L. *cornu*, horn; *copicus*, plenty. The inflorescence is a head encircled by a bell-shaped cover of apical leaves resembling the Horn of Plenty

cornucopiae L. *cornu*, horn; *copicus*, plenty. The inflorescence is partially included in the upper leaf-sheath

cornut-a, -um, -us L. horned. – (1) The long and shortly curved female spikelets resemble a horn. *Pharus cornutus* – (2) glumes with a horn-like appendage. *Phleum cornutum* – (3) leaf-blades with horn-like auricles. *Schizostachyum cornutum*

C

coroi In honor of M. Coro (fl. 1982) Bolivian ecologist

corollat-um, -us L. *corolla*, small crown; *-atum*, possessing. Glumes white, resembling petals

coromandelian-a, -us L. *-ana*, indicating connection. From the Coromandel, that is south-eastern India

coromandelin-a,-um L. *-ina*, indicating possession. See *coromandeliana*

coronalis L. *corona*, crown; *-alis*, pertaining to. The delicate branches of small leaf-blades, produced in whorls at the nodes resemble a crown

coronata L. *corona*, crown; *-ata*, possessing. Lemma densely appressed-villous with the hairs projecting above the apex to form a pappus-like crown

coronensis L. *-ensis*, denoting origin. From *Corona*, that is Krondstat now Brasov in the Carpathian Mountains, Romania

coronifera L. *corona*, crown; *fero*, carry or bear. The upper glume bears white or violet hairs towards its apex

coronulatum L. *corona*, crown; *-ula*, diminutive; *-ata*, possessing. There is a ring of cilia about the scar left by disarticulation of the racemes

corradii In honor of Bartolomeo Giacomo Rinaldo Corradi (1897–1976) who collected in Ethiopia

corralensis L. *-ensis*, denoting origin. From Corral, Chile

correlliae In honor of Helen B. Correll (1907–) United States botanist

corrugat-a, -um L. *corrugo*, wrinkle up. – (1) Lower glume transversely rugose. *Rottboellia corrugata* – (2) fertile lemma transversely rugose. *Panicum corrugatum*, *Setaria corrugata*

corsica From Corsica

Cortaderia Spanish *cortadera*, knife for cutting honey-comb. The margins of the leaf-blades are sharply scabrous

corvi L. *corvus*, crow. An abbreviation of *crus-corvi*

Corycarpus Gk *korys*, helmet; *karpos*, fruit. The fruit is free within the glumes

corymbifer-a, -um L. *corymbus*, cluster of grapes; *fero*, carry or bear. Inflorescence much branched with the racemes arising in clusters

corymbos-a,-um,-us L. *corymbus*, cluster of grapes. Inflorescence branches forming a false corymb

Corynephorus Gk *koryne*, club; *phero*, bear. The apices of the awns are swollen into club-shaped structures

Corynophorus See *Corynephorus*

corynotrich-a, -um Gk *koryne*, club; *thrix*, hair. Glumes bear capitate hairs

coryophorum Gk *korys*, helmet; *phero*, bear. Upper glume inflated

coryphaeum Gk *koryphe*, top of the head; L. *-eum*, resembling. Culms tall, freely branching from upper nodes

cossoniana L. *-ana*, indicating connection. In honor of Ernest Saint Charles Cosson (1819–1889) French botanist

costaricens-e, -is L. *-ense*, denoting origin. From *Costa* Rica

costat-a, -um L. *costa*, rib; *-ata*, possessing. – (1) Leaf-blades conspicuously longitudinally ribbed. *Eragrostis costata*, *Festuca costata* – (2) fertile lemma transversely rugose. *Chaetochloa costata*, *Chamaeraphis costata*, *Panicum costatum*, *Paspalum costatum*, *Setaria costata* – (3) upper glume prominently ribbed. *Rhynchelytrum costatum*

costatispiculum L. *costa*, rib; *-ata*, possessing; *spica*, a point; hence, in particular, an ear or spike of grain; *-ula*, diminutive. Spikelets with prickle hairs on the ribs of the glumes

costatum See *costata*

Costia In honor of Antonio Cypriano Costa of Cuxart (1817–1886) Spanish botanist

costiniana L. *-ana*, indicating connection. In honor of Alec Baillie Costin (1925–) Australian plant ecologist

Cottaea See *Cottea*

Cottea In honor of Heinrich Cotta (1763–1844) German botanist

cotulifer,-a,-um,-us Gk *kotyle*, small cup; *fero*, carry or bear. The pedicel tips are concave after the shedding of the spikelets

coursii In honor of D. Cours (fl. 1937) sometime Director of Agriculture in Madagascar

courtallumensis L. *-ensis*, denoting origin. From Courtallum, Madras State, India

cowanii In honor of Richard Sumner Cowan (1921–1997) United States botanist and bibliographer who migrated to Australia

coxii In honor of Felix Arthur Douglas Cox (1835–1916) amateur botanist and long-time resident of Chatham Islands, New Zealand

coytaei In honor of William Beeston Coyte (1740–1810) English cleric and botanist

Craepalia Gk *kraipale*, frenzy. Intoxication results from eating the diseased grain

craigii In honor of Barry William Charles Craig (1938–) Australian anthropologist

craiovensis L. *-ensis*, denoting origin. From Craiova, Romania

Craspedorhachis Gk *kraspedon*, border; *rhachis*, backbone. The appressed spikelets are borne on a flat rhachis, the whole resembling a winged backbone

crass-a, -um, -us L. thick. – (1) Racemes thick. *Andropogon crassus, Panicum crassum, Paspalum crassum* – (2) spikes thick. *Aegilops crassa, Roegneria crassa* – (3) spikelets inflated. *Eragrostis crassa*

crassiapiculatum L. *crassus*, thick; *apiculus*, small tip; *-atum*, possessing. Glumes and sterile lemma with thickened tips

crassicaudex L. *crassus*, thick; *caudex*, stem. Culms thick

crassicaulis L. *crassus*, thick; *caulis*, stem. Culms thick

crassiculmis L. *crassus*, thick; *culmus*, stalk. Culms thicker than those of related species

crassiflora L. *crassus*, thick; *flos*, flower. Spikelets thick

crassifolia L. *crassus*, thick; *folium*, leaf. Leaf-blades thick

crassinervis L. *crassus*, thick; *nervus*, nerve. Lemmas conspicuously nerved

crassinoda L. *crassus*, thick; *nodus*, knot. Culms with swollen nodes

Crassipes L. *crassus*, thick; *pes*, foot. The axes, pedicels and rhachillas are relatively thick

crassipes L. *crassus*, thick; *pes*, foot. – (1) As for *Crassipes*. *Andropogon crassipes, Panicum crassipes* – (2) culms swollen at the base. *Bromus crassipes, Koeleria crassipes*

crassiuscul-a, -us L. *crassius*, thicker; *-ula*, indicating a tendency. – (1) Leaf blades thicker than those of related species. *Chionochloa crassiuscula, Deyeuxia crassiuscula* – (2) spike relatively thicker than those of related species. *Leymus crassiusculus*

crass-um, -us See *crassa*

cratericola L. *craterus*, bowl; *-cola*, dweller. From craters in Central Anatolia, Turkey

crateriferum L. *crater*, bowl; *fero*, carry or bear. Bearing pit-like glands

crateriformis L. *craterus*, bowl; *forma*, appearance. With pit-like glands on culms, peduncles and pedicels

cravenii In honor of Lyndley Alan Craven (1945–) Australian botanist

crebra L. *creber*, pressed together. Racemes held erect adpressed to inflorescence axis

Cremopyrum See *Eremopyrum*

cretace-a, -um L. chalk-white. Spikelets white

cretense L. *-ense*, denoting origin. From Crete

cretic-um, -us L. *-icum*, belonging to. From Crete

crevostii In honor of Crevost who collected in Cochinchina, now part of Vietnam

Criciuma The vernacular name for members of the genus in Bahai, Brazil

criniforme L. *crinum*, hair; *forma*, appearance. Leaf-blades fine

Crinipes L. *crinum*, hair; *pes*, foot. The lower glume is drawn out into a long non-hygroscopic awn

crinit-a, -um, -us L. with long hair. Hairy as of leaf-blades or awns invested in long weak hairs

crinizonatum Gk *zone*, belt; L. *-atum*, possessing. Growing in the company of *Crinum* in places subject to inundation

crinum-ursi L. *crinum*, hair; *ursus*, bear. Plant densely hairy

crisp-a, -um L. curly. Leaf-blades very flexuous or coiled

crispatopilosa L. *crispo*, curl; *pilosa*, hairy. Panicle axis with short curly hairs

crispatum L. *crispo*, curl. Hairs on the spikelet conspicuously crisped

crispifolius L. *crispus*, curled, *folium*, leaf. Old leaves coiled like a watch spring

crispiseta L. *crispus*, curly; *seta*, bristle. Awn much curled in the upper part

crispulum L. *crispus*, curly; *-ulum*, tendency. Leaf-blade irregularly twisted

crispum See *crispa*

cristat-a, -um, -us L. crested. – (1) Inflorescence forming a crest. *Agropyron cristatum, Aira cristata, Anthephora cristata, Bromus cristatus, Cynosurus cristatus, Dactylis cristata, Festuca cristata, Koeleria cristata, Lophochloa cristata, Phalaris cristata, Poa cristata, Trisetum cristatum* – (2) glumes and sterile lemmas awned. *Oplismenus cristatus, Panicum cristatum, Paspalum cristatum*

cristatellum L. *crista*, crest; *-ata*, possessing; *-ellum*, diminutive. Upper lemma shortly apiculate

cristat-um, -us See *cristata*

Critesion Gk *krithe*, barley. Meaning uncertain other than referring to barley

Critho Gk *krithe*, barley. Classical Greek vernacular name for barley

Crithodium, crithodium Gk *krithe*, barley; *-odes*, resembling. The inflorescences resemble those of barley

Crithopsis Gk *krithe*, barley; *opsis*, resemblance. Similar to barley

Crithopyrum Gk *krithe*, barley; *pyros*, wheat. Resembling *Triticum* and *Hordeum* with respect to the inflorescence

croatica L. *-ica*, belonging to. From Croatia

crocata L. *crocus*, saffron crocus. Lemma apices similar in color to the saffron-yellow dye obtained from the stigmas of *Crocus sativus*

cromyorrhizon Gk *kromyon*, onion; *rhiza*, root. Culm-bases swollen

crossotos Gk *krossos*, tassle. Internodes of racemes purple and the pedicels with yellow hairs

Crossotropis Gk *krossos*, tassle; *tropis*, keel. Nerves of lemma fringed

crucensis L. *-ensis*, denoting origin. From Santa Cruz, Bolivia

cruciabile L. *crucio*, torment; *-abilis*, indicating capacity. The coarse hairs on the leaf-sheaths readily break off and may irritate the skin

crucianelloides Gk *-oides*, resembling. Similar to *Crucianella* in habit

crucianus L. *-ana*, indicating connection. From Santa Cruz, Bolivia

cruciat-a, -um L. *crux*, cross; *-ata*, possessing. Inflorescence of sub-alternate racemes appearing cross-like as viewed from above

crucinensis L. *-ensis*, denoting origin. From Santa Cruz, Bolivia

crupina An allusion to the anthoecia, with their long upwards-directed hairs, resembling the fruits of *Crupina vulgaris* (Asteraceae)

crusardeae, crus-ardeae L. *crus*, foot; *ardea*, heron. Inflorescence resembles a heron's foot

crus-corvi L. *crus*, foot; *corvus*, crow. Inflorescence resembles a crow's foot

crus-galli L. *crus*, foot; *gallus*, cock. Inflorescence resembles a cock's foot

cruspavonis, crus-pavonis L. *crus*, foot; *pavonis*, peacock. Inflorescence resembles a peacock's foot

crustarium L. *crusta*, anything baked; *-arium*, pertaining to. Anthoecia dark-brown

cruttwellii In honor of Norman Edward Gary Cruttwell (1916–) English-born clergyman who collected in New Guinea

crymophila Gk *krymos*, cold; *phileo*, love. Growing in high mountains

cryophila Gk *kryos*, frost; *phileo*, love. Growing on the shores of the Arctic Sea

Crypsinna Combining the characters of *Crypsis* and *Cinna*

Crypsis Gk *crypsis*, concealment. The inflorescences of some species are partially hidden

crypsoides Gk *-oides*, resembling. Resembling *Crypsis* in that part of the inflorescence is hidden in the leaf-bases

cryptacanthum L. *kryptos*, hidden; *akanthos*, prickly plant. The spikelets are subtended by one or more bristles, the whole being enveloped in a foliaceous bract

cryptandrus Gk *kryptos*, hidden; *aner*, man. The panicles sometimes remain enclosed

cryptanth-a, -um Gk *kryptos*, hidden; *anthos*, flower. – (1) Inflorescences partly hidden in the leaf-sheaths. *Panicum cryptanthum* – (2) glumes enclosing the florets. *Festuca cryptantha*

cryptatherus Gk *cryptos*, hidden; *ather*, spike or ear of wheat. Inflorescence enclosed in leaf-sheath

Cryptochloa Gk *kryptos*, hidden; *chloa*, grass. The inflorescences are hidden amongst the leaves

Cryptochloris Gk *kryptos*, hidden. Other than for the concealed inflorescences similar to *Chloris*

cryptolopha Gk *kryptos*, hidden; *lophos*, crest. Lemma awn projecting only slightly above the subtending glumes

cryptopodus Gk *kryptos*, hidden; *pous*, foot. Peduncles mostly hidden in subtending spathes

cryptopus See *cryptopodus*

Cryptopyrum Gk *kryptos*, hidden; *pyros*, wheat. Readily confused with *Triticum*

cryptosperma Gk *kryptos*, hidden; *sperma*, seed. The fruit is concealed by the leaves sheathing the cob

Cryptostachys Gk *kryptos*, hidden; *stachys*, spike as of an ear of wheat. Panicle not exserted until after anthesis

Crypturus Gk *kryptos*, hidden; *oura*, tail. The spikelets are sunk in the axis of the spike

csatoi In honor of János Csató (1883–1913) Transsylvanian botanist

ctenantha Gk *ktenos*, comb; *anthos*, flower. The margins of the lower lemma of the stalked spikelets bear widely separated, long stiff hairs

Ctenium Gk *ktenion*, a small comb. The inflorescence usually consists of a single one-sided spike

ctenodes Gk *ktenos*, comb; *-odes*, resembling. Racemes comb-like

Ctenopsis Gk *ktenos*, comb; *opsis*, resemblance. The inflorescence resembles a comb

ctenostachyum Gk *ktenos*, comb; *stachys*, spike as of an ear of wheat. Racemes are one-sided and so resemble combs

cualensis L. *-ensis*, denoting origin. From Minas del Cuale, Mexico

cuanzensis L. *-ensis*, denoting origin. From Cuanza River, Angola

cuatrecasasii In honor of José Cuatrecasas (1903–1996) Colombian botanist

cuban-a, -um L. *-ana*, indicating connection. From Cuba

cubens-e, -is L. *-ense*, denoting origin. From Cuba

cucaense L. *-ense*, denoting origin. From Morro do Cuca, Brazil

cucphuongensis L. *-ensis*, denoting origin. From Cuc Phuong, Ha Nam Ninh Province, Vietnam

cucullat-a, -um L. *cucullus*, hood; *-ata*, possessing. – (1) Sterile lemmas hood-like. *Chloris cucullata* – (2) sheathing leaves hood-like. *Cornucopiae cucullatum* – (3) apex of leaf-blade forming a hood. *Poa cucullata*

cugnacii In honor of Antoine de Cugnac (fl. 1928) French botanist

cuiabensis See *cuyabense*

cujabense L. *-ense*, denoting origin. From Cujaba, Brazil

culeou The vernacular name of the species in Chile

culiacanum L. *-anum*, indicating connection. From the mountains of Culiacan, Mexico

culicinum L. *culex*, mosquito; *-inum*, indicating resemblance. Latin form of the vernacular name "Mosquito grass"

culionensis L. *-ensis*, denoting origin. From Culión Island, Philippines

culmineus L. *culmen*, summit; Gk *-eus*, belonging to. Growing on mountain tops

cultratum L. shaped like a knife-blade. Leaf-blades linear-lanceolate with a slightly scalloped margin

cumbucana L. *-ana*, indicating connection. From river Cumbuca, Bahia, Brazil

cumingiana L. *-ana*, indicating connection. As for *cumingii*

cumingii In honor of Hugh Cuming (1791–1865) English naturalist and traveller

cumminsii In honor of Henry Alfred Cummins (1864–1939) Irish-born physician who collected in India

cundinamarcae, cundinamarce From Cundinamarca, Municipality of Guadalito, Colombia

cuneatifolia L. *cuneatus*, wedge-shaped; *folium*, leaf. Leaf-blade wedge-shaped

cunninghamii In honor of Allan Cunningham (1791–1839) English-born Australian and New Zealand botanist

cupaniana L. *-ana*, indicating connection. In honor of Francesco Cupani (1657–1710) Italian botanist, cleric and physician

cupanii As for *cupaniana*

cupressifolium L. *folium*, leaf. Leaf-blades only 1–1.5 mm long thereby resembling *Cupressus* leaves

cupre-um, -us L. coppery. – (1) The purple spikelets are reminiscent of burnished copper. *Panicum cupreum* – (2) cilia of leaf-blades copper-colored. *Himalayacalamus cupreus*

cuprina L. *cupreum*, coppery; *-ina*, indicating resemblance. Spikelets copper-colored

curamalalensis L. *-ensis*, denoting origin. From Mt Curámalál, Argentina

curassavic-um, -us L. *-icum*, belonging to. From Curassava, that is Curaçao

curicoana L. *-ana*, indicating connection. From Curicó Province, Chile

curranii In honor of Hugh McCollum Curran (1875–1960) United States forester

curt-a, -um L. short. Leaf-blades short

curtiaristat-a, -us L. *curtis*, short; *arista*, bristle; *-atus*, possessing. Awn short relative to the base of the lemma

curticoma L. *curtus*, short; *coma*, hair-tuft. With a short tuft of hairs at the apex of the lemma

curtifoli-a, -um L. *curtus*, short; *folium*, leaf. Leaf-blades short

curtigluma L. *curtus*, short; *gluma*, husk. Upper glume short

curtipedicellata L. *curtus*, short; *pedicellus*, stalk; *-ata*, possessing. Spikelets shortly stalked

curtipendula L. *curtus*, short; *pendeo*, hang down; *-ulus*, indicating tendency. With spikelets shortly twisted to one side of their slender axis

curtisetosa L. *curtus*, short; *seta*, bristle; *-osa*, abundance. Lemma awn very short

curtisianum L. *-anum*, indicating connection. In honor of Moses Ashley Curtis (1808–1872) who collected in Carolina, USA

curtisii In honor of William Curtis (1746–1799) English botanist and entomologist

curtissii In honor of Allen Hiram Curtiss (1845–1907) United States botanist

curtivaginatum L. *curtus*, short; *vagina*, sheath; *-atum*, possessing. Internodes much longer than the sheaths

Curtopogon Gk *kyrtos*, bent; *pogon*, beard. Seta subreflexed

curtum See *curta*

curva L. bent. Panicle branches more or less secund

curvat-a, -um, -us L. curved. Spikelets curved or gibbous

curvatiforme L. *forma*, appearance. Resembling *Agropyron curvatum*

curvat-um, -us See *curvata*

curviaristata L. *curvus*, bent; *arista*, bristle; *-ata*, possessing. Awn curved

curviflorum L. *curvus*, bent; *flos*, flower. Spikelets curved

curvifoli-a, -um L. *curvus*, bent; *folium*, leaf. – (1) The leaf-blades become spirally coiled in senescence, or on drying. *Chasmanthium curvifolia*, *Danthonia curvifolia*, *Ectrosiopsis curvifolia*, *Gouldochloa curvifolia*, *Panicum curvifolium*, *Pentaschistis curvifolia* – (2) leaf-blades convolute. *Aristida curvifolia*

curvinerv-e, -is L. *curvus*, bent; *nervus*, nerve. The nerves of the upper glume and sterile lemma are conspicuously bent towards the midrib

curvipes L. *curvus*, bent; *pes*, foot. Pedicels curved

curviseta L. *curvus*, bent; *seta*, bristle. Lemma with a curved awn

curvistachyus L. *curvus*, bent; Gk *stachys*, spike as of an ear of wheat. Racemes incurved

curvula L. *curvus*, bent; *-ula*, diminutive. Leaf-blades curved

cusickii In honor of William Caiklin Cusick (1842–1922) United States botanist

cuspidat-a, -um, -us L. *cuspis*, head of a spear; *-ata*, possessing. Glumes, lemmas or leaf-blades long tapering or terminating in a sharp tip

cuspidiglumis L. *cuspis*, head of a spear; *gluma*, husk. Glume apices sharply tapering

custurae From Mt Custura, Romania

Cutandia In honor of Vincente Cutanda (1804–1866) Spanish botanist

cuthbertii In honor of Alfred Cuthbert (1859–1932) United States botanist

Cuviera In honor of Georges Léopold Chrétien Frédéric Dagobert Cuvier (1769–1832) French biologist

cuyabens-e, -is L. *-ense*, denoting origin. From Cuyas, Brazil

cuzcoensis L. *-ensis*, denoting origin. From Cuzco, now Cusco Region, Peru

cyanantha Gk *kyanos*, blue; *anthos*, flower. Spikelets glaucous

cyanescens L. *cyanesco*, become blue. – (1) Fertile spikelets bluish-green. *Andropogon cyanescens, Cymbopogon cyanescens, Hyparrhenia cyanescens* – (2) foliage bluish-green. *Panicum cyanescens*

cyathopod-a, -us Gk *kyathos*, cup; *pous*, foot. Apex of peduncle is cup-shaped

Cyathopus Gk *kyathos*, cup; *pous*, foot. The spikelets arise from cup-shaped expanded pedicels

Cyathorhachis Gk *kyathos*, cup; *rhachis*, backbone. The spikelets are embedded in cavities along the inflorescence branches

cyatophora Gk *kyathos*, cup; *phero*, bear. Bearing cup-like glands

cycladum From Cyclades Islands, Greece

Cyclostachya Gk *kyklo*, wheel; *stachys*, spike as of an ear of wheat. The inflorescence is curved into an annulus and breaks away as a whole shedding grain as it rolls along

Cycloteria Gk *kyklo*, wheel. Origin uncertain, not given by the author

cygnorum L. *cygnis*, swan. Of the swans, that is, from the Swan River, Western Australia

Cylichnium Gk *chylichnion*, a small cup. Grain terminates in a small cup

cylindracea L. *cylindrus*, cylinder; *-acea*, indicating resemblance. Inflorescence cylindrical

cylindric-a, -um, -us L. *cylindrus*, cylinder; *-ica*, belonging to. Inflorescences cylindrical

cylindriflora L. *cylindrus*, cylinder; *flos*, flower. Spikes terete

cylindrispica L. *cylindrus*, cylinder; *spica*, a point; hence, in particular, an ear or spike of grain. Panicle narrow, almost cylindrical

Cylindropyrum L. *cylindrus*, cylinder; *pyros*, wheat. Spike cylindrical

cyllenaea From Mt Kyllene, Greece

cyllenica L. *-ica*, belonging to. See *cyllenaea*

Cymatochloa Gk *kyma*, anything swollen; *chloa*, grass. Aquatic perennial with swollen floating culms

Cymbachne, cymbachne Gk *kymbe*, boat; *achne*, scale. The glumes are boat-shaped when viewed from the side

Cymbanthelia Gk *kymbe*, boat; *anthele*, inflorescence. Racemes mostly included in spatheoles

cymbari-a, -us Gk *kymbe*, boat; *-aria*, pertaining to. Inflorescence with boat-like spathes

cymbiandra Gk *kymbe*, boat; *aner*, man. The lower floret is male and narrowly boat-shaped

cymbiferus Gk *kymbe*, boat; *fero*, carry or bear. Inflorescence partially enclosed in a spatheate leaf

cymbiform-e, -is Gk *kymbe*, boat; *forme*, appearance. Lower glume boat-shaped

Cymbopogon Gk *kymbe*, boat; *pogon*, beard. In most species, the hairy spikelets project from boat-shaped spathes

Cymbosetaria Gk *kymbe*, boat. The spikelets resemble those of *Setaria* but the fertile lemma has a well-marked keel

Cynochloris Hybrids between species of *Cynodon* and *Chloris*

D

Cynodon Gk *kyon*, dog; *odous*, tooth. From chiendent the French name for the commonest species

cynodontoides Gk *-oides*, resembling. Similar to *Cynodon*

cynosuroides Gk *-oides*, resembling. Inflorescence resembling that of *Cynosurus*

Cynosurus Gk *kyon*, dog; *oura*, tail. An allusion to the shape of the panicle of *Cynosurus cristata*

cynotis Gk *kyon*, dog; *ous*, ear. Upper lemma expanded into two wings at the base

Cyperochloa Gk *chloa*, grass. The inflorescence resembles that of *Cyperus*

cyperoides Gk *-oides*, resembling. Similar to *Cyperus* in the form of the inflorescence

Cyphochlaena Gk *kyphos*, bent; *chlaena*, cloak. The lemma is gibbous

Cypholepis Gk *kyphos*, bent; *lepis*, scale. Lemma keel curved concavely towards rachilla

cypricola L. *-cola*, dweller. From Cyprus

cyrenaica L. *-ica*, belonging to. From *Cyrenaica*, Libya

Cyrenia See *cyrenaica*

cyri In honor of Cyrus (c. 423–401 B.C.) Persian Emperor, within whose domain the species grows

cyrnea From Cyrneus, now Corsica

Cyrrococcum See *Cyrtococcum*

Cyrtochloa Gk *kyrto*, to bend forward; *chloa*, grass. Culms slightly geniculate

cyrtoclad-um, -us Gk *kyrtos*, bent; *klados*, branch. With curved ascending branches

Cyrtococcum Gk *kyrtos*, bent; *kokkos*, kernel. The mature spikelet is asymmetric in outline

Cyrtopogon See *Curtopogon*

czarnahorensis L. *-ensis*, denoting origin. From Czarna Hora, Romania

czekanovskii As for *czekanowskiana*

czekanowskiana L. *-ana*, indicating connection. In honor of Aleksander Piotr (Lawrentievitsh) Czekanowsky (1833–1876) Polish born Russian botanist

czerepanovii In honor of Sergei Kirillovich Czerepanov (1921–)

Czernaievia In honor of Basil Matvéievich Czerniaév (1796–1871) Russian botanist

Czernya In honor of Johann Czerny (1480–1530) Bohemian apothecary and botanist

cziczinii In honor of Cziczin (fl. 1960) Russian botanist

czilikens-e, -is L. *-ense*, denoting origin. From Czilik River, former Soviet Union

czimganic-a, -um, -us As for *tschimganica*

czirahica L. *-ica*, belonging to. From Czirah in Daghestan

czukczorum From the Chukchi Peninsula, Russian Far East

D

Dactilis See *Dactylis*

Dactilon See *Dactylis*

Dactylis Gk *daktylos*, finger. Inflorescence branches finger-like

Dactyloctenium Gk *daktylos*, finger; *ktenion*, a little comb. The finger-like inflorescence branches resemble small combs

Dactylodes Gk *daktylos*, finger; *-odes*, resembling. The inflorescence comprises finger-like racemes

Dactylogramma Gk *daktylos*, finger; *gramma*, drawing. Origin unclear

dactyloides Gk *daktylos*, finger; *-oides*, resembling. Inflorescence branches finger-like

dactylolepis Gk *daktylos*, finger; *lepis*, scale. Outer spines of burr sometimes surpassing the inner forming a structure resembling cupped hands

dactylon Gk *daktylos*, a finger. Inflorescence of finger-like racemes

Dactylopogon Gk *daktylos*, finger; *pogon*, beard. Inflorescence of finger-like hairy racemes

Dactylus See *Dactylis*

dafengdingensis L. *-ensis*, denoting origin. From Dafengding, Sichuan Province, China

dagana L. *-ana*, indicating connection. From Daga Dzang, Kingdom of Bhutan

daghestanic-a, -um L. *-ica*, belonging to. From Daghestan, Russian Federation

dagnae From Dagna Mountain, Arazdayan, Russian Federation

dagussa Amharic *dag*, kind; *-ussa*, an exclamation meaning "what about". The name serves to remind people of the kindness of the grass in providing food in times of famine

dahuric-a, -us L. *-ica*, belonging to. From Dahuria, a region of south-eastern Siberia

dakarensis L. *-ensis*, denoting origin. From Dakar, Senegal

Daknopholis Gk *dachno*, bite; *pholis*, scale of a snake. Upper glume truncate, as if bitten off

dakotense L. *-ense*, denoting origin. From South Dakota, USA

Dallwatsonia In honor of Michael John Dallwitz (1943–) Australian computer programmer and Leslie Watson (1938–) English-born Australian botanist

dalmatic-a, -um, -us L. *-ica*, belonging to. From Dalmatia, now a region of Croatia

daltonii In honor of Joseph Dalton Hooker (1817–1911) English botanist

Dalucum Meaning uncertain, origin not given by author

dalzellii In honor of Nicholas Alexander Dalzell (1817–1878) Scots-born Indian forester and botanist

dalzielii In honor of John McEwen Dalziel (1872–1942) Indian-born English botanist and medical officer who collected principally in West Africa

damarensis L. *-ensis*, denoting origin. From Damaraland in Namibia

damascena L. a citizen of Damascus. From Damascus, Syria

damazioanus L. *-anus*, indicating connection. In honor of Leonidus Botelho Damazio (1854–1922) Brazilian botanist

damiensiana L. *-ana*, indicating connection. In honor of M. Damiens

dandyana L. *-ana*, indicating connection. In honor of James Edgar Dandy (1903–1976) English botanist

danesii In honor of Jiri Vadav Daneš (1880–1928) Czech geologist who collected in Australia

danguyi In honor of P. Danguy, Museum assistant in Madagascar

Danthonia In honor of Étienne Danthione (fl. 1800–1815) French botanist

danthoniae Spikelets resemble those of *Danthonia*

Danthoniastrum Gk *-astrum*, incomplete resemblance. Resembling *Danthonia* in spikelet structure

Danthonidium Gk *-idium*, diminutive but here used as a name-forming suffix. Similar to *Danthonia* with respect to spikelets

danthonii Spikelets resemble those of *Danthonia*

danthonioides Gk *-oides*, resembling. Resembling *Danthonia* with respect to spikelets

Danthoniopsis Gk *-opsis*, appearance. Resembling *Danthonia* with respect to spikelets

Danthorhiza Gk *rhiza*, root. Origin uncertain, but may refer to a rhizomatous species with spikelets similar to those of *Danthonia*, as understood by the author

Danthosieglingia Presumed hybrids between species of *Danthonia* and *Sieglingia*

danubialis L. *-alis*, pertaining to. Growing near the river Danube

dardori In honor of Dardor

darfuricum L. *-icum*, belonging to. From Darfur Province, Sudan

darlacensis L. *-ensis*, denoting origin. From Darlac Province, Vietnam

darvelana L. *-ana*, indicating connection. From Darvel Bay, Sabah

darwiniana L. *-ana*, indicating connection. In honor of Charles Robert Darwin (1809–1882) English naturalist

Dasiola Gk *dasys*, hairy; *-ola*, diminutive. Lemma with short stiff hairs and a scabrid awn

dasyanth-a, -um, -us Gk *dasys*, hairy; *anthos*, flower. Glumes and lemmas densely hairy

dasycarpa Gk *dasys*, hairy; *karpos*, fruit. Lemma villous all over

dasyclada Gk *dasys*, hairy; *klados*, branch. Inflorescence branches pubescent

dasycoleos Gk *dasys*, hairy; *koleos*, sheath. Lower leaf-sheaths densely hairy

D

dasydesmis Gk *dasys*, hairy; *desmos*, mooring cable. Callus hairy

Dasyochloa Gk *dasys*, hairy; *chloa*, grass. Lemma bases woolly-hairy

dasyphyll-a, -um Gk *dasys*, hairy; *phyllon*, leaf. Leaf-blades hairy

dasypleurum Gk *dasys*, hairy; *pleuron*, rib. The lateral nerves of the subtending glumes and sterile lemmas bear long hairs

Dasypoa Gk *dasys*, hairy. Lower portion of lemma nerves invested in long hairs, otherwise resembling *Poa*

Dasypyrum Gk *dasys*, hairy; *pyros*, wheat. Plants villose

dasystachy-s, -um Gk *dasys*, hairy; *stachys*, spike as of an ear of wheat. Spikelets hairy

dasytrichium Gk *dasys*, hairy; *thrix*, hair. Culms densely villous

dasyurum Gk *dasys*, hairy; *oura*, tail. Inflorescence a spicate-panicle with spikelets subtended by bristles

Davidsea In honor of Gerrit Davidse (1942–) Netherlands-born United States botanist

davidseana As for *Davidsea*

davidsei As for *Davidsea*

daviesii In honor of John Griffiths Davies (1904–1969) Welsh-born Australian agriculturalist

davisii In honor of Peter Hadland Davis (1918–) Scots botanist

Davyella L. *-ella*, diminutive but here used as a name-forming suffix. In honor of Joseph Burtt Davy (1870–1940) Scots-born Californian and South African botanist

davyi As for *Davyella*

dawesense L. *-ense*, denoting origin. From Dawes County, Nebraska, USA

dayanandanii In honor of P. Dayanandan, Professor of Botany, Madras Christian College

dayongensis L. *-ensis*, denoting origin. From Dayong, Hunan Province, China

deamii In honor of Charles Clemon Deam (1865–1953) United States botanist

deasyi In honor of Henry Hugh Peter Deasy (1866–1947) Irish soldier and plant collector in Tibet and Chinese Turkestan

debil-e, -is L. weak. Culms decumbent often much branched at the base

debilior L. weaker. Habit lax compared with that of related species

debilis See *debile*

decaisnei In honor of Joseph Decaisne (1809–1882) Belgian-born French botanist

Decandolia In honor of Augustin Pyramus de Candolle (1774–1841) French botanist

decaryan-a, -us L. *-ana*, indicating connection. As for *Decaryella*

Decaryella L. *-ella*, diminutive but here used as a name-forming suffix. In honor of Raymond Decary (1891–1973) French botanist

Decaryochloa Gk *chloa*, grass. See *Decaryella*

deccanens-e, -is L. *-ense*, denoting origin. From the Deccan region of India

decempedal-e, -is L. *decem*, ten; *pes*, foot. Culms to about three meters tall

deceptorix L. *deceptor*, a deceiver; *-trix*, indicating femininity. Sspecies often mistaken for another

decidu-a, -um L. *deciduo*, cut off. Leaf-blades or spikelets deciduous

decipiens L. *decipio*, deceive. Resembling another species, or in some other way deceptive

deckeri In honor of S. Decker who collected in France

declinata L. *declino*, bend aside. Racemes one-sided

decolorans L. *decoloro*, deprive of its natural color. Glumes or lemmas pale-green to white

decolorat-a, -um L. *decoloro*, deprive of its natural color. Spikelets paler green than those of related species

decomposit-a, -um, -us L. much divided. Inflorescence much branched

decor-a, -us L. *decor*, elegance. Attractive, usually with respect to habit

decumbens L. *decumbo*, fall down. Culms prostrate

decus-sylvae L. *decus*, ornament; *silva*, wood. Woodland species with an attractive inflorescence

dedeccae In honor of D.M. Dedecca (fl. 1950–1957) Brazilian botanist

dedegenii In honor of A. De Degen (fl. 1906) Sicilian botanist

defectoides L. *deficio*, be wanting; *-oides*, resembling. Many spikelets but fails to produce grain

deficiens L. *deficio*, be wanting. Lacking in some respect

deflex-a, -um L. bent abruptly downwards. Inflorescence branches or spikelet pedicels reflexed

defracta L. *defringo*, break off. Sheaths break into short segments

degenii In honor of Arpád Degen (1866–1934) Hungarian botanist

deightonii In honor of Frederick Claude Deighton (1926–1949) British botanist who collected in West Africa

Deina Gk *deinos*, large. Segregated from *Triticum* on account of its robust habit and long grain

dekindtii In honor of Eugène De Kindt (fl. 1899–1902) who collected in Angola

delavayi In honor of Pierre Jean Marie Delavay (1834–1895) French cleric who collected in China

delavignei In honor of Gislain François de la Vigne (?–1805) sometime Professor of Botany, Kharkov, Ukraine

delawarense L. *-ense*, denoting origin. From Delaware State, USA

delawarica L. *-ica*, belonging to. As for *delawarense*

delfinii In honor of Fredericho Torres Delfin (1852–1904) who collected in Patagonia

delicat-a, -um, -us L. delicate. Dainty, usually of habit

delicatissima L. *delicata*, delicate; *-issima*, most. Very dainty, usually of habit

delicatul-a, -um, -us L. *delicatus*, dainty; *-ula*, diminutive. Small and dainty as of habit

delicat-um, -us See *delicata*

delilean-a, -us L. *-ana*, indicating connection. In honor of Alire Raffeneau Delile (1778–1850) French botanist

delilei As for *delileana*

delochei In honor of Deloche (fl. 1859) who collected in Uruguay

deltae L. *Delta*, originally *delta* of the Nile. From delta of river Paraná, Buenos Aires Province, Argentina

deltoide-a, -um Gk *delta*, shaped like a river delta; *-oidea*, resembling. Spikelets ovate-lanceolate

deludens L. *deludo*, deceive. Suppression of the midrib in the glumes incorrectly suggests the spikelets are diversely oriented

Demazeria See *Desmazeria*

demeusei In honor of Fernand Demeuse (fl. 1880–1892) Belgian botanist

demiss-a, -us L. drooping. Culms bent

demnatensis L. *-ensis*, denoting origin. From Demnat, Morocco

dendeniwae In honor of Goë Dendeniwa (fl. 1972) Papua-New Guinea botanist

Dendragrostis Gk *dendron*, tree; *agrostis*, name of an unknown grass of the Ancients. Woody grass from South America

Dendrocalamopsis Gk *opsis*, resemblance. Similar to *Dendrocalamus*

Dendrocalamus Gk *dendron*, tree; *kalamos*, reed. Culms tall and woody

Dendrochloa Gk *dendron*, tree; *chloa*, grass. Culms woody

dens-a, -um, -us L. dense. – (1) Growing in large clumps. *Bromus densus*, *Paspalum densum*, *Trachypogon densus* – (2) panicle with many spikelets. *Calamagrostis densus*, *Deyeuxia densa*, *Dissanthelium densum*, *Eragrostis densa*, *Stipa densa*

densiflor-a, -um, -us L. *densus*, dense; *flos*, flower. Inflorescences congested

densifolia L. *densus*, dense; *folium*, leaf. Leaves densely imbricate

densipaniculata L. *densus*, dense; *paniculus*, panicle; *-ata*, possessing. Inflorescence contracted

densispica L. *densus*, dense; *spica*, a point; hence, in particular, an ear or spike of grain. Panicle branches erect and closely adpressed to central axis of inflorescence

densissima L. *densus*, dense; *-issima*, most. Panicles contracted and almost spike-like

dens-um, -us See *densa*

D

dentat-a, -us L. *dens*, tooth; *-ata*, possessing. Glume apices obliquely truncate, unidentate, mucronate or rarely awned

dentatosulcatum L. *dens*, tooth; *-ata*, possessing. sulcus, furrow; *-atum*, possessing. Pedicels triquetrous and setulose ciliate on the keels

dentatus See *dentata*

denticulat-a, -um L. *dens*, tooth; *-ulus*, diminutive; *-ata*, possessing. Margin of any part of plant bearing small teeth

dentifera L. *dens*, tooth; *fero*, carry or bear. Lemma three-toothed

dentiflora L. *dens*, tooth; *flos*, flower. The lemma and palea apices are bifid

denudat-a, -um L. lacking in hairs. Foliage of spikelets glabrous or weakly hairy

depallens L. *de-*, very; *palleo*, be pale. Spikelets pale-colored

depauperat-a, -um, -us L. *depaupero*, impoverish. Stunted as if for want of nourishment

dependens L. *dependeo*, hang down. Culms drooping

Deplachne See *Diplachne*

deppeana In honor of Ferdinand Deppe (1794–1861) German botanist and traveller

depress-a, -us L. *depressio*, press down. – (1) Culms ascending from a spreading base. *Agrostis depressa* – (2) lower glume with a median pit. *Andropogon depressus*

deqenensis L. *-ensis*, denoting origin. From Deqen, Yunnan Province, China

derbyanus L. *-anus*, indicating connection. In honor of Orville Derby, United States geologist

dertonensis L. *-ensis*, denoting origin. From Derton, now Tortona, in the Italian Piedmont

derzhavinii In honor of Derzhavin, Russian botanist

Deschampsia, Deschampsie In honor of Louis Auguste Deschamps (1765–1842) French physician and naturalist

deschampsiiformis L. *forma*, appearance. Resembling *Deschampsia*

deschampsioides Gk *-oides*, resembling. Resembling *Deschampsia*

deserti L. *desertum*, desert. Species of the desert

deserticola L. *desertum*, desert; *-cola*, dweller. Growing in arid places

desertorum L. *desertum*, desert. Of deserts, that is, from arid places

desfontainii In honor of René Louiche Desfontaines (1750–1833) French botanist

desmantha Gk *desmos*, anything used for tying; *anthos*, flower. The three arms of the awn intertwine at maturity

Desmazeria In honor of Jean Baptiste Henri Joseph Desmazières (1787–1862) Belgian botanist

Desmostachya Gk *desmos*, anything used for tying; *stachys*, spike as of an ear of wheat. The stems and inflorescences are used for string in North Africa and the Middle East

desmoulinsii In honor of Charles Robert Alexandre Des Moulins (1798–1875) English-born French botanist

desolata L. *desolo*, abandon. A species previously overlooked

despreauxii In honor of Jean Despréaux (1794–1843) French-born traveller and plant collector

Despretzia In honor of C. Despretz (fl. 1831) French physician

Desvauxia In honor of Nicaise Auguste Desvaux (1784–1856) French botanist

desvauxii (1) As for *Desvauxia*. *Enneapogon desvauxii*, *Mibora desvauxii* – (2) In honor of Étienne-Émile Desvaux (1830–1854) French botanist who wrote on the subject of Chilean grasses. *Festuca desvauxii*

detinens L. *detineo*, detain. Climber forming thickets difficult to traverse

deustum L. *deuro*, burn up. The spikelet tips are purple or blackish, suggesting they have been severely burned

Devauxia A misspelling of *Desvauxia*

devia L. out of the way. The only species of the genus not Australian

devincenzii In honor of Garibaldi José Devincenzi (1882–1943) Uruguayan physician and museum director

dewaensis L. *-ensis*, denoting origin. From Dewa, Japan

dewevrei In honor of Alfred Dewèvre (1886–1897) Belgian botanist

dewildemanii In honor of Émile Auguste Joseph De Wildeman (1866–1947) Belgian botanist

dewinteri In honor of Bernard de Winter (1924–) South African botanist

Deyeuxia In honor of Nicolas Deyeux (1753–1837) French pharmacist

deyeuxioides Gk *-oides*, resembling. Resembling *Deyeuxia* in some respect

deyliana L. *-ana*, indicating connection. In honor of Milos Deyl (1906–1985) Czech botanist

deylii As for *deyliana*

dezhnevii In honor of Semen Ivanovich Dezhnev (1605?–1673) Russian traveller through Bering Strait

dhofariensis L. *-ensis*, denoting origin. From Dhofar, Oman

diaboli From Diablo Canyon, California, USA

diabolica L. devilish. Sharing characters with several other species

Diachroa Gk *dis*, twice; *chroia*, color. Lemmas greenish-purple or a reference to the contrast between the colors of the lemmas and stigmas

Diachyrium Gk *dis*, twice; *achyron*, chaff. Glumes of unequal length

Diacisperma A misspelling of *Disakisperma*

diadelpha Gk *dis*, twice; *adelphos*, brother. Florets with stamens arranged in two groups

diagonal-e, -is L. diagonal. Ligule apex oblique

dialytostachya Gk *dialytes*, a breaker up; *stachys*, spike as of an ear of wheat. Origin uncertain, not given by author

diamantinum From Diamantina, Serra de San Antonio, Minas Gerais, Brazil

diamesum Gk *dis*, by reason of; *mesos*, in the middle. Intermediate in appearance between two other species

dianae From Mt Diana, a peak on the island of St Helena in the South Atlantic

diand-er, -ra, -rum, -rus Gk *dis*, twice; *aner*, man. Florets with two stamens

diandr-a, -um, -us See *diander*

Diandrochloa Gk *dis*, twice; *aner*, man; *chloa*, grass. The florets have two stamens

Diandrolyra Gk *dis*, twice; *aner*, man. In contrast to *Olyra*, the male flowers have two instead of three stamens

Diandrostachya Gk *dis*, twice; *andros*, male; *stachys*, spike as of an ear of wheat. Floret has two stamens

diantha Gk *dis*, twice; *anthos*, flower. Spikelets with two florets

dianthemum Gk *dis*, twice; *anthemon*, flower. Spikelets with two florets

diaphora Gk *dia*, all through; *phero*, bear. Rhacilla not prolonged, that is spikelets lack a terminal sterile floret

Diarina See *Diarrhena*

Diarrhena, diarrhena Gk *dis*, twice; *arrhen*, male. The florets have two stamens

Diastemanthe Gk *diastema*, an interval; *anthos*, flower. At maturity the inflorescence breaks up into segments each bearing a spikelet

diatherus Gk *dis*, twice; *ather*, spike or ear of wheat. Inflorescence of paired racemes

Dicantheum See *Dichanthium*

Dichaetaria Gk *dis*, twice; *chaete*, bristle; *-aria*, pertaining to. Lateral lemma lobes long and bristle-like

Dichanthelium Gk *dicha*, in two; *anthele*, inflorescence; *-ium*, characteristic of. The spring and autumn panicles differ markedly, the former being exserted and the latter mostly enclosed in the upper leaf-sheaths

dichanthioides Gk *-oides*, resembling. Similar to *Dichanthium*

Dichanthium Gk *dicha*, in two; *anthos*, flower. Inflorescence with two kinds of spikelet, hermaphrodite and male or neuter

Dichelachne Gk *dichelos*, cloven-hoofed; *achne*, scale. Lemma apex two-lobed

dichotom-a, -um, -us L. with divisions in pairs. – (1) Inflorescence branching dichotomously. *Curtopogon dichotomus*, *Cutandia dichotoma*, *Panicum dichotomum*, *Pennisetum dichotomum* – (2) culms branching dichotomously. *Aristida dichotoma*

dichotomiflorum L. *dichotoma*, with divisions in pairs; *flos*, flower. Inflorescence an open panicle

dichotom-um, -us See *dichotoma*

dichro-a, -us Gk *dis*, twice; *chroia*, color. – (1) Two-colored as with callus hairs white, glume hairs fulvous. *Hyparrhenia dichroa* – (2) leaf-blades green above, reddish-purple below. *Guaduella dichroa* – (3) leaf-blades either green or reddish. *Andropogon dichrous*

Dichromus Gk *dis*, twice; *chromis*, color. Spikelets bicolored

dichrous See *dichroa*

diclina Gk *dis*, twice; *kline*, bed. Lower florets pistillate, upper florets staminate

dicoccoides Gk *-oides*, resembling. Resembling *Triticum dicoccum*

dicoccum Gk *dis*, twice; *kokkos*, a kernel. Mature spikelets mostly two-grained

Dictyochloa Gk *diktyon*, net; *chloa*, grass. Inflorescence subtended by an inflated bract with prominent reticulate venation

dictyoneur-a, -on Gk *diktyon*, net; *neuron*, nerve. Upper glume with prominent cross veins

didactyl-a, -um Gk *dis*, twice; *daktylos*, finger. Inflorescence of two racemes

Didactylon, didactylon Gk *dis*, twice; *daktylos*, finger. Inflorescence of two racemes

didactylum See *didactyla*

didistichum Gk *dis*, twice; *distichos*, two-rowed. Racemes with two rows of paired spikelets

Didymochaeta Gk *didymos*, twin; *chaete*, bristle. Both palea and lemma terminate in a short awn

didymostachyum Gk *didymos*, twin; *stachys*, spike as of an ear of wheat. Inflorescence of two branches

Diectomis Gk *dis*, twice; *ectemon*, castrate. Two of the spikelets in each cluster of three are sterile. The name has been given to two genera

diegoensis L. *-ensis*, denoting origin. From San Diego County, California, USA

diehlii In honor of I. E. Diehl (fl. 1903) United States editor and plant collector

dielsii As for *Dielsiochloa*

Dielsiochloa Gk *chloa*, grass. In honor of Friedrich Ludwig Emil Diels (1874–1945) German botanist

diemenica L. *-ica*, belonging to. From Tasmania, formerly Van Diemen's Land

diemii In honor of José Diem (1899–1986) German-born Argentinian botanist

dieterleniae In honor of Anna Dierterlen (1859–1945) French missionary and amateur botanist in Lesotho

dieterlenii See *dieterleniae*

dietrichiana L. *-ana*, indicating connection. In honor of Amalie Dietrich (1821–1891) German museum collector who lived for several years in Australia

difforme L. unlike what is usual. Leaf-blades with pellucid spots

diffus-a, -um, -us L. widely spreading. Inflorescence an open panicle

diffusissimus L. *diffusus*, widely spreading; *-issimus*, most. Panicle branches very widely spreading

diffusum See *diffusa*

Digastrium Gk *dis*, twice; *gaster*, belly. The internodes of the rhachis and the pedicels of the spikelets are swollen

digen-a, -ea Gk *dis*, twice; *genea*, descent. An interspecific hybrid

Digitaria L. *digitus*, finger; *-aria*, pertaining to. The ultimate inflorescence branches are often finger-like. The name has been applied to three distinct genera

digitaria Resembling *Digitaria* in inflorescence form

Digitariella L. *-ella*, diminutive but here used as a name-forming suffix. Resembling *Digitaria*

digitarioides Gk *-oides*, resembling. Resembling *Paspalum digitaria*

Digitariopsis Gk *opsis*, resemblance. The spikelets resemble those of *Digitaria*

digitat-a, -um, -us L. *digitus*, finger; *-ata*, possessing. Inflorescence branches finger-like

digitiformis L. *digitus*, finger; *forma*, appearance. Inflorescence branches whorled or sub-whorled

Dignathia Gk *dis*, twice; *gnathos*, jaw. The hardened glumes of the fertile spikelets resemble a pair of jaws or mandibles

Digraphis Gk *dis*, twice; *graphis*, brush. Both sterile lemmas have hairy tips

digyn-a, -us Gk *dis*, twice; *gyne*, woman. Pistil has two styles

Diheteropogon Gk *dis*, twice. Unlike *Heteropogon* which has solitary racemes, the racemes are paired

dilacerata Gk *dis*, twice; L. *lacero*, tear apart; *-ata*, possessing. Origin uncertain, not given by author

dilatat-a, -um L. *differo*, spread abroad; *-ata*, possessing. Racemes spreading in pseudoverticils

Dilepyrum Gk *dis*, twice; *lepyron*, shell or hulk. Both lemma and palea are indurated

Dileucaden Gk *dis*, twice; *leukos*, white; *aden*, gland. The florets have two fleshy, white lodicules

dillonii As for *quartiniana*

Dilophotriche Gk *dis*, twice; *lophos*, crest; *thrix*, hair. The lemma bears two hair tufts

diluta L. *diluo*, dissolve. Doubt as to the reality of the species dissolved with further collecting

Dimeiostemon Gk *dis*, twice; *meion*, less; *stemon*, thread. Florets possess only one stamen

Dimeria Gk *dis*, twice; *meros*, part. Racemes occur in pairs

Dimesia Gk *dis*, twice. Possibly a reference to the spikelets having two florets, one male, the other female

dimidiat-a, -um L. divided down the middle. – (1) The lemma sometimes with lateral hair tufts. *Danthonia dimidiata* – (2) lower glume leathery basally then abruptly thinner. *Sorghum dimidiatum* – (3) leaf-blade asymetric. *Olyra dimidiata* – (4) panicle with paired racemes. *Panicum dimidiatum*, *Paspalum dimidiatum*, *Rottboellia dimidiata*, *Stenotaphrum dimidiatum*

diminuta L. *diminuo*, make less. – (1) Upper glume much reduced. *Digitaria diminuta* – (2) lower glume minute. *Melinis diminuta* – (3) awn apparently single due to non development of the lateral arms. *Aristida diminuta* – (4) awn much reduced. *Stipa diminuta*

dimorpha Gk *dis*, twice; *morphe*, appearance. – (1) Spikelets of two kinds. *Festuca dimorpha* – (2) leaf-blade surfaces differently colored above and below. *Arundinaria dimorpha* – (3) annual or perennial in habit. *Brachiaria dimorpha*

dimorphantha Gk *dis*, twice; *morphe*, appearance; *anthos*, flower. Spikelets with hermophrodite and female florets in the same spikelet

Dimorphochloa Gk *dis*, twice; *morphe*, appearance; *chloa*, grass. Culms perennial, those in their first year differing greatly in appearance from those in their second year

dimorpholemma Gk *dis*, twice; *morphe*, appearance; *lemma*, scale. With two forms of lemma in the same panicle

Dimorphostachys Gk *dis*, twice; *morphe*, appearance; *stachys*, spike as of an ear of wheat. Torsion of its pedicel results in the fertile lemma of the upper spikelet facing the rhachis whereas that of the lower spikelet is turned away from the rhachis

dimyloides Gk *dis*, twice; *myle*, millstone; *-oides*, resembling. The pair of florets resemble millstones

Dinaeba See *Dinebra*

Dineba See *Dinebra* and a more faithful representation of the Arabic

Dinebra Arabic *danaiba*, a little tail. The apices of the glumes are prolonged

dinhensis L. *-ensis*, denoting origin. From Mt Dinh, near Baria, Vietnam

dinirica L. *-ica*, belonging to. From Dinira National Park, Venezuela

dinklagei In honor of Max Julius Dinklage (1864–1935) German factory manager and botanist who collected in Liberia

Dinochloa Gk *dinos*, whorl; *chloa*, grass. The inflorescence branches are whorled

dinteri In honor of Kurt Dinter (1868–1945) German botanist who collected widely in southern Africa

dioic-a, -um Gk *dis*, twice; *oikos*, house. The male and female spikelets occur on different plants or in different spikelets on the same plant

D

diomedarum See *diomedearum*
diomedearum An allusion to Diomedea, an albatross. Of the Galapagos Islands, home of the Ward Albatross (*Diomedea irrorata*)
Diperium Gk *dis*, twice; *pera*, leather pouch. Spikelets embedded in axis
Diplachne Gk *diploos*, double; *achne*, lobe. The lemma is two-lobed
diplachnoides Gk *-oides*, resembling. Resembling *Diplachne*
Diplachyrium Gk *diploos*, double; *achyron*, chaff. Spikelets with two glumes plus palea and lemma
diplandr-a, -us Gk *diploos*, double; *aner*, man. Bisexual floret with two anthers
Diplasanthum Gk *diplasios*, twice as many; *anthos*, flower. Spikelets paired, one sessile, the other stalked
Diplax Gk *diploos*, double. The spikelets have two florets
Diplocea Gk *dis*, twice; *ploche*, tissue. Lemma deeply bifid
Diplogon See *Diplopogon*
diploperennis Gk *diploos*, double; L. *perenne*, perennial. Diploid with respect to chromosome complement and perennial with respect to habit
Diplopogon Gk *diploos*, double; *pogon*, beard. A reference to there being awns on both the glumes and lemmas or to the palea being two-awned
diplostachya Gk *diploos*, double; *stachys*, spike as of an ear of wheat. The spikelets are sessile and arise in two rows from a central axis
diplotaphrum Gk *diploos*, double; *taphros*, ditch. Rhachis with a spikelet bearing groove on both sides instead of one as with related species
Dipogon Gk *dis*, twice; *pogon*, beard. Origin uncertain, not given by author
Dipogonia Gk *dis*, twice; *pogon*, beard. Palea two-awned
dipsacea Gk *dipsas*, thirst; *-ea*, possessed by. Growing in the spray of waterfalls
dipteros Gk *dis*, twice; *pteron*, wing or feather-like. Both glumes prominently winged on the keel
Diptychum Gk *dis*, twice; *ptyche*, fold. The lemma apex is bifid and plicate
Disakisperma Gk *dis*, twice; *akis*, point; *sperma*, seed. Apex of grain bicuspid
Disarrenum Gk *dis*, twice; *arrhen*, male. There are two male florets per spikelet
Dischlis See *Distichlis*
discifera L. *discus*, quoit; *fero*, carry or bear. The androecium is discus-shaped in the modern sporting sense
discolor L. not of the same color. Spikelets two-colored
discospor-um, -us L. *discus*, disc; *sporus*, seed. Grain the shape of a disc
discrepans L. *discrepo*, be different. A species of *Panicum* but with the habit of *Agrostis*
disjecta See *disiecta*
disjunct-um, -us L. *disjunco*, separate. – (1) Occurring in widely separated localities. *Paspalidium disjunctum* – (2) panicle interrupted. *Sporobolus disjunctus, Trisetum disjunctum*
dispar L. unequal. – (1) Glumes unequal in length or nerve number. *Holcolemma dispar, Isachne dispar, Microstegium dispar, Paspalum dispar, Pollinia dispar* – (2) glumes unequally awned. *Agrostis dispar*
disperma Gk *dis*, twice; *sperma*, seed. Both florets of spikelet with fertile seed
dispermis Gk *dis*, twice; *sperma*, seed. Spikelets producing two grains
dispersa L. *dispergo*, scatter around. Inflorescence an open panicle
dissanthelioides Gk *-oides*, resembling. Similar to *Dissanthelium*
Dissanthelium Gk *dissos*, double; *anthelion*, a small flower. The spikelets usually have only two fertile florets
dissectum L. cut up. Lodicules emarginate
dissimilis L. different. – (1) Male and female inflorescences and spikelets dissimilar. *Arundoclaytonia dissimilis* – (2) awn-branches of two lengths. *Aristida dissimilis*
dissimulator L. *dissimulator*, concealer. Species readily confused with others
dissita L. lying apart. With secondary inflorescence branches diverging widely from the central axis

dissitiflor-a, -um, -us L. *dissitus*, widely separated; *flos*, flower. – (1) Florets widely separated in spikelet. *Spartina dissitiflora* – (2) spikelets widely separated in inflorescence. *Agrostis dissitiflora, Paspalum dissitiflorum, Sasa dissitiflora* – (3) panicles few-flowered. *Bromus dissitiflorus, Festuca dissitiflora*

Dissochondrus Gk *dissos*, double; *chondros*, grain. Both florets of the spikelet are bisexual

dissoluta L. lax. The inflorescence consists of racemes pendant on filiform peduncles

distachia See *distachya*

distachy-a, -on, -os, -um, -us Gk *dis*, twice; *stachys*, spike as of an ear of wheat. Inflorescences with two branches or with regularly bifurcating branches

distachyoides Gk *-oides*, resembling. Resembling *Brachiaria distachya*

distachy-on, -os As for *distachya*

distachy-um, -us See *distachya*

distans L. *disto*, be apart. Spikelets widely separated in inflorescence

distantiflor-um, -us L. *disto*, be apart; *flos*, flower. – (1) Spikelets well separated in inflorescence. *Panicum distantiflorum* – (2) florets well separated in spikelet. *Myriocladus distantiflorus*

distich-a, -um, -us Gk *distichos*, in two rows. Plants with conspicuously two-rowed spikelets or leaves

Disticheia Gk *distichos*, in two rows. The many florets are conspicuously two-rowed in the spikelet

Distichlis Gk *distichos*, in two rows; *lisse*, smooth. Leaves conspicuously two-rowed and glabrous

distichon Gk *distichos*, in two rows. Spikelets borne in two rows on the spike

distichophyll-a, -um, -us Gk *distichos*, in two rows; *phyllon*, leaf. Leaves arranged conspicuously in two rows

distichophylloides Gk *-oides*, resembling. Similar to *Panicum distichophylla*

distichophyll-um, -us See *distichophylla*

distichovaginat-a, -us Gk *distichos*, in two rows; L. *vagina*, sheath; *-ata*, possessing. Leaf-sheaths conspicuously distichous

distich-um, -us L. *distichum*, in two rows. Leaves distinctly distichous

distichus See *disticha*

Distomomischus Gk *distomos*, double; *mischos*, husk. Glumes two instead of one as with the related *Lolium*

distort-a, -um L. *distanqueo*, distort. – (1) Culms recurved. *Bambusa distorta, Guadua distorta* – (2) sterile lemma irregularly crumpled. *Paspalum distortum*

distylum Gk *dis*, twice; *stylos*, column. Pistils with two styles

diuturnus L. long lasting. Densely tufted perennial

divaricat-a, -um, -us L. *divarico*, spread out. Inflorescence branches or culms regularly dividing into equal branches

divaricatissim-a, -um L. *divarico*, spread out; *-issima*, most. Panicle, large and diffuse

divaricat-um, -us See *divaricata*

divergens L. *divergo*, go in different directions. The panicle branches are not disposed as in an open panicle. – (1) Branches drooping. *Panicum divergens, Paspalum divergens* – (2) branches held erect. *Calamagrostis divergens*

divers-a, -us L. variable. – (1) Stamen number variable. *Apoclada diversa* – (2) leaf-blades variable in length. *Pleioblastus diversus*

diversiflor-a, -us L. *diversus*, variable; *flos*, flower. – (1) Spikelets of different types in the same inflorescence. *Andropogon diversiflorus, Eragrostis diversiflora, Polytrias diversiflora, Triticum diversiflora* – (2) inflorescences of two kinds on the same plant. *Eragrostis diversiflora*

diversifolius L. *diversus*, variable; *folium*, leaf. Lower leaf-blades of culm lanceolate, upper leaf-blades somewhat cordate

diversiglumis L. *diversus*, variable; *gluma*, husk. One glume of some spikelets long-awned

diversinerv-e, -is L. *diversus*, variable; *nervus*, nerve. The upper and lower glumes have different numbers of nerves

diversispicula L. *diversus*, variable; *spica*, a point; hence, in particular, an ear or spike of grain; *-ula*, diminutive. Spikelets polymorphic, varying according to position and sex

D

diversus See *diversa*
dives L. plentiful. Common grasses
divisum L. *divido*, divide. Differing from related species in some respect such as branched culms
divuls-a, -um L. *divello*, tear apart. Panicles or racemes interrupted
djalonicum L. *-icum*, belonging to. From Fouta Djalon, French Guinea
djimilensis L. *-ensis*, denoting origin. From valléee de Djimil, Algeria
djurdjurae From Kabylie Djuradjura Mts., Algeria
dmitrievae In honor of A. Dimitrieva (fl. 1962) Russian botanist
dobbelaerei In honor of Dobbelaere (fl. 1913) who collected in Zaire
dochna Latinized version of Arabic, *dochn*, the vernacular name of the species
dodii In honor of Anthony Hurt Wolley-Dod (1861–1948) English botanist who collected in South Africa
doelliana L. *-ana*, indicating connection. As for *Doellochloa*
doellii As for *Doellochloa*
Doellochloa Gk *chloa*, grass. In honor of Johann Christoph Doell (1808–1885) German botanist
doerfleri In honor of Ignaz Dörfler (1866–1950) Austrian botanist and publisher
doiyoshiwoana L. *-ana*, indicating connection. In honor of Yoshiwo Doi (fl. 1934) Japanese botanist
dokyoanus L. *-anus*, indicating connection. In honor of Dokyo, Japanese botanist
dolichachyra Gk *dolichos*, long; *achyron*, chaff. Spikelets with narrow glumes and lemmas
dolichanth-a, -us Gk *dolichos*, long; *anthos*, flower. Spikelets narrow
dolichathera Gk *dolichos*, long; *ather*, barb or spine. Spicate inflorescence longer than that of related species
dolichoadenotrichum Gk *dolichos*, long; *aden*, gland; *thrix*, hair. Inflorescence branches invested in long glandular hairs
dolichocarpus Gk *dolichos*, long; *karpos*, fruit. Grain terete
dolichochaeta Gk *dolichos*, long; *chaete*, bristle. Long lateral lemma bristles overtop the glumes
Dolichochaete Gk *dolichos*, long; *chaete*, bristle. Upper lemma with two lobes, each of which tapers into a long thin awn
dolichoclada Gk *dolichos*, long; *klados*, branch. Culms thin
dolicholepis Gk *dolichos*, long; *lepis*, scale. Glumes narrow-lanceolate
dolichomerithalla Gk *dolichos*, long; *meros*, part; *thallos*, young shoot. Internodes of young shoots long and thin
dolichophyll-a, -um Gk *dolichos*, long; *phyllon*, leaf. Leaf-blades long and narrow
dolichopus Gk *dolichos*, long; *pous*, foot. Origin unclear
dolichostachy-a, -um, -us Gk *dolichos*, long; *stachys*, spike as of an ear of wheat. Spikelets terete
dolichotrichum Gk *dolichos*, long; *thrix*, hair. The raceme bases bear numerous long hairs
dolos-a, -um L. deceitful. Easily confused with another species
dombeyana In honor of Joseph Dombey (1742–1794) French physician and naturalist who collected in Peru
dombeyi As for *dombeyana*
domingens-e, -is L. *-ense*, denoting origin. From Dominican Republic
dominii In honor of Karel Domin (1882–1953) Czech educator and botanist who collected widely in Australia
donacifoli-a, -um Gk *donax*, reed; L. *folium*, leaf. The leaf-blades resemble those of *Donax* (Marantaceae)
donaciformis L. *forma*, appearance. Resembling *Donax* (Poaceae) in habit and inflorescence
Donacium An alternative for *Donax* (Poaceae) if that is kept as the prior name for a genus of molluscs
Donax Gk *donax*, a type of reed in Classical literature, so called because the inflorescences move to and fro in the slightest breeze (Gk *doneo*, shake) and so the diaspores are easily dispersed by the wind

donax Reed-like in habit
dongicola L. *-cola*, dweller. Growing in shallow gullies, that is dongas (from Zulu *udonga*) in Western Australia
dongvanensis L. *-ensis*, denoting origin. From Dong Van, Ha Tien Province, Vietnam
don-hensonii In honor of Don Clifford Henson (1945–) United States artist and field botanist
donian-um, -us L. *-anum*, indicating connection. In honor of George Don (1798–1856) Scots-born botanist who collected in South America and Africa
dora See *durra*
doreana L. *-ana*, indicating connection. As for *dorei*
dorei In honor of William George Dore (1912–) Canadian botanist
dorsense L. *-ense*, denoting origin. From Dorse, Gamo Gofar Province, Ethiopia
douglasii In honor of David Douglas (1799–1834) Scots-born botanical explorer of the north-western Pacific
dowsonii In honor of Walter John Dowson (1887–1963) plant pathologist in East Africa and later Tasmania, Australia
dozei In honor of Frans Dozy (1807–1856) Dutch botanist
dozyi In honor of Jean Jacques Dozy (1908–) Dutch geologist
dracomontana L. *-ana*, indicating connection. From the Drakensbergs of Natal and Transvaal, South Africa
Drake-Brockmania In honor of Ralph Evelyn Drake-Brockman (1875–?) British Army medical officer who collected in Africa
drakensbergens-e, -is L. *-ense*, denoting origin. From Drakensberg, a range of mountains in South Africa
drarii In honor of Mohammed Drar (1894–1964) Sudanese taxonomist
dregean-a, -um, -us L. *-ana*, indicating connection. As for *Dregeochloa*
dregei As for *Dregeochloa*
Dregeochloa In honor of Johann Franz Drège (1794–1881) German horticulturalist and plant collector in South Africa
drepanophyllus Gk *drepanos*, sickle; *phyllon*, leaf. Leaf-blades sickle-shaped
Drepanostachyum Gk *drepanos*, sickle; *stachys*, spike as of an ear of wheat. Spikelets curved
drepanothrix Gk *drepanos*, sickle; *thrix*, hair. Hairs on sterile lemmas sickle-shaped
dressleri In honor of Robert Louis Dressler (1927–) United States botanist
drobovii In honor of Vasilii Petrovich Drobov (1885–1956) Russian botanist
drosocarpum Gk *drosos*, dew; *karpos*, fruit. Anthoecium milky-white
drucei In honor of Anthony Peter Druce (1920–1998) New Zealand botanist
drummondiana L. *-ana*, indicating connection. In honor of James Drummond. See *drummondii*
drummondii In honor of – (1) Thomas Drummond (c. 1790–1835) Scots-born naturalist and traveller in North America. *Andropogon drummondii, Dimorphostachys drummondii, Merisachne drummondii, Paspalum drummondii, Sorghum drummondii, Sporobolus drummondii, Triodia drummondii, Vilfa drummondii* – (2) James Drummond (c. 1784–1863) Scots-born Australian botanist. *Pentapogon drummondii, Plectrachne drummondii, Polypogon drummondii, Poroteranthe drummondii, Stipa drummondii*
drymea See *drymeia*
drymeia Gk *drymeios*, of oak woods. Growing in oak woods
Drymochloa Gk *drymos*, copse; *chloa*, grass. A genus comprising a single woodland species
Drymonaetes Gk *drymos*, copse; *naetes*, inhabitant. A genus of woodland species
Dryopoa Gk *dryos*, tree; *poa*, grass. A tall Australian forest-grass
dschungarica L. *-ica*, belonging to. From the Jungaria (Dzungaria)-Kashgaria Mountains, central Asia
dshilgensis L. *-ensis*, denoting origin. From Dzhilga, Kazakhstan

dshinalica L. *-ica*, belonging to. From Dzhinal, a mountain in the Caucasus

duartei In honor of Pedro Pablo Duarte-Bello (1922–) Cuban botanist who emigrated to the USA

dubi-a, -um, -us L. doubtful. – (1) Readily confused with other species. *Chloris dubia*, *Diplachne dubia*, *Elionurus dubius*, *Leersia dubia*, *Leptochloa dubia*, *Muhlenbergia dubia*, *Panicum dubium*, *Paspalum dubium*, *Saccharum dubium*, *Setaria dubia*, *Spodiopogon dubius* – (2) placed in the genus with misgiving. *Festuca dubia*

dubioides Gk *-oides*, resembling. Inflorescence similar to that of a species named *dubia* in the same genus

dubitatus L. *dubium*, doubtful; *-atus*, likeness. Resembling *Sorghum dubium*

duchaissingii In honor of Édouard Placide Duchassaing de Frontbressin (1818–1873) physician and naturalist, of Guadelupe in West Indies

duclouxii In honor of François Ducloux (1864–1945) French cleric and amateur botanist who collected in China

dudleyi In honor of William Russell Dudley (1849–1911) United States botanist

duernsteinensis L. *-ensis*, denoting origin. From the ruins of Dürnstein, near Donan River, Austria

dufourei In honor of Jean Marie Léon Dufour (1779–1865) French physician and botanist

duhamelianum L. *-anum*, indicating connection. In honor of Henri Louis Du Hamel du Monceau (1700–1781) French horticulturalist

dulcicaule L. *dulcis*, sweet; *caulis*, stem. Stems sweet to the taste

dullooa Latinized form of *dalu* the vernacular name of the species in Assam State, India

dumasiana L. *-ana*, indicating connection. In honor of Dumas

dumetorum L. *dumus*, thorn-bush; *-etum*, place of growth. Growing in thickets of thorn-bushes

dumetos-a, -us L. *dumus*, thorn-bush; *-osa*, abundance. Culms woody and much branched

dummeri In honor of Richard Arnold Duemmer (1887–1922) South African plant collector

dumosus L. *dumus*, thorn-bush; *-osus*, well developed. Habit decidedly bushy

dumus L. thorn-bush. Resembling a thorn bush

dunensis L. *-ensis*, denoting origin. From dunes on the shore of Lake Nyasa

dunlopii In honor of Clyde Robert Dunlop (1946–) Australian botanist

duongii From Duong huu Thai, Thai-binh Province, Vietnam

duparquetii In honor of Charles Victor Aubert du Parquet (1830–1888) French cleric and amateur botanist

duplicata L. *duplico*, double. Closely resembling another species

Dupoa Hybrids between species of *Dupontia* and *Poa*

Dupontia In honor of J. D. Dupont (fl. 1805–1813) French botanist. Author of a valuable essay on the "Sheath of the Leaves of Grasses"

Dupontopoa Hybrids between species of *Dupontia* and *Poa*

duquilioi See *quilioi*

dur-a, -um, -us L. hard. – (1) Lemmas hard or tough. *Brachiaria dura*, *Brachypodium durum*, *Cynosurus durus*, *Danthonia dura*, *Eragrostis dura*, *Hordeum durum*, *Merxmuellera dura*, *Poa dura*, *Sclerochloa dura*, *Scutachne dura*, *Thrixgyne dura*, *Trachypogon durus* – (2) culms tough. *Fargesia dura* – (3) leaf-blades tough. *Ehrharta dura*, *Stipa dura* – (4) grain flinty. *Triticum durum*

durandiana L. *-ana*, indicating connection. In honor of Elias Magliore Durand (1794–1873) French-born United States botanist

durandoi In honor of Gaètano Leone Durando (1811–1891) Italian-born Algerian botanist

durangens-e, -is L. *-ense*, denoting origin. From Durango, Mexico

duriaei In honor of Michel Charles Durieu de Maisonneuve (1796–1878) French botanist
durie See *duriaei*
durifoli-a, -um L. *durus*, hard; *folium*, leaf. Leaf-blades tough
duriotagana L. *-ana*, indicating connection. From the valleys of the Durio, now Douro and Tagus Rivers of Portugal
durissima L. *durus*, hard; *-issima*, most. Foliage harsh
duriuscula L. *durius*, harder; *-ula*, diminutive. Foliage somewhat harsh
durra Vernacular name of the species in Arabic
dur-um, -us See *dura*
d'urvillei See *urvilleana*
dusenii In honor of Per Karl Hjalmar Dusén (1855–1926) Swedish civil engineer and traveller who collected in Africa, Greenland and South America
Duthiea In honor of John Firminger Duthie (1845–1922) English botanist who collected in India
duthieana L. *-ana*, indicating connection. As for *Duthiea*
duthiei See *Duthiea*
dutillyanum L. *-anum*, indicating connection. In honor of Arthème Dutilly (1896–1973) Canadian cleric and botanist
duttoniana L. *-ana*, indicating connection. In honor of J. Dutton (1863–?) field assistant, Bathurst Agricultural Station, Australia
duvalii In honor of Joseph Duval-Jouve (1810–1883) French educator and botanist
Dybowskia In honor of Jan Dybowski (1858–1928) Polish botanist
dybowskii As for *Dybowskia*
Dydactylon See *Didactylon*
dyeri In honor of William Turner Thiselton-Dyer (1843–1928) English botanist
Dyneba See *Dinebra*
dyris From Dyris, a mountain peak in the High Atlas Mountains in Morocco
dzhilgensis See *dschilgensis*
dzongicola L. *-cola*, dweller. Growing on walls at Lingshi Dong, Tibet Autonomous Region, China

E

earlei In honor of Franklin Summer Earle (1856–1929) United States plant collector
eastwoodiae In honor of Alice Eastwood (1859–1953) United States botanist
Eatonia In honor of Amos Eaton (1776–1842) United States botanist
eatonii (1) As for *Eatonia*. *Poa eatonii* – (2) in honor of Alvah Augustus Eaton (1865–1908) United States botanist. *Panicum eatonii*
eberhardtii In honor of Philippe Albert Eberhardt (1874–1942) who collected in Annam, now included in Vietnam
ebingeri In honor of John E. Ebinger (1933–) United States botanist
ebracteata L. *e-*, without; *bractea*, bract; *-atus*, possessing. Sessile spikelets lacking subtending bracts
eburne-a, -um L. *ebur*, ivory; *-ea*, pertaining to. – (1) Fertile lemma pale-yellow. *Paspalum eburneum* – (2) Bulbous culm base resembles an ivory bead. *Ehrhartia eburnea*
eburonensis L. *-ensis*, denoting origin. From Eburones, in the Ardennes
ecarinata, ecarinatum L. *e-*, without; *carina*, keel; *-ata*, possessing. – (1) Palea keels reduced. *Eragrostis ecarinata* – (2) glumes lacking a keel. *Sorghum ecarinatum*
ecaudata L. *e-*, without; *cauda*, tail. Glume apices attenuated but not long and narrow
Eccoilopus Gk *ek*, out; *koilos*, hollow; *pous*, foot. Pedicel apices conspicuously cup-shaped after spikelets have been shed
Eccoptocarpha Gk *ekkoptos*, cut off; *karphos*, chaff. The upper part of the upper glume and lower lemma have hyaline apices abruptly differentiated from their heavily veined lower parts
echidnae Gk *echidna*, viper. Leaf-blades rigid and pointed, characters which they share with the teeth of a viper. Furthermore, in the vegetative state the plant resembles the echidna, a spiny Australian marsupial (*Echidna hystrix*) whose vernacular name is "echidna"

Echinalysium Gk *echinos*, hedgehog; *alysis*, chain. The spikelets are arranged in clusters along a central axis thereby resembling a string of hedgehogs

Echinanthus Gk *echinos*, hedgehog; *anthos*, flower. The upper glume bears hooked spines

Echinaria Gk *echinos*, hedgehog; *-aria*, pertaining to. The lemmas and paleas terminate in hardened subulate awns and the spikelets are clustered in capitate inflorescences which thereby resemble hedgehogs

echinat-a, -um, -us L. spiny. – (1) Inflorescence a very condensed panicle and the spikelets or auxillary structures are awned the whole thereby resemble a hedgehog. *Aristida echinata, Bathratherum echinatum, Cenchrus echinatus, Cynosurus echinatus, Lasiochloa echinata, Phleum echinata, Sesleria echinata, Tribolium echinatum* – (2) spikelets with bristly glumes. *Panicum echinatum* – (3) plants forming spiny tussocks. *Stipa echinata* – (4) lower glume with abundant spicular hairs. *Andropogon echinatus, Arthraxon echinatus*

Echinochlaena See *Echinolaena*

Echinochloa, echinochloa Gk *echinos*, hedgehog; *chloa*, grass. The glumes are often awned and the inflorescence congested, thereby resembling a hedgehog

echinochloidea Gk *-oides*, resembling. Inflorescences resembling those of *Echinochloa*

echinoides Gk *echinos*, hedgehog; *-oides*, resembling. Spikelets subtended by an involucre of spiny bristles

Echinolaena Gk *echinos*, hedgehog; *chlaena*, cloak. The lower glume bears numerous acute or shortly barbed bristles

echinolaenoides Gk *-oides*, resembling. Similar to *Echinolaena*

Echinopogon Gk *echinos*, hedgehog; *pogon*, beard. The inflorescence is a capitate condensed panicle with bristle like glumes, the whole thereby resembling a hedgehog

echinotrichum Gk *echinos*, hedgehog; *thrix*, hair. Inflorescence digitate, racemes short and the glumes invested in long tuberculate-based hairs resembling the quills of a hedgehog

echinulat-a, -um, -us L. *echinus*, hedgehog; *-ulus*, diminutive; *-ata*, possessing. – (1) Leaf-blades hispid with small prickles. *Andropogon echinulatus, Chrysopogon echinulatus, Raphis echinulata* – (2) spikelets hispid with small prickles. *Brachiaria echinulata, Loudetia echinulata, Panicum echinulatum*

echinurus Gk *echinos*, hedgehog; *oura*, tail. Spike-like inflorescence very bristly

echinus Gk *echinos*, hedgehog. Spikes very short and the spikelets awned

Echistachys Gk *echinos*, hedgehog; *stachys*, spike as of an ear of wheat. The inflorescence is contracted and the upper glume of each spikelet bears hooked spines

eckloniana L. *-ana*, indicating connection. As for *ecklonii*

ecklonii, ecklonis In honor of Christian Frederick Ecklon (1795–1868) German-born South African apothecary and plant collector

Ectosperma Gk *ektos*, free from; *sperma*, seed. The caryopsis falls readily from its subtending palea and lemma

Ectrosia Gk *ektrosis*, miscarriage. The spikelets have one-two hermaphrodite proximal florets with one-two male or neuter distal florets

Ectrosiopsis Gk *opsis*, resemblance. Similar to *Ectrosia*

ecuadorens-e, -is L. *-ense*, denoting origin. From Ecuador

ecuadoriana As for *ecuadorense*

ecuadoricus L. *-icus*, belonging to. From Ecuador

ecuadoriens-e, -is L. *-ense*, denoting origin. As for *ecuadorense*

edelbergii In honor of Lennart Edelberg (1915–1982) Danish botanist

editissima L. *editus*, high place; *-issima*, most. Growing on high mountains

editorum L. *editus*, high place. Of high places, that is growing on mountains

edlundiae In honor of Sylvia A. Edlund (fl. 1984–1997) Canadian botanist

edmondii In honor of Brother Edmond, a colleague of Brother Léon, who collected in Cuba. See *Saugetia* for details of the latter

eduardii In honor of Eduard Hackel (1850–1926) Austrian agrostologist

edul-e, -is L. edible. Young shoots edible

edwardii In honor of Edward E. Terrell (1923–) United States botanist

edwardsian-a, -us L. *-ana*, indicating connection. From Edwards County, Texas, USA

effus-a, -um L. spread out. Inflorescence an open panicle

effusiflora L. *effusus*, spread out; *flos*, flower. Inflorescence an open panicle

effusum See *effusa*

egena Gk *e-*, without; *genys*, under jaw. Lower leaf-sheaths glabrous whereas those of related species hairy

eggersii In honor of Heinrich Franz Alexander Eggers (1844–1908) German-born Danish military officer and plant collector in the West Indies and South America

eggertii In honor of Heinrich Karl Daniel Eggert (1841–1904) German-born United States botanist

eggleri In honor of Joseph Eggler (1896–1963) Austrian phytosociologist

egleri In honor of Walter A. Egler (?–1961) who collected in Brazil

eglumis L. *e-*, without; *gluma*, husk. The spikelets lack a lower glume

egmontiana L. *-ana*, indicating connection. From Egmont Bay, New Zealand

egregi-a, -um L. extraordinary. Differing markedly in some respect from closely related species

Ehrartha See *Ehrharta*

ehrenbergiana L. *-ana*, indicating connection. In honor of Christian Gottfried Ehrenberg (1795–1876) German physician and naturalist

ehrenbergii As for *ehrenbergiana*

Ehrhardta See *Ehrharta*

Ehrharta, Ehrhartia In honor of Jakob Friederick Ehrhart (1742–1795) Swiss botanist

eichingeri In honor of Alfons Eichinger (1883–?) German botanist

eickii In honor of Emil Eick (fl. 1898–1899) who collected in Usumbara, Tanzania

eigiana L. *-ana*, indicating connection. In honor of Alexander Eig (1895–1938) Russian-born Palestine botanist

eigii As for *eigiana*

eitenii In honor of Georg Eiten (1923–) United States-born Brazilian botanist and Liene Teixeira Eiton (1925–1979) Brazilian botanist

ejubata L. *e-*, without; *juba*, mane; *-ata*, possessing. Glumes and sterile lemma almost glabrous

ekeanum L. *-anum*, indicating connection. From Mt Eke on Maui one of the Hawaiian Islands

ekmanian-a, -um, -us L. *-ana*, indicating connection. As for *Ekmanochloa*

ekmanii See *Ekmanochloa*

Ekmanochloa Gk *chloa*, grass. In honor of Erik Leonard Ekman (1888–1931) Swedish botanist

elanata L. *e-*, without; *lana*, wool; *-ata*, possessing. Lacking woolly hairs on lemma

elat-a, -um L. tall. Culms taller than those of most species of the genus

elati-or, -us L. taller. Culms taller than those of related species

elatiuscula L. *elatius*, taller; *-ula*, tending towards. The culms taller than those of related taxa

elatoides Gk *-oides*, resembling. Resembling *Nastus elatus*

elatum See *elata*

elbrusica See *elbrussica*

elbrussica L. *-ica*, belonging to. From Elbrus, a mountain in Republic of Georgia

elbursensis L. *-ensis*, denoting origin. From Elburs, near Keredj, Iran

eleanoriae In honor of Eleanor Constance Bor (?–1957)

Electra Gk *electron*, amber. The grains are transparent and yellow

elegans L. elegant. – (1) Inflorescence attractive. *Acroceras elegans, Aira elegans, Andropogon elegans, Anthephora elegans, Aristida elegans, Briza elegans, Burmabambus elegans, Ceresia elegans, Chloris elegans, Cymbopogon elegans, Deyeuxia elegans, Digitaria elegans, Elionurus elegans, Enneapogon elegans, Fargesia elegans, Gymnothrix elegans, Isachne elegans, Loudetia elegans, Mesosetum elegans, Neoschischkinia elegans, Panicum elegans, Paspalum elegans, Pennisetum elegans, Poa elegans, Podosaemum elegans, Sericura elegans, Thelepogon elegans, Trichoneura elegans, Trichopteryx elegans, Tristachya elegans, Vilfa elegans, Zenkeria elegans* – (2) culms slender. *Arundinaria elegans, Leleba elegans, Sasa elegans*

elegantissim-a, -um, -us L. *elegans*, elegant; *-issima*, most. Very attractive in some respect, usually the inflorescence

elegantul-a, -um, -us L. *elegans*, elegant; *-ulus*, diminutive. Of attractive appearance

elephantin-a, -us L. *elephantus*, elephant; *-ina*, indicating resemblance. – (1) A gigantic species. *Erianthus elephantinus* – (2) a translation of the vernacular name Olifant grass. *Danthonia elephantina*

elephantipes L. *elephantus*, elephant; *pes*, foot. Culms thick and fleshy with black nodes and so resembling the feet of elephants

Eleusine The Ancient Greek town of Ceres, where the Goddess of the harvest was worshipped

eleusine Resembling *Eleusine*

eleusinoides Gk *-oides*, resembling. Similar to *Eleusine* with respect to the inflorescence

elevata L. *elevo*, raise up. Growing at high altitudes

elevatissimus L. *elevo*, raise up; *-issimus*, most. Sheaths of culm leaves elongated

elevatovenulosa L. *elevo*, raise up; *venulus*, small-vein; *-osa*, abundance. Veins of the sterile lemma conspicuously raised

elgonensis L. *-ensis*, denoting origin. From Mt Elgon on the border of Kenya and Uganda

Elicotrichum Gk *helisso*, turn around; *thrix*, hair. The awn is strongly twisted

elidis From the Peleponnesis, situated in the Ancient Greek Province of Elis

eligulat-a, -um L. *e-*, without; *ligula*, small tongue; *-ata*, possessing. Ligule lacking or very much reduced

Elimus See *Elymus*

elionuroides Gk *-oides*, resembling. Similar to *Elionurus*

Elionurus Gk *eloios*, doormouse; *oura*, tail. The cylindrical inflorescence resembles the tail of a doormouse

elisabethiae In honor of Elisabeth, nothing further given by author

elisabethpolense L. *-ense*, denoting origin. From Elisavetpol, Azerbaijan

elisabethvilleana L. *-ana*, indicating connection. From Elisabethville, now Lubumbashi, Zaire

elliotea As for *elliottii* (2)

elliotiana As for *elliotii* (2)

elliotii (1) As for *elliottii* (1) *Sorghum elliotii* – (2) as for *elliottii* (2). *Cenchrus elliotii*, *Festuca elliotii*

elliottian-a, -um L. *-ana*, indicating connection. As for *elliottii* (2)

elliottii In honor of – (1) George Francis Scott Elliott (1861–1934) Indian-born Scottish botanist. *Agrostis elliottii*, *Festuca elliottii*, *Sorghum elliottii*, *Triraphis elliottii* – (2) Stephen Elliott (1771–1830) United States planter, legislator and amateur botanist. *Andropogon elliottii*, *Chrysopogon elliottii*, *Eragrostis elliottii*, *Poa elliottii*, *Sorghastrum elliottii*, *Sorghum elliottii*, *Triodia elliottii*

elliptic-a, -um Gk *elluipsis*, ellipse; *-ica*, belonging to. – (1) Spikelets elliptical in outline. *Echinochloa elliptica*, *Paspalum ellipticum* – (2) culms semiterete. *Aristida elliptica*

elmeri In honor of Adolph Daniel Edward Elmer (1870–1942) United States botanist

elongat-a, -um, -us L. elongated. – (1) Inflorescence elongated. *Agropyron elongatum*, *Agrostis elongata*, *Andropogon elongatus*, *Anthephora elongata*, *Avenastrum elongatum*, *Chrysopogon elongatus*, *Danthonia elongata*, *Elymus elongatus*, *Elytrigia elongata*, *Eragrostis elongata*, *Gymnopogon elongatus*, *Helictotrichon elongatum*, *Hordeum elongatum*, *Hyparrhenia elongata*, *Lasiagrostis elongata*, *Megastachya elongata*, *Mesosetum elongatum*, *Muhlenbergia elongata*, *Panicum elongatum*, *Poa elongata*, *Polypogon elongatus*, *Sesleria elongata*, *Sporobolus elongatus*, *Stipidium elongatum*, *Triticum elongatum*, *Uralepis elongata* – (2) leaf-sheath elongated. *Axonopus elongatus*, *Rhynchelytrum elongatum* – (3) culms elongated. *Nastus elongatus*, *Pleioblastus elongatus*

elongatiform-e, -is L. *elongatus*, elongated; *forma*, appearance. Spikelets longer than those of related species

elongat-um, -us See *elongata*

eludens L. *eludo*, avoid. – (1) Not to be confused with a series of closely related species. *Muhlenbergia eludens*, *Reederochloa eludens* – (2) Growing amongst the branches and leaves of shrubs thereby avoiding grazers. *Aristida eludens*

elwendiana L. *-ana*, indicating connection. From Elwend or Alwand, a mountain in south-west Iran

elygantha Gk *elyge*, darkness; *anthos*, flower. Anthoecia dark-colored

Elyhordeum Hybrids between species of *Elymus* and *Hordeum*

Elyleymus Hybrids between species of *Elymus* and *Leymus*

Elymandra Gk *elymos*, sheath; *aner*, man. The ultimate units of the inflorescence comprise a single terminal hermaphrodite spikelet with several male florets below

Elymohordeum Hybrids between species of *Elymus* and *Hordeum*

elymoides Gk *-oides*, resembling. Resembling *Elymus* with respect to the inflorescence

Elymopyrum Hybrids between species of *Elymus* and *Agropyron*

Elymordeum Hybrids between species of *Elymus* and *Hordeum*

Elymostachys Hybrids between species of *Elymus* and *Psathyrostachys*

Elymotrigia Hybrids between species of *Elymus* and *Elytrigia*

Elymotritichum See *Elymotriticum*

Elymotriticum Hybrids between species of *Elymus* and *Triticum*

Elymus Gk *elymos*, an ancient Greek name for an unidentified cereal

Elynorus See *Elionurus*

Elyonurus See *Elionurus*

Elysitanion Hybrids between species of *Elymus* and *Sitanion*

elythrophyllus Gk *elythron*, cover; *phyllon*, leaf. Fertile floret sessile and protected by leaf-like reduced stalk

Elytrigia Gk *eletryon*, cover; *tryge*, a crop of corn. Some species are weedy and grow, that is, seek cover in corn crops

elytrigioides Gk *-oides*, resembling. Similar to *Elytrigia*

elytroblephar-a, -um Gk *elytron*, cover; *blepharis*, eye-lash. The nerves of the upper glume and sterile lemma often bear long hairs

elytrochaet-a, -um Gk *elytron*, cover; *chaete*, bristle. Lower glume awned

elytrophoroides Gk *-oides*, resembling. Inflorescence an interrupted spicate-panicle as with *Elytrophorus*

Elytrophorum, Elytrophorus Gk *elytron*, cover; *phero*, bear. Lower glume of outer fascicle of spikelets enlarged

Elytrordeum Hybrids between species of *Elytrigia* and *Hordeum*

Elytrostachys Gk *elytron*, cover; *stachys*, spike as of an ear of wheat. Bracts resembling the wing-covers (elytra) of beetles cover the main rhachis thereby concealing the real structure of the inflorescence from the casual observer

emaculata L. *e-*, without; *macula*, spot; *-ata*, possessing. Culm-sheaths without spots

emarginat-a, -um L. *emargino*, deprive of its edge. Lemma apex notched

embergeri In honor of Louis Emberger (1897–1969) French botanist

emeiensis L. *-ensis*, denoting origin. From Emei Shan, Sichuan Province, China

emeinica L. *-ica*, belonging to. From Emei Shan, Sichuan Province, China

emergens L. *emergo*, emerge. Panicle strongly exserted

emersleyi In honor of J. D. Emersley, United States plant collector

emersus L. *emergo*, appear. Origin uncertain as name not accompanied by a description

emeryi In honor of Emery, English botanist who collected in Nepal

eminens L. *emineo*, stand out. – (1) Conspicuous with respect to inflorescence. *Agrostis eminens*, *Calamagrostis eminens*, *Festuca eminens*, *Stipa eminens*, *Stylagrostis eminens* – (2) conspicuous with respect to stature. *Digitaria eminens*, *Panicum eminens*, *Paspalum eminens*, *Poa eminens*

E

eminii In honor of Emin Pasha the adopted name of Eduard Sennritzer (1840–1892) German physician and traveller in service of the Egyptian Government

emirnensis L. *-ensis*, denoting origin. From Emern, Madagascar

emodensis L. *-ensis*, denoting origin. From Emodi Montes, Latin name of Himalayas

emodi See *emodensis*

emsonii In honor of H. E. Emson (1931–1937) who collected in Tanzania

encaustiomarginata L. *encaustus*, burned in; *margo*, border; *-ata*, possessing. Margins of leaf-blades white

Endallex Gk *endon*, within; *alloios*, of another sort; *hex*, six. Within the glumes there are six quite different structures

endlichii In honor of Rudolf Endlich (?–1915) who collected in Kilimanjaro region of East Africa

Endodia Gk *endon*, within; *dis*, double. Florets with two stamens

endomelas Gk *endon*, within; *melas*, black. Inner surface of glumes blackish at the base

engleri In honor of Heinrich Gustav Adolph Engler (1844–1930) German botanist

englishii In honor of Carl Schurz English (1904–1976) United States botanist and horticulturalist

Enneapogon Gk *ennea*, nine; *pogon*, beard. The lemma has nine hairy awns

enod-e, -is L. *e-*, without; *nodus*, knot. Internodes very short

Enodium L. *e-*, without; *nodus*, knot. Culm with a single node at its base and hence easily overlooked

ensifoli-a, -um L. *ensis*, sword; *folium*, leaf. Leaf-blades sword-like

ensiform-e, -is L. *ensis*, sword; *forma*, appearance. Leaf-blades sword-like

enslinii In honor of Aloysius Enslin who collected in North America

Enteropogon Gk *enteron*, bowel; *pogon*, beard. Lemma-awn long and twisted

entolaseum As for *Entolasia*

Entolasia Gk *entos*, within; *lasios*, hairy. Fertile lemma hairy

Entoplocamia Gk *entos*, within; *plokamis*, a curl of hair. The gynoecium has very long styles

entrerriensis L. *-ensis*, denoting origin. From Entre Réos Province, Argentina

enysii In honor of John Davis Gilbert Enys (1837–1912) English-born amateur botanist and long serving New Zealand magistrate

epacridifoli-a, -um L. *folium*, leaf. The leaf-blades resemble the leaves of certain species of Epacridaceae

epactia Gk *epaktios*, coastal. Latinized form of Greek word for the usual habitat of the species

epaleata L. *e-*, without; *palea*, scale; *-ata*, possessing. The lower floret lacks a palea

epauneroi See *pauneroi*

epectinata L. *e-*, without; *pecten*, comb; *-atus*, possessing. – (1) The lemma lacks long hairs present on the keel of a closely related species. *Loxodera epectinata* – (2) differs from a related species by the absence of teeth on the lower glume margins. *Lasiurus epectinatus*

Ephebopogon Gk *ephebos*, arrived at puberty; *pogon*, beard. Possibly a reference to white pedicels of the stalked spikelets

ephemeroblepharis Gk *ephemeros*, short-lived; *blepharis*, eye-lash. At the base of the upper floret there are two membranous lateral wings which may shrivel at maturity

ephemeroides L. *-oides*, resembling. Similar to *Panicum ephemerum*

ephemerum Gk *ephemeros*, short lived. Completing life-cycle in a few weeks

Epicampes Gk *epikampe*, bend. Lemma bears a short, curved awn

epigeios Gk *epi*, on; *ge*, earth. Growing on land whereas related species grow in swamps

epigejos See *epigeios*

epil-e, -is L. *e-*, lacking; *pilus*, hair. Glabrous in some respect

epileuca Gk *epi*, on; *leukos*, white. Upper surface of leaf-blade glaucous-white

epilifolium L. *e-*, without; *pilus*, hair; *folium*, leaf. Lacking cilia on the margins of the leaf-blades

epilis See *epile*
episetosus Gk *epi*, on; L. *seta*, bristle; *-osus*, abundance. Upper surface of leaf-blade densely hairy
epitrich-a, -us Gk *epi*, on; *thrix*, hair. Upper leaf-surface hairy
epitrichoides Gk *-oides*, resembling. Similar to *Sasa epitricha*
epitrichus See *epitricha*
equilaterale L. *aequus*, equal; *laterus*, side; *-ale*, pertaining to. Lower glume triangular in outline
equinum L. pertaining to horses. From the grasslands of Bahia and Paraguay
equitans L. *equito*, ride. Leaves distinctly equitant (the lower leaf when turned horizontally seeming to ride upon the one above it)
eradii In honor of N. A. Erady (fl. 1953) Indian botanist
Eragrostiella L. *-ella*, diminutive but here used as a name-forming suffix. Similar to *Eragrostis*
eragrostiformis L. *forma*, appearance. Resembling *Eragrostis*
Eragrostis Several meanings have been proposed, of which three follow. – (1) Gk *eros*, loving, together with *Agrostis* the Greek name of an indeterminate herb. – (2) Gk *er*, early. *agrostris*, wild. Species of *Eragrostis* are commonly early invaders of arable land. – (3) Gk *eri*, an inseparable particle used as a prefix to strengthen a word in the sense of very much, that is a many-floreted *Agrostis*
eragrostis Resembling *Eragrostis*
eragrostoides Gk *-oides*, resembling. Similar to *Eragrostis* with respect to spikelet structure
erect-a, -um, -us L. held upright. – (1) Spikelets or inflorescence branches held erect. *Brachyelytrum erectum, Briza erecta, Bromus erectus, Cathestecum erectum, Cynodon erectus, Ehrharta erecta, Glyceria erecta, Panicularia erecta, Paspalum erectum, Stipa erecta* – (2) with stems erect from a rhizome. *Setosa erecta* – (3) with blades of culm-leaves erect. *Dinochloa erecta*
erectiflora L. *erectus*, tending towards being erect; *flos*, flower. Panicle branches adpressed to central axis
erectifoli-a, -um L. *erectus*, tending towards being erect; *folium*, leaf. Leaf-blades held erect
erect-um, -us See *erecta*
eremaeum Gk *eremos*, desert; *-eum*, belonging to. Desert species
Eremitis Gk *eremos*, solitary; *-itis*, close connection. The male florets have a single stamen and their spikelets form a single whorl below the solitary female spikelet
Eremium Gk *eremos*, desert. An allusion to the ability of the only known species to withstand drought
Eremocaulon Gk *eremos*, solitary; *kaulos*, stem. The culms are widely separated along the rhizome
Eremochlamys Gk *eremos*, solitary; *chlamys*, cloak. The spikelets usually have only one subtending glume, the lower being undeveloped
Eremochloa Gk *eremos*, solitary; *chloa*, grass. The inflorescence is a solitary spike
Eremochloe, Eremochloë See *Eremochloa* but a different genus
eremophil-a, -us Gk *eremos*, desert; *phileo*, love. Desert species
Eremopoa Gk *eremos*, desert; *poa*, grass. From the deserts of Central Asia
Eremopogon Gk *eremos*, solitary; *pogon*, beard. Spikelets with a single conspicuous awn
eremopyroides Gk *-oides*, indicating resemblance. Inflorescences resemble those of *Eremopyrum*
Eremopyrum Gk *eremos*, desert; *pyros*, wheat. Desert grasses related to wheat
Eriachne Gk *erion*, wool; *achne*, scale. Lemma bears long hairs
eriachne As for *Eriachne*
erianth-a, -um Gk *erion*, wool; *anthos*, flower. With woolly glumes, lemmas or awns
Erianthecium Gk *erion*, wool; *anthos*, flower; *oikos*, house. Both the paleas and lemmas are hairy

erianthoides Gk *-oides*, resembling. – (1) Similar to *Erianthus* with respect to the inflorescence. *Bothriochloa erianthoides* – (2) similar to *Paspalum erianthum*. *Paspalum erianthoides*

erianthum See *eriantha*

Erianthus Gk *erion*, wool; *anthos*, flower. The subtending glumes are covered with long hairs

ericetorum L. *-etum*, place of growth. Growing amongst *Erica*

erigens L. *erigo*, lift up. Meaning obscure

erinacea L. hedgehog. Plants caespitose with rigid, pungent foliage

eriobasis Gk *erion*, wool; *basis*, bottom. Lower leaf-sheaths densely hairy

Erioblastus Gk *erion*, wool; *blastos*, shoot. Panicle branches and leaf-blades shortly pilose

eriocaulis Gk *erion*, wool; *kaulos*, stem. Culms hairy

Eriochaeta Gk *erion*, wool; *chaete*, bristle. The bristles subtending the spikelet clusters are densely hairy

Eriochloa Gk *erion*, wool; *chloa*, grass. Spikelets woolly

eriochryseoides Gk *-oides*, resembling. Similar to *Eriochrysis*

Eriochrysis Gk *erion*, wool; *chrysos*, gold. The spikelets are invested with golden-yellow hairs

eriocoleus Gk *erion*, wool; *koleos*, sheath. Leaf-sheath softly hairy

Eriocoma Gk *erion*, wool; *kome*, head of hair. The lemma carries a fastigiate tuft of silky hair

eriogon-a, -um Gk *erion*, wool; *gony*, knee. Nodes hairy

eriolepis Gk *erion*, wool; *lepis*, scale. Glumes, lemmas or paleas densely hairy

Eriolytrum Gk *erion*, wool; *elytron*, cover. Glumes densely woolly

Erioneuron Gk *erion*, wool; *neuron*, nerve. Palea keels ciliate

eriophorum Gk *erion*, wool; *phero*, bear. Glumes and sterile lemma densely hairy

eriophylla Gk *erion*, wool; *phyllon*, leaf. Leaves densely woolly

eriopod-a, -um Gk *erion*, wool; *pous*, foot. Basal leaf-sheaths woolly

Eriopodium Gk *erion*, wool; *pous*, foot. The raceme joints are densely hairy

eriopodum See *eriopoda*

eriostachy-a, -um, -us Gk *erion*, wool; *stachys*, spike as of an ear of wheat. Inflorescence branches or spikelets densely hairy

eriostoma Gk *erion*, wool; *stoma*, mouth. Orifice of leaf-sheath woolly-tomentose

eriwanense L. *-ense*, denoting origin. From Eriwan, but origin uncertain, not given by the author

Erochloe, Erochloë Gk *eros*, loving; *chloa*, grass. A transliteration of "love grass", the vernacular name of several *Eragrostis* species

erosa L. *erodo*, grow into. Apices of glumes irregularly toothed

erosiglumis L. *erodo*, grow into; *gluma*, husk. Glume apices irregularly toothed

Erosion Gk *-ion*, diminutive here employed as a name-forming suffix. In honor of Eros, the Greek God of Love

errabundum L. wandering. Culms decumbent and spreading

erratica L. wandering to and fro. Weedy and easily dispersed

erroneus L. *erro*, wanderer; *-eus*, pertaining to. Origin of the name uncertain as not given by author but may refer to the species having a creeping habit

erubescens L. *erubesco*, blush. Inflorescences pinkish

erucaeformis See *eruciforme*

Erucaria L. *eruca*, caterpillar; *-aria*, pertaining to. The inflorescence resembles a caterpillar

eruciferum L. *eruca*, caterpillar; *fero*, carry or bear. The inflorescence resembles a caterpillar

eruciform-e, -is L. *eruca*, caterpillar; *forma*, shape. Racemes bear a fanciful resemblance to caterpillars

erythrae-a, -um Gk *erythros*, red; *-ea*, pertaining to. Inflorescences reddish

Erythranthera Gk *erythros*, red; *antheros*, blooming. The inflorescence is red

erythrocarpon Gk *erythros*, red; *karpos*, fruit. Anthoecia red

erythrochaetum Gk *erythros*, red; *chaete*, bristle. Margins of glumes bear red hairs

erythrogona Gk *erythros*, red; *gony*, knee. Nodes reddish-brown

erythrolepis Gk *erythros*, red; *lepis*, scale. Glumes of staminate florets red

erythropoda Gk *erythros*, red; *pous*, foot. Sheath of lower leaves red

erythrosperm-a, -um Gk *erythros*, red; *sperma*, seed. Anthoecia red

erythrostachya Gk *erythros*, red; *stachys*, spike as of an ear of wheat. Panicle purple-red

esculent-a, -um L. edible. Grain edible

esenbeckii In honor of Christian Gottfried Daniel Nees von Esenbeck (1776–1858) German botanist

eskia Vernacular name of the species in Spain

euadenia Gk *eu-*, well; *aden*, gland. Margins of leaf-blades fringed with long-stalked glands

eucampta Gk *eu-*, well; *campto*, bend. Leaf-blades twisted

euchaetus Gk *eu-*, well; *chaete*, bristle. The lower glume of the sessile spikelet has a long bifid awn and that of the stalked spikelet a long single awn

Euchlaena, Euchlaene Gk *eu-*, well; *chlaena*, cloak. The female spikelets are ensheathed

Euchlaezea Hybrids between species of *Euchlaena* and *Zea*

euchlamydeum Gk *eu-*, well; *chlamys*, cloak; L. *-eum*, pertaining to. Glumes well developed

Euclasta Gk *eu-*, well; *klastos*, broken in pieces. At maturity, the raceme freely falls apart as segments

euclaston Gk *eu-*, well; *klastos*, broken in pieces. The spike breaks up readily at maturity

eucnemis Gk *eu-*, well; *knemis*, leg armour. Well armoured, lower glume chartaceous with rigid hairs

eucom-um, -us Gk *eu-*, well; *kome*, head of hair. – (1) Raceme joints silky with soft, long hairs. *Andropogon eucomus* – (2) glumes and sterile lemmas invested in long hairs. *Paspalum eucomum*

Eudonax Gk *eu-*, good. Proposed as an alternative name for *Donax* if it were reserved for a genus of molluscs

Eufournia Gk *eu-*, a name-forming prefix. See *Fourniera* which is invalid being a later homonym

eugenii In honor of Eugene B. Alexeev (1946–1987) Russian botanist who specialized on the taxonomy of *Festuca*

Euklastaxon Gk *eu*, well; *klaston*, broken in pieces; *axon*, axle. Inflorescence readily fragmenting into small segments

Eulalia In honor of Eulale Delile (fl. 1825–1840) French artist and illustrator of Kunth's Révision des Graminées

eulalioides Gk *-oides*, resembling. Resembling *Eulalia*

Eulaliopsis Gk *opsis*, resemblance. Resembling *Eulalia*

Eupogon Gk *eu-*, well; *pogon*, beard. Similar to *Andropogon* but differing in that the sterile spikelets are sessile

euprepes Gk *eu-*, well; *prepo*, catch the eye. In contrast to *Andropogon* lacks pedicellate sterile spikelets

Euraphis Gk *eu-*, well; *rhaphis*, needle. The upper florets are reduced to awns

europae-um, -us L. *-eum*, belonging to. From Europa now Europe

eurylemma Gk *eurys*, broad; lemma, scale. Lemma of lower spikelet broad

euryphyllum Gk *eurys*, wide; *phyllon*, leaf. Leaf-blades broad-lanceolate

Eustachys Gk *eu-*, well; *stachys*, spike as of an ear of wheat. Inflorescence of one sided racemes bearing two rows of spikelets on very short pedicels

Euthryptochloa Gk *eu-*, well; *thrypto*, break in pieces; *chloa*, grass. The spikelets fall entire

Eutriana Gk *eu-*, well; *triaena*, trident. Lemma of terminal sterile floret three-awned

eutuldoides Gk *eu-*, well. Closely resembles *Bambusa tuldoides*

euxina From about the Black Sea, know to the Romans as *Mare Euxinum*

evagans L. *evago*, wander around. Rhizomatous in habit

evenkiensis L. *-ensis*, denoting origin. From Evenkia, Siberia

evolut-a, -um L. *evolvo*, unroll. Leaf-blades short

evrardii In honor of Charles Marie Evrard (1926–) Belgian botanist

ewartian-a, -um, -us L. *-ana*, indicating connection. In honor of Albert James Ewart (1872–1937) English-born Australian botanist

Exagrostis Gk *ex*, outside. The spikelets resemble those of *Agrostis* but differ in possessing several florets

exaltat-a, -um, -us L. lofty. Culms tall

exarat-a, -um, -us L. *exaro*, plough up. – (1) With furrows between the nerves of the glumes. *Andropogon exaratus, Mesosetum exaratum* – (2) of the sterile lemma. *Panicum exaratum, Paspalum exaratum* – (3) of the lemmas. *Agrostis exarata, Phleum exarata*

exaristat-a, -us L. *ex*, without; *arista*, bristle; *-ata*, possessing. Lemmas awnless or almost so

exarmata L. *ex*, without; *arma*, instrument of war; *-ata*, possessing. Lemma unawned

exasperat-a, -um, -us L. rough. – (1) Glumes scabrous. *Agrostis exasperata* – (2) sterile lemmas scabrous. *Digitaria exasperata* – (3) leaf-blade margins scabrous. *Panicum exasperatum, Paspalum exasperatum* – (4) inflorescence branches scabrid. *Eragrostis exasperata* – (5) leaf-blade scabrid. *Chusquea exasperata*

exaurita L. *ex*, without; *aurita*, eared. Leaf-blade without a pair of basal lobes

excavat-um, -us L. hollowed out. – (1) Apices of raceme joints cup-shaped. *Andropogon excavatus, Cymbopogon excavatus* – (2) spikelets sunk in the rhachis. *Axonopus excavatus, Paspalum excavatum* – (3) fertile lemma-base concave. *Panicum excavatum*

excels-a, -um, -us L. tall. – (1) Culms tall. *Arundinaria excelsa, Elymus excelsus, Eragrostis excelsa* – (2) culms high climbing. *Arthrostylidium excelsum*

excurrens L. *excurro*, run out. Lemma of upper floret mucronate

exelliana L. *-ana*, indicating connection. In honor of Arthur Wallis Exell (1901–1993). English botanist

exigu-a, -um, -us L. wanting in size and number. – (1) Spikelets or inflorescence branches few. *Agrostis exigua, Oryzopsis exigua, Panicum exiguum, Pennisetum exiguum, Tripogon exiguus* – (2) Spikelets few-flowered. *Eragrostis exigua*

exiguiflorum L. *exiguus*, wanting in size and number; *flos*, flower. Panicle weakly developed

exigu-um, -us See *exigua*

exil-e, -is L. slender. Culms slender

eximi-a, -us L. exceptional. Readily distinguished from related species

Exotheca, -um, -us Gk *exo*, outside; *theke*, box. The homogamous spikelets form an involucre below the heterogamous triad

expans-a, -um L. *expando*, expand. – (1) Inflorescence an open panicle. *Agrostis expansa, Dissanthelium expansum, Eragrostis expansa, Melica expansa, Muhlenbergia expansa, Panicum expansum, Pentaschistis expansa* – (2) an orthographic error for *inexpansa*. *Calamagrostis expansa*

explicata L. *ex*, without; *plico*, fold up. Leaf-blades flat

exsaniosa L. *ex*, without; *saniosa*, bloody matter. Origin uncertain, not given by author

exsert-a, -um, -us L. exserted. – (1) Rhachilla longer than in related species. *Agrostis exserta, Cymbopogon exsertus* – (2) peduncle longer than in related species. *Aristida exserta* – (3) lateral awns of the glumes are exserted. *Pentaschistis exserta*

exstans L. *exsto*, stand out or project. The lower glume bears distinctive apical keels with protruding stump-like thickenings

extensa L. widespread. Culms widely separated on the rhizome

extenuatum L. *extenuo*, make thin. Inflorescence branches narrow compared with those of related species

extremioriental-e, -is L. *extremus*, extreme; *oriens*, the east; *-alis*, pertaining to. From the Far East, that is Eastern Asia

Exydra Gk *exo*, outside; *hydor*, water. Growing around the margins of pools

eylesii In honor of Frederick Eyles (1864–1937) English-born botanist in Zimbabwe

F

faberi In honor of Ernst Faber (1839–1899) English cleric who collected in China

fabrei In honor of G. Fabre (fl. 1980) French botanist

fabrisii In honor of Umberto Antonio Fabris (1924–1976) Argentinian botanist

factorovskyi In honor of Eliezer Faktorovsky (1897–1926) Russian-born Palestine botanist

fagetorum L. *-etum*, place of growth. Growing in beech (*Fagus*) woods

falcat-a, -um L. *falx*, sickle; *-ata*, possessing. – (1) Inflorescence branches curved. *Arundinaria falcata, Chloris falcata, Dimeria falcata, Drepanostachyum falcatum, Ischaemum falcatum, Leptochloa falcata, Panicum falcatum, Paspalum falcatum, Pogonarthria falcata, Stipa falcata* – (2) spikelets curved. *Chusquea falcata, Eragrostis falcata*

falcatiaurita L. *falx*, sickle; *aurita*, long-eared. Auricles sickle-shaped

falcatum See *falcata*

falcifer-a, -um L. *falx*, sickle; *fero*, carry or bear. Racemes sickle-shaped

falcifolia L. *falx*, sickle; *folium*, leaf. Leaf-blade sickle-shaped

falcipes L. *falx*, sickle; *pes*, foot. Beneath each triad of spikelets the subtending axis is recurved so the whole bears a fanciful resemblance to a baited hook

falcis L. *falx*, sickle. Of sickles, that is, the leaf-blades are often sickle-shaped

falconeri In honor of Hugh Falconer (1808–1865) Scots-born Indian physician and botanist

falcula L. *falx*, sickle; *-ula*, diminutive. Racemes sub-falcate

Falimiria In honor of Stephanek Falimierz, 16th century Polish naturalist

falklandica L. *-ica*, belonging to. From Falkland Islands, United Kingdom Territory in the South Atlantic

fallax L. deceptive. Closely resembling another species

fallens L. *fallo*, to escape notice. A replacement name for a species which if transferred to its correct genus would generate a homonym

Falona Gk *phalos*, ridge on a Greek helmet. The subtending glumes are strongly ridged

falsum L. *falsus*, false. Resembling species of another genus

falx L. sickle. Inflorescence a curved spike

famatinensis L. *-ensis*, denoting origin. From Sierra Famatina, Rioja Province, Argentina

familiar-e, -is L. *familia*, family; *-aris*, pertaining to. Of wide-spread distribution

fangiana L. *-ana*, indicating connection. In honor of Fang Wen-pei (1899–1983) Chinese botanist

fansipanensis L. *-ensis*, denoting origin. From Mt Fansipan, Vietnam

farct-a, -um, -us L. solid with centre softer than outside. Culms solid except in the very centre

farcticaulis L. *farctus*, solid with centre softer than outside; *caulis*, stem. Culms solid

farct-um, -us See *farcta*

Fargesia In honor of Paul Guillaume Farges (1844–1912) French cleric and amateur botanist who collected in China

fargesii As for *Fargesia*

farinacea L. *farina*, flour; *-acea*, resembling. Culms white with wax

farinos-a, -us L. *farina*, flour; *-osa*, abundance. Surface of culms very waxy

Farrago L. hotchpotch. The genus has an unusual combination of characters

farrum L. *far*, spelt. Vernacular name for the species in Classical Rome, maintained as farro in contemporary Italian

Fartis Origin uncertain, not given by author

fascicular-e, -is L. *fascis*, bundle; *-ula*, diminutive; *-aris*, pertaining to. – (1) Spikelets or racemes clustered in the inflorescence. *Agrostis fascicularis, Andropogon fascicularis, Bromus fascicularis, Digitaria fascicularis, Diplachne fascicularis, Festuca fascicularis, Leptochloa fascicularis, Pseudosorghum fasciculare* – (2) with clusters of hairs on the lemma. *Chaetobromus fascicularis*

F

fasciculat-a, -um, -us L. *fascis*, bundle; *-ulus*, diminutive. *-ata*, possessing. With spikelets or branches clustered in the inflorescence

fasciculiflorum L. *fascis*, bundle; *-ula*, diminutive; *flos*, flower. Inflorescence branches in erect, digitate fascicles

fasciculiforme L. *fascis*, bundle; *-ula*, diminutive; *forma*, appearance. Inflorescences arising in clusters from upper leaf-nodes

Fasciculochloa L. *fasces*, bundle; *-ulus*, diminutive; Gk *chloa*, grass. The spikelets are arranged in bundles on the inflorescence branches

fascinata L. *fascino*, bewitch. Sufficiently different from related species to draw attention to itself

fastigiat-a, -um, -us L. *fastigium*, gable or vertex; *-ata*, possessing. Panicle branches or culms held erect rather than diverging

fastuos-a, -um L. *fastus*, proud; *-osa*, abundance. Habit more luxuriant than that of related species

fatmens-e, -is L. *-ense*, denoting origin. From Wadi Fatmima in Arabia

fatua L. tasteless. Grain not favored as food

faucicola L. *fauces*, abyss; *-cola*, dweller. Growing on the edge of a volcanic crater

fauriei In honor of Urbain Jean Faurie (1847–1914) French cleric and amateur botanist

fawcettiae In honor of Stella Grace Maisie Fawcett (1912–1988) Australian botanist

fax L. a torch. With an inflorescence bearing a fanciful resemblance to a torch with ascending flames

faxonii In honor of Charles Edward Faxon (1846–1918) United States botanist

fazoglensis L. *-ensis*, denoting origin. From Fazogl, Sudan

featherstonei In honor of William Featherstone (fl. 1922) United States botanist who collected in Peru

fecund-a, -um L. fruitful. – (1) Producing abundant spikelets. *Dichanthium fecundum* – (2) producing abundant culms. *Bambusa fecunda*

fedtschenkoi In honor of Boris Alexsevitch Fedtschenko (1872–1947) German-born Russian botanist

feekesiana L. *-ana*, indicating connection. In honor of Feekes

feensis L. *-ensis*, denoting origin. From Santa Fe, Mexico

feildingianum L. *-anum*, indicating connection. In honor of I. B. Feilding (fl. 1892–1893) who collected in Malaysia

felix L. fruit-bearing. Known in fruit

felliana L. *-ana*, indicating connection. In honor of David Gregory Fell (1962–) Australian botanist

fenas Vernacular name for the species in Spain

Fendleria In honor of August Fendler (1813–1883) German-born United States botanical collector

fendleriana L. *-ana*, indicating connection. As for *Fendleria*

fenestrat-a, -um L. *fenestra*, window; *-ata*, possessing. – (1) The lower lemma has two hyaline patches at the base. *Sacciolepis fenestrata* – (2) fertile lemma with rectangular raised ornamentation. *Digitaria fenestrata*, *Panicum fenestratum*

fenixii In honor of Eugenio Fenix (1883–1939) Philippine botanist

fenshamii In honor of Roderick John Fensham (1960–) Australian ecologist

fenzliana L. *-ana*, indicating connection. In honor of Eduard Fenzl (1808–1879) Austrian botanist

fera L. uncultivated. Not cultivated

feratiana L. *-ana*, indicating connection. In honor of Férat (fl. 1818) who collected in Pyrenees

ferax L. fruitful. Readily setting grain

feresetacea L. *fere*, nearly. Similar to *Stipa setacea*

ferganens-e, -is L. *-ense*, denoting origin. From Fergana Range, Kyrgyzstan

fergusonii In honor of William Ferguson (1820–1887) plant collector in Sri Lanka

ferioliana L. *-ana*, indicating connection. In honor of Feriol

fernaldiana L. *-ana*, indicating connection. As for *fernaldii*

fernaldii In honor of Merritt Lyndon Fernald (1873–1950) United States botanist

fernandesii In honor of Rosette Mercedes Saraiva Batarda Fernandes (1916–) Spanish botanist

fernandezian-a, -us L. *-ana*, indicating connection. From the Juan Fernández Archipelago, Chile

fernandopoanum L. *-anum*, indicating connection. From Fernando Po, now Bioko, Equatorial Guinea

ferreyrae In honor of Ramón Alejandro Ferreyra (1910–2005) Peruvian botanist

ferrilateris L. *ferreus*, iron; *laterus*, side. In honor of John Richard Ironside Wood (1944–) English plant collector and botanist

Ferrocalamus L. *ferreus*, iron; *calamus*, reed. The culms are solid at the base

ferronii In honor of Henri de Ferron

ferrugine-a, -um, -us L. light-brown. Glumes and/or lemmas invested in light-brown hairs

fertilis L. fruitful. Producing abundant grain

ferventicola L. *ferveo*, boil; *-cola*, dweller. Growing in warm soil close to boiling springs

fessum L. exhausted. Leaf-blades inrolled as if plant had been subjected to drought

festivus L. pretty. Plant of attractive appearance

Festuca The name of a weed in Pliny and the Latin term for a stem or straw. Based on the Celtic *fest*, pasture or food

festucace-a, -um, -us L. *-acea*, resembling. Resembling *Festuca* in some respect, usually the habit or inflorescence

festucaeformis, festuciformis L. *forma*, appearance. Resembling *Festuca* in habit or inflorescence

Festucaria L. *-aria*, pertaining to. Resembling *Festuca*

Festucella L. *-ella*, diminutive. Here a suffix to form a generic name for a group of species previously included in *Festuca*

festuciformis See *festucaeformis*

festucoides Gk *-oides*, resembling. Resembling *Festuca*, especially with respect to the inflorescence

Festucopsis Gk *opsis*, resemblance. Similar to *Festuca*

Festulolium Hybrids between species of *Festuca* and *Lolium*

Festulpia Hybrids between species of *Festuca* and *Vulpia*

Fibichia In honor of Johann Fiebig (?–1792) German botanist

fibrata L. *fibra*, fibre; *-ata*, possessing. Lower leaf-sheaths fibrous

fibrifera L. *fibra*, fibre; *fero*, carry or bear. Leaf-sheaths fibrous at the base

fibros-a, -um, -us L. *fibra*, fibre; *-osa*, abundance. Leaf-sheaths disintegrating at length into copious fibres

fibrovaginata L. *fibra*, thread; *vagina*, sheath; *-ata*, possessing. Leaf-sheath fibrous

fiebrigii In honor of Karl Fiebrig Gertz (1869–1951) German-born South American botanist

fieldingii In honor of Henry Barron Fielding (1805–1851) English botanist

figarei See *figarii*

figarian-a, -us L. *-ana*, denoting connection. As for *figarii*

figarii In honor of Antonio bey Figari (1804–1870) Italian physician and naturalist

figertii In honor of Ernst Figert (1848–1925)

figueirae In honor of Figueira who collected in Uruguay

fiherenensis L. *-ensis*, denoting origin. From Fiherenana, Madagascar

filabrensis L. *-ensis*, denoting origin. From Sierra de Filabres, Spain

filamentosum L. *filamentus*, filament; *-osa*, abundance. Pedicels long and thin

Filgueirasia In honor of Tarisco S. Filgueiras (1950–) Brazilian botanist

filgueirasii As for *Filgueirasia*

filicaul-e, -is L. *filum*, thread; *caulis*, stem. Culms slender

filiculm-e, -is L. *filum*, thread; *culmus*, stalk. Culms slender

filifera L. *filum*, thread; *fero*, carry or bear. Blades of basal leaves very long and narrow

filifoli-a, -um, -us L. *filum*, thread; *folium*, leaf. Leaf-blades very narrow

F

filiform-e, -is L. *filum*, thread; *forma*, shape. – (1) Leaf-blades narrow. *Agrostis filiformis, Arundinella filiformis, Atheropogon filiformis, Bouteloua filiformis, Ehrharta filiformis, Eragrostis filiformis, Festuca filiformis, Gymnopogon filiformis, Lachnagrostis filiformis, Leptosaccharum filiforme, Milium filiforme, Parapholis filiformis, Psilostachys filiformis, Reynaudia filiformis, Tripogon filiformis* – (2) pedicels or peduncles thread-like. *Avena filiformis, Olyra filiformis* – (3) culms thin. *Andropogon filiformis, Saccharum filiforme* – (4) racemes thin. *Aira filiformis, Digitaria filiformis, Leptochloa filiformis, Panicum filiforme*

Filipedium L. *filum*, thread; *pes*, foot. Inflorescence branches thread-like

filipendul-a, -us L. *filum*, thread; *pendo*, hang down; *-ula*, indicating tendency. Spikelet borne on slender pedicels or peduncles

filipendulinus L. *filum*, thread; *pendula*, pendulous; *-inus*, indicating resemblance. Racemes borne on slender peduncles

filipendulus See *filipendula*

filipes L. *filum*, thread; *pes*, foot. – (1) Pedicels of spikelets slender. *Agrostis filipes, Deyeuxia filipes, Panicum filipes* – (2) peduncles of racemes slender. *Vetiveria filipes*

filiramum L. *filum*, thread; *ramus*, branch. Culms very slender

filostachyum L. *filum*, thread; Gk *stachys*, spike as of an ear of wheat. Inflorescence branches with spikelets only at the base

fimbriat-a, -um, -us L. *fimbriae*, fringe; *-ata*, possessing. – (1) With fringed glumes or lemmas. *Arundinaria fimbriata, Chimonocalamus fimbriatus, Chusquea fimbriata, Coridochloa fimbriata, Digitaria fimbriata, Eulalia fimbriata, Panicum fimbriatum, Paspalum fimbriatum, Piptochaetium fimbriatum, Sporobolus fimbriatus, Stipa fimbriata, Syntherisma fimbriatum* – (2) with fringed leaf-blades. *Pollinia fimbriata* – (3) with bristles at orifice of leaf-sheath. *Arthrostylidium fimbriatum* – (4) with ligule fringed. *Festuca fimbriata, Himalayacalamus fimbriatus, Melocalamus fimbriatus*

Fimbribambusa L. *fimbriae*, fringe and *Bambusa*. Origin of name uncertain, possible referring to crested nodes

fimbriligula L. *fimbriae*, fringe; *ligula*, small tongue. Ligule a fringe of hairs

fimbriligulata L. *fimbriae*, fringe; *ligula*, small tongue; *-ata*, possessing. Ligule margin with long hairs

fimbrillata L. *fimbriae*, thread; *-illum*, diminutive; *-ata*, indicating likeness. Lemma bears slender hairs

fimbrinodum L. *fimbriae*, fringe; *nodum*, knot. Nodes bearing a skirt of reflexed hairs

Fingerhuthia In honor of Karl Anton Finger-huth (1798–1876) German physician and amateur botanist

finitim-a, -us L. neighboring. Readily confused with another species

fiorii In honor of Adriano Fiori (1865–1950) Italian botanist

Fiorinia In honor of Elisabetta Fiorini-Mazzanti (1799–1879) Italian botanist

firm-a, -um, -us L. *firm*, in the sense of opposite to frail. – (1) Culms stout. *Eragrostis firma, Panicum firmum* – (2) glumes cartilaginous. *Heteropogon firmus*

firmandus L. *firmo*, declare. Worthy of recognition

firmiculm-e, -is L. *firmus*, stout; *culmus*, stalk. Culms robust

firmior L. stouter. Culms stouter than those of related species

firmul-a, -um L. *firmus*, stout; *-ula*, diminutive. More robust in habit or spikelet size than related species

firm-um, -us See *firma*

fischeri In honor of – (1) Cecil Ernest Claude Fischer (1874–1950) Indian botanist born of Europaean parents. *Arundinaria fischeri, Dimeria fischeri* – (2) Henri Fischer, French professor. *Bromus fischeri* – (3) Alexander Fischer (fl. 1820s) British naval surgeon. *Dupontia fischeri, Graphephorum fischeri*

fischerianus L. *-anus*, indicating connection. As for *fischeri* (1)

fisheri In honor of Alexander Fisher (fl. 1820) naval surgeon who collected in the Arctic

fissa L. *fissum*, cleft. Lemma apex bifid

fissifoli-um, -us L. *fissum*, cleft; folium leaf. Apex of leaf-blade sometimes bifid

fissura L. a cleft made by splitting. Growing in rock fissures

fitzgeraldii In honor of William Vincent Fitzgerald (1867–1929) Western Australian forest botanist

flabellat-a, -um, -us L. *flabella*, fan; *-ata*, possessing. – (1) Spikelets or inflorescences fan-shaped. *Agrostis flabellata, Avenastrum flabellatum, Bromus flabellatum, Chloris flabellata, Panicum flabellatum, Parodiochloa flabellata, Tetrapogon flabellata* – (2) culms fan-shaped at the base. *Aristida flabellata, Muhlenbergia flabellata*

flabelliformis L. *flabella*, fan; *forma*, appearance. The crowded equitant basal leaves resemble a fan

flaccid-a, -um, -us L. unable to support its own weight. – (1) Inflorescence branches long and thin and so droop. *Agrostis flaccida, Andropogon flaccidus, Arberella flaccida, Aristida flaccida, Deyeuxia flaccida, Digitaria flaccida, Eragrostis flaccida, Olyra flaccida, Pennisetum flaccidum* – (2) leaf-blades drooping. *Festuca flaccida*

flaccidula L. *flaccidus*, unable to support its own weight; *-ula*, diminutive. Inflorescence branches drooping

flaccid-um, -us See *flaccida*

flacciflorum L. *flaccidus*, unable to support its own weight; *flos*, flower. Panicle long exserted, branches pendulous

flaccifolia L. *flaccidus*, unable to support its own weight; *folium*, leaf. Leaf-blades pendulous

flacourtii In honor of Etienne de Flacourt (1607–1660) French colonial administrator and linguist

flagellifer, -a L. *flagellum*, whip; *fero*, carry or bear. Tip of leaf-blade thread-like

flamignii In honor of Agosto Flamigni (1907–1934) who collected in Zaire

flammida L. *flammo*, blaze; *-ida*, becoming. Panicle large and yellow

flav-a, -um L. yellow. – (1) Spikelets straw-colored. *Agrostis flava, Chaetochloa flava, Melica flava, Panicum flavum, Paspalum flavum, Poa flava* – (2) bristles subtending spikelets yellow. *Setaria flava*

flavens L. *flaveo*, be yellow. Spikelets yellow-brown

flavescens L. *flavesco*, become yellow. – (1) Foliage yellowish. *Agrostis flavescens, Avena flavescens, Bromus flavescens, Chionochloa flavescens, Danthonia flavescens, Enneapogon flavescens, Eragrostis flavescens, Erianthus flavescens, Panicum flavescens, Pappophorum flavescens, Polypogon flavescens* – (2) spikelets yellowish. *Stipa flavescens, Trisetum flavescens* – (3) involucral bristles yellowish. *Pennisetum flavescens*

Flavia L. *flavus*, yellow. Spikelets yellow-green

flavicans L. *flaveo*, be yellow; *-icans*, becoming. Plants overall with yellow to greenish-yellow foliage

flavicomum L. *flavus*, yellow; *coma*, head of hair. Inflorescence yellow

flavid-a, -um L. *flavidus*, pale yellow. Spikelets yellow

flavidodula L. *flavidus*, pale yellow; *-ula*, diminutive. Spikelets yellow

flavidula L. *flavidus*, pale yellow; *-ula*, diminutive. Spikelets yellow

flavidum See *flavida*

flavovirens L. *flavus*, yellow; *virens*, green. Spring foliage pale yellow-green

flavum See *flava*

fleckii In honor of Eduard Fleck (fl. 1890) German geologist and plant collector in South Africa

fleuryi In honor of François Fleury (fl. 1948) French collector in tropical Africa

flex-a, -um, -us L. *flecto*, bend. – (1) Rhachis flexuose. *Brachypodium flexum, Yushania flexa* – (2) spike slightly bent. *Leymus flexus*

flexibarbata L. *flecto*, bend; *barba*, beard; *-ata*, possessing. Lemma with a hygroscopic awn

flexil-e, -is L. *flecto*, bend; *-ile*, property. Culm geniculate at base and slender

F

flexispica L. *flecto*, bend; *spica*, a point; hence, in particular, an ear or spike of grain. Inflorescence spike-like with a tendency to bend

Flexularia L. *flecto*, bend; *-ula*, diminutive; *-aria*, pertaining to. Awns and pedicels flexuose

flex-um, -us See *flexa*

flexuos-a, -um, -us L. *flecto*, bend; *-osa*, abundance. – (1) Inflorescence branches lax and drooping or bent in a zigzag fashion. *Andropogon flexuosus, Aristida flexuosa, Arundinaria flexuosa, Arundarbor flexuosa, Avenella flexuosa, Bambusa flexuosa, Cymbopogon flexuosus, Digitaria flexuosa, Eragrostis flexuosa, Erioblastus flexuosus, Imperata flexuosa, Phyllostachys flexuosa, Poa flexuosa, Roegneria flexuosa, Sorghastrum flexuosum, Sporobolus flexuosus* – (2) awn flexuous. *Aira flexuosa, Avena flexuosa, Deschampsia flexuosa, Muhlenbergia flexuosa, Stipa flexuosa* – (3) stolons arching. *Axonopus flexuosus* – (4) culms weak. *Uniola flexuosa*

flexuosissimum L. *flecto*, bend; *-osa*, abundance; *-issima*, most. Inflorescence spike-like and very flexible

flexuos-um, -us See *flexuosa*

flocciculmis L. *floccus*, lock of wool; *culmus*, stem. Leaf-sheaths woolly

floccifoli-a, -us L. *floccus*, lock of wool; *folium*, leaf. Leaf-margins bear tufts of hairs

floccos-a, -us L. *floccus*, lock of wool; *-osus*, abundance. – (1) Basal leaf-sheaths densely hairy. *Apocopis floccosa, Aristida floccosa* – (2) racemes densely villous. *Andropogon floccosus, Cymbopogon floccosus* – (3) leaf-blade adjacent to ligule densely villous. *Eragrostis floccosa*

flodmanii In honor of Julius Hjalmar Flodman (fl. 1859–1896) Swedish-born United States botanist

floresii In honor of Antonio Jijon Flores (1833–1915) Ecuadoran novelist and statesman

floribund-a, -um L. *floreo*, bloom; *-bunda*, indicating action. Inflorescence of many flowers

florid-a, -us L. *floreo*, bloom; *-idus*, becoming. Profusely flowering

floridan-a, -um, -us L. *-ana*, indicating connection. From Florida State, USA

floridulus L. *floridus*, profusely flowering; *-ulus*, diminutive. Inflorescence of abundant small florets

floridus See *florida*

florissanti From Florissant, Colorado, USA

florulenta L. *flos*, flower; *-ulenta*, indicating abundance. Panicle large with many spikelets

fluitans L. *fluito*, float. Leaves or rhizomes floating

fluminens-e, -is L. *flumen*, a river; *-ense*, denoting origin. Pertaining to Rio de Janeiro, Brazil

flumineum L. relating to a river. Growing near water

Fluminia L. *flumen*, flowing or flooding water. Growing in swampy places

fluviatile, fluviatilis L. *fluvius*, river; *-atilis*, place of growth. Growing along riverbanks

fluviicola L. *fluvius*, river; *-cola*, dweller. Growing along river banks

fockei In honor of Hendrik Charles Focke (1802–1858) who collected in Suriname

foena L. hay. The upper glumes bear piliferous glands which give off courmarin

Foenodorum L. *foenus*, hay; *odorus*, sweet smelling. Fragrant as of hay

foermerianum L. *-anum*, indicating connection. In honor of Rudolf Förmer (fl. 1900–1901) German botanist

foetid-um, -us L. evil smelling. Crushed foliage is strongly scented

foexiana L. *-ana*, indicating connection. In honor of Étienne Edmond Foëx (1876–1944) French plant pathologist or of Gustav Louis Emile Foëx (1844–1906) viticulturalist of Montepellier

foliacea L. *folium*, leaf; *-acea*, indicating resemblance. The racemes have a leaf-like winged rhachis

foliat-a, -us L. *folium*, leaf; *-ata*, possessing. Panicle with many leafy bracts

foliiforme L. *folium*, leaf; *forma*, appearance. Inflorescence branches winged

foliis-variegatis L. *folium*, leaf; *variegatis*, variegated. Leaf-blades variegated

folios-a, -um, -us L. *folium*, leaf; *-osa*, abundance. Culms more leafy than those of related species

fominii In honor of Aleksandr Vasilievich Fomin (1869–1935) Russian botanist

fonkii In honor of Fr. Fonk (fl. 1857–1858) who collected in Chile

fontanale L. *fontanus*, spring; *-ale*, pertaining to. From Steyermark Falls, on Río Tirica, Venezuela

fontanesianum L. *-anum*, indicating connection. As for *fontanesii*

fontanesii In honor of René Louiche Desfontaines (1750–1833) French botanist

fonticola L. *fons*, spring; *-cola*, dweller. Growing in the spray of waterfalls

fontismagni L. *fons*, spring; *magnus*, large. The latinized name of the type locality, Grootfontein, Namibia

font-queri In honor of Pes Font-Quer (1888–1964) Spanish botanist

fontqueriana L. *-ana*, indicating connection. As for *font-queri*

Forasaccus From the Italian vernacular name *forasacco*, which is given to several species of *Vulpia*, *Festuca*, *Bromus* and *Hordeum*

forbesian-a, -um L. *-ana*, indicating connection. In honor of John Forbes Royle (1799–1858) English physician in service of East India Company

fordeana L. *-ana*, indicating connection. In honor of Helena Forde (1830–1910) New South Wales plant collector

forficulata L. *forficula*, small scissors; *-ata*, possessing. Apex of lower glume resembling a pair of shears

formicarum L. *formica*, ant; *-arum*, belonging to. The twisting of the hygroscopic awns causes the dispersed floret to move across the ground with irregular ant-like movements

formos-a, -um, -us L. handsome. Attractive in appearance

formosae From Formosa, now Taiwan

formosan-a, -um, -us L. *-ana*, indicating connection. For Formosa, now Taiwan

formosensis L. *-ensis*, denoting origin. See *formosae*

formosulum L. *formosus*, beautiful; *-ulus*, tendency. The purple rhachis contrasts sharply with the white-haired spikelets

formos-um, -us See *formosa*

forrestii L. *-ana*, indicating connection. In honor of George Forrest (1873–1932) who collected in China

forskalii, forskalei, forskålei, forskålii, forskeelii, forskhalei, forskohlii, forskolii, forsskalii In honor of Pehr Forsskål (1736–1768) Swedish botanist

forsteri In honor of – **(1)** Johann Georg Adam Forster (1754–1794) German explorer and botanist. *Agrostis forsteri*, *Deyeuxia forsteri* – **(2)** Paul Irwin Forster (1961–) Australian botanist. *Aristida forsteri*

forsterianum L. *-anum*, indicating connection. As for *forsteri* **(1)**

fortis L. *fortis*, strong. Culms robust

fortunae-hibernae L. luck of the Irish. The type specimen was grown at Royal Botanic Gardens, Kew, from soil brought to England from Tasmania by Lord Talbot de Malahide (1912–1973) an Irish peer

fortunei In honor of Robert Fortune (1812–1880) English botanist

fosbergii In honor of Francis Raymond Fosberg (1908–1993) United States botanist

fossae-rusticorum L. *fossa*, ditch; *rus*, the country; *-icus*, belonging to. Of country ditches, that is growing alongside ditches in fields

foucaudii In honor of Julien Foucaud (1847–1904) French botanist

fouilladeana L. *-ana*, indicating connection. As for *fouilladei*

fouilladei In honor of Amédée Fouillade (1870–) French botanist

foulkesii In honor of Thomas Foulkes (fl. 1855–1860) English cleric who collected in India

fourcadei In honor of Georges Henri Fourcade (1866–1948) French-born South African forester and plant collector

Fourniera In honor of Eugène Pierre Nicolas Fournier (1834–1884) French botanist

F

fournieriana L. *-ana*, indicating connection. As for *Fourniera*

foveolat-a, -um, -us L. *fovea*, pit; *-olus*, minute. *-atum*, possessing. Lower glume has a conspicuous circular depression

fractus L. weak. The spikelets are pendulous because they terminate long thin panicle branches

fragil-e, -is L. weak. – (1) Inflorescences readily disarticulating. *Agropyron fragile, Andropogon fragilis, Bambusa fragilis, Digitaria fragilis, Garnotia fragilis, Gaudinia fragilis, Homozeugos fragile, Luziola fragilis, Paspalum fragile, Schizachyrium fragile, Triticum fragile, Tuctoria fragilis* – (2) rhachilla readily disarticulating shortly after maturity. *Asthenatherum fragile, Avena fragilis, Bromus fragilis, Danthonia fragilis, Digastrium fragile, Helleria fragilis, Hordeum fragile, Ischaemum fragile, Tricholaena fragilis*

fragiliflora L. *fragilis*, weak; *flos*, flower. Rhachilla readily disarticulating

fragilis See *fragile*

fragilissimus L. *fragile*, weak; *-issima*, most. Racemes readily disarticulating

fragrans L. *fragro*, smell sweet. Foliage possessing an agreeable odour

francavillean-um, -us In honor of Albert Franqueville (?–1891)

franchetianum L. *-anum*, indicating connection. In honor of Adrien René Franchet (1834–1900) French botanist

franchetii As for *franchetianum*

francoi In honor of Felix Franco (1892–?) who collected in Mexico

frankii In honor of Joseph C. Frank (1782–1835) German botanist and physician

franksiae In honor of Millicent Franks (1886–1961) South African botanical artist

frappieri See *benoistii*

fratercula Origin unclear

fraudulentum L. *fraus*, deceit; *-ulentum*, filled with. The mature inflorescence may be mistaken for that of another genus

frederici In honor of Friedrich Martin Josef Welwitsch (1806–1872) Austrian-born mainly Angolan botanist

frederikseniae In honor of Signe Frederiksen (1942–) Danish botanist

fredscholzii In honor of Fred Scholz, outstanding expert on traditional land use in Oman

freita From Freitas, Portugal

Fremya In honor of Pierre Frémy (1880–1944) French cleric and algologist

freticola L. *fretum*, channel; *-cola*, dweller. From the Straits of Magellan

friesianum L. *-anum*, indicating connection. In honor of Elias Magnus Fries (1794–1878) Swedish botanist

friesii In honor of – (1) Thore Christian Elias Fries (1886–1930) Swedish botanist. *Eragrostis friesii, Leersia friesii, Panicum friesii, Sorghastrum friesii, Sorghum friesii* – (2) Robert Elias Fries (1876–1966) Swedish botanist. *Aristida friesii*

friesiorum In honor of Thore Christian Elias Fries (1886–1930) and Robert Elias Fries (1876–1966) Swedish botanists

frigid-a, -us L. cold. Growing at high altitudes

frigidis See *frigida*

frigidus See *frigida*

froesianum L. *-anum*, indicating connection. As for *Froesiochloa*

Froesiochloa Gk *chloa*, grass. In honor of Richardo de Lemos Fróes (1891–1960) Brazilian plant collector

frondescens L. *frondesco*, become leafy. Culms leafy, ascending from a creeping base

frondos-a, -us L. *frons*, leaf; *-osa*, abundance. Freely branching from the nodes and so habit bushy

frumentace-a, -um, -us L. *frumentum*, pertaining to grain; *-acea*, resembling. Species serving as cereals or suspected of being suitable as cereals

Frumentum L. relating to grain. A nomenclatural synonym of *Secale* and *Triticum*

frutescens L. *frutesco*, become bushy. Shrubby in habit

fruticans L. *frutesco*, become bushy. Branching from the base

fruticosa L. *frutex*, shrub; *-osa*, abundance. Plant shrubby

fruticulos-a, -us L. *frutex*, shrub; *-ulus*, diminutive; *-osa*, abundance. Culms rigidly erect and somewhat woody

fuegian-a, -um L. *-ana*, indicating connection. From Fuegia

fuegina L. *-ina*, indicating possession. From Fuegia, that is Tierra del Fuego, the southern most part of Chile and Argentina

fugax L. ephemeral. Short-lived species often from inhospitable habitats

fugeshiensis L. *-ensis*, denoting origin. From Fugeshigunn, Ishikawa Prefecture, Japan

fujianica L. *-ica*, belonging to. From Fujian Province, China

fukuchiyamensis L. *-ensis*, denoting origin. From Fukuchiyama, Kyoto Prefecture, Japan

fukuyamae In honor of K. Fukuyama, Japanese botanist

fulgens L. *fulgeo*, gleam. Anthoecium glossy

fulgid-a, -um L. *fulgeo*, gleam; *-idum*, becoming. – (1) Spikelets glossy. *Calamagrostis fulgida* – (2) anthoecia glossy. *Panicum fulgidum*

fulgor L. lightning. Culms rapidly growing

fultum L. *fulgeo*, gleam. Anthoecium glossy white

fulv-a, -um, -us L. *brown*, deep yellow. Usually a reference to spikelet color

fulvescens L. *fulvesco*, become brown. Panicle pale-brown

fulvibarbis L. *fulvus*, yellowish-brown; *barba*, beard. Callus fulvously bearded on the sides

fulvicom-a, -us L. *fulvus*, brown; *coma*, head of hair. Racemes densely clothed with brown hairs

fulvispica L. *fulvus*, brown; *spica*, ear of spike of grain. Inflorescence branches invested with brown hairs

fulv-um, -us See *fulva*

fumida L. smoky. Spikelets purple to black

fumigata L. *fumigo*, fumigate. Inflorescence dark-grey

funaensis L. *-ensis*, denoting origin. From the banks of the Funa, probably a river, in Zambia

funckianum L. *-anum*, indicating connection. In honor of Heinrich Christian Funck (1771–1839) German botanist and apothecary

funckii As for *funckianum*

funereum L. relating to a funeral. Lemma-awn black

funghomii In honor of Fung Hom also known as H. L. Fung (fl. c. 1931–1941) Chinese plant collector

funiculata L. *funis*, rope; *-ula*, diminutive; *-ata*, possessing. The twisted column of the awn resembles a rope

funiushanensis L. *-ensis*, denoting origin. From Mt Yunjushan, Hunan Province, China

funstonii In honor of Frederick Funston (1865–1917) United States botanist

furcat-um, -us L. *furca*, fork; *-atus*, possessing. Inflorescence branches arising in pairs

furfurosa L. brown. Spikelets pale-brown

furtiv-a, -um L. *secret*, hidden. – (1) Lemma partially hidden. *Rhytachne furtiva* – (2) species long overlooked. *Panicum furtivum*

furv-a, -um L. *dusky*, almost black. Lemmas darkish purple-brown

fusc-a, -um, -us L. dark, swarthy. Glumes or lemmas dark-brown

fuscata L. *fuscus*, dark; *-ata*, possessing. Spikelets invested with dark hairs

fuscescens L. *fuscesco*, become dark. Lemma brown

fuscoviolaceum L. *fuscus*, dark; *violaceus*, violet. Inflorescence invested in brownish-purple hairs

fusc-um, -us See *fusca*

fusiform-e, -is L. spindle-shaped. Spikelets long-pointed

Fussia In honor of Johann Michály Fuss (1814–1883) Transsylvanian botanist

futadensis L. *-ensis*, denoting origin. From Futada, Niigata Prefecture, Japan

G

gabelii In honor of Mark L. Gabel (1950–) United States palaeobotanist

gabesensis L. *-ensis*, denoting origin. From near de Gabès, Tunisia

gabonens-e, -is L. *-ense*, denoting origin. From Gabon

gabrieliae In honor of Gabriel Domin, wife of Karel Domin (1882–1953); see *dominii*

gabunense See *gabonense*

gaditan-a, -um, -us L. *-ana*, indicating connection. From Gades, now Cadiz, Spain

gaertnerianum L. *-anum*, indicating connection. In honor of Joseph Gaertner (1732–1791) German physician and botanist

gaetula Belonging to the Gaetulians, in Roman times a people of northwestern Africa

Gaimardia In honor of Joseph Paul Gaimard (1793–1858) French naturalist

galapageium L. *-ium*, indicating connection. From the Galapagos Islands

Galeottia In honor of Henri Galeotti (1814–1858) French botanist

galeottiana As for *Galeottia*

galeottii See *Galeottia*

galicicae From Galicica Planina, Macedonia

gallaensis L. *-ensis*, denoting origin. From the region of the Galla tribe, Arussi Province, Ethiopia

gallatlyi In honor of G. Gallatly (fl. 1876) who collected in Myanmar

gallecic-a, -um L. *-ica*, belonging to. From Gallecia, now southern France

galli A contraction of *crus-galli*

gallica L. *-ica*, belonging to. From Gallia, now France

galloinsulanus L. *-anus*, indicating connection. From Ile de France, now Republic of Mauritius, Indian Ocean

galmarra In honor of Galmarra (fl. 1848) an Aboriginal from Patrick Plains, New South Wales, Australia

galpinii In honor of Ernest Edward Galpin (1858–1941) banker and amateur botanist

gambicum L. *-icum*, belonging to. From Gambia

gambiense L. *-ense*, denoting origin. From Gambia

gamblei In honor of James Sykes Gamble (1846–1925) English-born Indian forester and botanist

Gamelythrum See *Gamelytrum*

Gamelytrum Gk *gamos*, wedding; *elytron*, cover. Lemma completely invests the palea

gamisansii In honor of Jacques Gamisans (1944–) Catalonian botanist

gammieana L. *-ana*, indicating connection. In honor of James Alexander Gammie (1839–1924) Scottish botanist

gammiei As for *gammieana*

ganaensis L. *-ensis*, denoting origin. From Gana, Zaire

gandogeri In honor of Michel Gandoger (1850–1926) French botanist

gandreanszkyi See *andreanszkyi*

ganeschinii In honor of Sergej Sergejewitsch Ganeschin (1879–1930) Russian botanist

gangangalaensis L. *-ensis*, denoting origin. From Gangangala, Zaire

gangetica L. *-ica*, belonging to. From Ganges River, India

gangitis Gk *-itis*, close connection. From Ganges in southern France

Gaoligongshania Type species collected by the 1978 Gaoligong Expedition to Yunnan Province, China

garamas From Garamas, Libya

gardneri In honor of – (1) George Gardner (1812–1849) Scots-born physician and botanist, sometime Director of Peradeniya Gardens, Sri Lanka. *Digitaria gardneri*, *Isachne gardneri* – (2) Charles Austin Gardner (1896–1970) English-born Western Australian botanist. *Eriachne gardneri*

gardnerian-a, -um L. *-ana*, indicating connection. As for *gardneri* (as for *Digitaria*)

garhwalensis L. *-ensis*, denoting origin. From Garhwal, India

garipensis L. *-ensis*, denoting origin. From the Garip River, South Africa

Garnotia In honor of Prosper Garnot (1794–1838) French surgeon-naturalist

Garnotiella L. *-ella*, diminutive here used as a name-forming suffix. Allied to *Garnotia*

garubensis L. *-ensis*, denoting origin. From Garub, Namibia

gasparricensis L. *-ensis*, denoting origin. From Gaspar Rico, a former name of Pokak Atoll, one of the Marshall Islands

gaspensis L. *-ensis*, denoting origin. From Gaspe Peninsula, Canada

gasteenii In honor of Wrixon James Gasteen (1922–) Australian agriculturalist and naturalist

Gastridium Gk *gaster*, paunch; *-idium*, diminutive. The glumes are gibbously swollen

Gastropyrum Gk *gaster*, belly; *pyros*, wheat. The inflorescence is moniliform and disintegrates into individual spikelets at maturity

gatacrei In honor of William Forbes Gatacre (1843–1906) Scottish-born British army officer

gatineauensis L. *-ensis*, denoting origin. From Gatineau Road, Eardley, Canada

gattingeri In honor of Augustin Gattinger (1825–1903) United States physician and botanist

gaubae In honor of Erwin Gauba (1891–1964) Austrian-born Australian botanist

gaudichaudii In honor of Charles Gaudichaud-Beaupré (1789–1854) French pharmacist and naturalist

Gaudinia In honor of Jean François Gottlieb Philippe Gaudin (1766–1833) Swiss cleric and botanist

gaudinian-a, -um L. *-ana*, indicating connection. As for *Gaudinia*

Gaudinopsis Gk *opsis*, resemblance. Similar to *Gaudinia*

gaumeri In honor of George Franklin Gaumer (1850–1929) who collected in the Americas

gausum Gk *gausos*, bent. The spikelets are curved

gautieri In honor of Marie Clément Gaston Gautier (1841–1911) French botanist

gayan-a, -um, -us L. *-ana*, indicating connection. In honor of – (1) Claude Gay (1800–1873) French natural historian and writer who spent much of his adult life teaching in Chile and Peru. *Agrostis gayana, Andropogon gayanus, Arundo gayana, Chloris gayanus, Digitaria gayana, Elymus gayanus, Panicum gayanum, Paspalus gayanus, Poa gayana* – (2) Jacques Gay (1786–1864) French civil servant and botanist. *Holcus gayanus*

Gazachloa Gk *chloa*, grass. See *gazensis*

gazensis L. *-ensis*, denoting origin. Of Gazaland, formerly a Territory extending from coastal Mozambique to the mountains in eastern Zimbabwe at about latitude 20° S. Now largely included in Mozambique

gedrosianus L. *-anus*, indicating connection. From Gedrosia, the name in Classical times for the coastal region of southeast Iran and south-west Pakistan

gegarkunii In honor of Gegarkun, Russian botanist

geibiensis L. *-ensis*, denoting origin. From Geibi, Hiroshima Prefecture, Japan

gelida L. icy cold. Growing at high altitudes

Gelidocalamus L. *gelidus*, icy cold; *kalamos*, reed. Reed-like grasses growing on high mountains

geminat-a, -um, -us L. *gemini*, twins; *-ata*, possessing. – (1) Inflorescence of paired branches. *Agropyron geminatum, Agrostis geminata, Andropogon geminatus, Arthrostylidium geminatum, Chloris geminata, Coelorachis geminata, Dactyloctenium geminatum, Mnesithea geminata, Pentarrhaphis geminata, Poecilostachys geminatus, Pollinia geminata, Rhipidocladum geminatum, Rottboellia geminata* – (2) spikelets paired. *Lophatherum geminatum, Panicum geminatum, Paspalidium geminatum, Sporobolus geminatus*

geminiflor-a, -um, -us L. *gemini*, twins; *flos*, flower. – (1) Spikelets with a pair of staminate or neuter florets. *Aegopogon geminiflorus* – (2) spikelets in pairs on a common peduncle. *Aristida geminiflora, Avena geminiflora, Paspalum geminiflorum*

geminifolia L. *gemini*, twins; *folium*, leaf. Only two of the culm leaves have blades

geminiramula L. *gemini*, twins; *ramus*, branch; *-ula*, diminutive. Inflorescence branches arising in pairs

gemmeum L. *gemma*, jewel; *-eum*, indicating resemblance. Upper lemma and palea with conspicuous wart-like outgrowths

gemmosum L. *gemma*, jewel; *-osum*, well developed. Papillae on upper lemma bear a fanciful resemblance to jewels

G

genalensis L. *-ensis*, denoting origin. From the valley of the Genale Wenz River, Ethiopia

Genea Gk offspring. A group of species segregated from *Bromus*

geneschinii In honor of S. Geneschin (fl. 1930)

genevensis L. *-ensis*, denoting origin. From Geneva, Switzerland

geniculat-a, -um, -us L. *genus*, knee; *-ulus*, diminutive. *-ata*, possessing. Plants with bent culms or awns

gentilis L. of the same clan. Belonging in the same Section of the genus

gentryi In honor of Howard Scott Gentry (1903–1993) United States botanist

genuensis L. *-ensis*, denoting origin. From Genua, now Genoa, Italy

genuflexum L. *genus*, knee; *flexum*, bend. Culms repeatedly geniculate

geoffreyi In honor of Geoffrey Thomas Jacobs (1980–) Australian information technologist

geometra Italian, map maker. In honor of Ettori Bovone (1880–1922) pioneer traveller and plant collector in Zaire

Geopogon Gk *ge*, earth; *pogon*, beard. The lower florets only of the spikelet are awned

georgian-a, -um L. *-ana*, indicating connection. – (1) From Georgia, USA. *Panicum georgianum* – (2) from Republic of Georgia. *Avena georgiana*

georgic-a, -um L. *-ica*, belonging to. From Republic of Georgia

georgii In honor of George Forrest (1873–1932) Scottish-born traveller and plant collector

gerardii In honor of – (1) John Gerard (1545–1612) English botanist. *Andropogon gerardii*, *Alopecurus gerardii*, *Colobachne gerardii*, *Festuca gerardii*, *Phleum gerardii*, *Schedonorus gerardii* – (2) Louis Gérard (1733–1819) French botanist. *Crypsis gerardii*, *Phleum gerardii*

gerdesii In honor of J. F. Gerdes who collected in Brazil

Germainea See *Germainia*

Germainia In honor of Jacques Nicolas Ernest Germain de Saint Pierre (1815–1882) French botanist

germanic-a, -um L. *-ica*, belonging to. From Germania, that is Germany

gerontogaea Gk *gerontos*, old; *ge*, earth. Old World, that is American species

gerrardii In honor of William Tyrer Gerrard (?–1866) who collected in Natal

Gerritea In honor of Gerrit Davidse (1942–) Netherlands-born United States botanist

gervaisii In honor of Camille Gervais (1933–) Canadian botanist

geyeri In honor of Carl Andreas Geyer (1809–1853) German-born United States botanist

geyeriana L. *-ana*, indicating connection. As for *geyeri*

ghatica L. *-ica*, belonging to. From Western Ghats, India

ghiesbreghtii In honor of August Ghiesbreghtii (1810–1893) Belgian botanist

gibb-a, -um L. swelling. – (1) Spikelets gibbous. *Ischaemum gibbum*, *Panicum gibbum*, *Phleum gibbum*, *Piptochaetium gibbum*, *Pseudophleum gibbum*, *Sacciolepis gibba* – (2) nodes gibbous. *Bambusa gibba*

gibboides Gk *-oides*, resembling. Similar to *Bambusa gibba*

gibbos-a, -um L. *gibba*, swelling; *-osa*, indicating abundance. – (1) Spikelets swollen asymmetrically. *Aristida gibbosa*, *Chaetaria gibbosa*, *Digitaria gibbosa*, *Indosasa gibbosa*, *Mesosetum gibbosum*, *Panicum gibbosum*, *Pennisetum gibbosum*, *Stipa gibbosa* – (2) culm-sheaths asymmetrical. *Sinobambusa gibbosa*

gibbsiae In honor of Lilian Suzette Gibbs (1870–1925) English traveller and botanist

gibbum See *gibba*

gidarba Origin uncertain, not given by the author but probably a vernacular name

giessii In honor of J. W. H. Giess (fl. 1971) Namibian botanist

Gigachilon Gk *gigas*, large; *chilos*, green fodder. Segregated from *Triticum* on account of its robust habit

gigante-a, -um, -us L. very large. Culms tall compared with those of related species

gigantissima L. *gigantea*, very large; *-issima*, most. Culms very tall

Gigantochloa L. *gigantea*, large; Gk *chloa*, grass. Tall, woody grasses

gigas L. giant. Culms taller than most other species in the genus

gilbertiana L. *-ana*, indicating connection. In honor of Michael George Gilbert (1943–) English botanist

gilesii In honor of Ernest Giles (1835–1897) English-born Australian explorer

gilgiana L. *-ana*, indicating connection. As for *Gilgiochloa*

Gilgiochloa Gk *chloa*, grass. In honor of Ernst Friedrich Gilg (1867–1933) German botanist

gilgitica L. *-ica*, belonging to. From Gilgit, north-east Pakistan

gillettii In honor of Jan Bevington Gillett (1911–1995) English-born East African and Iraqi botanist

gilliesii In honor of John Gillies (1747–1836) who collected in Argentina

gillii In honor of Gill, South African plant collector

gilvohirsutus L. *gilvum*, dull yellow; *hirsutus*, hairy. Leaf-sheath invested in dull yellow hairs

gilvum L. dull-yellow. Panicle dull-yellow

gimmae From Jimma, a district in Eritraea

ginae In honor of Gina Luzzato (fl. 1937) who collected in North Africa

Ginannia In honor of Giuseppe Ginnani (1692–1753) Italian botanist at Ravena

gintlii In honor of O. Gintl, Bohemian botanist

giovanninii In honor of Melchior Giovannini who collected in Mexico

gisekeanus L. *-anus*, indicating connection. In honor of Paul Dietrich Gesike (1741–1796) German physician and amateur botanist

giulianettii In honor of Amadeo Giulianetti (?–1901) who collected in Papua-New Guinea

glab-er, -ra, -rum L. smooth. – (1) Leaf-blades lacking hairs. *Agrostis glabra, Amphilophis glabra, Andropogon glaber, Arundinaria glabra, Avena glabra, Bothriochloa glabra, Deyeuxia glabra, Digitaria glabra, Dimeria glabra, Elionurus glaber, Enneapogon glaber, Gymnothrix glabra, Heteropogon glaber, Hierochloe glabra, Hordeum glabra, Lepargochloa glabra, Melinis glabra, Microcalamus glaber, Panicum glabrum, Paspalum glabrum, Pennisetum glabrum, Pharus glaber, Rottboellia glabra, Stenotaphrum glabrum, Syntherisma glabrum, Tricholaena glabra, Trichopteryx glabra, Tristachya glabra, Trisetum glabrum* – (2) lemmas lacking hairs. *Danthonia glabra* – (3) ligules of culm-sheaths and leaf-blades lacking hairs. *Neololeba glabra*

glaberrima L. most free of hairs. Plant glabrous

glabra See *glaber*

glabrat-a, -um, -us L. *glaber*, smooth; *-ata*, possessing. Plant glabrous in whole or in part

glabrescens L. *glabresco*, becoming glabrous. Quite glabrous with respect to the whole plant or one or more of its parts

glabriflor-a, -is L. *glaber*, smooth; *flos*, flower. Lemmas glabrous

glabrifoli-a, -um L. *glaber*, smooth; *folium*, leaf. Leaf-blades glabrous

glabriglaucum L. *glaber*, smooth; *glaucum*, bluish-green. Culms with glabrous nodes and when young, bluish-green

glabrinodis L. *glaber*, smooth; *nodus*, knot. Nodes glabrous

glabripoda L. *glaber*, smooth; Gk *pous*, foot. Callus of spikelet glabrous

glabrissimum L. *glaber*, smooth; *-issimum*, most. Plant quite glabrous

glabriuscul-a, -us L. *glabrius*, smoother; *-ula*, tendency. Tending towards being glabrous

glabrovagina L. *glaber*, smooth; *vagina*, sheath. Culm-sheaths glabrous

glabrum See *glaber*

glacial-e, -is L. frozen. Growing at high altitudes

G

gladiatum L. *gladius*, sword; *-atum*, possessing. Leaf-blade lanceolate to subcordate

Glandiloba L. *glans*, gland; *lobus*, lobe. The reduced lower glume and swollen pedicel fused to form a small gland-like swelling at the base of the spikelet

glandulopaniculatum L. *glandulosa*, with abundant small glands; *paniculus*, panicle; *-atum*, possessing. Panicle branches glanduliferous

glandulosa L. *glans*, gland; *-ula*, diminutive; *-osa*, abundance. – (1) With small glands especially on the leaf-blades. *Danthonia glandulosa*, *Erucaria glandulosa* – (2) with short teeth or hair cushions mistaken for glands. *Coelorachis glandulosa*, *Manisuris glandulosa*, *Rottboellia glandulosa*

glandulosipedata L. *glandulosa*, possessed of abundant glands; *pes*, foot; *-ata*, possessing. Pedicels with abundant glands

glanvillei In honor of R. R. Glanville, who collected in Sierra Leone

glareae L. *glarea*, shingle. Growing on shingle beds

glareosa L. *glarea*, shingle; *-osa*, abundance. From gravelly habitats

glauc-a, -um, -us L. *glauca*, bluish-green. Whole plant or any of its parts glaucous

glaucantha Gk *glaukos*, bluish-green; *anthos*, flower. Spikelets glaucous

glaucescens L. *glaucesco*, become glaucous. Foliage and/or other parts bluish-green

glaucidulum L. *glaucus*, bluish-green; *-idus*, becoming; *-ulum*, diminutive. Plant tinged with violet

glaucifoli-a, -um, -us L. *glaucus*, bluish-green; *folium*, leaf. With bluish-green foliage

glaucina Gk *glaukos*, bluish-green; *-ina*, indicating resemblance. Foliage glaucous

glaucispicula L. *glaucus*, bluish-green; *spica*, a point; hence, in particular, an ear or spike of grain; *-ula*, diminutive. Spikelets glaucous to pruinose

glaucissim-a, -um, -us L. *glaucus*, bluish-green; *-issimum*, most. Plant whole or in part quite glaucous

glaucocladum Gk *glaukos*, bluish-green; *klados*, branch. – (1) Lower internodes coated with a whitish wax. *Panicum glaucocladum*, *Pennisetum glaucocladum* – (2) shoots covered by white wax. *Schizostachyum glaucocladum*

glaucoides Gk *glaukos*, bluish-green; *-oides*, resembling. Plant glaucous

glaucophyll-a, -um, -us Gk *glaukos*, bluish-green; *phyllon*, leaf. – (1) Leaf-blades glaucous. *Andropogon glaucophyllus*, *Dactyloctenium glaucophyllum* – (2) leaf-blades green with longitudinal white stripes. *Bambusa glaucophylla*

glaucopsis Gk *glaukos*, bluish-green; *opsis*, appearance. Leaf-blades glaucous

glaucopurpureus L. *glaucus*, bluish-green; *purpureus*, dull-red tinted with blue. Plant glaucous with a red tinge

glaucostachyum Gk *glaukos*, bluish-green; *stachys*, spike as of an ear of wheat. Racemes bluish-green

glaucovirens L. *glaucus*, bluish-green; *virens*, green. Plant in whole or in part glaucous

glauc-um, -us See *glauca*

Glaziophyton Gk *phyton*, plant. In honor of Auguste François Marie Glaziou (1828–1906) French-born artist and Brazilian botanist

glaziovii, glaziowii As for *Glaziophyton*

gleasonii In honor of Henry Allan Gleason (1882–1975) United States botanist who collected in British Guiana, now Guyana

glischra Gk *glishros*, sticky. Plant viscid

globifera L. *globus*, sphere; *fero*, carry or bear. Inflorescence a congested globular panicle

globoideum L. *globus*, sphere; Gk *-oideum*, resemblance. Spikelets globose

globos-a, -um, -us L. *globus*, sphere; *-osa*, abundance. – (1) Spikelets spherical. *Aira globosa*, *Airopsis globosa*, *Isachne globosa*, *Lasiacis globosa*, *Milium globosum*, *Phaenosperma globosa* – (2) inflorescence spherical. *Andropogon globosus*, *Cymbopogon globosus*

globular-e, -is L. *globus*, sphere; *-ulus*, diminutive; *-aris*, pertaining to. – (1) Spikelets spherical. *Panicum globulare*, *Setaria globularis* – (2) spikelets clustered into ball-like aggregations. *Elytrophorus globularis*

globuliferum L. *globus*, sphere; *-ulus*, diminutive; *fero*, carry or bear. The spicate inflorescence is interrupted to produce clusters of spikelets

globulosum L. *globus*, sphere; *-ulus*, diminutive; *-osum*, abundance. Spikelets spherical

glochidiatus Gk *glochis*, arrow head; L. *-atus*, possessing. Dorsal apex of lower glume has barbed hairs

gloeoclados Gk *gloios*, anything sticky; *klados*, branch. Leaf-blades bearing an abundance of sticky hairs

gloeodes Gk *gloios*, sticky; *-odes*, resembling. Sticky at the nodes

glomerat-a, -um, -us L. *glomus*, ball of thread; *-ata*, possessing. Spikelets crowded and forming clusters in the inflorescence

glumace-a, -um L. *gluma*, husk; *-acea*, belonging to. Glumes conspicuous

glumaepatul-a, -um L. *gluma*, husk; *patula*, standing open. Glumes spreading at maturity

glumar-e, -is L. *gluma*, husk; *-aris*, pertaining to. Glumes well formed

glumos-a, -um, -us L. *gluma*, husk; *-osa*, abundance. Spikelets with conspicuous glumes

glutinos-a, -um L. *gluten*, glue; *-osa*, abundance. – (1) Leaves sticky. *Agrostis glutinosa, Eragrostis glutinosa, Poa glutinosa, Tristegis glutinosa* – (2) spikelets sticky. *Homolepis glutinosa, Panicum glutinosum* – (3) grain sticky. *Oryza glutinosa*

glutinoscabrum L. *gluten*, glue; *-osa*, abundance; *scaber*, rough. Wart-like secreting glands abound on the leaf-blades and leaf-sheaths

glutinosum See *glutinosa*

Glyceria Gk *glykeros*, sweet. The grain of the type species is sweet to the taste

glyceriantha Gk *anthos*, flower. Spikelets resemble those of *Glyceria*

glycerioides Gk *-oides*, resembling. Similar to *Glyceria*

Glyphochloa Gk *glypho*, carver; *chloa*, grass. Lower glume often elaborately sculptured

gmelinii In honor of – (1) Johan Friedrich Gmelin (1748–1804) German botanist. *Melica gmelinii, Poa gmelinii* – (2) Karl Christian Gmelin (1762–1837) German physician and botanist. *Agropyron gmelinii, Avena gmelinii, Elymus gmelinii, Roegneria gmelinii, Trisetum gmelinii*

gnaphalioideum Gk *-oideum*, resembling. Foliage densely woolly like that of *Gnaphalium*

gnezdilloi In honor of Gnezdillo

Gnomonia Gk *gnomon*, pointer, as of rod at centre of a sundial. An allusion to fescue which in English may refer either to a grass or to a sundial. The connection between the two arises from the usage in Latin of *festuca* for both straw and the rod by which slaves were touched during the ceremony of manumission. The double meaning of *festuca* enabled the author to hint obliquely that the new genus incorporated species previously included in *Festuca*

goaensis L. *-ensis*, denoting origin. From Goa State, India

goalparensis L. *-ensis*, denoting origin. From Goalpara District, Assam State, India

gobariensis L. *-ensis*, denoting origin. From Gobari, Zaire

gobica L. *-ica*, belonging to. From *Gobi* Desert

gobicola Mandarin *gobi*, a stony desert; *-cola*, dweller. Growing in cold stony deserts at the base of Mt Muztagata, southwest China

godefroyi In honor of Jules Godefroy (fl. 1895) sometime Director of the Agricultural College at Grand-Jouan, Réunion

goebelii In honor of Karl Immanuel Eberhard von Goebel (1855–1932) German botanist

goeldii In honor of Émil Andreas Goeldi (1859–1917) Swiss-born, Brazilian botanist

goeppertii In honor of Heinrich Robert Goeppert (1880–1884) German physician and botanist

goeringii In honor of Philip Friedrich Wilhelm Goering (1809–1879) German botanist

goetzenii In honor of Adolf Graf Goetzen (fl. 1894) who collected in Tanzania

goiasensis L. *-ensis*, denoting origin. From Goias, Brazil

goiranicum L. *-icum*, belonging to. In honor of Agostino (Augustin) Goiran (1835–1909) Italian botanist

goiranii As for *goiranicum*

golae In honor of Giuseppe Gola (1877–1956) Italian botanist

Goldbachia In honor of Karl Ludwig Goldbach (1793–1824) German-born Russian botanist

golestanensis L. *-ensis*, denoting origin. From Golestan National Park, Iran

goloskokovii In honor of Vitaliy Petrovich Goloskokov (1913–) Russian botanist

gombeiana L. *-ana*, indicating connection. From Gombei-toge, Yamagata Prefecture, Japan

gonatodes Gk *gony*, knee; *-odes*, resembling. Culms conspicuously geniculate

gonatostachys Gk *gony*, knee; *stachys*, spike as of an ear of wheat. Culms short, geniculate at the base

gongshanensis L. *-ensis*, denoting origin. From Gongshan Xian, Yunnan Province, China

gonopodus Gk *gony*, knee; *pous*, foot. Culms geniculate and rooting at the lower nodes

gonyrrhizum Gk *gony*, knee; *rhiza*, root. Culms rooted at the nodes

gonzalezii In honor of Angel Custodio González (1943–) Venezuelan botanist

gonzaloi In honor of Gonzalo (fl. 1925) who collected in Spain

gooddingii In honor of Leslie Newton Goodding (1880–1967) United States botanist

gorbunovii In honor of Mikhail Grigorievich Gorbunov (1912–) Russian geologist

gorodkovii, gorodkowii In honor of Boris Nikolaevich Gorodkov (1890–1953) Russian botanist

gossweileri In honor of John Gossweiler (1873–1952) Swiss-born Angolan botanist

Gossweilerochloa Gk *chloa*, grass. See *gossweileri*

gossypin-a, -um L. *gossipion*, cotton tree; *-ina*, indicating resemblance. Densely covered with long spreading white hairs

gouanii In honor of Antoine Gouan (1733–1821) French botanist

gougerotiana L. *-ana*, indicating connection. In honor of Mariane Gougerot friend of Aimée Camus (see *Camusia*)

goughensis L. *-ensis*, denoting origin. From Gough Island in the South Atlantic

Gouinia In honor of Gouin (fl. 1864–1867) who collected in Mexico

gouinii In honor of Antoine Gouinia (1733–1821) French botanist

Goulardia In honor of Pierre Etienne Goulard (?–1909) French botanist

Gouldochloa Gk *chloa*, grass. In honor of Frank Walter Gould (1913–1981) United States agrostologist

goyanum L. *-anum*, indicating connection. As for *goyazense*

goyasense L. *-ense*, denoting origin. From Goyás State, Brazil

goyazens-e, -is L. *-ense*, denoting origin. From Goyaz Province, Brazil

gozadakensis L. *-ensis*, denoting origin. From Mt Gozadake, Nishiomate Island, Japan

gracei In honor of Marvin Grace (c. 1935–) United States cattle rancher

gracil-e, -is L. slender. Culms or inflorescences slender

Gracilea L. *gracilis*, slender. Very slender annual

gracilenta L. *gracilesco*, become slender. Culms very slender

gracilescens L. *gracilesco*, become slender. Culms slender

gracilicaule L. *gracilis*, slender; *caulis*, stem. Culms slender

graciliflor-a, -um L. *gracilis*, slender; *flos*, flower. Primary inflorescence branches filiform

gracilifolia L. *gracilis*, slender; *folium*, leaf. Leaf-blades less than 0.5 mm broad

gracililaxa L. *gracilis*, slender; *laxa*, loose. Culms subcapillary, flexuose

gracilior L. more slender. In some respect more slender than related species

gracilipes L. *gracilis*, slender; *pes*, foot. Pedicels slender

gracilis See *gracile*

gracilissimum L. *gracilis*, slender; *-issimum*, most. Culms very slender

gracillim-a, -um, -us L. very delicate. Of slender habit

graec-a, -um From Graecia now Greece

grafiana L. *-ana*, indicating connection. In honor of Graf

grahamii In honor of R. J. Graham, economic botanist who worked in India

grallata L. *gralla*, stilt; *-ata*, possessing. Culms erect with abundant stilt-roots

Gramen L. grain. Meaning uncertain but has been applied to a single species of *Digitaria*. The name has been used twice as a *nomen nudum*

Gramerium L. *gramen*, grain; *-ium*, indicating connection. Meaning obscure

Graminastrum L. *-astrum*, indicating inferiority. Meaning obscure except in that the species are inferior in some respect

gramine-a, -us L. *gramen*, grain. In some respect resembling a cereal

Graminocarpon L. *gramen*, grain; Gk *karpos*, fruit. Form genus for fossils resembling anthoecia

Graminophyllum L. *gramen*, grain; Gk *phyllon*, leaf. Form genus for fossil leaves resembling those of grasses

granatensis L. *-ensis*, denoting origin. From Granata now Granada, Spain

grand-e, -is L. tall. – (1) Plants robust and vigorous, often with tall culms. *Agrostis grandis, Andropogon grandis, Avena grandis, Bromus grandis, Calamagrostis grandis, Dendrocalamopsis grandis, Glyceria grandis, Koeleria grandis, Muhlenbergia grandis, Panicum grande, Poa grandis, Puccinellia grandis, Roegneria grandis, Setaria grandis, Sorghum grande, Stipa grandis* – (2) anthoecia large. *Stipidium grande*

grandiaristata L. *grandis*, large; *arista*, bristle; *-ata*, possessing. Lemma long-awned

grandiflor-a, -um, -us L. *grandis*, large; *flos*, flower. – (1) Spikelets with more florets than those of related species. *Andropogon grandiflorus, Arundinella grandiflora, Bromus grandiflorus, Danthonia grandiflora, Diheteropogon grandiflorus, Festuca grandiflora, Germainia grandiflora, Helopus grandiflorus, Heteropogon grandiflorus, Holcus grandiflorus, Homalocenchrus grandiflorus, Leersia grandiflora, Melica grandiflora, Ottochloa grandiflora, Pennisetum grandiflorum, Rhynchelytrum grandiflorum, Saccharum grandiflorum, Tricholaena grandiflora, Triodia grandiflora* – (2) florets large. *Calamagrostis grandiflora, Gymnopogon grandiflorus, Poa grandiflora*

grandifoli-a, -um, -us L. *grandis*, large; *folium*, leaf. Leaf-blades large

grandiglumis L. *grandis*, large; *gluma*, husk. Glumes and/or lemmas large

grandis See *grande*

grandispic-a, -um L. *grandis*, large; *spica*, a point; hence, in particular, an ear or spike of grain. Spikelets large

grandispiculatum L. *grandis*, large; *spica*, a point; hence, in particular, an ear or spike of grain; *-ula*, diminutive; *-atum*, possessing. Spikelets large

granditectoria L. *grandis*, large; *tectorius*, of a cover. Leaf-blades broader than those of *Sasa tectoria*

granditectorius L. *grandis*, large. Resembling *Sasa tectorius* but having larger leaf-blades

graniflorum L. *granum*, grain; *flos*, flower. The anthoecium is smooth and glossy resembling a grain

granitica English granite; L. *-ica*, belonging to. Growing on granitic soils

graniticola L. *-cola*, dweller. See *granitica*

grantii In honor of D. K. S. Grant (fl. 1922–1923) who collected in Tanzania

granular-e, -is L. *granum*, grain; *-aris*, pertaining to. – (1) Segments of the inflorescence resemble beads. *Cenchrus granularis, Hackelochloa granularis, Manisuris granularis* – (2) spikelets resemble grain. *Digitaria granularis, Panicum granulare, Paspalum granulare, Rytilix granularis, Sporobolus granularis*

granulat-a, -um L. *granum*, grain; *-ata*, possessing. Lemma warty at the base

granulifera L. *granum*, grain; *-ula*, diminutive; *fero*, carry or bear. Second glume and sterile lemma surfaces granular

granulosa L. *granum*, grain; *-ula*, diminutive; *-osa*, abundance. Surface rough as if covered in small beads

Graphephorum Gk *graphis*, style for writing on wax tablets; *phero*, bear. Rhachilla extended between fertile and sterile floret

grat-a, -um, -us L. pleasing. Attractive in appearance

gravius L. *gravis*, heavier. Species overall more robust than related species

Graya In honor of Asa Gray (1818–1888) United States botanist

grayana L. *-ana*, indicating connection. From Gray's Peak, Colorado, USA

grayi See *Graya*

grayumii In honor of Michael Howard Grayum (1949–) United States botanist

gredensis L. *-ensis*, denoting origin. From Sierra de Gredos, Spain

greenei In honor of Edward Lee Greene (1843–1915) United States botanist

Greenia In honor of Benjamin Daniel Greene (1793–1862) Guyanan-born, United States botanist

greenwayi In honor of Percy James Greenway (1897–1980) English botanist

gregalis L. *grex*, flock; *-alis*, pertaining to. Growing everywhere in the region from which described

greggii In honor of Josiah Gregg (1806–1850) United States physician, explorer and botanical collector

gregoriense L. *-ense*, denoting origin. From San Gregoris, Peru

grenieri In honor of Jean Charles Marie Grenier (1808–1875) French botanist

gresicola French *grès*, sandstone; *-cola*, dweller. Growing on sandstones

Greslania In honor of Évenor de Greslan (1839–1900) French agriculturalist who was born on Réunion Island and died on New Caledonia

grevillensis L. *-ensis*, denoting origin. From Mt Greville, southeast Queensland, Australia

griffithian-a, -um L. *-ana*, indicating connection. In honor of William Griffith (1810–1845) English-born surgeon-botanist in India and southeast Asia

griffithii As for *griffithiana*

griffithsiae As for *griffithiana*

griffithsii See *Griffithsochloa*

Griffithsochloa In honor of David Griffiths (1867–1935) United States agronomist and botanist

griffonii In honor of Griffon du Bellay (fl. 1864) a French Naval surgeon and explorer who collected in Gabon

grigorjevii In honor of Jury Sergeyevich Grigoreiv (1905–) Soviet botanist

grillus See *Gryllus*

grimburgii In honor of Karl Grimburg (fl. 1898) who collected in Greece

griquensis L. *-ensis*, denoting origin. From Griqualand West, South Africa

grise-a, -um L. grey. Plant in whole or in part grey

grisebachian-a, -um L. *-ana*, indicating connection. As for *grisebachii*

grisebachii In honor of August Heinrich Rudolf Grisebach (1814–1879) German botanist

griseum See *grisea*

groenlandic-a, -um L. *-ica*, belonging to. From Groenland, that is Greenland

gross-a, -um, -us L. large. – (1) Culms talls. *Fargesia grossa* – (2) spikelets large. *Brachiaria grossa, Bromus grossus, Panicum grossum*

grossarium L. *grossus*, large; *-arium*, pertaining to. Large in some respect

grossheimiana L. *-ana*, indicating connection. In honor of Alexander Alfonsovich Grossheim (1888–1948)

gross-um, -us See *grossa*

grumosum L. broken into grains or small tubercules. Lemma surface irregularly sculptured in to tile-like areas

Gryllus, gryllus Gk *gryllus*, cricket. The spikelets are in clusters of three which together bear a fanciful resemblance to a cricket. Furthermore, the spikelets are shed as triads which move erratically in response to the twisting and untwisting of the hygroscopic awn on the lemma of the sessile spikelet and so resemble jumping crickets

guadaloupens-e, -is L. *-ense*, denoting origin. From Island of Guadaloupe

Guadella See *Guaduella*

guadeloupens-e, -is See *guadaloupense*

guadinii As for *Gaudinia*

Guadua, guadua Chibcha *gua-uba*, water flower. Vernacular name of the species in Colombia

Guaduella L. *-ella*, diminutive here used as a name-forming suffix. Resembling *Guadua* in some respect

guamanensis L. *-ensis*, denoting origin. From the páramo of Guamani, Ecuador

guangdongensis L. *-ensis*, denoting origin. From Guangdong Province, China

guangxiens-e, -is L. *-ense*, denoting origin. From Guangxi Province, China

guaramacalana L. *-ana*, indicating connection. From Guaramacal National Park, Venezuela

guaraniticum L. *-icum*, belonging to. From the land of the Guarani in Argentina

guaricense L. *-ense*, denoting origin. From Guarico, Venezuela

guatemalens-e, -is L. *-ense*, denoting origin. From Guatemala

guatemalica L. *-ica*, belonging to. From Guatemala

guayanerum From La Guayanera, Sinola State, Mexico

guayaquilense L. *-ense*, denoting origin. From Guayaquil, Ecuador

guenoarum In honor of the Guenoas, a people who lived on Isla Vizcaíno, Uruguay

guestphalica L. *-ica*, belonging to. From Guestphalia, Westfalia, Germany

guetrotii In honor of Guétrot (fl. 1944)

guianens-e, -is L. *-ense*, denoting origin. – (1) From British Guiana, now Guyana. *Dinebra guianensis*, *Eragrostis guianensis*, *Heteranthoecia guineensis*, *Isachne guineensis*, *Ischaemum guianense*, *Manisuris guianensis*, *Panicum guianense*, *Paspalum guianense*, *Rhytachne guianensis*, *Thrasya guianensis* – (2) from French Guyana. *Strephium guianense*

guidenensis L. *-ensis*, denoting origin. From Guide County, Qinghai Province, China

guillarmodiae In honor of Amy Jacot Guillarmod (1911–) South African botanist

guineens-e, -is L. *-ense*, denoting origin. For Guinea Coast, West Africa

guingensis L. *-ensis*, denoting origin. From Pedras de Guinga, Angola

guizhouensis L. *-ensis*, denoting origin. From Guizhou Province, China

gulliveri In honor of Thomas A. Gulliver, botanical collector in northern Australia

gummiflua L. *gummius*, containing gum; *fluo*, flow. Leaf-sheaths sticky

gunckelii In honor of H. Gunckel (fl. 1931) who collected in Chile

gunnian-a, -us L. *-ana*, indicating connection. In honor of Ronald Campbell Gunn (1808–1881) South African-born Tasmanian botanist

gunnii As for *gunniana*

gusindei In honor of Martin Gusinde (1886–1969) who collected in Chile

gussonei In honor of Giovanni Gussone (1787–1866) Italian botanist

gussonianum As for *gussonei*

gussonii As for *gussonei*

gussonis As for *gussonei*

gusuleacii In honor of Gusuleac (1904–1937) Romanian botanist

guthrie-smithiana L. *-ana*, indicating connection. In honor of William Herbert Guthrie-Smith (1861–1940) New Zealand author, farmer and naturalist

guttatum L. *gutta*, spot; *-atum*, possessing. Sterile lemma with red spots

guzmanii In honor of Raphael Guzman Mejía (1950) Mexican botanist

gyganteus See *gigantea*

gyirongensis L. *-ensis*, denoting origin. From Gyirong, China

Gymnachne Gk *gymnos*, naked; *achne*, scale. The lemma is glabrous

Gymnandropogon Gk *gymnos*, naked. Similar to *Andropogon* but lacking bracts in the inflorescence

gymnantha Gk *gymnos*, naked; *anthos*, flower. Lemmas glabrous

Gymnanthelia Gk *gymnos*, naked; *anthele*, inflorescence. Lacking conspicuous spathes in the inflorescence

gymnocarpon Gk *gymnos*, naked; *karpos*, fruit. The palea and lemma gape at maturity, exposing the grain

Gymnopogon Gk *gymnos*, naked; *pogon*, beard. The rhachilla lacks hairs and projects well beyond the terminal floret

gymnostachys Gk *gymnos*, naked; *stachys*, spike as of an ear of wheat. The spikelets lack a lower glume and the upper is much reduced

Gymnostichum Gk *gymnos*, naked; *stichos*, row. The glumes are minute or wanting thereby leaving the lemmas exposed

gymnostyla Gk *gymnos*, naked; *stylos*, column. Stigma base glabrous passing imperceptibly into hairy stigmas

gymnotheca Gk *gymnos*, bare; *theke*, cup. Anthoecium exposed because subtending glumes are very small

Gymnothrix, gymnothrix See *Gymnotrix*

Gymnotrix Gk *gymnos*, naked; *thrix*, hair. The bristles subtending the spikelets are scabrid rather than feathery

gynerioides Gk *-oides*, resembling. Similar to *Gynerium* in habit

Gynerium Gk *gyne*, woman; *erion*, wool. The glumes of the female florets are invested with long hairs

gynoglossa Gk *gyne*, woman; *glossa*, tongue. In addition to the two styles the apex of the gynoecium bears a deltoid appendage that may be likened to a tongue

gypsacea L. *gypsum*, gypsum; *-acea*, belonging to. Growing on gypsum soils

gypsophila Gk *gypsos*, gypsum; *phileo*, love. Growing on gypsum soils

gyrans L. *gyro*, turn round in a circle. Awns forming loose spirals on drying

H

haareri In honor of Alec Ernest Haarer (1894–1970) English-born Tanzanian plant ecologist

habahenensis L. *-ensis*, denoting origin. From Habahe, Xinjiang Uyghur Autonomous Region, China

habrantha Gk *habros*, delicate; *anthos*, flower. Spikelets minute

Habrochloa Gk *habros*, delicate; *chloa*, grass. Dwarf annual

habrothrix Gk *habros*, pretty; *thrix*, hair. Plant invested with a mixture of long and short, glandular or non-glandular hairs

Habrurus Gk *habros*, delicate; *oura*, tail. Inflorescence a single spike-like raceme

hachadoensis L. *-ensis*, denoting origin. From Pino Hachado, Neuquén Province, Argentina

Hackelia In honor of Eduard Hackel (1850–1926) Bohemian born Austrian botanist

hackelian-a, -um L. *-ana*, indicating connection. As for *Hackelia*

hackelii As for *Hackelia*

Hackelochloa Gk *chloa*, grass. See *Hackelia*

hadjikyriakou In honor of Georgios N. Hadjikyriakou (also as Chatzikyriakou) (fl. 1999) Cypriot botanist

haemacarpon Gk *haima*, blood; *karpos*, fruit. Anthoecia red

Haemarthria See *Hemarthria*

haematodes Gk *haima*, blood; *-odes*, resembling. Leaf-blades blotched with red

haemi From Haemus now Bulgarian Mountains, Bulgaria

haenkean-a, -um, -us L. *-ana*, indicating connection. In honor of Thaddaeus Peregrinus Xaverius Haenke (1761–1816) Bohemian botanist who travelled widely in the Pacific

haenkei As for *haenkeana*

hagenbeckian-um, -us L. *-anum*, indicating connection. In honor of C. F. Hagenbeck (fl. 1898 or earlier) German botanist who collected in Bolivia, Chile and Argentina

hagerupii In honor of Olaf Hagerup (1889–1961) Danish botanist

haifense L. *-ense*, denoting origin. From Haifa, Israel

hainanens-e, -is L. *-ense*, denoting origin. From Hainan Province, China

Hainardia In honor of Pierre Hainard (1936–) Swiss botanist and ecologist

Hainardiopholis Hybrids between *Hainardia* and *Pholiurus*

hait The vernacular name of this species in Sumatra, Indonesia. It means hooked to other neighbouring plants

haitiens-e, -is L. *-ense*, denoting origin. From Haiti

hajastanicum L. *-icum*, belonging to. From Hajastan, Armenia

hajrae In honor of P. K. Hajra (1940–) Indian botanist

Hakonechloa Gk *chloa*, grass. See *hakonensis*

hakonensis L. *-ensis*, denoting origin. From Hakone, Kanagawa Prefecture, Japan

hakusanensis L. *-ensis*, denoting origin. From Hakusan, Japan

halei In honor of Josiah Hale (?–1856) United States botanist

halepens-e, -is L. *-ense*, denoting origin. From Halab, Arabic for Aleppo, Syria

halleriana L. *-ana*, indicating connection. In honor of Albrecht Haller (1708–1777) Swiss botanist, physiologist and poet

halleridis As for *halleriana*

hallianus L. *-anus*, indicating connection. In honor of Hall

hallieri In honor of Johann Gottfried Hallier (1868–1932) who collected in the Philippines

hallii In honor of Elihu Hall (1822–1882) United States farmer and amateur botanist

halmaturina Gk *halme*, sea water that has dried; *-ina*, indicating possession. Growing in salt marshes

halmyris Gk salt water. Growing on coastal dunes

Halochloa Gk *halos*, salt; *chloa*, grass. Growing in saltmarshes or along sea shores

halophil-a, -um, -us Gk *halos*, salt; *phileo*, love. Growing on saline soils

Halopyrum Gk *halos*, salt; *pyros*, wheat. Wheat-like and growing in coastal habitats

Hamalocenchrus See *Homalocenchrus*

hamat-a, -um L. *hamus*, hook; *-ata*, possessing. Lemma awn slender with an abruptly deflexed bristle

hamatulus L. *hamus*, hook; *-atus*, possessing; *-ulus*, diminutive. The short, paired racemes are reflexed causing the inflorescence branches to resemble a series of small anchors

hamatum See *hamata*

hamhungensis L. *-ensis*, denoting origin. From Hamhung, Korea

hamiensis L. *-ensis*, denoting origin. From el Ham, Arabia

hamiltoniana L. *-ana*, indicating connection. As for *hamiltonii* (2)

hamiltonii In honor of – (1) Augustus Hamilton (1853–1913) English-born New Zealand educator and amateur botanist. *Poa hamiltonii* – (2) Francis Buchanan, later known as Francis Hamilton or Francis Buchanan-Hamilton (1762–1829) Scots-born Indian physician and botanist. *Andropogon hamiltonii*, *Chrysopogon hamiltonii*, *Dendrocalamus hamiltonii*, *Pennisetum hamiltonii* – (3) William Hamilton (1783–1856) British physician and traveller. *Panicum hamiltonii*

hamosum L. *hamus*, hook; *-osum*, abundance. Lemma awn strongly recurved

hamulatus L. *hamus*, hook; *-atus*, possessing; *-ulus*, diminutive. The short, paired racemes are reflexed causing the inflorescence branches to resemble a series of small anchors

hamulosa L. *hamus*, hook; *-ula*, diminutive; *-osa*, abundance. Lemma base and awn with reflexed barbs

hancei In honor of Henry Fletcher Hance (1827–1886) English botanist

hanningtonii In honor of James Hannington (1847–1885) English-born Ugandan cleric and plant collector

H

hannonensis L. *-ensis*, denoting origin. From Hannô, Musashi Province, now Tokyo Prefecture and parts of Saitama and Kanagawa Prefectures, Japan

hannoverianus L. *-anus*, indicating connection. From Hannover, Germany

hansenii In honor of – (1) George Hansen (1863–1908) United States plant collector. *Elymus hansenii, Poa hansenii, Sitanion hansenii* – (2) Bertel Hansen (1932–2005) Danish botanist. *Ischaemum hansenii*

hansiana L. *-ana*, indicating connection. From Hansi, India

hans-meyeri In honor of Hans Meyer (fl. 1907) who collected in Ecuador

hantu Malay *hantu*, ghost. A contraction of *buluk hantu*, the vernacular name of the species in Sarawak

hapalantha Gk *hapaloos*, soft; *anthos*, flower. Lemmas membranous with weakly developed nerves

hapalotricha Gk *hapaloos*, soft; *thrix*, hair. Rhachilla densely pubescent

Haplachne Gk *haploos*, single; *achne*, scale. The floret lacks a palea

haplocaulos Gk *haploos*, single; *kaulos*, stem. Culms unbranched

haploclad-a, -um Gk *haploos*, single; *klados*, branch. Panicle comprising racemes

haplodurum Gk *haploos*, single. Resembles *Triticum durum* in morphology but has only the haploid chromosome complement of that species

harae In honor of Hiroshi Hara (1911–1986) Japanese botanist

hararensis L. *-ensis*, denoting origin. From Harar, Ethiopia

harfordii In honor of William George Washington Harford (1825–1911) United States botanist

harimensis L. *-ensis*, denoting origin. From Harima Province, now part of Hyogo Prefecture, Japan

harmandii In honor of Jules Harmand (1845–1921) French naval physician who collected in Thailand, Cambodia and Laos

harmensiana L. *-ana*, indicating connection. In honor of Harmsen

harmonicum Gk *harmonia*, a skilfull blending of sounds; *-icum*, belonging to. Flutes are made from the internodes

Harpachne Gk *harpe*, sickle; *achne*, scale. The lemma is sickle-shaped

harpachnoides Gk *-oides*, resembling. Similar to *Harpachne*

Harpechloa Gk *harpe*, sickle; *chloa*, grass. Terminal spikelets sickle-shaped

Harpochloa See *Harpechloa*

Harpostachys Gk *harpe*, sickle; *stachys*, spike as of an ear of wheat. Inflorescence sometimes a single curved raceme

harrisii In honor of William Harris (1860–1920) Jamaican plant collector

harsukhii In honor of Harsukh (fl. c. 1900) Indian plant collector

hartmanniana L. *-ana*, indicating connection. In honor of Karl Johann Hartmann (1790–1849) Scandinavian physician and botanist

hartmannii In honor of Hans Hartmann (fl. 1962) who collected in Karakoram Range straddling the boundary separating China from Pakistan and India

hartwegianum L. *-anum*, indicating connection. In honor of Carl Theodor Hartweg (1812–1871) German botanical explorer

hartzii In honor of Nikolaj Eg Kruse Hartz (1867–1937) Danish botanist

hashimotoi In honor of C. Hashimoto (fl. 1930) Japanese botanist

hassei In honor of Hermann Edward Hasse (1846–1915) German-born United States botanist

hasskarliana L. *-ana*, indicating connection. In honor of Justus Karl Hasskarl (1811–1894) German-born Dutch botanist

hassleri In honor of Emil Hassler (1861–1939) Swiss-born Paraguayan botanist

hatchoensis L. *-ensis*, denoting origin. From Lake Hatchoike, Idzu or Izu Province, now part of Shizuoka and Tokyo Prefectures, Japan

hatenashiensis L. *-ensis*, denoting origin. From Hatenashi, Nara Prefecture, Japan

hatico From Verado hatico, Colombia

hatschbachii In honor of Gert Hatschbach (1923–) Brazilian botanist

hatsuroana L. *-ana*, indicating connection. In honor of Hatsuro, Japanese botanist

hatsusimanus L. *-ana*, indicating connection. In honor of Hatsusima-Gumihiho (fl. 1935) Japanese botanist

hattorian-a, -us L. *-ana*, indicating connection. In honor of Yasuyoshi Hattori (fl. 1934) Japanese botanist

hatusimae In honor of Sumihiko Hatsusima (1906–) Japanese botanist

haughtii In honor of Oscar Haught, also known as Oscar Lee Haught (1893–1975) who collected in North and South America

haumanii In honor of Lucien Leon Hauman, also Hauman-Merck (1880–1965) Belgian botanist who collected in Africa and South America

hauptiana L. *-ana*, indicating connection. In honor of Ernest Gottfried Haupt (1795–1862) who collected in Siberia

haussknechtianus L. *-anus*, indicating connection. As for *haussknechtii*

haussknechtii In honor of Heinrich Carl Haussknecht (1838–1903) German botanist

havanensis L. *-ensis*, denoting origin. From Havana, Cuba

havardii In honor of Valery Havard (1846–1927) United States physician and amateur botanist

hawaiiensis L. *-ensis*, denoting origin. From Hawaii

hayachinecola L. *-cola*, a dweller. See *hayachinensis*

hayachinensis L. *-ensis*, denoting origin. From Hayachinesan, a mountain in Rikuchiu Province, Japan

hayatae In honor of Bunzô Hayata (1874–1934) Japanese botanist

Haynaldia In honor of Stefan Franz Lajos Haynald (1816–1891) cleric and botanist who was born in Transylvania, now included in Hungary

haynaldiana L. *-ana*, indicating connection. As for *Haynaldia*

Haynaldoticum Hybrids between species of *Haynaldia* and *Triticum*

hebechlamys Gk *hebe*, pubic hair; *chlamys*, cloak. Upper leaves of culms softly pubescent

hebestachyum Gk *hebe*, pubic hair; *stachys*, spike as of an ear of wheat. Inflorescence shortly hairy

hebotes Gk *hebos*, youthful; *-otes*, denoting condition. Panicle branches with soft downy hair

hedbergii In honor of Karl Olov Hedberg (1923–) Swedish botanist

hedgei In honor of Ian Charleson Hedge (1928–) Scots botanist

hegetschweileri In honor of Johannes Hegetschweiler-Bodmer (1789–1839) Swiss physician and botanist

heidemaniae In honor of T. Heideman (fl. 1932–1934) who collected in Nakhichevan, Azerbaijan

heidenreichii In honor of Ferdinand Albert Heidenreich (1819–1901)

hejiangensis L. *-ensis*, denoting origin. From Hejiang Xian, Guizhou Province, China

Hekaterosachne Gk *hekateros*, each of two; *achne*, scale. Both glumes are awned

heldreichii In honor of Theodor Heldreich (1822–1902) German botanist

helenae (1) In honor of Helena, Duchess of Aosta, the collector. *Tristachya helenae* – (2) from St. Helena, an island in the South Atlantic. *Agrostis helenae*

heleniae In honor of the collector, Helen Collingwood Fortune Hopkins (1953–) English botanist

Heleochloa Gk *helos*, marsh; *chloa*, grass. Growing in swamps and mudflats

heleochloides Gk *-oides*, resembling. Resembling *Heleochloa* in some respect

helferi In honor of Johan Wilhem Helfer (1810–1840) Bohemian physician and traveller who collected in Myanmar

helgolandica L. *-ica*, belonging to. From Helgoland, an island off the mouth of the river Elbe, Germany

heliconia Leaf-blades resembling those of *Heliconia*

helicophylla Gk *helix*, twisted; *phyllon*, leaf. Old leaf-blades curled in the manner of a watch-spring

Helictotrichon, Helictotrichum Gk *helictos*, twisted; *thrix*, hair. The column of the lemma awn is twisted

heliochloides Gk *-oides*, resembling. With the habit of *Heleochloa schoenoides*
hellenica L. *-ica*, belonging to. Belonging to the Hellenes, that is from Greece
Hellera See *Helleria*
helleri As for *Helleria*
Helleria In honor of Amos Arthur Heller (1867–1944) United States botanist
Hellerochloa Gk *chloa*, grass. See *helleri*
helmsii In honor of Richard Helms (1842–1914) German born New Zealand and Australian botanist
helobium Gk *helos*, marsh meadow; *bios*, manner of living. Swamp species
helodes Gk *helos*, marsh meadow. Swamp species
helophilus Gk *helos*, marsh; *phileo*, love. Growing in marshes
Helopus, helopus Gk *helos*, swamp; *pous*, foot. Growing in damp soil
helvol-a, -um, -us L. *helvolus*, yellow-green. – (1) Bristles subtending spikelets yellow-green. *Pennisetum helvolum, Setaria helvola, Oplismenus helvolus* – (2) anthoecium yellow. *Panicum helvolum, Sporobolus helvolus, Vilfa helvola*
Hemarthria Gk *hemi-*, half; *arthron*, segment. The spikelets are sessile and embedded in the inflorescence axis which readily disarticulates into segments at maturity
Hemibromus Gk *hemi-*, half. Resembling *Bromus* in spikelet structure
hemignostum Gk *hemi-*, half; *gnosis*, enquiry. Species little known
Hemigymnia Gk *hemi-*, half; *gymnos*, naked. The subtending glumes are shorter than the spikelet
Hemimunroa Gk *hemi-*, half. Similar to *Munroa*
hemipoa Gk *hemi-*, half. Resembling *Poa*
hemipogon Gk *hemi-*, half; *pogon*, beard. Lemma hairy only towards the apex
Hemipus Gk *hemi-*, half; *pous*, foot. The upper glume is missing
Hemisacris Gk *hemi-*, half; *akris*, sharp. The lemma acute instead of bifid
Hemisorghum Gk *hemi-*, half. Near to *Sorghum*
hemisphericum Gk *hemi-*, half; *sphaera*, sphere; *-icum*, belonging to. Spikelets subhemispheric
hemitomon Gk *hemi-*, half; *tomon*, cutting. Aquatic culms rigid but soft and flaccid about the water line so liable to lean as if partly severed
hemmingii In honor of Hemming (fl. 1958) the collector
hendersonii In honor of Louis Fourniquet Henderson (1853–1942) United States botanist
hengshanica L. *-ica*, belonging to. From Hengshan, China
henonis In honor of Jacques Louis Hénon (1802–1872) French botanist
Henrardia In honor of Jan Theodor Henrard (1881–1974) Dutch botanist
henrardiana L. *-ana*, indicating connection. As for *Henrardia*
henrardii As for *Henrardia*
henriettae In honor of Henrietta Ippolitovna Poplavskaja (1885–1956) Russian botanist
henriquezii In honor of Julio Augusto Henriquez (1838–1928) Portuguese botanist
henryanum L. *-anum*, indicating connection. In honor of Charles Henry (fl. 1921) French official in the Marquesas
henryi In honor of – (1) Augustine Henry (1857–1936) Irish physician and forester. *Deyeuxia henryi, Digitaria henryi, Stipa henryi, Trisetum henryi* – (2) James N. Henry (fl. 1940) Provost of Lingnan University, China. *Phyllostachys henryi*
hensii In honor of Frans Hens (1856–1928) Belgian cleric who collected in Zaire
henslowian-a, -um L. *-ana*, indicating connection. In honor of John Stevens Henslow (1796–1861) English cleric and botanist
hentyi In honor of Edward Ellis Henty (1915–2002) Australian botanist who for many years worked in Papua New Guinea
hepburnii In honor of A. John Hepburn
hephaestophila Gk *Hephaistos*, God of fire; *phileo*, love. Growing on the slopes of the crater of Volcan de Agna, Guatemala

heptamera Gk *hepta*, seven; *meros*, part. Lemma with seven nerves, each terminating in a bristle

heptaneuron Gk *hepta*, seven; *neuron*, nerve. Upper glume seven-nerved

heptantha Gk *hepta*, seven; *anthos*, flower. Spikelets with seven florets

heptapotamica Gk *hepta*, seven; *potamos*, river; L. *-ica*, belonging to. From "Land of the Seven Streams" in Dzungaria region of Central Asia

Heptaseta, heptaseta Gk *hepta*, seven; L. *seta*, bristle. Florets with seven bristles

herbacea L. *herba*, herb; *-acea*, indicating resemblance. Culms less robust than those of related species

hercegovinica L. *-ica*, belonging to. From Herzegovina (Hercegovina), formerly Yugoslavia, now the country of Bosnia and Herzegovina

hercynica L. *-ica*, belonging to. From Hercynia, now Hartz region of Germany

hereroensis L. *-ensis*, denoting origin. From Heroro District, South Africa

heribaudii In honor of Heribaud Joseph otherwise Jean Baptiste Caumel (1841–1918) French cleric and botanist

herjedalica L. *-ica*, belonging to. From Herjedalen Province, Sweden

herklotsii In honor of Geoffrey Alton Craig Herklots (1902–1986) British-born Colonial Officer of Hong Kong

hermannii In honor of P. Hermann, plant collector in S.W. Africa

hermaphrodit-a, -um L. bisexual. Spikelets each with a single bisexual floret

herminieri In honor of Ferdinand l'Herminier (1802–1866) French botanist who collected in Guadeloupe

hermonis From Mount Hermon, now Jebel esh Sheikh, on the border of Syria and Lebanon

herpoclados Gk *herpo*, creep; *klados*, branch. Culms decumbent

herrerae In honor of Gerado Herrera Chacón (fl. 1980–1989) who collected in Costa Rica

hervieri In honor of Jean Hervier (1847–1900) French cleric and botanist

herzogiana L. *-ana*, indicating connection. In honor of Theodor Herzog (1880–1961) German botanist

herzogii As for *herzogiana*

hesperia L. *Hesperus*, the West. Grows mainly on the west coast of the South Island, New Zealand

hesperica L. *Hesperus*, the West; *-ica*, belonging to. From Hesperis, that is of the west. In this instance, from Spain and Portugal

hesperidium Gk *Hesperides*, daughters of the evening, who dwelt on a western island. From north-west Africa

Hesperochloa L. *Hesperus*, the West; *chloa*, grass. Endemic to western USA

Hesperostipa L. *Hesperus*, the West. Restricted to North America, that is the Western Hemisphere, in contrast to Eurasian species of *Stipa*

Heterachne Gk *heteros*, different; *achne*, scale. The glumes are of different lengths

heteranth-a, -um Gk *heteros*, different; *anthos*, flower. – (1) Spikelets paired and dissimilar. *Digitaria heterantha*, *Panicum heteranthum* – (2) florets dissimilar in the same spikelet. *Isachne heterantha*

Heteranthelium Gk *heteros*, variable; *anthele*, inflorescence. Spikelets of two kinds, fertile and sterile in the same inflorescence

heteranther-a, -us Gk *heteros*, different; *antheros*, blooming. – (1) Anthers of the sessile spikelets much smaller than those of the pedicelled. *Andropogon heterantherus* – (2) lemmas of the two florets much smaller than those of the spikelet. *Pogonatherum heteranthera*

Heteranthoecia Gk *heteros*, different; *anthos*, flower; *oikos*, house. Spikelets with two florets, the lower hermaphrodite, the upper pistillate

heteranth-um, -us Gk *heteros*, different; *anthos*, flower. The terminal spikelets of the inflorescence are either fertile or sterile

Heteranthus Gk *heteros*, different; *anthos*, flower. In the same spikelet the lower lemmas may be awnless or have straight awns whereas the upper lemmas have geniculate awns

H

Heterelytron, Heterelytrum Gk *heteros*, different; *elytron*, cover. The glumes of the staminate and hermaphrodite florets are dissimilar

Heterocarpha Gk *heteros*, different; *karphos*, any dry body. The subtending glumes differ in morphology, one being symmetric, the other asymmetric

Heterochaeta Gk *heteros*, different; *chaete*, bristle. Awns in lower and upper lemmas of spikelet differ in length

heterochaeta Gk *heteros*, different; *chaete*, bristle. Awns variable in the species

heterochlamys Gk *heteros*, different; *chlamys*, cloak. The glumes differ markedly in venation and indumentum

Heterochloa Gk *heteros*, different; *chloa*, grass. The florets are paired and dissimilar

heterochroa Gk *heteros*, different; *chroia*, color. Spikelets variously pigmented

heteroclada Gk *heteros*, different; *klados*, stem. Fertile and sterile culms morphologically quite different

heteroclit-a, -um, -us Gk *heteros*, different; *klitus*, hillside. Species with disjunct distributions

heterocrasped-a, -um Gk *heteros*, different; *kraspedon*, fringe. Leaf-margins serrate with hairs of varying lengths

heterocycla Gk *heteros*, different; *kyklos*, circle. Successive circular leaf-scars not parallel but obliquely directed to the left and right at successive nodes

heterogama Gk *heteros*, different; *gamos*, marriage. – (1) Sessile spikelet sterile and pedicellate fertile. *Apocopis heterogama* – (2) lower florets and spikelet bisexual or male and upper florets female. *Poa heterogama*

heteroglossa Gk *heteros*, different; *glossa*, tongue. The ligules of the lower rosette leaves and those higher up on the culm have ligules of different lengths

heterolepis Gk *heteros*, different; *lepis*, scale. The subtending glumes differ markedly in length

heteromalla Gk *heteros*, different; *mallon*, more. Very different in some respect from other species

heteromera Gk *heteros*, different; *meros*, part. Subtending glumes very different in length

heteromorpha Gk *heteros*, different; *morphe*, shape. Spikelets dimorphic

heteroneuron Gk *heteros*, different; *neuron*, nerve. Lower glume one-nerved, upper glume nine-nerved

heteropachys Gk *heteros*, different; *pachys*, thick. Adjacent leaf-blades often of different diameters

Heteropholis Gk *heteros*, different; *pholis*, scale as of snake. The upper and lower glumes are quite different

heterophyll-a, -um Gk *heteros*, different; *phyllon*, leaf. Culms with two types of leaf-blade. Usually the basal leaves are fine-bladed and those of the culm are conspicuously broader

heteropodium Gk *heteros*, different; *pous*, foot. Florets borne on pedicels of differing lengths

Heteropogon Gk *heteros*, different; *pogon*, beard. The lemma of the hermaphrodite floret has a well developed awn in contrast to the setae developed on the awns of the male florets

heterostachyum Gk *heteros*, different; *stachys*, spike as of an ear of wheat. Inflorescence has spikelets of two kinds

Heterosteca Gk *heteros*, different; *theke*, box. Lower floret of spikelet fertile, upper sterile

Heterostega, heterostega See *Heterosteca*

heterotrich-a, -um Gk *heteros*, different; *thrix*, hair. Glumes and sterile lemma bear hairs of various lengths

Heuffelia In honor of Johann A. Heuffel (1800–1857) Hungarian physician and botanist

heufleriana L. *-ana*, indicating connection. In honor of Ludwig Samuel Joseph David Alexander Heufler zu Rasen (1817–1885) Austrian botanist

hexaflorus Gk *hexa*, six; L. *flos*, flower. Inflorescence a spike bearing six burrs, each of which is a group of one or more spikelets enclosed in a ring of bristles

hexandr-a, -us Gk *hexa*, six; *aner*, man. The florets possess six anthers

Hexarrhena Gk *hexa*, six; *arrhen*, a male. The spikelets are in clusters of seven of which the central spikelet is female or bisexual and the six laterals male

hexastachyon Gk *hexa*, six; *stachys*, spike as of an ear of wheat. The spicate inflorescence has six vertical rows of spikelets

hexastachy-um, -us Gk *hexa*, six; *stachys*, spike as of an ear of wheat. Inflorescence of six branched

hexastichon Gk *hexa*, six; *stichon*, row. The spicate inflorescence has six rows of spikelets

heydei In honor of Enrique Téophila (also Heinrich Theophil) Heyde (fl. 1892) Guatemalan cleric and plant collector

heymannii In honor of A. L. Heymann

heynei In honor of Benjamin Heyne (1770–1819) German-born Indian, geologist, botanist and physician

heynii In honor of Benjamin Heyne (1770–1819) German-born Indian, geologist, botanist and physician

hians L. *hio*, gape. Glumes spreading at anthesis or when anthoecium mature

hiascens L. *hiasco*, tending to gape. Anthoecium exposed at maturity due to spreading of sterile lemma and glumes

hibaconuca From Hibagun and Onukamura, Hiroshima Prefecture, Japan

Hibanobambusa A woody grass resembling *Bambusa* from Hibasan, a mountain in Shimane Prefecture, Japan

hibernaculum L. *hiberno*, pass the winter; *-aculum*, indicating capacity. Sown in the autumn and overwinters as young plants

hibernans L. *hiberno*, pass the winter. Overwintering as seedlings

hibernum L. belonging to the winter. Sown in the autumn and so overwinters as young plants

Hickelia In honor of Paul Robert Hickel (1865–1935) English-born French botanist

hidaensis L. *-ensis*, denoting origin. From Hida Province, now part of Gifu Prefecture, Japan

hidakanus L. *-anus*, indicating connection. From Hidaka Province, now Hidaka Subprefecture, Hokkaido, Japan

hidejiroana L. *-ana*, indicating connection. In honor of Kato Hidejiro, Japanese botanist

hideoi In honor of Hideo Koidzumi (1886–1945) Japanese botanist

hiegaeri Orthographic variant of *higegaweri*

hiemalis L. *hiems*, winter; *-alis*, pertaining to. Winter flowering

hiemata L. *hiems*, cold; *-ata*, possessing. A component of high alpine grassland in south-eastern Australia, a region with a cold climate

hieminflatum L. *hiems*, winter; *inflo*, inflate. Spikelets swelling out in the winter

hierniana L. *-ana*, indicating connection. In honor of Walter Philip Hiern (1839–1925) English botanist

Hierochloa See *Hierochloe*

Hierochloe, Hierochloë Gk *hieros*, sacred; *chloa*, grass. From the custom of strewing plants of certain species before Church doors on Saints Days

Hierocloe See *Hierochloe*

hieronymi In honor of George Hans Emmo Wolfgang Hieronymus (1846–1921) German botanist, sometime resident of Argentina

hieronymusii As for *hieronymi*

higegaweri Vernacular name of some species of *Polypogon* in Japan

higoensis L. *-ensis*, denoting origin. From Higo Province, now Kumamoto Prefecture, Japan

hikosanensis L. *-ensis*, denoting origin. From Hikosan, a mountain in Buzen Province, now part of Fukuoka and Oita Prefectures, Japan

Hilarei As for *Hilaria*

Hilaria In honor of Auguste de St. Hilaire (1779–1853) French naturalist who travelled widely in South America

hilariae As for *Hilaria*

hildebrandtii In honor of Johann Maria Hildebrandt (1847–1881) German-born traveller and plant collector

hillebrandian-um, -us L. *-anum*, indicating connection. In honor of Wilhelm Hillebrand (1821–1886) German physician and botanist

H

hillebrandii As for *hillebrandianum*

hillmanii In honor of Frederick Hebard Hillman (1863–1954) United States botanist

himalaic-a, -um, -us L. *-ica*, belonging to. From the Himalayas

Himalayacalamus Gk *kalamos*, reed. A woody genus from the Himalayas

himalayan-a, -um, -us L. *-ana*, indicating connection. From the Himalayas

himalayens-e, -is L. *-ense*, denoting origin. From the Himalayas

hindsii In honor of – (1) Richard Brinsley Hinds (c. 1812–c. 1847) British naval surgeon and naturalist. *Arundinaria hindsii*, *Pleioblastus hindsii* – (2) J. Hinds (fl. 1947) who collected in Ghana. *Chrysochloa hindsii*

Hinterhuberia In honor of Georg Hinterhuber (1768–1850) Austrian pharmacist and amateur botanist

hintoniana L. *-ana*, indicating connection. As for *hintonii*

hintonii In honor of George Bode Hinton (1882–1943) United States botanist

Hippagrostis Gk *hippos*, anything coarse; *agrostis*, an unidentified fodder plant. The leaf-blades are broad and non grass-like from a European perspective

hippothrix Gk *hippos*, coarse; *thrix*, hair. Hairs gathered, beard-like, towards the base of the leaf-blade

hippuris Panicle branches erect, the plants thereby resembling those of *Hippuris*

hirstii In honor of Frank Hirst (fl. 1959)

hirsut-a, -um, -us L. hairy. Plant hairy in respect to all or some parts

hirsutissim-a, -um, -us L. *hirsutus*, hairy; *-issima*, most. Plant in whole or in part very hairy

hirsutulum L. *hirsutus*, hairy; *-ulum*, denoting tendency. Plant hirsute throughout

hirsut-um, -us See *hirsuta*

hirt-a, -um, -us L. hairy. Hairy in part or extensively

hirtell-a, -um, -us L. *hirtus*, hairy; *-ella*, diminutive. Plants with slightly hairy leaves or spikelets

hirthii In honor of Adolphus Hirth (fl. 1885) who collected in Chile

hirticaul-e, -is L. *hirtus*, hairy; *caulis*, stem. Culms hairy

hirticulmis L. *hirtus*, hairy; *culmus*, stem. Culm bases densely hirsute

hirtiflor-a, -um, -us L. *hirtus*, hairy; *flos*, flower. Spikelets with hairy glumes and or lemmas

hirtifoli-a, -us L. *hirtus*, hairy; *folium*, leaf. Leaf-blades hairy

hirtigluma L. *hirtus*, hairy; *gluma*, husk. Glumes hairy

hirtiglumis See *hirtigluma*

hirtinoda L. *hirtus*, hairy; *nodus*, noded. Young culms having hairy nodes

hirtinodes L. *hirtus*, hairy; *nodes*, knot. Nodes hairy

hirtinoides A misspelling of *hirtinodes*

hirtissima L. *hirtus*, hairy; *-issima*, most. Leaf-sheath densely hairy

hirtiusculum L. *hirtius*, more hairy; *-ulum*, denoting tendency. Somewhat hairy

hirtivaginat-a, -us L. *hirtus*, hairy; *vagina*, sheath; *-ata*, possessing. Leaf-sheath hairy

hirtivaginum L. *hirtus*, hairy; *vagina*, sheath. Leaf-sheaths hirsute

hirtovaginatus A misspelling of *hirtivaginatus*

hirtul-a, -um L. *hirtus*, hairy; *-ula*, diminutive. Somewhat hairy with respect to some or all parts

hirt-um, -us See *hirta*

hisauchii In honor of Kiyotaka Hisauchi or Hisauti (1884–1981) Japanese botanist

hispanic-a, -um L. *-ica*, belonging to. From Hispania, now Spain

hispid-a, -um, -us L. bearing coarse stiff hairs. Plant wholly or partly rough to the touch

hispidifolium L. *hispidus*, bearing coarse stiff hairs; *folium*, leaf. Leaves hairy

hispidissim-um, -us L. most hispid. Plant densely invested with stiff hairs

hispidul-a, -um, -us L. *hispidus*, bristly; *-ula*, diminutive. With minutely hispid glumes

hispid-um, -us See *hispida*

hissaric-a, -um, -us L. *-ica*, belonging to. From Hissar District, Turkestan

hitachiensis L. *-ensis*, denoting origin. From Hitachi Province, now Ibaraki Prefecture, Japan

Hitchcockella L. *-ella*, diminutive but here used as a name-forming suffix. In honor of Albert Spear Hitchcock (1865–1935) United States agrostologist

hitchcockian-a, -um L. *-ana*, indicating connection. As for *Hitchcockella*

hitchcockii (1) As for *Hitchcockella*. *Axonopus hitchcockii, Digitaria hitchcockii, Paleoericoma hitchcockii, Paspalum hitchcockii, Redfieldia hitchcockii, Styppeiochloa hitchcockii, Thrasya hitchcockii, Trichachne hitchcockii, Tristachya hitchcockii, Valota hitchcockii* – (2) in honor of Hitchcock who collected in Zimbabwe. *Loudetia hitchcockii*

hiugensis L. *-ensis*, denoting origin. From Hiuga or Hyuga Province, now Miyazaki Prefecture, Japan

hiyamana L. *-ana*, indicating connection. In honor of Kôzô Hiyama (1905–) Japanese botanist

hiyeiana L. *-ana*, indicating connection. From Mt Hiyeizan, Japan

hizaoriensis L. *-ensis*, denoting origin. From Hizaori, Musashi Province, now Tokyo Prefecture and parts of Saitama and Kanagawa Prefectures, Japan

hizenensis L. *-ensis*, denoting origin. From Hizen Province, now much of Saga and Nagasaki Prefectures, Japan

hobdyi In honor of Robert Warner Hobdy (1942–) United States botanist

hochreutineri In honor of Benédict Pierre Georges Hochreutiner (1873–1959) Swiss botanist

hochstetterian-a, -um L. *-ana*, indicating connection. In honor of Christian Gottlob Ferdinand Hochstetter (1829–1884) German anthropologist and geologist

hockii In honor of Adrien Hock (fl. 1910) who collected in Zaire

hodgsonii In honor of Harlow James Hodgson (1917–) United States agronomist

hoehnei In honor of Fredrico Carlos Hoehne (1882–1959) Brazilian botanist

hoffmannii In honor of Carl Hoffmann (?–1859) who collected in Costa Rica

hoffmannseggii In honor of Johannes Centurius, Graf von Hoffmannsegg (1766–1849) German museum curator and traveller

hoggarensis L. *-ensis*, denoting origin. From the Hoggar, also known as Ahaggar Mountains of southern Algeria

hoggariensis As for *hoggarensis*

hogoensis L. *-ensis*, denoting origin. From Hogo, a mountain in Taiwan

hohenackeriana L. *-ana*, indicating connection. In honor of Rudolf Friedrich Hohenacker (1798–1874) Swiss-born German botanist

hoi In honor of Y. Y. Ho (fl. 1957) Chinese plant collector

hoiensis L. *-ensis*, denoting origin. From Hoia, Romania

hokianum L. *-anum*, indicating connection. From Hoki Province, now part of Tottori Prefecture, Japan

holathera Gk *holos*, wholly; *ather*, barb or spine. Awn not disarticulating from base of lemma at maturity

Holboellia In honor of Cave Peter Holbøll (1795–1856) Danish botanist

holciform-e, -is L. *forma*, appearance. Inflorescence a dense panicle as with *Holcus*

holcoides Gk *-oides*, resembling. Similar to *Holcus* with respect to the inflorescence

Holcolemma Gk *holcos*, strap; *lemma*, scale. Lemma of lower floret strap-like

Holcus Gk *holco*, draw. Used in Classical Times to remove hairs from the body

holgateana L. *-ana*, indicating connection. In honor of Martin Wyatt Holdgate (1931–) British biologist

hollei In honor of G. Holle (1825–1893)

holmbergii In honor of Rudolf Holmberg (1874–1930) Swedish botanist

holmesii In honor of G. E. Holmes (fl. 1940–1944) Australian cleric and plant collector

holmii In honor of Herman Theodor Holm (1854–1932) Danish-born United States botanist

holochrysum Gk *holos*, entire; *chrysos*, yellow. Leaf-blades and spikelets invested with yellow hairs

Hologamium Gk *holos*, entire; *gamo*, marriage. Spikelets both heterogamous and polygamous

H

hololeuca Gk *holos*, entire; *leukos*, white. Densely pubescent with long white hairs

holoserice-a, -um Gk *holos*, entire; *sericea*, silky. Plant or spikelets totally invested in dense hairs

Holosetum Gk *holos*, entire; L. *seta*, bristle. Upper glume hairy overall

holotricha Gk *holos*, entire; *thrix*, hair. Plants greyish-tomentose

holstii In honor of Carl Hugo Ehrenfried Wilhelm Holst (1865–1894) German-born botanist who collected in East Africa

holttumiana L. *-ana*, indicating connection. As for *Holttumochloa*

Holttumochloa Gk *chloa*, grass. In honor of Richard Eric Holttum (1895–1990) English botanist

holubii In honor of – (1) Emil Holub (1847–1902) Bohemian naturalist, explorer and physician. *Echinochloa holubii*, *Panicum holubii* – (2) Josef Holub (1930–1999) Czech botanist. *Festuca holubii*

holwayi As for *holwayorum*

holwayorum In honor of Edward Willet Dorlan Holway (1853–1923) and Mary Ellen Holway (1872) United States plant collectors in North and South America

Homalachna, Homalachne Gk *homalos*, of like degree; *achne*, scale. Glumes approximately equal in length

Homalocenchrus Gk *homalos*, of like degree; *kegchros*, millet. The spikelets resemble those of millet (*Panicum miliaceum*) as understood in Classical times

homblei In honor of Henri Antoine Homblé (1883–1921) Belgian botanist

Homeoplitis See *Homoplitis*

homochlamys Gk *homos*, alike; *chlamys*, cloak. Glumes similar

Homoeantherum See *Homoeatherum*

Homoeatherum Gk *homos*, alike; *ather*, barb or spine. Lateral nerves of lower glume extend as a pair of similar awns

homogamus Gk *homos*, alike; *gamos*, marriage. The sessile and stalked spikelets are alike

Homoiachne See *Homalachna*

Homolepis Gk *homos*, alike; *lepis*, scale. The glumes are similar in size, shape and texture

homomalla Gk *homos*, alike; *mallos*, stem of onion. Culms with only two leaves

homonym-a, -um Gk *homos*, alike; *onoma*, name. Species that have synonyms which are also homonyms

Homopholis Gk *homos*, alike; *pholis*, scale as of a snake. Upper glume and proximal lemma similar

Homoplitis Gk *homos*, alike; *hoplitis*, armed. One of the glumes and one of the lemmas is very long awned

Homopogon Gk *homos*, alike; *pogon*, beard. The bases of both the rhachis segments and pedicels bearded with long fulvous hairs

Homozeugos Gk *homos*, alike; *zeugos*, pair. The sessile and stalked spikelets are similar

hondae As for *hondana*

hondana L. *-ana*, indicating connection. In honor of Masaji Honda (1897–1984) Japanese botanist

hondoensis L. *-ensis*, denoting origin. From Hondo, Japan

hondurensis L. *-ensis*, denoting origin. From Honduras

hongyuanensis L. *-ensis*, denoting origin. From Hongyuan, Sichuan Province, China

honokowaiense L. *-ense*, denoting origin. From Honokowaion, Maui, one of the Hawaiian islands

hooglandii In honor of Ruurd Dirk Hoogland (1922–1994) Dutch botanist

hookeri In honor of – (1) Joseph Dalton Hooker (1817–1911) English botanist, traveller and Garden's Director. *Achnatherum hookeri*, *Andropogon hookeri*, *Anthistiria hookeri*, *Anthoxanthum hookeri*, *Arthraxon hookeri*, *Arundinaria hookeri*, *Ataxia hookeri*, *Avena hookeri*, *Avenula hookeri*, *Bambusa hookeri*, *Cymbopogon hookeri*, *Deschampsia hookeri*, *Erianthus hookeri*, *Hierochloe hookeri*, *Oplismenus hookeri*, *Saccharum hookeri*, *Stipa hookeri*, *Themeda hookeri*, *Trikeraia hookeri* – (2) William Jackson Hooker (1785–1865) English botanist. *Helictotrichon hookeri*, *Imperata hookeri*, *Vilfa hookeri*

hookeriana L. *-ana*, indicating connection. As for *hookeri*, usually in honor of Joseph Dalton Hooker

Hookerochloa Gk *chloa*, grass. In honor of Joseph Dalton Hooker (1817–1911) English botanist, traveller and Garden's Director

hooveri In honor of Robert Francis Hoover (1913–1970) United States botanist

hooverianus L. *-anus*, indicating connection. As for *hooveri*

Hoplismenus See *Oplismenus*

hoppeana L. *-ana*, indicating connection. In honor of David Heinrich Hoppe (1760–1846) German apothecary and botanist

Hordale Hybrids between species of *Hordeum* and *Secale*

hordeace-a, -us L. *-acea*, resembling. Inflorescence as with *Hordeum*

hordeiform-e, -is L. *forma*, appearance. Inflorescence resembling that of *Hordeum*

Hordeleymus Hybrids between species of *Hordeum* and *Elymus*

hordeoides Gk *-oides*, resembling. Similar to *Hordeum*

Hordeopyrum Hybrids between species of *Hordeum* and *Agropyron*

Horderoegneria Hybrids between species of *Hordeum* and *Roegneria*

Hordeum Roman name of barley

horizontal-e, -is L. horizontal. Primary panicle branches horizontal

hornemanniana L. *-ana*, indicating connection. As for *hornemannii*

hornemannii In honor of Jens Wilken Hornemann (1770–1841) Danish botanist

horneri In honor of – (1) Robert M. Horner (fl. 1896–1897) United States botanist. *Poa horneri* – (2) Ludwig Horner (1811–1838) Swiss surgeon-geologist who collected in Sumatra. *Eremochloa horneri, Paspalum horneri*

hornungiana L. *-ana*, indicating connection. In honor of Ernest Gottfried Hornung (1795–1862)

horrens L. *horro*, stand on end, as of hair. The leaf-blades are stiff and disposed more or less at right angles to the culm

horribilis L. terrible. Origin uncertain, not given by author

horridula L. *horridus*, prickly; *-ula*, diminutive. Leaf-blades stiff erect

horrifolia L. *horreo*, stand erect; *folium*, leaf. Leaf-blades, short and sub-pungent

horsfieldii In honor of Thomas Horsfield (1773–1859) United States physician and naturalist

horsfordianum L. *-anum*, indicating connection. In honor of Horsford

horstianum L. *-anum*, indicating connection. In honor of Horst

hortensis L. *hortus*, garden; *-ensis*, denoting origin. Commonly cultivated in pots

horticola L. *hortus*, garden; *-cola*, dweller. Described from plants cultivated at the Botanic Garden, Berlin

horvatiana L. *-ana*, indicating connection. In honor of Ivor Horvat (fl. 1937–1974) Yugoslav botanist

hosakae In honor of Edward Yataro Hosaka (1907–1961) Hawaiian botanist

hosidaikitiana L. *-ana*, indicating connection. In honor of Daikichi Hoshi (fl. c. 1936) Japanese botanist

hosomiana L. *-ana*, indicating connection. In honor of Hosomi, Japanese botanist

hosseana L. *-ana*, indicating connection. As for *hosseusii*

hosseusii In honor of Carl Curt Hosséus (1878–1950) German-born botanist and traveller who in later life lived in Argentina

hosteanum L. *-anum*, indicating connection. As for *hostii*

hostii In honor of Nicolaus Thomas Host (1761–1834) Austrian physician and botanist

hostilis L. enemy. Habit cushion-like and leaf-blades needle-like

hothamensis L. *-ensis*, denoting origin. From Mt Hotham, Victoria, Australia

houttuynii In honor of Maarten Houttuyn (1720–1798) Dutch naturalist and physician

Houzeaubambus In honor of Jean Houzeau de Lehaie (1820–1888) Belgian botanist

H

howellii In honor of – (1) Thomas Jefferson Howell (1842–1912) United States botanist. *Agrostis howellii, Alopecurus howellii, Calamagrostis howellii, Festuca howellii, Poa howellii* – (2) John Thomas Howell (1903–1994) United States botanist. *Puccinellia howellii, Trisetum howellii*

howensis L. *-ensis*, denoting origin. From Lord Howe Island, part of New South Wales, Australia

hozuensis L. *-ensis*, denoting origin. From Hozu, Hida Province, now part of Gifu Prefecture, Japan

hsuehana In honor of Chi-Ju Hsueh (1921–) Chinese botanist

huachucae L. of Huachuca Mountains, Arizona, USA

huallancaensis L. *-ensis*, denoting origin. From Huallanca, also called Huánuco, Peru

huamachucensis L. *-ensis*, denoting origin. From Huamachuco, Peru

huancavelicae From Huancavelica, Peru

huantensis L. *-ensis*, denoting origin. From Huanta Province, Argentina

huashanica L. *-ica*, belonging to. From Hua Shan, China

huatensis See *huantensis*

Hubbardia In honor of Charles Edward Hubbard (1900–1980) English agrostologist

hubbardiana L. *ana*, indicating connection. As for *Hubbardia*

hubbardii As for *Hubbardia*

Hubbardochloa Gk *chloa*, grass. See *Hubbardia*

hubeiensis L. *-ensis*, denoting origin. From Hubei Sheng, China

huberi In honor of Otto Huber (1944–) Venezuelan botanist

huber-morathii In honor of Arthur Huber-Morath (1901–1990) Swiss merchant and amateur botanist

hubsugulica L. *-ica*, belonging to. From Lake Khubsugal, Mongolia

huebneriana L. *-ana*, indicating connection. In honor of Hübner (fl. 1930) who collected in Saxony, Germany

huecu The vernacular name *huecú* means "intoxicator" in the Araucanian language of Chile and western Argentina. Plants host an ergot fungus toxic to grazing animals

huegelii In honor of Carl Alexander Anselm Huegel (1794–1870) German botanist

hugelii See *huegelii*

hugeninii In honor of Auguste Huguenin (1780–1860) French teacher and botanist

hughii In honor of Pietro Ugo Marchese delle Favare (1827–1898)

hugoniana L. *-ana*, indicating connection. In honor of Hugh (fl. 1898) cleric and botanical collector in China

hui In honor of Hsen Hsu Hu (1894–1968) Chinese botanist

huillens-e, -is L. *-ense*, denoting origin. From Huilla, Angola

hukudaeana L. *-ana*, indicating connection. In honor of Yutaka Hukuda, Japanese botanist

hukudana See *hukudaeana*

hulettii In honor of Garry K. Hulett (1936–) United States ecologist

hultenii In honor of Eric Oskar Gunnar Hultén (1894–1980) Swedish botanist

humbertian-a, -um, -us L. *-ana*, indicating connection. As for *Humbertochloa*

humbertii As for *Humbertochloa*

Humbertochloa Gk *chloa*, grass. In honor of Jean Henri Humbert (1887–1967) French botanist who collected in Madagascar

humboldtian-a, -um, -us L. *-ana*, indicating connection. In honor of Friedrich Heinrich Alexander Humboldt (1769–1859) German botanist and traveller

humboldtii As for *humboldtiana*

humidicola L. *humidus*, moist; *-cola*, dweller. – (1) Growing in damp meadows. *Brachiaria humidicola, Panicum humidicola* – (2) growing on the margins of lakes. *Eragrostis humidicola*

humidorum L. *humidus*, moist. Growing in moist places

humifusa L. procumbent. Culms prostrate

humila See *humile*

humil-e, -is L. low growing. Short-statured in comparison with related species and often prostrate

humilior L. more dwarfed. Species dwarf for the genus
humilis See *humile*
humillima L. most low growing. Low growing with respect to related species
hunanensis L. *-ensis*, denoting origin. From Hunan, China
hungarica L. *-ica, belonging to. From Hungary*
hunzikeri *In honor of Armando Teodoro Hunziker (1919–2001) Argentinian botanist*
huonii In honor of A. Huon (fl. 1961–1966) French botanist
hupehensis L. *-ensis*, denoting origin. From Hupeh, China
huppenthalii In honor of Huppenthal
husnotii In honor of Pierre Tranquilla Husnot (1840–1929) French botanist
hutatabiensis L. *-ensis*, denoting origin. From Hutatabiyama, Settsu Province, now part of Hyogo and Osaka Prefectures, Japan
huttonensis L. *-ensis*, denoting origin. In honor of J. H. Hutton of the Indian civil service
huttoniae In honor of Caroline Atherstone Hutton (1826–?) South African plant collector
hyachinensis L. *-ensis*, denoting origin. From Mt Hyachine, Japan
hyalin-a, -um Gk *hyalos*, glass; *-ina*, indicating resemblance. – (1) Margins of glumes, lemmas and sterile lemmas hyaline. *Digitaria hyalina*, *Melica hyalina*, *Paspalum hyalinum* – (2) glumes hyaline. *Stipa hyalina*
Hyalopoa Gk *hyalos*, glass. Similar to *Poa* but glumes membranous
hyaloptera Gk *hyalos*, glass; *pteron*, wing or feather-like. Lemma has two large hyaline wings
hybernum See *hibernum*
hybrid-a, -us L. of mixed parentage. Sharing the characters of two or more species and not necessarily genetic hybrids
hydaspicum L. *-icum*, belonging to. From Hydaspes, now the Behut or Djelun River, India
Hydrochloa Gk *hydor*, water; *chloa*, grass. Grasses of marshes and stream banks
hydrolithica Gk *hydor*, water; *lithos*, stone; *-ica*, belonging to. Growing on rocks in a seasonal stream
hydrophil-a, -um Gk *hydor*, water; *phileo*, love. Growing in or close to water
hydrophylla Gk *hydor*, water; *phyllon*, leaf. Growing in water
hydrophylum See *hydrophila*
Hydropoa Gk *hydor*, water; *poa*, grass. Growing in swamps
Hydropyrum Gk *hydor*, water; *pyros*, wheat. The species grow in fresh and brackish water
Hydrothauma Gk *hydor*, water; *thauma*, wonder. The leaves have long slender pseudopetioles allowing the blades to float upon water
hyemalis See *hiemalis*
hygrocharis Gk *hygros*, moisture; *charis*, favour. Growing in water or very moist habitats
Hygrochloa Gk *hygros*, moisture; *chloa*, grass. Swamp grasses
hygrometric-a, -um Gk *hygros*, water; *metron*, measure; *-ica*, belonging to. Aquatic or swamp species
hygrophila Gk *hygros*, water; *phileo*, love. Swamp or stream bank plants
Hygrorhiza See *Hygroryza*
Hygroryza Gk *hygros*, moisture. A swamp grass similar to *Oryza* but floating instead of rooted
hylaeicum Gk *hyle*, woodland; *-icum*, belonging to. Growing on forest margins
Hylebates Gk one who haunts the woods. Growing in shady places
hylobates See *Hylebates*
Hymenachne Gk *hymen*, membrane; *achne*, scale. The fertile lemma is membranous
hymeniochilum Gk *hymen*, membrane; *chilos*, green fodder. Spikelets green except for lemma of lower floret that has a hyaline margin
hymenoglossa Gk *hymen*, membrane; *glossa*, tongue. Ligule membranous
hymenoides Gk *hymen*, membrane; *-oides*, resembling. Spikelets with papery glumes

H

Hymenothecium Gk *hymen*, membrane; *anthos*, flower; *oikos*, house. Glumes membranous

hypanica L. *-ica*, belonging to. From Hypanis, the Classical Greek name for the river Bug in the Ukraine

Hyparrhenia Gk *hypo*, below; *arrhen*, male. Inflorescence with male homogamous spikelets are at the base of the raceme

hyperarctic-a, -us Gk *hyper*, over. Growing within the Arctic circle

hyperborea Gk *hyper*, over; *boreas*, north wind. Growing within the Arctic circle

Hyperthelia Gk *hyper*, above; *thele*, a female. The fertile spikelets occur above the homogamous pair

hypnoides Gk *hypnos*, moss; *-oides*, resembling. Habit moss-like

hypogona Gk *hypo*, below; *gony*, knee. Origin uncertain, not given by author

Hypogynium Gk *hypo*, below; *gyne*, woman. The stalked members of the paired spikelets are male and arise from below the sessile hermaphrodite floret

hypogynus See *hypogyna*

hypomegas Gk *hypo*, below; *megas*, large. Lower glume much longer than the upper

hypopsila Gk *hypo*, below; *pilos*, bare. Lemma with a few hairs only at the base

hypsenephis Gk *hypsi*, aloft; *nephos*, cloud. Alpine species

Hypseochloa Gk *hypsi*, aloft; *chloa*, grass. The genus is endemic at high altitudes on Mt Cameroon, West Africa

hypsophila Gk *hypsi*, aloft; *phileo*, love. Alpine species

Hypudaerus Origin uncertain, not given by author. Forming thick bushes

Hystericina Gk *hystrix*, hedgehog; *-ina*, indicating resemblance. Inflorescence resembles a hedgehog

hystrichoides Gk *-oides*, resembling. Similar to *Hystrix*

hystricina Gk *hystrix*, hedgehog; *-ina*, indicating resemblance. Plant tufted with setaceous-juncoid leaf-blades which resemble the quills of a hedgehog

hystricula L. *-ula*, diminutive. Resembling *Hystrix* in the form of the inflorescence

Hystringium Gk *hystrinx*, hedgehog; L. *-ium*, resembling. Spikelets bristly

Hystrix Gk *hystrix*, hedgehog. The long awned spikelets of the type species resemble a hedgehog

hystrix Resembling *Hystrix*. Usually a reference to the inflorescence resembling that of *Hystrix*

I

ianthina L. *iantha*, violet; *-ina*, indicating resemblance. Inflorescence violet-colored

ianthoides L. *iantha*, violet; Gk *-oides*, resembling. Spikelets pale-purple

ianthum L. violet. Spikelets pale-purple

iaponica See *japonica*

ibarii In honor of Enrique Ibar (fl. 1877–1878) who collected in Patagonia

ibarrens-e, -is L. *-ense*, denoting origin. From Villa de Iberra, near Quito, Ecuador

iberica L. *-ica*, belonging to. – (1) From Iberia, a province of the Republic of Georgia. *Calamagrostis iberica*, *Poa iberica* – (2) from the Iberian Peninsula, that is Spain and Portugal. *Stipa iberica*

ibiramae From Ibirama, Caterina Province, Brazil

ibitense L. *-ense*, denoting origin. See *ibityensis*

ibityensis L. *-ensis*, denoting origin. From Mt Ibity, Madagascar

ibizensis L. *-ensis*, denoting origin. From Ibiza, now Ivaza, one of the Balearic Islands

ibukiana L. *-ana*, indicating connection. From Mt Ibuki, Shiga Prefecture, Japan

iburua Hausa *iburu*, local name. A cereal grown in Upper Guinea and Northern Nigeria

ichnanthoides Gk *-oides*, resembling. Similar to *Ichnanthus*

Ichnanthus, Ischnanthus Gk *ichnos*, vestige; *anthos*, flower. Lower floret of spikelet incomplete

ichnodes Gk *ichnos*, vestige; *-odes*, resembling. Fertile floret with two ligular appendages at its base
ichu Quechua, straw or grass-like plant. Name for several grass species in Peru
ichunense L. *-ense*, denoting origin. From Río Ichun, Venezuela
ichyostachyum Gk *ichthyos*, fish; *stachys*, spike as of an ear of wheat. The overlapping spikelets resemble fish-scales
iconia From Icona, now Konia, Turkey
idahoensis L. *-ensis*, denoting origin. From Idaho, USA
idjenensis L. *-ensis*, denoting origin. From Idjin, Java, Indonesia
idukkiensis L. *-ensis*, denoting origin. From the Idukki District, Kerala state, India
igaensis L. *-ensis*, denoting origin. From Iga Plateau, Japan
igagoyeana L. *-ana*, indicating connection. From Iga Province, now western Mie Prefecture, Japan
ignoratum L. *ignoro*, mistake. Previously included in another species
igoschinae In honor of K. Igoshina (fl. 1958) who collected in Urals
ihosyense L. *-ense*, denoting origin. From Ihosy, Madagascar
ikegamii In honor of Yoshinobu Ikegami (fl. 1934)
ikomanum L. *-anum*, indicating connection. In honor of Y. Ikoma (fl. 1929) Japanese botanist
ikopense L. *-ense*, denoting origin. From the Ikopa River basin, Madagascar
ilgazensis L. *-ensis*, denoting origin. From Ilgaz Dagh, a mountain in Turkey
iliensis L. *-ensis*, denoting origin. From river Ili, Kazakhstan
iljinii In honor of Modesta Michailovich Iljin (1889–1967) Polish-born Russian botanist
illimanica L. *-ica*, belonging to. From Mt Illiman, Bolivia
illinoniense L. *-ense*, denoting origin. From Illinois State, USA
illyrica L. *-ica*, belonging to. From Illyria, nowadays Dalmatia and Albania
imadatensis L. *-ensis*, denoting origin. From Imadategun, Yetizan Province, Japan
imatongensis L. *-ensis*, denoting origin. From Imatong Mountains, Sudan
imatophylla Gk *imas*, leather-strop; *phyllon*, leaf. Leaf-blades membranous to subchartaceous
imbaburensis L. *-ensis*, denoting origin. From Imbabura, Ecuador
imbecill-a, -is L. feeble. – (1) Habit creeping or forming lax tufts. *Oplismenus imbecillus* – (2) inflorescence slender. *Agrostis imbecilla, Eragrostis imbecilla, Poa imbecilla*
imberb-e, -is L. beardless. Glumes and/or lemmas glabrous
imbricat-a, -um L. *imbricare*, overlap like roof-tiles. The shorter branches bear densely overlapping spikelets
imeretica L. *-ica*, belonging to. From Imeretia, a mountain range in the Republic of Georgia
imerinensis L. *-ensis*, denoting origin. From Imerin, Madagascar
immers-um, -us L. sunken. Spikelets on very short pedicels borne on one side of a winged rhachis
impeditum L. *empedio*, hinder, hence not completely formed. Lower floret sterile
Imperata In honor of Ferrante Imperato (1550–1625) an apothecary from Naples
imperatoides Gk *-oides*, resembling. Similar to *Imperata*
imperfect-a, -us L. incomplete. – (1) Only one floret of spikelet fertile. *Melica imperfecta* – (2) only one spikelet of cluster developed. *Aegopogon imperfectus*
imperialis L. *imperium*, rule; *-alis*, pertaining to. The finest of all Himalayan species of *Poa*
implexa L. *implecto*, interweave. – (1) Awn-branches long and intertwined. *Aristida im-plexa* – (2) leaf-blades intertwining. *Poa implexa*
implicat-a, -um L. *implico*, entangle. Inflorescence branches intertwined
importunus L. troublesome. A troublesome weed difficult to eradicate
impress-a, -um, -us L. *imprimo*, press in. Lower glume of sessile spikelet deeply concave

imrinum L. *-inum*, indicating possession. From the island of Imbros in the Aegean Sea

inaequal-e, -is L. *in-*, not; *aequalis*, equal. – (1) The glumes differ in length. *Deyeuxia inaequalis, Muhlenbergia inaequalis* – (2) the spikelet pedicels are of different length. *Panicum inaequalis* – (3) the upper glume and sterile lemma markedly dissimilar. *Digitaria inaequale, Panicum inaequale*

inaequiglum-e, -is L. *inaequalis*, unequal; *gluma*, husk. Glumes differing in length and/or shape

inaequilateralis L. *-alis*, pertaining to. See *inaequilaterus*

inaequilaterus L. *in-*, not; *aequus*, equal; *latus*, side. Leaf-blades asymmetric with respect to midrib

inaequiloba L. *in-*, not; *aequus*, equal; *lobus*, lobe. Median lobe of lemma longer than the laterals

inaequivalve L. *in-*, not; *aequus*, equal; *valvus*, leaf of a folding door. Glumes differing in size and nervation

inamaena See *inamoena*

inamoena L. *in-*, not; *amoena*, beautiful. Appearance unattractive in comparison with related species

inarmata L. *in-*, not; *armo*, arm. Lemma apex blunt

inaurita L. *in-*, not; *aurita*, eared. Auricles and ligules inconspicuous

incan-a, -um, -us L. *grey*, hoary. Leaf-blades or spikelets villous

incanellus L. *-ellus*, diminutive. Similar to but smaller than *Andropogon incanus*

incan-um, -us See *incana*

incis-a, -um L. cut deeply. Lower glume deeply bifid

inclusum L. *includo*, enclose. – (1) Racemes enclosed within spathes. *Schizachyrium inclusum* – (2) spike partially enclosed in subtending leaf-sheath. *Pennisetum inclusum*

incomplet-um, -us L. imperfect. – (1) Terminal floret incomplete. *Cynodon incompletus* – (2) pedicellate florets sterile. *Andropogon incompletus, Sorghastrum incompletum, Sorghum incompletum*

incomptum L. unadorned. Anthoecial surface dull rather than glossy

incomtum L. unadorned. Glumes and sterile lemma glabrous except for a few hairs at their apices

inconspicu-a, -us L. inconspicuous. Easily overlooked

inconstans L. variable. Leaf-blades vary in outline from sub-cordate to linear-lanceolate in outline

incrassat-a, -us L. thickened. Culm base swollen

increscens L. *incresco*, grow. Similar to *Andropogon fulvus* but with bigger spikelets

incumbens L. *incumbo*, lie upon. Lateral inflorescence branches appressed to central axis

incurv-a, -us L. bowed. – (1) Inflorescences curved spikes. *Aegilops incurva, Nardus incurva, Parapholis incurva, Psilurus incurvus, Sacciolepis incurva* – (2) leaf-blades recurved. *Poa incurva*

incurvat-a, -us L. *incurvo*, bend. Inflorescence an incurved spike

incurvus See *incurva*

indandamanica L. *-ica*, belonging to. From southern India and the Andaman Islands

indeprensa L. *in-*, not; *deprendo*, detect. Segregated from a closely related species

indetonsus L. unshorn. Pedicels of reduced spikelets densely hairy

indic-a, -um, -us L. *-ica*, belonging to. From India

indigesta L. *in-*, not; *digero*, dissolve. The rigid and pungent leaf-blades are not edible

Indocalamus L. *Indus*, India; *kalamos*, reed. Small reed-like Indian bamboo

Indochloa L. *Indus*, India; Gk *chloa*, grass. From India

Indopoa L. *Indus*, India. From India and resembling *Poa*

Indoryza L. *Indus*, India. Resembling *Oryza* and from India

Indosasa Similar to *Sasa* with the type species described from Tonkin, now Vietnam

indum L. *Indus*, India. From India

indurat-a, -um, -us L. *induro*, make hard. – (1) Palea of lower floret rigid and woody at maturity. *Gilgiochloa indurata* – (2) glumes hardened. *Coelachyrum induratum* – (3) leaf-sheaths woody. *Bromus induratus*

indut-a, -um L. *induo*, clothe. Glumes and/or lemmas densely hairy

inebrians L. *inebrio*, intoxicate. When grazed by cattle in Mongolia, they exhibit symptoms of intoxication

ineptum L. unsuitable. Meaning obscure, origin not given by the author

inerm-e, -is L. unarmed. Apices of lemmas or glumes rounded

inexpectans L. *in-*, not; *expecto*, expect. Unexpected in the sense of segregated from another species

infecunda L. *in-*, not; *fecundus*, fertile. No grain found after persistent searching

infest-a, -um L. hostile. Meaning uncertain but may refer to growing in a hostile environment

infirm-a, -um, -us L. *lax*, weak. Culms decumbent

inflat-a, -um L. swollen. – (1) Pedicels or inflorescence internodes inflated. *Agrostis inflata*, *Thyrsia inflata* – (2) lemmas swollen. *Berriochloa inflata*, *Chloris inflata*, *Melica inflata*, *Panicum inflatum*, *Triticum inflatum*

inflex-a, -us L. bent inwards. Panicle branches held erect

infuscum L. dusky-brown. Anthoecium light-brown

ingens L. enormous. Exceeding in size that which is usual for related species

ingrat-a, -us L. disagreeable. The sharp callus enables the spikelets to catch on to clothing

inguschetica L. *-ica*, belonging to. From Inguschetia, the upper reaches of the Shon-don River, Caucasus, a mountain range separating the Black and Caspian Seas

innominata L. *in-*, not; *nomen*, name; *-ata*, possessing. The species was recognized but remained unnamed for many years

innovatus L. *innovo*, renew. Readily regenerating from creeping root stalks

inopia L. scarcity. Locally restricted to seashores around Sea of Okhotsk

inops Gk weak. Culms slender

inordinatus L. *in-*, not; *ordino*, arrange. Lower inflorescence branches not whorled

inscalpt-um, -us L. *inscalptus*, engraved. Upper glume transversely ribbed

insculpt-a, -um, -us L. engraved. Glumes ridged or with a round depression

inserta L. *insero*, place among. Inflorescence overtopped by upper leaves

insign-e, -is L. outstanding. Culms tall for genus

insolit-a, -us L. uncommon. – (1) The species is rare. *Poa insolita* – (2) the species is geographically restricted although locally abundant. *Andropogon insolitus*

insperata L. unexpected. A newly recognized genus in an otherwise taxonomically well studied genus

inspersum L. *insergo*, scatter. Lower glume with scattered hairs

insubrica L. *-ica*, belonging to. From the land of the Insubres, now included in Northern Italy and Southern Switzerland

insulae-cypri L. *insula*, island. From Cyprus

insular-e, -is L. *insula*, island; *-are*, pertaining to. Island species

insularum L. *insula*, island. Of islands of the Lesser Antilles, Netherlands islands in the Caribbean

insulatlantica L. *insula*, island. From Cape Verde Islands, a Republic in the Atlantic Ocean

insulicola L. *insula*, island; *-cola*, dweller. The species is from the island of Java, Indonesia

intect-a, -um L. *in-*, without; *tectum*, cover. – (1) Spikelets lacking glumes. *Digitaria intecta* – (2) lacking subtending bristles. *Pennisetum intectum*

integ-er, -a L. *integer*, entire. Upper apex of lemma not divided

intercedens L. *intercedo*, come between. Intermediate between two other species

interceptus L. *intercipio*, interrupt. Pedicels hairy at their bases and apices but not in between

interi-or, -us L. interior. From inland areas such as the central part of the United States

interjacens L. *interjaceo*, lie between. A presumed hybrid

interjectum L. *interjicio* or *interjacio*, put between or intermix. Not stated by the author, but probably either because the species is known from two widely separated areas or because some diagnostic characters are intermediate in nature

intermedi-a, -um, -us L. intermediate. Having affinities with but distinct from other species

interrupt-a, -um, -us L. not continuous. Spikelets or inflorescence branches clustered at intervals along an axis

intersita L. *inter*, between; *situs*, place. Intermediate between its putative parents

interstipitata L. *inter*, between; *stipes*, stalk; *-ata*, possessing. Glumes separated by a conspicuous internode

intons-um, -us L. *in-*, not; *tonsus*, shaven. Plant densely hairy

intrans L. *intro*, enter. With their long awns and short calli, the detached spikelets readily penetrate animal skins and clothing

intricata L. *intrico*, entangle. – (1) Inflorescence branches entangled. *Agrostis intricata*, *Stipa intricata* – (2) distinguished with difficulty from another species. *Aristida intricata*

intrusa L. *intrudo*, thrust in. The geographical distribution of the species is included within that of a related taxon

intumescens L. *intumesco*, swell up. Pedicel of stalked spikelet club-shaped

inukamiensis L. *-ensis*, denoting origin. From Inukamigun, Shiga Prefecture, Japan

inundat-a, -um, -us L. *inundo*, flooded. Growing in places subject to flooding

inutilis L. *in-*, not; *utilis*, useful. Not useful for domestic grazing animals

invaginata L. *in-*, not; *vagina*, sheath; *-ata*, possessing. Intravaginal shoots numerous

invalida L. *in-*, not; *validus*, valid. Spikelet structure not fully characteristic of the genus

invers-a, -um L. *inverto*, turn upside down. – (1) Contrary to expectation, the lower glume is larger than the upper glume. *Aristida inversa*, *Panicum inversum* – (2) in contrast to a related species, the culms are glabrous. *Pleioblastus inversus*

involucrat-a, -um, -us L. *involucrum*, cover; *-ata*, possessing. – (1) Inflorescence subtended by bracts. *Alopecurus involucratus*, *Ammophila involucrata*, *Andropogon involucratus*, *Chaetobromus involucrata*, *Cornucopiae involucratum*, *Hyparrhenia involucrata*, *Periballia involucrata* – (2) subtended by bristles. *Panicum involucratum*

involut-a, -um, -us L. *involvo*, inroll. Leaf-blades or sheaths inrolled

Ioackima See *Joachimia*

ioclados Gk *ion*, the violet; *klados*, branch. Panicle branches violet

iodostachys Gk *ion*, the violet; *-oides*, resembling; *stachys*, spike as of an ear of wheat. Inflorescence purple

ionanthum Gk *ion*, the violet; *anthos*, flower. Stigmas and anthers deep purple

iowense L. *-ense*, denoting origin. From Iowa, USA

ipamuensis L. *-ensis*, denoting origin. From Ipamua, Zaire

Ipnum Gk *hypnos*, moss. Habit moss-like

iranic-a, -um L. *-ica*, belonging to. From Iran

iraten Vernacular name of the species in Java, Indonesia, and translating as "split bamboo"

irazuens-e, -is L. *-ense*, denoting origin. From Volcán Irazú, Costa Rica

ircutensis See *irkutensis*

ircutica L. *-ica*, belonging to. From Irkutsk Province, Siberia

irianensis L. *-ensis*, denoting origin. From Irian Jaya, now Papua, Indonesia

iridaceum L. *-aceum*, resembling. Leaf-sheaths strongly compressed the plant thereby resembling an *Iris* in habit

iridenscens Possibly a misspelling of *iridescens*

iridifoli-a, -um L. *folium*, leaf. Leaf-blades rather wide and flat, the plants thereby resembling *Iris* species

iringense, iringensis L. *-ense*, denoting origin. From Iringa, Tanzania

irkutensis L. *-ensis*, denoting origin. From Irkut River, southern Siberia

irkutica L. *-ica*, belonging to. From Irkutsk Province, Siberia

irratun See *iraten*

irregulare L. irregular. Spikelets arranged on one side of the rhachis instead of an open panicle

irrigata L. *irrigo*, conduct water. Growing in swamps, meadows and the sides of drains

irritans L. *irrito*, irritate. – (1) Leaf-blades rigid, pungent. *Triodia irritans* – (2) callus sharp. *Pollinia irritans, Pseudopogonatherum irritans, Saccharum irritans*

irtyshensis L. *-ensis*, denoting origin. From Irtysh, western Siberia

Irulia Vernacular name for species of the genus in Travancore, India

isabelensis L. *-ensis*, denoting origin. From Isla Isabel, Nayarit, Mexico

Isachne Gk *isos*, equal; *achne*, scale. Glumes more or less similar

isachne Gk *isos*, equal; *achne*, scale. Upper glume and sterile lemma similar

isachnoides Gk *-oides*, resembling. Resembling *Isachne* with respect to spikelets

isalens-e, -is L. *-ense*, denoting origin. See *Isalus*

Isalus From the Isalo Range, Madagascar

ischaemoides Gk *-oides*, resembling. Similar to *Ischaemum*

Ischaemopogon Gk *pogon*, beard. Resembling *Ischaemum* but base of fertile spikelet invested with hairs

Ischaemum Gk *ischaemon*, styptic. The hairy spikelets of *Ischaemum* are reputed to staunch bleeding

ischaemum Gk *ischaemon*, styptic. Spikelets hairy

Ischnanthus See *Ichnanthus*

ischnocaulon Gk *ischnos*, meagre; *kaulos*, stems. Culms terete, slender

Ischnochloa Gk *ischnos*, meagre; *chloa*, grass. The inflorescence is poorly developed

Ischnurus Gk *ischnos*, meagre; *oura*, tail. The inflorescence is a spike of small spikelets

Ischoemum See *Ischaemum*

Ischurochloa Gk *ischyros*, great; *chloa*, grass. Culms tall and woody

ischyranthus Gk *ischyros*, great; *anthos*, flower. The spikelets bear awns up to 20 cm in length

ischyroneura Gk *ischyros*, strong; *neuron*, nerve. Lemmas conspicuously seven-nerved

Iseilema Gk *isos*, equal; *eilema*, covering. The involucral male spikelets surround and so protect the central hermaphrodite spikelet

ishiharae In honor of Ishihara, Japanese botanist

ishizuchiana L. *-ana*, indicating connection. From Ishizuchiyama, a mountain in Ehime Prefecture, Japan

isiaca In honor of Isis, an Egyptian Goddess from Egypt

isingiana L. *-ana*, indicating connection. In honor of Ernest Horace Ising (1884–1973) Australian civil servant and amateur botanist

isocalycin-a, -um Gk *isos*, equal; *kalyx*, cup; *-ina*, indicating resemblance. Glumes similar in size and shape

isoldeae In honor of Isolde Hagemann (1944–) German botanist

isolepis Gk *isos*, equal; *lepis*, scale. Glumes and lemmas similar

isopholis Gk *isos*, equal; *pholis*, scale of a snake. Glumes equal or subequal

isostachyus Gk *isos*, equal; *stachys*, spike as of an ear of wheat. Ultimate racemes of the inflorescence fasciculate and of the same length

ispahanicum L. *-icum*, belonging to. From Esfahan, Iran

ispanicum A misspelling of *hispanica*

issatchenkoi In honor of Boris Laurentiewicz Issatchenko (1871–?) Russian botanist

issongense L. *-ense*, denoting origin. From Isongo, East Africa

itaboense L. *-ense*, denoting origin. From Itabo, Isla de la Juventud, Cuba

italic-a, -um, -us L. *-ica*, belonging to. From Italia, that is Italy

itatiaiae From Serra Itatiaia, Brazil

ithaburense L. *-ense*, denoting origin. From Mount Ithaburum, now Mt Tabor, Israel

itieri In honor of Jules Itier (fl. 1843–1874) French naturalist and traveller

iuncl-um, -us See *juncea*

ivakoanyensis L. *-ensis*, denoting origin. From Massif de l'Ivakoany, Madagascar

ivanovae In honor of Valentina Ivanova (1928–) Russian botanist

ivingense L. *-ense*, denoting origin. From Ivinga, Malawi

ivohibens-e, -is L. *-ense*, denoting origin. From Pic d'Ivohibé, Madagascar

ivorensis L. *-ensis*, denoting origin. From Ivory Coast, in particular from Togo

iwabuchiana L. *-ana*, indicating connection. As for *iwabuchii*

iwabuchii In honor of Hatsuro Iwabuchi, Japanese botanist

iwakiana L. *-ana*, indicating connection. From Iwaki Province, now part of Fukushima Prefecture, Japan

iwakiensis L. *-ensis*, denoting origin. From Iwaki Province, now part of Fukushima Prefecture, Japan

iwamatoi In honor of Hidenobu Iwamato (fl. 1932) Japanese botanist

iwamiana L. *-ana*, indicating connection. From Iwami Province, now part of Shimane Prefecture, Japan

iwarancusa Sanskrit *jwara*, fever; *khusa*, grass. Used medicinally to control fever

iwateana L. *-ana*, indicating connection. From Mt Iwateyana, Hondo Prefecture, Japan

iwatekensis L. *-ensis*, denoting origin. From Iwate-ken, Japan

iwayae In honor of K. Iwaya (fl. 1932) Japanese botanist

Ixalum Gk *ixalos*, bounding. The detached spherical female inflorescences bowl along the beach in response to the slightest breeze

Ixophorus Gk *ixos*, birdlime; *phero*, bear. Bristle subtending spikelet sticky

iyasakaensis L. *-ensis*, denoting origin. From Iyasaka, Hiroshima Prefecture, Japan

iyomontana L. *mons*, mountain; *-ana*, indicating connection. From the mountains of Iyo Province, now Ehime Prefecture, Japan

J

jaboncillo Origin unknown, not given by author

jacobinae From Jacobina, Brazil

jacobsiana L. *-ana*, indicating connection. In honor of Surrey Wilfred Laurance Jacobs (1946–) Australian botanist

jacobsii In honor of Marius Jacobs (1929–) Dutch botanist

jacquemontii In honor of Victor Jacquemont (1801–1832) French naturalist and traveller

Jacquesfelixia In honor of Henri Jacques-Félix (1907–) French botanist and tropical agronomist

jacquiniana L. *-ana*, indicating connection. As for *jacquinii*

jacquinii In honor of Nikolaus Joseph Jacquin (1727–1817) Dutch-born Austrian botanist

jaculatorium L. *jaculatorius*, for throwing. The young inflorescences resemble spears

jacutens-e, -is L. *-ense*, denoting origin. See *jacutica*

jacutica L. *-ica*, belonging to. From Jacutia, now Yakutsk, eastern Siberia

jacutorum See *jacutica*

jaegeri In honor of Fritz and Oehler Eduard Jaeger (fl. 1906–1907) who collected in East Africa

jaegerian-a, -us L. *-ana*, indicating connection. – (1) As for *jaegeri*. *Hyparrhenia jaegeriana*, *Parahyparrhenia jaegeriana* – (2) in honor of Paul Jaeger (Strasbourg). *Tripogon jaegerianus*

jaffuelii In honor of P. Félix Jaffuel (1874–1939) Chilean plant collector

jagnobica L. *-ica*, belonging to. From Yagnoba Valley, Central Asia

jaguaense L. *-ense*, denoting origin. From Castillo de Jagua, Cuba

jahandiezii In honor of Émile Jahandiez (1876–1938) who wrote about the grasses of North Africa

jahnii In honor of Alfredo Jahn (1867–1940) who collected in Venezuela

jaime-hintonii In honor of Jaime (James C.) Hinton (fl. 1940) who collected in Mexico; collections by Hinton family members were often, as in this case, attributed solely to George Boole Hinton (1882–1943) the father of Jaime
jainiana L. *-ana*, indicating connection. In honor of Sudhanshu Kumar Jain (1926–) Indian botanist
jainii As for *jainiana*
jakubzineri In honor of Jakubziner (fl. 1958) Russian cereal breeder
jakutens-e, -is L. *-ense*, denoting origin. See *jacutica*
jalapense L. *-ense*, denoting origin. From Jalapa, Guatemala
jaliscan-a, -um L. *-anum*, indicating connection. From Jalisco, Mexico
jaliscoanum L. *-anum*, indicating connection. From Jalisco, Mexico
jamaicens-e, -is L. *-ense*, denoting origin. From Jamaica
jamesensis L. *-ensis*, denoting origin. From Baie James, Quebec, Canada
jamesii In honor of Edwin James (1797–1861) United States physician and botanist
jamesoniana L. *-ana*, indicating connection. As for *jamesonii*
jamesonii In honor of William Jameson (1796–1873) who collected in South America
jaminianum L. *-anum*, indicating connection. In honor of Pierre Jamin (?–1866)
janczewskii In honor of Edward Franciszek Janczewski-Glinka (1846–1918) Polish botanist
jankae In honor of Victor Janka (1837–1890) Austrian-born Hungarian botanist
Jansenella L. *-ella*, diminutive here used as a name-forming suffix. in honor of Pieter Jansen (1882–1955) Dutch agrostologist
jansenii As for *Jansenella*
januarium L. *-ium*, belonging to. From Rio de Janiero, Brazil
japonensis L. *-ensis*, denoting origin. From Japan
japonic-a, -um, -us L. *-ica*, belonging to; *Japan*, a modified spelling of *Zhapan* introduced into Europe by Marco Polo as a transliteration for the Chinese name for the large islands to the east of that country. From Japan
Jarapha Variant spelling of *Jarava*
Jarava, jarava In honor of Juan de Jarava (fl. 1557) Spanish physician and naturalist
Jardinea In honor of Désiré Edeleston Stanilus Aimé Jardin (1822–1896) French naval officer and amateur botanist
jardinii As for *Jardinea*
jarenskianum L. *-anum*, indicating connection. In honor of Jarenski
jauaensis L. *-ensis*, denoting origin. From Cerra Jaua, Venezuela
jauanum L. *-anum*, indicating connection. From Jaua Plateau, Bolívar State, Venezuela
jaucensis L. *-ensis*, denoting origin. From Pedaleros del Jauco, Cuba
jaunsarensis L. *-ensis*, denoting origin. From Jaunsar Hills in north-west Himalayas
javan-a, -um L. *-anum*, indicating connection. From Java, Indonesia
javanic-a, -um, -us L. *-ica*, belonging to. From Java, Indonesia
javensis L. *-ensis*, denoting origin. See *javanica*
javorkae In honor of Sándor Jávorka (1883–1961) Hungarian phytogeographer
jayachandranii In honor of V. Jayachandran Nair (1940–) Indian botanist
jeanpertii In honor of "Jeanpeart", friend of A. St-Yves, French agrostologist
jeanyae In honor of Jeany Vander Neut Davidse (1945–) United States research assistant at Missouri Botanical Garden
jeffreysii In honor of Jeffreys (fl. 1907) plant collector in Rhodesia, now Zimbabwe
jeholensis L. *-ensis*, denoting origin. From Jehol, China
jejunum L. unproductive. A weedy species
jelskii In honor of Constantin von (Konstanty) Jelski (1837–1896) Polish ornithologist and collector who worked in Lima, Peru in the 1870s
jemenic-a, -us L. *-ica*, belonging to. From the Yemen
jemensis L. *-ensis*, denoting origin. From the Yemen
jemtlandica L. *-ica*, belonging to. From Jemtland, Sweden

J

jenisseiensis L. *-ensis*, denoting origin. From lower reaches of the Enesei, also known as the Yenisey or Jenisseisk River, in the Russian Far East

jensenii In honor of J. A. D. Jensen (fl. 1879) Danish lieutenant

jeremiadis In honor of Jeremy Michael Bayliss Smith (1945–) English born Australian ecologist and an acknowledgement by the author "that any study of the genus may be a jeremiad"

jerichoensis L. *-ensis*, denoting origin. From Jericho, Queensland, Australia

jesuitic-um, -us L. *-icum*, belonging to. From Mission areas under the jurisdiction of the Jesuit Order

jimenezii In honor of Otón Jiménez (1895–?) Costa Rican plant collector

jingpoense L. *-ense*, denoting origin. From Jingpo, Yunnan Province, China

jinshaensis L. *-ensis*, denoting origin. From Jinsha Jiang, Yunnan Province, China

jinshaicola L. *-cola*, dweller. From the banks of the Jinsha River, Yuanmou Xian, Yunnan Province, China

jiulongensis L. *-ensis*, denoting origin. From Jiulong Xian, Sichuan Province, China

jivarancusa See *jwarancusa*

Joachimia In honor of Joachim Murat (1771–1815) brother-in-law of Napoleon

Joannegria In honor of Giovanni Negri (1877–1960) Italian paleobotanist

joannis From St. Joansthale, Czech Republic

johannae From Johanna, one of the islands of the Comoros Republic, Indian Ocean

johannense L. *-ense*, denoting origin. As for *johannae*

johnii In honor of John Correia Alphonso, College Principal, Bombay, India

johnstonii In honor of – (1) Ivan Murray Johnston (1898–1960) United States botanist. *Bouteloua johnstonii, Nassella johnstonii* – (2) Henry Hamilton Johnston (1858–1927) who collected in Africa. *Sacciolepis johnstonii*

jonesii In honor of Marcus Eugene Jones (1852–1934) United States mining engineer, teacher and botanist

joorii In honor of Joseph Finley Joor (1849–1892) United States botanist

jordalii In honor of Louis Henrik Jordal (1919–1951) United States botanist

jorullensis L. *-ensis*, denoting origin. From Playas de Jorullo, Mexico

josephii In honor of J. Joseph (fl. 1964–1979) Indian botanist

jouldosensis L. *-ensis*, denoting origin. From Yao ér Du Si, Xin Yang Province, China

Jouvea In honor of Joseph Duval-Jouve (1810–1883) French agrostologist

Joycea In honor of Joyce Winfred Vickery (1908–1979) Australian botanist

joyceae As for *Joycea*

jubaensis L. *-ensis*, denoting origin. From Jubbada Hoose, Somalia

jubat-a, -um, -us L. *jubum*, mane; *-ata*, possessing. The inflorescence or awn resembles a fox tail

jubiflor-um, -us L. *jubum*, mane; *flos*, flower. Inflorescence mane-like

jucunda L. pleasant. Attractive in appearance

judziewiczii In honor of Emmet J. Judziewicz (1953–) United States botanist

juergensii In honor of Carlo Juergens (fl. 1905) who collected in Brazil

jugicola L. *jugum*, mountain ridge; *-cola*, dweller. From the Central Highlands of Tasmania

jugorum L. *jugum*, mountain ridge. From the ridges of the Witten Bergen in South Africa

jujuyense L. *-ense*, denoting origin. From Jujuy Province, Argentina

juldusicola L. *-cola*, dweller. From Mt Juldus, Turkestan region of Central Asia

juliae In honor of Julia but origin unclear, not given by author

julietii In honor of Cárlos Juliet, who collected in Chile

jumentorum L. *jumentum*, a yoke-beast. Of bullocks, and possibly a reference to the grass being used as a fodder

junatovii In honor of Alexander Afanasievich Junatov (1909–) Russian botanist

junce-a, -um, -us L. *juncea*, rush-like. Leaf-blades convolute resembling those of certain *Juncus* species
junceiform-e, -is L. *junceus*, rush-like; *forma*, appearance. In habit resembling certain *Juncus* species
junce-um, -us See *juncea*
juncifoli-a, -um, -us L. *juncea*, rush-like; *folium*, leaf. With rush-like leaf-blades or culms
junciformis L. *forma*, appearance. Resembling *Juncus* especially in habit
juncoides Gk *-oides*, resembling. Similar to *Juncus*
junghuhnian-a, -um L. *-ana*, indicating connection. In honor of Franz Wilhelm Junghuhn (1809–1864) German botanist who collected in Java, Indonesia
junghuhnii As for *junghuhniana*
juniperinum L. juniper; *-inum*, indicating resemblance. Leaf-blades ovate-lanceolate resembling those of *Juniperus*
junnarensis L. *-ensis*, denoting origin. From Junnar, Maharashta State, India
junodii In honor of Henri Alexandre Junod (1863–1934) Swiss missionary doctor and amateur botanist in Mozambique
jurassica L. *-ica*, belonging to. From Jura on the French-Swiss border
juressi L. from Serra de Gerez, Portugal
jurtzevii In honor of Jurtzev (fl. 1969)
juruana L. *-ana*, indicating connection. From the river Jurua, Department Amazonas, Brazil
juvenal-e, -is L. *-ale*, pertaining to. From Port Juvénal, near Montpellier, France

K

kaalaense L. *-ense*, denoting origin. From Mt Kaala, Hawaiian Islands
kachinensis L. *-ensis*, denoting origin. From Mt Kachen, Myanmar
Kaeleria See *Koeleria*
kafuroense L. *-ense*, denoting origin. From Kafuro(a), Tanzania
kagamiana L. *-ana*, indicating connection. In honor of Jasunosuké Kagami, Japanese agriculturalist and forester
kagerensis L. *-ensis*, denoting origin. From Kagera River, Mozambique
kahiliense L. *-ense*, denoting origin. From Kahil, a mountain on Kaui, one of the Hawaiian Islands
kahoolawense L. *-ense*, denoting origin. From Kahoolawe, one of the Hawaiian Islands
kaialpina L. *alpes*, high mountain; *-ina*, indicating possession. From Kai, Honshu Island, Japan
kaiensis L. *-ensis*, denoting origin. From Kai Province, now Yamanashi Prefecture, Japan
kaieteurana L. *-ana*, indicating connection. As for *kaietukense*
kaietukens-e, -is L. *-ense*, denoting origin. From Kaietuka Fall, sometimes mispelt Kaieteur, a raised area in the valley of the Potaro River, Guyana
kaini Vernacular name for the species on Japen Island, Papua, Indonesia
kajkaiense L. *-ense*, denoting origin. From Kajakai, Afghanistan
kakakton Gk breakable. Rhachis fragile at the nodes
kakao Maori *kakaho*, a batten for carrying thatching. The culms are used as roof poles
kakudensis L. *-ensis*, denoting origin. From Kakudemura, Yetsigo Province, Japan
kalaharens-e, -is L. *-ense*, denoting origin. From Kalahari Desert, south-west Africa
kalarica L. *-ica*, belonging to. From Kalar Mountains, Transbaikal region, Russian Federation
kalavoorensis L. *-ensis*, denoting origin. From Kalavoor, Kerala State, India
kalbica L. *-ica*, belonging to. From the eastern Kalba Mountains, Kazakhstan
kalininae In honor of J. V. Kalinina, the collector
kallimorphon Gk *kallion*, more beautiful; *morphe*, appearance. Attractive in appearance
kalmii In honor of Pehr Kalm (1715–1779) Swedish botanist
kalnikensis L. *-ensis*, denoting origin. From Mt Kalnik, Croatia

kalpongianum L. *-anum*, indicating connection. From Kalpong, Andaman Islands, India
kalugense L. *-ense*, denoting origin. From Kaluga, Russian Federation
kamczadalorum From Kamchatka, Russian Far East
kamczatensis L. *-ensis*, denoting origin. See *kamtschatica*
kamerunense L. *-ense*, denoting origin. From the Cameroons, a mountain range in West Africa
kammurensis L. *-ensis*, denoting origin. From Kammura, Mino Province, now part of Gifu Prefecture, Japan
kamoji In honor of Kamoj, Japanese botanist
Kampmannia In honor of Frédéric Edouard Kampmann (1830–1914) Swiss botanist
Kampochloa Gk *kampe*, caterpillar; *chloa*, grass. The inflorescence bears a fanciful resemblance to a caterpillar
kamtschatica L. *-ica*, belonging to. From Kamchatka, Russian Far East
kanaii In honor of Hiroo Kanai (1930–) Japanese botanist
kanaioense L. *-ense*, denoting origin. From Kanaio on Maui, one of the Hawaiian Islands
kanashiroi In honor of Tetsuo Kanashiro (fl. 1912) Japanese botanist, also known as Tetsuo Amano
kanayamensis L. *-ensis*, denoting origin. From Kanayamamura, Fukushima Prefecture, Japan
kanboensis L. *-ensis*, denoting origin. From Kanboho, Korea
kanehirae In honor of Ryôzô Kanehira (1882–1948) Japanese botanist
kangeanensis L. *-ensis*, denoting origin. From Kangean Islands, Indonesia
kanijirapallilana See *kanjirapallilana*
kanjirapallilana L. *-ana*, indicating connection. From Kanjirapallil, Travancore, India
kansasens-e, -is L. *-ense*, denoting origin. From Kansas, USA
kansuensis L. *-ensis*, denoting origin. From Kansu, China
Kaokochloa Occurring on the Kaokoveld in south-west Africa
kaonohuaense L. *-ense*, denoting origin. From Kaonohua Gulch on Maui, one of the Hawaiian Islands
kapandensis L. *-ensis*, denoting origin. From Kapanda, Zaire
kapiriensis L. *-ensis*, denoting origin. From Kapiri Valley, Zaire
kappleri In honor of August Kappler (1815–1887) German soldier and naturalist
karadagensis L. *-ensis*, denoting origin. From Kara Dag, a mountain in Turkey
karadaghense As for *karadagensis*
karakabinic-a, -us L. *-ica*, belonging to. From Karakabin Basin, Kazakhstan
karamyschevii In honor of Alexander Karamyschev, Russian botanist
karasbergensis L. *-ensis*, denoting origin. From Karasberg, Angola
karatavica L. *-ica*, belonging to. From Karatau Mountain range, Turkestan
karataviense L. *-ense*, denoting origin. See *karatavica*
karateginensis L. *-ensis*, denoting origin. From the Karategin Range, Central Asia
karavajevii In honor of Mikhail Nikolaevich Karavajev (Karavaev) (1903–?) Soviet botanist
karelinii In honor of Grigorij Silych Karelin (1801–1872) Russian botanist
kariwaensis L. *-ensis*, denoting origin. From Karihagun, Yetsigo province, Japan
kariyosensis L. *-ensis*, denoting origin. From Mt Kariyose, Musashi Province, now Tokyo Prefecture and parts of Saitama and Kanagawa Prefectures, Japan
karka Origin obscure, possibly the corruption of an Indian vernacular name referring to its white inflorescence
karkaralens-e, -is L. *-ense*, denoting origin. From the Karkaraly Mountains, Kazakhstan
karlobagensis L. *-ensis*, denoting origin. From Karlobag, Yugoslavia
Karroochloa Gk *chloa*, grass. From the Karroo of southern Africa
karsiana L. *-ana*, indicating connection. From Kars, a Province of Turkey

karstenii In honor of Gustav Karl Wilhelm Hermann Karsten (1817–1908) German botanist
karwinskiana L. *-ana*, indicating connection. In honor of Wilhelm Friedrich Karwinski Karwin (1780–1855) Hungarian-born German botanist
karwinskii As for *karwinskiana*
karwinskyan-a, -us L. *-ana*, indicating connection. As for *karwinskiana*
karwinskyi As for *karwinskiana*
karwynskii As for *karwinskiana*
karzinianum L. *-anum*, indicating connection. In honor of Karzin but origin unclear, not given by author
kasamaensis L. *-ensis*, denoting origin. From Kasama, Northern Province, Zambia
kashidensis L. *-ensis*, denoting origin. From Kashidamura, Japan
kashmiriana L. *-ana*, indicating connection. From Kashmir
kasimontana L. *mons*, mountain; *-ana*, indicating connection. Growing on Mount Kasi, Fukushima Prefecture, Japan
kassiana L. *-ana*, indicating connection. From Mt Kashizan, Fukushima Prefecture, Japan
kassizanensis L. *-ensis*, denoting origin. From Kashizan, Fukushima Prefecture, Japan
kastalskyi In honor of G. Kastalsky (fl. 1826–1829) who collected in Kamchatka, Russia
kasteki From Kastek River, near Tashkent, Uzbekistan
kasumense L. *-ense*, denoting origin. From Kazuma Range, Zimbabwe
katakton Gk *katakton*, capable of being broken. The spikelets deciduous from the base, falling entire or with accessory branch structures attached
katangens-e, -is L. *-ense*, denoting origin. From Katanga Province, Zaire
katentaniense L. *-ense*, denoting origin. From Katentania, Republic of Congo
kathaensis L. *-ensis*, denoting origin. From Katha Mountains, Myanmar
katsuragiana L. *-ana*, indicating connection. In honor of Lord Katsuragi
kattegatensis L. *-ensis*, denoting origin. From the shores of the Kattegat, the seaway separating northern Denmark from Sweden
kauaiense L. *-ense*, denoting origin. From Kauai, one of the Hawaiian Islands
kavanayense L. *-ense*, denoting origin. From Kavanayen, Venezuela
kawakamii In honor of Takiya Kawakamii (1871–1915) Japanese botanist
kawanoyuensis L. *-ensis*, denoting origin. From Kawanoyu, Kushiro Province, Hokkaido, Japan
kayi In honor of Omar Lamar Kay (1920–2001) United States soil scientist
kazachstanica L. *-ica*, belonging to. From Kazakhstan
keckii In honor of David Daniels Keck (1903–1995) United States botanist
keenanii In honor of J. Keenan (fl. 1961) who collected in Myanmar
kegelii In honor of Hermann Aribert Heinrich Kegel (1819–1856) German botanist
kelibiae From Kelibia, Tunisia
kelleri In honor of A. Keller (1873–1945) Swiss botanist
kelloggii In honor of Albert Kellogg (1813–1887) United States physician and amateur botanist
kelungens-e, -is L. *-ense*, denoting origin. From Kelung, Taiwan
kemerovensis L. *-ensis*, denoting origin. From Kemerovskya Oblast, Russian Federation
kempffii In honor of Noel Kempff Mercado (1924–1986) Bolivian biologist
kempirica L. *-ica*, belonging to. From Kempirbulach Range, Kazakhstan
Kengia In honor of Keng Yi-li (1894–1975) Chinese agrostologist
kengiana As for *Kengia*
kengii As for *Kengia*
Kengyilia In honor of Keng Yi-Li (1897–1975) Chinese botanist
keniensis L. *-ensis*, denoting origin. From Kenya

Keniochloa Gk *chloa*, grass. From Kenya
kennedyae In honor of Mary Bozzom Kennedy (1838–1915) of Wonnaminta station near Broken Hill, Australia
kennedyana In honor of Patrick Beveridge Kennedy (1874–1930) United States agronomist
kenteica L. *-ica*, belonging to. From Kentei-chan, a mountain in Mongolia
kentii Possibly in honor of Adolphus Henry Kent (1828–1913) English nurseryman
kentrophyllus Gk *kentron*, spur; *phyllon*, leaf. Leaf-blades involute and rigid
kentuckense, kentuckiense L. *-ense*, denoting origin. From Kentucky, USA
keralae See *keralensis*
keralensis L. *-ensis*, denoting origin. From Kerala, India
kerguelensis L. *-ensis*, denoting origin. From Kerguelen Island in the Antarctic Ocean
Kerinozoma Gk *kerinos*, wax; *zone*, girdle. Pedicel with a cartilaginous band just below the spikelet
kermesinum L. *-inum*, belonging to. From Kerma, Sudan
kerneri In honor of Anton Joseph Kerner von Marilaun (1831–1898) Austrian botanist
kerriana As for *Kerriochloa*
kerrii As for *Kerriochloa*
Kerriochloa Gk *chloa*, grass. In honor of Arthur Francis George Kerr (1877–1942) Irish-born medical officer and government botanist, Thailand
kersteniana L. *-ana*, indicating connection. In honor of Kersten who collected in East Africa
kerstingii In honor of Otto Kersting (1863–?) German botanist
kesenensis L. *-ensis*, denoting origin. From Kesengun, Rikuchiu Province, Japan
ketoiensis L. *-ensis*, denoting origin. From Keto, one of the Kuril or Chishima Islands
ketzchovelii In honor of Nikoloy Nikolaevich Ketzchoveli (1897–1982) Russian botanist
keyense L. *-ense*, denoting origin. From the Keys, Florida, USA
keysseri In honor of Christian Keysser (1877–1961) German missionary, linguist and ethnographer
khasian-a, -um, -us L. *-ana*, indicating connection. From the Khasia Hills, India
khasyana As for *khasiana*
khoonmengii In honor of Khoon Meng Wong (1954–) Malaysian botanist and educator
kialaensis L. *-ensis*, denoting origin. From Kiala, Zaire
kiarchanum L. *-anum*, indicating connection. From Kiarch but origin unclear, not given by the author
kibambeleensis L. *-ensis*, denoting origin. From Kibambele, Zaire
kiboensis L. *-ensis*, denoting origin. From Kibôsan, a mountain in Kumamoto Prefecture, Japan
kidumaensis L. *-ensis*, denoting origin. From Kiduma, Zaire
Kielboul Meaning uncertain, origin not given by author
Kielbul See *Kielboul*
kiensieleense L. *-ense*, denoting origin. From Kinsélé, Republic of Congo
Kiharapyrum Gk *pyros*, wheat. In honor of Hitoshi Kihara (1893–1986) Japanese botanist and resembling wheat
kila Vernacular name of the species in southern Chile
kilimandscharic-a, -us L. *-ica*, belonging to. From Mt Kilimandjaro, East Africa
kilimanjarica See *kilimandscharica*
killeenii In honor of Timothy John Killeen (1952–) collector of the species
killickii In honor of Donald Joseph Boomer Killick (1926–) South African botanist
killipii In honor of Ellsworth Paine Killip (1890–1968) United States botanist
kimayalaensis L. *-ensis*, denoting origin. From Kimayala, Zaire
kimberleyensis L. *-ensis*, denoting origin. From the Kimberley Region of north-western Australia
kimpasaensis L. *-ensis*, denoting origin. From Kimpasa, Zaire
kimuinguensis L. *-ensis*, denoting origin. From Kimuingua, Zaire

kimurae In honor of Arika Kimura (1900–1996) Japanese botanist
Kinabaluchloa Gk *chloa*, grass. From Mt Kinabalu, Borneo
kinabaluensis L. *-ensis*, denoting origin. From Mt Kinabalu, Borneo
kindunduensis L. *-ensis*, denoting origin. From Kindundu, Zaire
kingesii In honor of Heinrich Kinges (1912–) German botanist
kingiana L. *-ana*, indicating connection. In honor of Philip Parker King (1791–1856) British naval officer, born on Norfolk Island, died Sydney, Australia
kingii In honor of – (1) Clarence King (1842–1901) United States geologist and explorer. *Blepharidachne kingii*, *Eremochloe kingii*, *Festuca kingii*, *Hesperochloa kingii*, *Oryzopsis kingii*, *Poa kingii* – (2) Philip Parker King (1791–1856) British naval officer, born on Norfolk Island, died Sydney, Australia. *Aira kingii*, *Deschampsia kingii* – (3) George King (1840–1909) Scots-born physician and Indian botanist. *Ischaemum kingii*
kingundaensis L. *-ensis*, denoting origin. From Kingunda, Zaire
kinkiensis L. *-ensis*, denoting origin. From Kinki botanical region, Japan
kinshasaensis L. *-ensis*, denoting origin. Kinshasa, Zaire
kinsudiensis L. *-ensis*, denoting origin. From Kinsude, Zaire
kirelowii In honor of Ivan Petrovich Kirilov (1821–1843) Russian botanist
kirghisorum From Karakirghizica in the Terskej Alatau, Kyrgyzstan
kirghizica See *kirghisorum*
kirishimensis L. *-ensis*, denoting origin. From Kirishimayama, a mountain in Miyazaki Prefecture, Japan
kirisimensis See *kirishimensis*
kirkii In honor of – (1) Thomas Kirk (1828–1898) English-born New Zealand forester and amateur botanist. *Agropyron kirkii*, *Poa kirkii* – (2) John Kirk (1833–1922) Scots physician and botanist who worked largely in East Africa. *Pennisetum kirkii*
kirstingii See *kerstingii*
kisantuense L. *-ense*, denoting origin. From Kisantu, Zaire
kishinoana L. *-ana*, indicating connection. In honor of Yorisaburo Kishino (fl. 1933) Japanese botanist
kisoensis L. *-ensis*, denoting origin. From Kiso, Nagano Prefecture, Japan
kitadakens-e, -is L. *-ense*, denoting origin. From Kita Dake, a mountain in Kumamoto Prefecture, Japan
kitagawae In honor of Masao Kitagawa (1909–) Japanese botanist
kitaibeliana L. *-ana*, indicating connection. In honor of Paul Kitaibel (1757–1817) German botanist
kitaibelii As for *kitaibeliana*
kitamiana L. *-ana*, indicating connection. From Kitami Province, Hokkaido, Japan
kitanoensis L. *-ensis*, denoting origin. From Kitano, Niigata Prefecture, Japan
kiusian-a, -us L. *-ana*, indicating connection. From Kyusha, Japan
kiwuensis L. *-ensis*, denoting origin. From Lake Kivu, Yemen
kiyalaens-e, -is L. *-ense*, denoting origin. From Kiyala, Zaire
kjellmanii In honor of Frans Reinhold Kjellman (1846–1907) Swedish botanist
klagha Vernacular name for the species in Java, Indonesia
klasterskyi In honor of Ivan Klástersky (1901–1979)
kleinianum L. *-anum*, indicating connection. In honor of Jakob Theodor Klein (1685–1759) German botanist
kleinii In honor of Robert Miguel Klein (1923–1992) Brazilian botanist
Klemachloa Gk *klema*, an unknown plant referred to by Pliny, but generally assumed to have possessed jointed stems with swollen nodes; *chloa*, grass. Culm nodes conspicuous
klemenzii In honor of Elisabet Nikolaevna Klementz (fl. 1883–1898) Russian botanist
klingii In honor of Eric Kling (?–1892) German Army officer who collected in Togo

klossii In honor of Cecil Boden Kloss (1877–1949) English zoologist and museum administrator who collected in Malaysia

Knappia In honor of John Leonard Knapp (1767–1845) English writer on British grasses

kneuckeri In honor of Johann Andreas Kneucker (1862–1946) German botanist

knudsenii In honor of Valdemar Emil Knudsen (1819–1898) Norwegian born United States publisher, merchant and sugar cane farmer on Hawaii

knuthii In honor of Reinhard Knuth (1874–1957) German botanist

kobayashii In honor of – (1) Sumiko Kobayashi (1922–) Japanese botanist. *Poa kobayashii* – (2) M. Kobayashi. *Puccinellia kobayashii*

kobemontana L. *mons*, mountain; *-ana*, indicating connection. From the mountains near Kobe, Japan

koboi In honor of Kobo, Japanese botanist

kochii In honor of Karl Heinrich Emil Koch (1809–1879) German physician, botanist and traveller

kodzumae In honor of Masayuki Kôdzuma

koeiean-a, -us L. *-ana*, indicating connection. In honor of Mogens Ergell Køie (1911–2000) Danish botanist

Koelera See *Koeleria*

koeleri As for *Koeleria*

Koeleria In honor of Georg Ludwig Koeler (1765–1807) German botanist

koeleriiformis L. *forma*, appearance. With the habit of *Koeleria*

koelerioides Gk *-oides*, resembling. Similar to *Koeleria* especially with respect to the inflorescence

koelzii In honor of Walter Norman Koelz (1895–?) United States botanist

koenigii In honor of Johan Gerhard Koenig (1728–1784). Born in Duchy of Courland between present-day Poland and former Soviet Union; missionary-surgeon and economic botanist in India

koestlinii In honor of Köstlin

kogasensis L. *-ensis*, denoting origin. From Kogashi, Tochigi Prefecture, Japan

kogensis L. *-ensis*, denoting origin. From Kôga, Shiga Prefecture, Japan

kohautianum L. *-anum*, indicating connection. In honor of Franz Kohaut (?–1822)

kohyafoemina L. *foemina*, female. Meaning uncertain, not given by the author

kohzegawana L. *-ana*, indicating connection. From Kohzegawa, Japan

koibalensis L. *-ensis*, denoting origin. From the Koibal or Kaibal homelands in Northern Russia

koidzumian-a, -um L. *-ana*, indicating connection. In honor of Gen'ichi Yonezawa Koidzumi (1883–1953) Japanese botanist

koidzumii As for *koidzumiana*

koiyeana L. *-ana*, indicating connection. In honor of Gihachiro Koiye, Japanese botanist

kokanica L. *-ica*, belonging to. From Kokankishlak, Uzbekistan

kokeeense L. *-ense*, denoting origin. From Kokee State Park on Kauagi, one of the Hawaiian islands

kokonorica L. *-ica*, belonging to. From Koko Nor, now Ching Hai Su, China

koksuensis L. *-ensis*, denoting origin. From Koksu River, Kazakhstan

kolakovskyi In honor of Alfred Alekseevich Kolakovsky (1906–) Russian botanist

koleopodum Gk *koleos*, sheath; *pous*, foot. Peduncle mostly invested by upper leaf-sheath

koleostachys Gk *koleos*, sheath; *stachys*, spike as of an ear of wheat. Inflorescence partly enclosed in its sheathing leaf

koleotricha Gk *koleos*, sheath; *thrix*, hair. Leaf-sheaths densely hairy

kolesnikovii In honor of Boris Pavlovich Kolesnikov (1909–) Russian forester and plant collector

kolgujewensis L. *-ensis*, denoting origin. From Kolgujew, an island in the Russian Arctic

kollimalayana L. *-ana*, indicating connection. From Kollimala, South India

kolymaensis See *kolymense*

kolymens-e, -is L. *-ense*, denoting origin. From the Kolyma Basin, north-eastern Siberia

komarovii In honor of Vladimir Leontievitch Komarov (1869–1946) Russian botanist

komiyamana L. *-ana*, indicating connection. In honor of Komiyama

komoriana L. *-ana*, indicating connection. In honor of Hikotaro Komori, Japanese botanist

konaense L. *-ense*, denoting origin. From the Kona district, formerly the Kingdom of Kona, Hawaii

kongocacuminis L. *cacuminis*, high point. From Kongô-san, a mountain in Osaka Prefecture, Japan

kongosanensis L. *-ensis*, denoting origin. From Kongô-san, a mountain in Osaka Prefecture, Japan

koolauense L. *-ense*, denoting origin. From Koolau Range, Oahu one of the Hawaiian Islands

Koordersiochloa In honor of Sijfert Hendrik Koorders (1863–1919) Dutch botanist

kopetdagensis L. *-ensis*, denoting origin. From Kopet Dag Khrebet Mountains, Turkmenskaja (Turkmenistan)

kora Hindi *kodu*. A corruption of the Hindi name for the species

korabensis L. *-ensis*, denoting origin. From Mt Korab, Serbia

korbuensis L. *-ensis*, denoting origin. – (1) From Gunong Korbu, Malaysia. *Holttumochloa korbuensis* – (2) from the Kubor Range, Papua New Guinea. *Poa korbuensis*

korczaginii In honor of Aleksandr Alexandrovich Korczagin (1900–1987) Russian botanist

kordofana L. *-ana*, indicating connection. From Kordofan, Sudan

koreana L. *-ana*, indicating connection. From Korea

koreano-alpina L. *alpes*, mountain; *-ina*, indicating possession. From the mountains of Korea

koretrostachys Gk *koris*, crab; *stachys*, spike as of an ear of wheat. Inflorescence branches subverticillate

koritnicensis L. *-ensis*, denoting origin. From Koritnik, Albania

koriyamensis L. *-ensis*, denoting origin. From Kôriyama, Fukushima Prefecture, Japan

korotkyi In honor of M. F. Korotkij (?–1915) Russian botanist

korovinii In honor of E. Korovin (fl. 1929) Russian botanist

korschinskyana L. *-ana*, indicating connection. In honor of Sergei Ivanovich Korzhinskii (1860–1900) Russian botanist

korschinskyi As for *korschinskyana*

korshinskianum See *korschinskyana*

korshinskyi As for *korschinskyana*

korshunensis L. *-ensis*, denoting origin. From Korzhun River near Alma Ata, Kazakhstan

Korycarpus See *Corycarpus*

koryoens-e, -is L. *-ense*, denoting origin. From Koryo-shikenrin-ippan, Japan

kosakensis L. *-ensis*, denoting origin. From Kosaka, Ugo Province, now the major part of Akita and Yamagata Prefectures, Japan

kosaninii In honor of Nedelyko Kosanin (1874–1934) Serbian botanist

koshaninii As for *kosaninii*

koshiensis L. *-ensis*, denoting origin. From Koshi, Japan

koshinaiana L. *-ana*, indicating connection. From Koshinai, Sakhalin Island, Russian Far East

koshisimonii From a place name in Japan

kossinskyi In honor of Ekaterina Konstantinova Kosinskaja (1874–1928) Russian botanist

kostermansiana L. *-ana*, indicating connection. In honor of André Joseph Guillaume Henri Kostermans (1907–1994) Indonesian botanist of Dutch parentage

kotovii In honor of Michael Ivanovich Kotov (1896–1978) Russian botanist

kotschyan-a, -um L. *-ana*, indicating connection. In honor of Karl Georg Theodor Kotschy (1813–1866) Austrian botanist and traveller

kotschyi As for *kotschyana*

kottoensis L. *-ensis*, denoting origin. From Basse Kotto, Republic Central Africa

kotulae In honor of Bolestaw Kotula (1849–1892) Polish botanist

kotzebuensis L. *-ensis*, denoting origin. From Kotzebue Sound, Alaska

koyana L. *-ana*, indicating connection. From Kôyasan, a mountain in Kii Province, now Wakayama and part of Mie Prefectures, Japan

kozanensis L. *-ensis*, denoting origin. From Kozani District in northern Greece

kozasa Japanese *ko*, small; *sasa*, small bamboo. Dwarf bamboo

krajinae In honor of Vladimir Joseph Krajina (1905–1993) who collected in Slovakia

kralifii From Djebel Sidi-Kralif, Tunisia

Kralikia In honor of Jean Louis Kralik (1813–1892) French botanist

Kralikiella L. *-ella*, diminutive but here a name forming suffix. Resembling *Kralikia* in some respect, but smaller

kransei In honor of Kranse who collected at Tschotkol, Turkestan region of Central Asia

krapovickasii In honor of Antonia Krapovickas (1921–) Argentinian botanist

krascheninnikovii In honor of Ippolit Mikhailovich Krascheninnikov (1884–1945) Russian botanist

Kratzmannia In honor of Emil Kratzmann (1814–1867) Czech botanist

kraussii In honor of Christian Ferdinand Friedrich von Krauss (1812–1890) who collected in South Africa

kreczetoviczii In honor of V. I. Krechetovich (1901–1942) Russian botanist

krivotulenkoae In honor of U. F. Krivotulenko (fl. 1955) Russian botanist

Krombholzia In honor of Julius Vincenz Krombholtz (1782–1842) Bohemian lawyer, surgeon and mycologist

kronenbergii In honor of A. Kronenburg (fl. 1903–1904) plant collector in Central Asia

kronokens-e, -is L. *-ense*, denoting origin. From Lake Kronotzkoe, Kamtchatka, Russian Far East

krusemaniana L. *-ana*, indicating connection. In honor of Gideon Kruseman (1904–) Dutch entomologist and plant ecologist

krylovian-a, -um L. *-ana*, indicating connection. In honor of Porfirij Nikitic Krylov (1850–1931) Russian botanist

krylovii As for *kryloviana*

Ktenosachne Gk *ktenion*, small comb; *achne*, scale. The glumes have comb-like keels

kuborensis L. *-ensis* denoting origin. From Kubor Range, Papua New Guinea

kuchariana L. *-ana*, indicating connection. In honor of Kuchar (fl. 1984) who collected in Somalia

kudoi In honor of Yûshun Kudô (1887–1932) Japanese botanist

kuenlunica L. *-ica*, belonging to. From Kuen-Lun, Inner Mongolia

kuhlmannii In honor of Joao Geraldo Kuhlmann (1882–1958) who collected in Brazil

kukaiwaaense L. *-ense*, denoting origin. From Kukaiwaa on Molokaione, one of the Hawaiian islands

kumaensis L. *-ensis*, denoting origin. From Kumagunn, Kumamoto Prefecture, Japan

kumarakodiense L. *-ense*, denoting origin. From Kumarakodi, Kerala State, India

kumasasa Japanese *kuma*, bear; *sasa*, dwarf bamboo. Growing in places frequented by bears

kumasoana Origin uncertain but probably in honor of Kumaso

kumgansani From Kongosan, a mountain in Korea

kundjuana L. *-ana*, indicating connection. From Kundju, Japan

kungeica L. *-ica*, belonging to. From Kungei Alatau, Kazakhstan

kuniense L. *-ense*, denoting origin. From Kunie, now Île de Pins off New Caledonia

kunimiana L. *-ana*, indicating connection. From Kunimiyama, a mountain in Rikuchiu Province, Japan

kunishii In honor of Kunish, Japanese botanist

kunmingensis L. *-ensis*, denoting origin. From Kunming, Yunnan Province, China

kuntaensis L. *-ensis*, denoting origin. From Kuntagun, Shiga Prefecture, Japan

kunthian-a, -um L. *-ana*, indicating connection. In honor of Karl Sigismund Kunth (1788–1850) German botanist

kunthii As for *kunthiana*

kuntzean-a, -us L. *-ana*, indicating connection. In honor of Carl Ernst Otto Kuntze (1843–1907) German botanist

kuntzei As for *kuntzeana*

kuoi In honor of Pung (Pen) Chao Kuo (fl. 1980–1987) Chinese botanist

kuprijanovii In honor of Andrei Nikolayevich Kuprijanov (fl. 1972) Russian botanist

kuramense L. *-ense*, denoting origin. From Kurrum Valley, Afghanistan

kurdica L. *-ica*, belonging to. From country inhabited by the Kurds, presently included in eastern Turkey, north-eastern Iraq and north-western Iran

kurdistanica L. *-ica*, belonging to. From Kurdistan, country of the Kurds. See *kurdica*

kurehaensis L. *-ensis*, denoting origin. From Kurehayama, Yettsui Province, Japan

kurilensis, kurillensis L. *-ensis*, denoting origin. Of the Kuriles

kuring The vernacular name of the species in Sumatra, Indonesia

kuriyamensis L. *-ensis*, denoting origin. From Kuriyama, Tochigi Prefecture, Japan

kurokawana L. *-ana*, indicating connection. In honor of Takao Kurokawa (fl. 1931–1934) Japanese botanist

kurtczumica L. *-ica*, belonging to. From the Kurtczum saddle in the Altai Mountains, Kazakhstan

kurtschumica L. *-ica*, belonging to. From Kurtschum

kurtziana L. *-ana*, indicating connection. As for *kurtzii*

kurtzii In honor of Fritz (Federico) Kurtz (1854–1920) who collected in Europe and South America

kurumthotticalana L. *-cola*, dweller; *-ana*, indicating connection. From Kurrumthotti, South India

kurzii In honor of Wilhelm Sulpiz Kurz (1834–1878) German botanist

kusirensis L. *-ensis*, denoting origin. From Kushiro Province, Hokkaido, Japan

kutaiensis L. *-ensis*, denoting origin. From West Kutai, Borneo

kutcharoensis L. *-ensis*, denoting origin. From Lake Kutcharo, Kushiro Province, Hokkaido, Japan

kuzakaina L. *-ina*, indicating possession. From Kuzakaitoge, Rikuchiu Province, Japan

kuznetzovii In honor of Nicolai Ivanovitch Kuznetsov (1864–1932)

kwaiensis L. *-ensis*, denoting origin. From Kwaihu, Kenya

kwamouthensis L. *-ensis*, denoting origin. From Kwamouth, Zaire

kwangsiensis L. *-ensis*, denoting origin. From Guangxi Province, China

kwashotensis L. *-ensis*, denoting origin. From Kwashoto, Taiwan

kwiluense L. *-ense*, denoting origin. From Moyen-Kwilu, Zaire

kyathaungtu A district in Pégu State, Myanmar

kyberi In honor of D. Kyber (fl. 1820–1825) who collected in Kamchatka, Russian Far East

kyongsongensis L. *-ensis*, denoting origin. From Kyongsong, Korea

kyzlkiensis L. *-ensis*, denoting origin. From Kyzylk-kuga, Kazakhstan

L

laagei In honor of Louise de Laage de Meux, mother of Antoine de Cugnac (1898–?) who described the species

labillardierei In honor of Jacques Julian Houtlan de Labillardière (1755–1834) French botanist and explorer

labradoric-a, -um L. *-ica*, belonging to. From Labrador, Canada

lacei In honor of John Henry Lace (1857–1918) English-born Indian forester

lachenalii In honor of Werner de la Chenal (1736–1800) Swiss botanist

lachenensis L. *-ensis*, denoting origin. From Lachen, Sikkim State, India

Lachnagrostis Gk *lachnos*, wool. Like *Agrostis* but lemma hairy and rhachilla prolongated

lachnanth-a, -um, -us Gk *lachnos*, wool; *anthos*, flower. Spikelets woolly
lachne-a, -um Gk *lachnos*, wool; L. *-ea*, pertaining to. Leaf-blades and leaf-sheaths densely hairy
Lachnochloa Gk *lachnos*, wool; *chloa*, grass. Lemma pubescent
lachnophyll-a, -um Gk *lachnos*, wool; *phyllon*, leaf. Leaf-blades densely hairy
lachnorrhachis Gk *lachnos*, wool; *rhachis*, backbone. Rhachis densely covered with short soft hairs
Lachryma-job, Lacryma-job As for *lacryma-jobi*
Lachrymaria L. *lacryma*, tear-drop; *-aria*, pertaining to. The cupule resembles a giant tear-drop
laciniatus L. *lacinia*, flap on fringe or edge of a garment; *-atus*, possessing. Lemma margin ciliate
lacmonicus L. *-icus*, belonging to. From Lacmon an area on the northern slopes of Mt Pindus, Greece
laconicum From Laconica, Greece
Lacryma, lacryma L. tear-drop. The terminal racemes project from a grey or white glistening cupule
Lacryma-job g See *lacryma-jobii*
lacryma-jobii, lacrymajobii L. *lacryma*, tear-drop; *jobi*, of Job a Biblical character who experienced much suffering. The bead-like bract surrounding the base of each inflorescence unit resembles a tear-drop
lacte-a, -um L. milky. Spikelets milky-white
lactiflorum L. *lactius*, milk-white; *flos*, flower. Spikelets white
lactistriata L. *lactius*, milk-white; *stria*, furrow; *-ata*, possessing. Leaf-sheaths with white stripes
lacunaria L. *lacuna*, cavity; *-aria*, pertaining to. Surface of grain pitted
lacunis L. *lacuna*, pond. Growing around the margins of ponds
lacunos-a, -us L. *lacuna*, cavity; *-osus*, well developed. – (1) Lower glume pitted. *Andropogon lacunosus* – (2) fertile lemma pitted. *Setaria lacunosa*
lacustr-e, -is L. *lacus*, lake; *-estre*, place of growth. Growing in or around lakes
ladakhensis L. *-ensis*, denoting origin. From the Ladakh Range, India
ladyginii In honor of V. Ladygin (fl. 1901) Russian botanist
laegaardii In honor of Simon Laegaard (1933–) Danish botanist
laersii As for *Leersia*
Laertia In honor of Laertes, King of Ithaca, an ancient Greek state
laestadii In honor of Lars Levi Laestadius (1800–1861) Swedish cleric and botanist
laet-a, -um L. fruitful. Setting abundant grain
laetevirens L. *laetum*, bright; *virens*, green. Foliage bright-green
laeteviridis L. *laetum*, bright; *viridis*, green. Foliage bright-green
laetum See *laeta*
laev-e, -is L. smooth. Lacking hairs or roughness, usually of leaf-blades or lemmas
laevifolium L. *laevis*, smooth; *folium*, leaf. Leaf-blades glabrous
laevigat-a, -um L. smooth and polished. – (1) Spikelets glabrous. *Avena laevigata*, *Spartina laevigata* – (2) culm leaf-sheaths glabrous. *Yushania laevigata* – (3) anthoecia glabrous. *Panicum laevigatum* – (4) plant generally glabrous. *Aristida laevigata*
laeviglumis L. *laevis*, smooth; *gluma*, husk. Glumes glabrous
laevipaleatum L. *laevis*, smooth; *palea*, chaff; *-atum*, possessing. Glumes and lemmas glabrous
laevipes L. *laevis*, smooth; *pes*, foot. Pedicels smooth
laevis See *laeve*
laevispica L. *laevis*, smooth; *spica*, a point; hence, in particular, an ear or spike of grain. Inflorescence smooth
laevissim-a, -um L. *laevis*, smooth; *-issima*, most. Plants usually quite glabrous
laeviuscula L. *laevius*, smoother; *-ula*, tendency. Plants almost glabrous
lagascae In honor of Mariano de la Lagasca (1776–1839) Spanish botanist

lagopoides Gk *lagos*, hare; *pous*, foot; *-oides*, resembling. The inflorescence resembles a hare's foot

lagostachyum Gk *lagos*, hare; *stachys*, spike as of an ear of wheat. Inflorescence a spike-like panicle

lagotis Gk *lagos*, hare; *ous*, ear. Upper lemma expanded into two ear-like wings at the base

lagunculiforme L. *lagunus*, flask; *-ula*, diminutive; *forma*, appearance. Diaspores narrow flask-shaped

laguriformis L. *forma*, appearance. Inflorescence like that of *Lagurus*

laguroides Gk *-oides*, resembling. Resembling *Lagurus* in respect of the inflorescence

laguroideum Gk *lagos*, hare; *-oideum*, resembling. Inflorescence resembling a hare's tail

Lagurus Gk *lagos*, hare; *oura*, tail. The panicle is a densely ovate spike resembling a hare's tail

lahittei In honor of Raul Lahitte (fl. 1930–1940) Argentinian botanist

lahonderei In honor of Christian La Hondère (fl. 1987) French botanist

lahulensis L. *-ensis*, denoting origin. From Lahul District, India

laidlawii In honor of William Laidlaw (?–1935) Scots-born Australian botanist and Garden's Director

laki-a, -um From the Laksii region of Daghestan, Russian Federation

lako Vernacular name for the species in Tetun (Tetum) language of East Timor

Lamarckia In honor of Jean Baptiste Antoine Pierre Monet de Lamarck (1744–1829) French biologist

lamarckian-a, -um L. *-ana*, indicating connection. As for *Lamarckia*

lamarckii As for *Lamarckia*

Lamarkia See *Lamarckia*

lamarkiana L. *-ana*, indicating connection. As for *Lamarckia*

lambinonii In honor of Jacques Ernest Joseph Lambinon (1936–) Belgian botanist

lamiatile L. *lama*, bog; *-atile*, place of growth. Bog dweller

lamii In honor of Hermann Johannes Lam (1892–1977) Dutch botanist

laminarum L. *lamina*, blade. Awns flat not twisted into columns

laminata L. *lamina*, blade; *-ata*, indicating possession. Known only from a fragment of fossil leaf blade

lampranthus Gk *lampros*, splendid or brilliant; *anthos*, flower. Spikelets shiny and olive-green

lamprocaryon Gk *lampros*, splendid or brilliant; *karyon*, nut. Anthoecium glossy and chestnut brown

lamproparia Gk *lampros*, splendid or brilliant; *pareia*, cheek. Lemma of the fertile floret has glabrous, shiny flanks

lamprophylla Gk *lampros*, splendid or brilliant; *phyllon*, leaf. Foliage attractive

lamprospicula Gk *lampros*, splendid or brilliant; *spica*, a point; hence, in particular, an ear or spike of grain; *-ula*, diminutive. Spikelets shiny

Lamprothyrsus Gk *lampros*, splendid or brilliant; *thyrsos*, an ornamental wand. The inflorescence is a compact silvery panicle

lanaiense L. *-ense*, denoting origin. From Lanai, one of the Hawaiian Islands

lanat-a, -um, -us L. lana, wool; *-ata*, possessing. – (1) Leaf-blades densely pubescent. *Agrostis lanata*, *Bromus lanatus*, *Holcus lanatus*, *Leptocoryphium lanatum*, *Loudetia lanata*, *Muhlenbergia lanata*, *Navicularia lanata*, *Panicum lanatum*, *Paspalum lanatum*, *Pennisetum lanatum*, *Pleioblastus lanatus*, *Trichopteryx lanata*, *Triodia lanata* – (2) lemmas and/or glumes densely pubescent. *Anthaenantia lanata*, *Ischaemum lanatum*, *Poa lanata* – (3) ribs of lowermost leaf-sheaths woolly. *Stipa lanata*

lanatiflor-a, -um L. *lana*, wool; *-ata*, possessing. *flos*, flower. Lemma invested with cottony hairs

lanatifolia L. *lana*, wool; *-ata*, possessing; *folium*, leaf. Upper surface of leaf-blade densely woolly

lanatipes L. *lana*, wool; *-ata*, possessing; *pes*, foot. Pedicels hairy

lanat-um, -us See *lanata*

lancangensis L. *-ensis*, denoting origin. From Lancang, Yunnan Province, China

lancea L. *lanceus*, lance. Shape lanceolate as of glumes, leaf-blades or inflorescences

lanceari-um, -us L. *lanceus*, lance; *-arium*, pertaining to. Leaf-blades lanceolate

lancearum As for *lancearium*

lanceolat-a, -um, -us L. *lanceus*, lance; *-ola*, diminutive; *-ata*, possessing. Mostly a reference to lanceolate leaf-blades

lanceolatiformis L. *lanceolatus*, lanceolate; *forma*, appearance. Glumes lanceolate

lanceolat-um, -us See *lanceolata*

lanciflorum L. *lanceus*, lance; *flos*, flower. Spikelets lanceolate in outline

lancifoli-a, -um, -us L. *lanceus*, lance; *folium*, leaf. Leaf-blade lanceolate

landbeckii In honor of Maximilian Landbeck who collected in Chile

lanea L. *lana*, wool; *-ea*, resembling. The sheath-hairs are fleece-like

langbianense L. *-ense*, denoting origin. From Lang-bian Plateau, Annam

langeana L. *-ana*, indicating connection. – (1) In honor of Karl Heinrich Lang (1800–1843) German cleric and naturalist. *Poa langeana* – (2) see *langei*. *Koeleria langeana*, *Puccinellia langeana*

langei In honor of Johann Martin Christian Lange (1818–1898) Danish botanist

langkawiensis L. *-ensis*, denoting origin. From Langkawi Islands, off the Malay Peninsula

langloisii In honor of Auguste Barthelemy Langlois (1832–1900) French-born United States cleric and plant collector

Langsdorffia In honor of Georg Heinrich von Langsdorff (1774–1852) German surgeon and naturalist

langsdorffian-a, -us L. *-ana*, indication connection. See *Langsdorffia*

langsdorffii As for *Langsdorffia*

langsdorfianus See *Langsdorffia*

langsdorfii As for *Langsdorffia*

langtangensis L. *-ensis*, denoting origin. From Langtang, Nepal

languid-a, -um L. weak. Culms spreading

languidior L. weaker. Culms weak and forming less dense tussocks than related species

languidum See *languida*

lanicaulis L. *lana*, wool; *caulis*, stem. Culms woolly-hairy at their base

lanifera L. *lanos*, wool; *fero*, carry or bear. Panicle branches densely woolly

laniflora L. *lana*, wool; *flos*, flower. Lemma and palea hairy

laniger, -a L. *lana*, wool; *gero*, carry or bear. – (1) Glumes invested with long hairs. *Andropogon laniger* – (2) leaf-blades or leaf-sheaths woolly. *Agrostis lanigera*, *Aristida lanigera*, *Neurachne lanigera*, *Triodia lanigera*

lanipes L. *lana*, wool; *pes*, foot. – (1) Lemmas woolly at the base. *Eragrostis lanipes* – (2) leaf-sheaths woolly at the base. *Aristida lanipes*, *Germainia lanipes* – (3) rhizomes woolly. *Panicum lanipes*

lanos-a, -um L. *lana*, wool; *-osa*, abundance. Leaf-sheath woolly

lanshanensis L. *-ensis*, denoting origin. From Lanshan, Hunan Province, China

lanuginos-um, -us L. *lanuginus*, woolly; *-osum*, abundance. – (1) Leaf-blades densely woolly. *Andropogon lanuginosus*, *Dichanthelium lanuginosum*, *Panicum lanuginosum* – (2) involucres subtending spikelets densely woolly. *Pennisetum lanuginosum*

lapalmae From La Palma, Canary Islands

lapidea L. *lapis*, stone; *-idea*, resembling. Latin form of the vernacular name "Stone Bamboo" by which the species is known in Southern China

lapidosa L. *lapis*, stone; *-osa*, abundance. Growing in rocky mountains

lappace-a, -us L. *lappa*, burr; *-acea*, indicating resemblance. – (1) Lemmas of the upper floret bearing reflexed bristles. *Centotheca lappacea* – (2) lemma awns shortly recurved. *Astrebla lappacea* – (3) involucral bristles barbed. *Cenchrus lappaceus*

Lappago L. *lappa*, burr; *-ago*, indicating resemblance. Glumes stiffly fringed forming a burr

Lappagopsis Gk *opsis*, resemblance. Resembling *Lappago*

lapponic-a, -um L. *-ica*, belonging to. From Lapponia, now Lapland

lappula L. *lappa*, burr; *-ula*, diminutive. Spikelets burr-like because of tubercule-based hairs on lemma

lappulaceus L. *lappa*, burr; *-ula*, diminutive; *-aceus*, indicating resemblance. Lemma densely clothed with hooked hairs and forming a burr at maturity

larcomianum L. *-anum*, denoting connection. From Mt Larcom, Queensland, Australia

larentii See *lorentii*

larranagae In honor of Dámaso Antonio Larrañaga (1771–1848) Uruguayan cleric and plant collector

larsenii In honor of Kai Larsen (1926–) Danish botanist

Lasiacis Gk *lasios*, shaggy; *akis*, point. Lemmas and palea apices terminate in tufts of hairs

Lasiagrostis Gk *lasios*, shaggy. Lemmas and bases of awns densely hairy, otherwise resembling *Agrostis*

lasianth-a, -um Gk *lasios*, shaggy; *anthos*, flower. Spikelets hairy

Lasingrostis See *Lasiagrostis*

Lasiochloa Gk *lasios*, shaggy; *chloa*, grass. The lemmas are hairy

lasioclada Gk *lasios*, shaggy; *klados*, branch. Culms pilose

lasiocole-os, -um Gk *lasios*, shaggy; *koleos*, sheath. Leaf-sheath hairy

lasiogon-um, -us Gk *lasios*, shaggy; *gony*, knee. Nodes and leaf-sheath bases densely hairy

lasiolepis Gk *lasios*, shaggy; *lepis*, scale. Glumes and lemmas hairy

Lasiolytrum Gk *lasios*, shaggy; *elytron*, cover. Glumes with abundant short hairs

lasionodosa Gk *lasios*, shaggy; L. *nodus*, knot; *-osa*, abundance. Nodes densely hairy

lasiophyll-a, -us Gk *lasios*, shaggy; *phyllon*, leaf. Leaf-blades or sheaths hairy

Lasiopoa Gk *lasios*, shaggy; *poa*, grass. Plants invested with long hairs

lasiopodium Gk *lasios*, shaggy; *pous*, foot. Pedicels hairy

Lasiorhachis Gk *lasios*, shaggy; *rhachis*, backbone. The inflorescence branches are ciliate

lasiorrhachis See *Lasiorhachis*

Lasiostega Gk *lasios*, shaggy; *stegos*, roof. Meaning uncertain, not given by author

lasiostoma Gk *lasios*, shaggy; *stoma*, mouth. Orifice of leaf-sheath densely hairy

lasiothyrsa Gk *lasios*, shaggy; *thyrsos*, ornamental wand. Inflorescence loosely, long hairy

Lasiotrichos Gk *lasios*, shaggy; *thrix*, hair. Lemmas pubescent

Lasiurus Gk *lasios*, shaggy; *oura*, tail. The inflorescence resembles a shaggy tail

lassenianum L. *-anum*, indicating connection. From Lassen Peak, California, USA

lasseri In honor of Tobias Lasser (1911–) Venezuelan botanist

lat-a, -um L. broad. Leaf-blades broad

lateral-e, -is L. *laterus*, side; *-ale*, pertaining to. Inflorescence forming as a lateral shoot

lateriflora L. *laterus*, side; *flos*, flower. Panicles arising from lateral shoots

lateritectoria L. *latus*, broad; *tectorius*, of a cover. Leaf-blades broader than those of *Sasa tectoria*

lateritic-a, -um L. *-ica*, belonging to. Growing on laterites

latichino L. *latus*, broad. Resembling *Pleioblastus chino* but with broader leaf-blades

laticomum L. *latus*, broad; *coma*, head of hair. Panicle very lax and erect, with filiform to capillary branches arranged in fascicles and so resembling a head of hair

laticulmum L. *latus*, broad; *culmus*, stalk. Culms strongly compressed

latiflor-a, -us L. *latus*, broad; *flos*, flower. Spikelets broad

latifoli-a, -um, -us L. *latus*, broad; *folium*, leaf. Leaf-blades broad or relatively broad with respect to related species

latifrons L. *latus*, broad; *frons*, leaf. Leaf-blades broad

latiglum-e, -is L. *latus*, broad; *gluma*, husk. Glumes broad

Latipes L. *latus*, broad; *pes*, foot. Pedicels broad, flattened

latipes L. *latus*, broad; *pes*, foot. Spikelets broad-based

latispicea L. *latus*, broad; *spica*, a point; hence, in particular, an ear or spike of grain. Spikelets broad

latispicula L. *latus*, broad; *spica*, a point; hence, in particular, an ear or spike of grain; *-ula*, diminutive. Spikelets broad

latisquamea L. *latus*, broad; *squama*, scale. Glumes or lemmas broad

latissimifolia L. *latus*, broad; *-issima*, most; *folium*, leaf. Leaf-blades very broad

latissimum L. *latus*, broad; *-issimum*, most. Leaf-blades very broad

latitectoria L. *latus*, broad. Resembling *Sasa tectoria* but leaves broader

latronum L. *latro*, bandit. Replacement name for a homonym which had, as it were, stolen the identity of a previously described species

latum See *lata*

latzii In honor of Peter Kenneth Latz (1941–) Australian plant ecologist

laudanensis L. *-ensis*, denoting origin. From Laudan Pass in the Pamirs, a mountain range mostly in Tajikistan

lauriolii In honor of J. Lauriol (fl. 1934) who collected in the Sahara

lautum L. *lavo*, wash and by implication thereby elegant. Panicle long and slim

lautumia L. *lautumia*, quarry. The first collected was from a disused limestone quarry

lavrenkoanum L. *-anum*, indicating connection. In honor of Eugen(y) M. Lavrenko (fl. 1925) Russian botanist

lavrenkoi In honor of A. N. Lavrenko, Russian agrostologist

lawii In honor of John Sutherland Law (1810–1885) Indian Civil Servant and amateur botanist

lawrencei In honor of Robert Williams Lawrence (1807–1833) plant collector in Tasmania

lawsonii In honor of Marmaduke Alexander Lawson (1840–1896) English botanist

lax-a, -um, -us L. loose. – (1) Inflorescence much branched either as a single panicle or from branching of the culms. *Agropyron laxum, Agrostis laxa, Aira laxa, Andropogon laxus, Anthistiria laxa, Aristida laxa, Axonopus laxus, Diectomis laxa, Ectrosia laxa, Eragrostis laxa, Gastridium laxum, Hemarthria laxa, Holcus laxus, Hordeum laxum, Ischaemum laxum, Matudacalamus laxa, Orthoclada laxa, Panicum laxum, Pennisetum laxum, Setaria laxa, Simplicia laxa, Sporobolus laxus, Themeda laxa, Tripsacum laxum, Trisetum laxum, Tristachya laxa* – (2) lateral branches lax and flexuose ascending. *Erianthus laxus, Eriochrysis laxa* – (3) culms overarched and reaching the ground. *Bambusa laxa, Triniochloa laxa*

laxatus L. *laxus*, loose; *-atus*, possessing. Common axis of racemes slightly flexuous or nodding

laxiflor-a, -um, -us L. *laxus*, loose; *flos*, flower. – (1) Inflorescence an open panicle. *Achnatherum laxiflora, Agrostis laxiflora, Alopecurus laxiflorus, Atropis laxa, Dichanthelium laxiflorum, Distichlis laxiflora, Festuca laxiflora, Panicum laxiflorum, Roegneria laxiflora, Sorghum laxiflorum, Stipa laxiflora, Trichodium laxiflora* – (2) spikelets with widely separated florets. *Poa laxiflora*

laxinodis L. *laxus*, loose; *nodus*, knot. Culms geniculate at the base and so the plant not densely tufted

laxior L. more lax. Racemes more lax than those of related species

laxispica L. *laxus*, loose; *spica*, spike. Spikelets widely separated on spike

laxissima L. *laxus*, lax; *-issima*, most. Panicle very open

laxiuscula L. *laxius*, looser; *-ula*, diminutive. Spikelets more lax than those of related species

lax-um, -us See *laxa*

laysanensis L. *-ensis*, denoting origin. From Laysan Island, one of the Hawaiian Islands

lazaridis In honor of Mike Lazarides (1928–) Australian botanist

lazic-a, -um L. *-ica*, belonging to. From Lazica, now Lazistan, Turkey

lazistanica L. *-ica*, belonging to. See *lazica*

leandri, leandrii In honor of – (1) Jacques Désiré Leandri (1903–1982) Corsican-born French botanist. *Poecilostachys leandrii* – (2) Leandro do Sacramento (?1779–1829) Carmelite friar who collected in Brazil. *Panicum leandri*

lebrunii In honor of Jean-Paul Antoine Lebrun (1906–1985) Belgian botanist

lecardii In honor of Th. Lécard (1834–1880) French botanist

Lechlera In honor of Willibald Lechler (1814–1856) German apothecary and traveller who collected in South America

lechleri See *Lechlera*

lechleriana L. *-ana*, indicating connection. As for *Lechlera*

leckenbyi In honor of A. B. Leckenby (fl. 1898) United States botanist

lecomtei See *Lecomtella*

Lecomtella L. *-ella*, diminutive but here used as a name-forming suffix. In honor of Paul Henri Lecomte (1856–1934) French botanist

leconteanum L. *-anum*, indicating connection. As for *Lecomtella*

ledebouri In honor of Carl Friedrich Ledebour (1785–1851) German botanist

ledermannii In honor of Carl Ludwig Ledermann (1875–1958) Swiss-born West African plant collector

leekei In honor of George Gustav Paul Leeke (1883–1933) German botanist

Leersia In honor of Johann Daniel Leers (1727–1774) German apothecary and botanist

leersianum L. *-anum*, indicating connection. As for *Leersia*

leersii As for *Leersia*

leersiiformis L. *forma*, appearance. Resembling *Leersia*

leersioides Gk *-oides*, resembling. Resembling *Leersia* in inflorescence form or habit

legei In honor of Émile Legé who collected in France

legrandii In honor of Carlos Diego (Carlos María Diego Enrique) Legrand (1901–1982) Uruguayan zoologist, botanist and Museum director

lehmannian-a,-um L. *-ana*, indicating connection. In honor of Friedrich Karl Lehmann (1850–1903) who collected in South America

lehmannii (1) As for *Paspalum lehmanniana*. *Agrostis lehmannii*, *Andropogon lehmannii*, *Chusquea lehmannii*, *Dimeria lehmannii*, *Eragrostis lehmannii*, *Pterygostachyum lehmannii* – (2) in honor of Friedrich Carl Lehmann (1850–1903) German-born botanist who collected in central America. *Agrostis lehmannii*

leianth-a, -um Gk *leios*, smooth; *anthos*, flower. Spikelets with glabrous glumes or lemmas

leiarthria Gk *leios*, smooth; *arthron*, joint. Unlike related species, the rhachilla internodes lack short hairs

leibergii In honor of John Bernhard Leiberg (1853–1913) United States forester and plant collector

leichhardtiana L. *-ana*, indicating connection. In honor of Friedrich Wilhelm Ludwig Leichhardt (1813–1848) German-born physician and Australian explorer

leiocalycina Gk *leios*, smooth; *kalyx*, cup; *-ina*, indicating resemblance. Glumes glabrous

leiocarp-a,-on,-us Gk *leios*, smooth; *karpos*, fruit. Spikelets glabrous

leioclad-a, -um Gk *leios*, smooth; *klados*, branch. Panicle branches smooth

leiocladium See *leioclada*

leiocladum See *leioclada*

leiocolea Gk *leios*, smooth; *koleos*, sheath. Leaf-sheath glabrous

leiogonum Gk *leios*, smooth; *gony*, knee. Nodes glabrous

leiophylla Gk *leios*, smooth; *phyllon*, leaf. Leaf-blades glabrous

Leiopoa Gk *leios*, smooth; *poa*, grass. The lemmas are glabrous, the spikelets otherwise resembling those of *Poa*

leiopoda Gk *leios*, smooth; *pous*, foot. Callus is glabrous

L

leioptera Gk *leios*, smooth; *pteron*, wing or feather-like. Lemma keels glabrous

leiostachya Gk *leios*, smooth; *stachys*, spike as of an ear of wheat. Glumes glabrous

leiotropis Gk *leios*, smooth; *tropis*, keel. The midribs of the glumes and lemmas are glabrous

leishanensis L. *-ensis*, denoting origin. From Leishan County, Guizhan Province, China

lejeunii In honor of Alexandre Louis Simon Lejeune (1779–1852) Belgian physician and botanist

lejocarpa See *leiocarpa*

lejocolea See *leiocolea*

lejophylla See *leiophylla*

lejopoda See *leiopoda*

Leleba Vernacular name for a species from the Moluccan Islands, Indonesia

lelievrei In honor of Le Lièvre de la Morinière (?–1845) French botanist

lemanii In honor of Dominique Sébastien Leman (1781–1829) French botanist

lembaensis L. *-ensis*, denoting origin. From Lemba, Zaire

lemeean-a, -um L. *-ana*, indicating connection. In honor of Albert Marie Victor Lemée (1872–1961) French botanist

lemmonii In honor of John Gill Lemmon (1832–1908) United States forester and botanist

Lemstrix Hybrids between species of *Hystrix* and *Leymus*

lencoranicum L. *-icum*, belonging to. See *lenkoranensis*

lendiger-a, -um L. *lens*, lentil; *gero*, carry or bear. The base of the caryopsis resembles a lentil

lenens-e, -is L. *-ense*, denoting origin. From Lena River Basin, Siberia

lengguanii In honor of Leng-guan Saw (fl. 1997) Malaysian botanist

leninogorica L. *-ica*, belonging to. From the Leninogor depression in the Altai Mountains, Kazakhstan

lenis L. soft. Leaf-blades softly hairy

lenkoranensis L. *-ensis*, denoting origin. From Lencoran, Republic of Georgia

Lenormandia In honor of Sébastian René Lenormand (1776–1871) French lawyer and botanist

lensaei In honor of Adrian Jacques de *Lens* (fl. 1828) who collected in southern France

lenta L. flexible. Culms wiry

lenticularis L. *lens*, lentil; *-ulus*, diminutive; *-aris*, pertaining to. – (1) Spikelets nearly orbicular. *Leersia lenticularis* – (2) grains lenticular. *Sporobolus lenticularis*

lentiferum L. *lens*, lentil; *fero*, carry or bear. The anthoecia resemble lentils

lentiginos-a, -um, -us L. *lentigo*, lentil-shaped spot; *-osum*, abundance. Spikelets ovate-orbicular resembling lentils

lentigiosus See *lentiginosa*

leonardii In honor of Emery Clarence Leonard (1892–1968) United States botanist

leonardiorum L. *-orum*, indicating possession in the plural. In honor of Thomas and Ann Leonard and their family, United States botanical benefactors

leonii In honor of Rolando J.C. León (fl. 1962) Argentinian ecologist

leonin-a, -um L. *-ina*, indicating possession. – (1) From places inhabited by lions. *Avena leonina*, *Helictotrichon leoninum* – (2) in honor of Brother, Frère or Hermano Léon (also known as Joseph Sylvestre Sauget-Bargier). As for *Saugetia*. *Eragrostis leonina*, *Paspalum leoninum*

leonis (1) From Sierra Leone. *Pennisetum leonis* – (2) see *Saugetia*. *Panicum leonis*, *Paspalidium leonis*

lepageana L. *-ana*, indicating connection. In honor of Ernest Lepage (1905–1981) Canadian cleric and botanist, major explorer in northern Canada and Alaska, hybrid specialist (sedges, Triticeae)

lepagei As for *lepageana*

Lepargochloa Gk *lepargos*, with white feathers; *chloa*, grass. The inflorescence is densely hairy

Lepeocercis Gk *leipo*, lack; scale; *kerkis*, arm or leg bone. Unlike related species the pedicels are bone-shaped with a translucent, that is, semisolid centre

lepid-a, -um, -us L. fine, elegant. Plant attractive in appearance
Lepideilema Gk *lepis*, scale; *eilema*, cover. The base of the spikelets is covered with bracts
lepidobasis Gk *lepis*, scale; *basis*, that which supports something. Laminae of basal leaves much reduced
Lepidopironia Gk *lepis*, scale; *pyros*, wheat; *oon*, egg; *-ia*, characteristic of. The seed is enclosed in a transparent utricle and the genus was regarded by the author as related to *Triticum* (wheat)
lepidopoda Gk *lepis*, scale; *pous*, foot. Rhizomes clothed in scales
lepidul-a, -um L. *lepidus*, pretty; *-ula*, diminutive. Plant attractive in appearance
lepid-um, -us See *lepida*
lepidura Gk *lepis*, scale; *oura*, tail. Racemes solitary, terminal
Lepidurus Gk *lepis*, scale; *oura*, tail. The spicate inflorescence with its sessile spikelets resembles the tail of a rat
lepidus See *lepida*
Lepitoma Gk *lepis*, scale; *tome*, stump. The glumes are truncate
Lepiurus Gk *lepis*, scale; *oura*, tail. The inflorescence is a cylindrical spike with the spikelets embedded and sealed into cavities by the glumes
leporin-a, -um L. *lepus*, hare; *-ina*, indicating resemblance. Inflorescence resembles a hare's tail
leprodes Gk *lepros*, rough; *-odes*, indicating resemblance. Lower glume of sessile spikelet densely scabrid
leprosulum L. *leprosus*, leprous; *-ulum*, diminutive. Lemma and palea somewhat shiny-white
lepta Gk *leptos*, narrow. Leaf-blades thread-like
leptacanthus Gk *leptos*, narrow; *acantha*, spine. Involucral bristles slender
leptachne Gk *leptos*, narrow; *achne*, scale. Glumes narrow-lanceolate
leptachyrium Gk *leptos*, narrow; *achyron*, chaff; *-ium*, characteristic of. Glumes hyaline
Leptagrostis Gk *leptos*, narrow. Lemma acuminate in contrast to that of *Agrostis*
leptalea Gk *leptaleos*, delicate. Habit of plant slender
leptanth-a, -us Gk *leptos*, narrow; *anthos*, flower. Spikelets narrow
Leptaspis Gk *leptos*, narrow; *aspis*, shield. The lemma of the female spikelet resembles a narrow shield
Leptatherum Gk *leptos*, narrow; *ather*, barb or spine. The lemma of the upper floret long attenuate
Leptocanna Gk *leptos*, narrow; *kanna*, cane. Thin-stemmed woody grasses
leptocarpa Gk *leptos*, narrow; *karpos*, fruit. Grains elongated
Leptocarydion Gk *leptos*, narrow; *karyon*, nut; *-ion*, diminutive. *Caryopsis* linear, trigonous, resembling a small nut
leptocaulon Gk *leptos*, narrow; *kaulos*, stem. Culms narrow
Leptocercus, Leptocereus Gk *leptos*, narrow; *kerkis*, tapering rod. Inflorescence narrow, cylindrical
leptochaeta Gk *leptos*, narrow; *chaete*, bristle. Lemma awn thread-like
Leptochloa Gk *leptos*, narrow; *chloa*, grass. Inflorescence a slender spike
Leptochloe, Leptochloë See *Leptochloa*
leptochlooides Gk *-oides*, resembling. Similar to *Leptochloa*
Leptochloopsis Gk *opsis*, resemblance. Similar to *Leptochloa*
Leptochloris Gk *leptos*, narrow. Similar to *Chloris* but with narrow spikelets
leptoclad-a, -us Gk *leptos*, narrow; *klados*, branch. Culms slender
leptocom-a, -um, -us Gk *leptos*, narrow; *kome*, hair of the head. – (1) Lemmas with a basal tuft of long hairs. *Poa leptocoma* – (2) pedicels of stalked spikelets hairy. *Anadelphia leptocoma*, *Andropogon leptocomus* – (3) lemma awn shortly hairy. *Hypogynium leptocomum*
Leptocoryphium Gk *leptos*, delicate; *koryphe*, summit; *-ium*, characteristic of. The fertile lemma has a delicate white apex

leptogluma Gk *leptos*, narrow; L. *gluma*, husk. Glumes narrow-lanceolate
leptolepis Gk *leptos*, narrow; *lepis*, scale. Lemmas narrow-lanceolate
Leptoloma Gk *leptos*, delicate; *loma*, border. The lemma of the hermaphrodite floret has a narrow hyaline margin
leptolomoides Gk *-oides*, resembling. Resembling *Leptoloma* with respect to inflorescence
Leptoma See *Lepitoma*
leptomerum Gk *leptos*, narrow; *meros*, portion. Slender in all its parts
leptophyll-a, -um, -us Gk *leptos*, narrow; *phyllon*, leaf. Leaf-blades narrow
Leptophyllochloa Gk *leptos*, narrow; *phyllon*, leaf; *chloa*, grass. Leaf-blades very narrow
leptophyll-um, -us See *leptophylla*
leptopoda Gk *leptos*, narrow; *pous*, foot. Pedicel slender
Leptopogon Gk *leptos*, narrow; *pogon*, beard. Racemes slender and bearded like those of *Andropogon*
leptopogon Gk *leptos*, narrow; *pogon*, beard. Awn thin and flexuous
leptopus Gk *leptos*, narrow; *pous*, foot. Spikelet with narrow, acute callus
Leptopyrum Gk *leptos*, narrow; *pyros*, wheat. Inflorescence a narrow spike
leptorhachis Gk *leptos*, narrow; *rhachis*, backbone. Inflorescence with a slender central axis
leptorrhachis See *leptorhachis*
leptorrhiza Gk *leptos*, narrow; *rhiza*, root. Rhizomes thin
leptos Gk delicate. Inflorescence a contracted panicle
Leptosaccharum Gk *leptos*, narrow. Culms slender but in many other respects similar to *Saccharum*
leptostachy-a, -um, -us Gk *leptos*, narrow; *stachys*, spike as of an ear of wheat. – (1) Spikelets long and narrow. *Glyceria leptostachya, Hymenachne leptostachya, Stipa leptostachya* – (2) inflorescence branches slender. *Andropogon leptostachyus, Chloris leptostachya* – (3) culms slender. *Lasiacis leptostachya, Panicum leptostachyum*
Leptostachys, leptostachys Gk *leptos*, narrow; *stachys*, spike as of an ear of wheat. Inflorescences narrow
leptostachy-um, -us See *leptostachya*
leptothera Gk *leptos*, narrow; *ather*, ear or spike of wheat. Inflorescence a spicate panicle
Leptothrium Gk *leptos*, narrow; *thrix*, hair. The glumes are subulate
Leptothrix Gk *leptos*, narrow; *thrix*, hair. Glumes awn-like
leptothrix Gk *leptos*, narrow; *thrix*, hair. Lemmas terminating in long, thin awns
leptotricha Gk *leptos*, narrow; *thrix*, hair. – (1) Inflorescence branches very thin. *Agrostis leptotricha* – (2) lemmas invested with long slender hairs. *Eragrostis leptotricha*
leptour-a, -um Gk *leptos*, narrow; *oura*, tail. Inflorescence a thin spike or spike-like panicle
leptura, Lepturus Gk *leptos*, narrow; *oura*, tail. – (1) Inflorescence a cylindrical spike. *Setaria leptura* – (2) awns filiform. *Aristida leptura*
Lepturella L. *-ella*, diminutive but here used as a name-forming suffix. Similar to *Lepturus*
Lepturidium Gk *-idium*, resembling. Similar to *Lepturus*
lepturoides Gk *-oides*, resembling. Inflorescence a spike as with *Lepturus*
Lepturopetium Combining the characters of *Lepturus* and *Oropetium*
Lepturopsis Gk *opsis*, resemblance. Similar to *Lepturus*
lepusnica L. *-ica*, belonging to. From Lake Napusnicul, Romania
Lepyroxis Gk *lepyron*, husk; *oxis*, vinegar cruet. Spikelets shaped like a vinegar cruet of Greek times
Lerchenfeldia In honor of Josef Radnitzky von Lerchenfeld (1753–1812) Austrian-born cleric, educator and botanist
leschenaultian-a, -us L. *-ana*, indicating connection. In honor of Jean-Baptiste Louis-Claude-Théodore, Leschenault de la Tour (1773–1826) French botanist and traveller

Lesourdia In honor of E. le Sourd, French physician and amateur botanist

lessingiana L. *-ana*, indicating connection. In honor of Christian Friedrich Lessing (1809–1862) Polish-born German botanist in Siberia

lessoniana L. *-ana*, indicating connection. In honor of René Primivère (1794–1849) or his brother Pierre Adolphe Lesson (1805–1888) French botanists

letestui In honor of Georges Marie Patrice Charles le Testu (1877–1967) French plant collector

letourneuxii In honor of Tacite Letourneux (1804–1880) or Aristide Horace (1820–1890) French botanists

letouzeyi In honor of René Letouzey (fl. 1972) who collected in Republic of Cameroon

lettermanii In honor of George Washington Letterman (1841–1913) United States teacher and botanist

leucacranth-a, -um Gk *leukos*, white; *akros*, at the tip; *anthos*, flower. Anthoecium whitish in contrast to the glumes which are whitish with green veins

leucanth-a, -um Gk *leukos*, white; *anthos*, flower. Spikelets invested in long silky white hairs

leucites Gk *leukos*, white; *-ites*, indicating connection. Leaf-sheath invested in dazzling white hairs

leucoblepharis Gk *leukos*, white; *blepharis*, eye-lash. Leaf-blades bearing long white hairs

leucocephala Gk *leukos*, white; *kephale*, head. Inflorescences pale as if blanched

leucocom-a, -um Gk *leukos*, white; *kome*, hair of head. Glumes and sterile lemma invested with long erect hairs

leucogluma Gk *leukos*, white; L. *gluma*, husk. Glumes hyaline to green

leucolepis Gk *leukos*, white; *lepis*, scale. Spikelets greenish-white

leucophae-a, -um Gk *leukos*, white; *phaeos*, grey. Panicles pale-colored

Leucophrys Gk *leukos*, white; *ophrys*, eyebrow. The lemma of the lower floret bears a line of stiff erect hairs below the middle

leucopila Gk *leukos*, white; *pilos*, felt. Plant invested with short white hairs

Leucopoa Gk *leukos*, white; *poa*, grass. The leaf-blades are glaucous and the spikelets bluish-white

leucopogon Gk *leukos*, white; *pogon*, beard. Inflorescence with abundant white hairs

leucorhod-a, -us Gk *leukos*, white; *rhodon*, rose. Oral setae white

leucosperma Gk *leucos*, white; *sperma*, seed. Lemma investing grain, white

leucostachy-a, -um, -us Gk *leukos*, white; *stachys*, spike as of an ear of wheat. The spikelets are white and sometimes invested with copious white hairs

leucosticta Gk *leukos*, white; *stictos*, spotted. Leaf-blade with white spots

leucothrix Gk *leukos*, white; *thrix*, hair. Leaf-sheaths invested with long white hairs

leucotricha Gk *leukos*, white; *thrix*, hair. Apex of lemma white with a ring of hairs about the base of the awn

lev-e, -is L. smooth. As for *laeve*

leviculme L. *levis*, smooth; *culmus*, stalk. Culms smooth

levigatus L. smooth. Culms smooth with polished internodes

levingei In honor of Henry Corbin Levinge (1828–1896) Irish botanist

levipes L. *levis*, smooth; *pes*, foot. Pedicels glabrous

levis See *leve*

leyboldtii In honor of Frederico Leyboldt (1827–?) who collected on Más Afuera, an island in the Juan Fernández Archipelago, Chile

Leydeum Hybrids between species of *Leymus* and *Hordeum*

Leymopyron Hybrids between species of *Leymus* and *Agropyron*

Leymostachys Hybrids between species of *Leymus* and *Psathyrostachys*

Leymotrigia Hybrids between species of *Leymus* and *Elytrigia*

Leymotrix Hybrids between species of *Leymus* and *Hystrix*

Leymstrix Hybrids between species of *Hystrix* and *Leymus*

Leymus An anagram of *Elymus*

leysseri In honor of Friedrich Wilhelm Leysser (1731–1815) German soldier and botanist

Leytesion Hybrids between species of *Critesion* and *Leymus*

lhasaensis L. *-ensis*, denoting origin. From Lhasa, Tibet Autonomous Region, China

l'Herminieri See *herminieri*

lhotskyi In honor of Johann Lhotzky (1795–1866) Polish-born of Czech parents, explorer and naturalist

lianatherus French *liana*, tropical twining twine; *ather*, barb or spine. Lemmas with hygroscopic awns to 20 cm long

liangshanensis L. *-ensis*, denoting origin. From Liangshan, Sichuan Province, China

libanoticum L. *libanos*, rosemary; *-icum*, belonging to. Growing amongst rosemary (*Rosemarinus officinalis*)

Libertia In honor of Anna Maria Libert (1782–1865) French botanist

libyca L. *-ica*, belonging to. From the Libyan Desert

Libyella L. *-ella*, diminutive but here used as a name-forming suffix. From Libya

licentiana L. *-ana*, indicating connection. In honor of Eugène Licent (fl. 1930) cleric who collected in China

lichiangensis L. *-ensis*, denoting origin. From Mt Lichiang, Yunnan Province, China

liebigiana L. *-ana*, indicating connection. In honor of Manfred Liebig (fl. 1912–1974) German cleric and plant collector in Togo

liebmannian-a, -um L. *-ana*, indicating connection. As for *liebmannii*

liebmannii In honor of Frederik Michael Leibmann (1813–1856) Danish botanist

lignosa L. *lignum*, wood; *-osa*, abundance. Culms woody

ligular-e, -is L. *ligula*, small tongue; *-aris*, pertaining to. Ligule conspicuous

ligulat-a, -us L. *ligula*, small tongue; *-ata*, possessing. – (1) Ligule conspicuous. *Agrostis ligulata, Andropogon ligulatus, Calamagrostis ligulata, Deyeuxia ligulata, Festuca ligulata, Isachne ligulata, Lasiacis ligulata, Leptochloa ligulata, Poa ligulata, Sporobolus ligulatus* – (2) leaf-blades short and strap-shaped. *Muhlenbergia ligulata*

ligustic-a, -um, -us From Ligusticus, now Liguria, part of the Italian Piedmont

lihauense L. *-ense*, denoting origin. From Lihau Peak on Maui one of the Hawaiian Islands

liliana In honor of Liliana Zimmermann, sister of R. C. Zimmermann whose generosity supported the field work which led to the description of the species

lilloi In honor of Miguel Lillo (1862–1931) Argentinian botanist

lima In honor of Abelardo Rodriques Lima who collected in Brazil

limbat-a, -um L. *limbus*, border; *-ata*, possessing. Glumes or lemmas colored differently on their margins and centres

limensis L. *-ensis*, denoting origin. From Lima, Peru

limicola L. *limus*, mud; *-cola*, dweller. Growing around swamps

limitanea L. *limes*, pathway; *-anea*, relating to. Growing in railway reserves, South Australia

Limnas Gk *limnas*, swamp. Swamp plants

Limnetis Gk *limnas*, swamp; L. *-etis*, place of growth. Plants of sea-coast salt-marshes

Limnodea Gk *-odea*, resembling. Similar to *Limnas*

Limnopoa Gk *limnas*, swamp; *poa*, grass. Forms mats on water

limonias Gk *leimon*, meadow. Inhabiting meadows

limos-a, -um, -us L. *limus*, mud; *-osa*, abundance. Growing in muddy places or swamp species

limprichtii In honor of Hans Wolfgang Limpricht (1877–?) German botanist who collected in Japan, China and Tibet as well as in Europe

lincangensis L. *-ensis*, denoting origin. From Lincang, Yunnan Province, China

linczerskii See *linczevskyi*
linczevskii In honor of Linczevsky
Lindbergella L. *-ella*, diminutive but here used as a name-forming suffix. See *Lindbergia*
Lindbergia In honor of Harold Lindberg (1871–1963) Finnish botanist
lindenbergian-a, -um L. *-ana*, indicating connection. As for *Lindbergia*
lindenian-a, -um L. *-ana*, indicating connection. In honor of Jean Jules Linden (1817–1898) Luxembourg-born Belgian botanist
lindenii As for *lindeniana*
lindheimeri In honor of Ferdinand Jakob Lindheimer (1801–1879) German-born United States botanist
lindiens-e, -is L. *-ense*, denoting origin. From Lindi, a district in Tanzania
lindigii In honor of Alexandro M. Lindigio (fl. 1862) who collected in Colombia
lindleyan-a, -um L. *-ana*, indicating connection. In honor of John Lindley (1799–1865) English botanist
lindleyi As for *lindleyana*
lindmanii In honor of Carl Axel Magnus Lindman (1856–1928) Swedish botanist
lindsayi In honor of William Lauder Lindsay (1829–1880) Scots botanist and physician
lineale L. *linea*, linen thread; *-ale*, pertaining to. Leaf-blades long and narrow
linear-e, -is L. *linea*, linen thread; *-are*, pertaining to. – (1) Leaf-blades narrow. *Agrostis linearis, Andropogon linearis, Arundinaria linearis, Cynodon linearis, Digitaria linearis, Panicum lineare, Paspalum lineare, Sporobolus linearis, Trisetaria linearis, Trisetum lineare* – (2) inflorescence a spike-like panicle. *Polypogon linearis*
linearifoli-a, -us L. *linea*, linen thread; *-aris*, pertaining to; *folium*, leaf. Leaf-blades long and thin
linearis See *lineare*
lineat-a, -um, -us L. *linea*, linen thread; *-ata*, possessing. – (1) Leaf-blades, glumes or lemmas marked by fine parallel lines. *Andropogon lineatus, Bambusa lineata, Panicum lineatum* – (2) leaf-blade narrow. *Axonopus lineatus*
lineicus L. *-icus*, belonging to. From Linieski Pass, western Altai Mountains, Kazakhstan
lineispatha L. *linea*, linen thread; *spatha*, sheathing base and false petiole of a palm leaf. Rhachis winged with conspicuous veins
lineolata L. *linea*, linen thread; *-ola*, diminutive; *-ata*, possessing. – (1) Leaf-sheath marked by fine parallel lines. *Yushania lineolata* – (2) leaf-blade filiform. *Stipa lineolata*
Lingnania Commemorating Lingnan University, China
lingnanioides Gk *-oides*, resembling. Similar to *Lingnania*
lingua L. *lingua*, tongue. Ligule long-fimbriate
lingulata L. *lingua*, tongue; *-ula*, diminutive; *-ata*, possessing. – (1) Ligule conspicuous. *Coix lingulata, Sasa lingulata* – (2) spikelets tongue-shaped in outline. *Eragrostis lingulata*
linicola L. *-cola*, dweller. Growing in fields of *Linum*, that is amongst flax
linifoli-a, -us L. *linum*, thread; *folium*, leaf. Leaf-blades linear
Linkagrostis Segregated from *Agrostis* and honoring Link as in *linkii*
linkian-a, -um L. *-ana*, indicating connection. As for *linkii*
linkii In honor of Johann Heinrich Friedrich Link (1767–1851) German botanist
linnaei In honor of Carl Linnaeus (1707–1778) Swedish botanist
linnean-a, -um, -us As for *linnaei*
Linosparton Gk *linon*, linen; *spartine*, cord. Used for rope making
linozodes Gk *linon*, flax; *-odes*, resembling. Culms resemble the stems of flax
Lintonia In honor of A. Linton (fl. 1904–1906) who collected in Kenya
lintonii As for *Lintonia*
liouae In honor of Lian(g) Liou (1933–) Chinese botanist
Lipeocercis See *Lepeocercis*
Lipeoceris See *Lepeocercis*
lipskyi In honor of Vladimir Hippolitowitsch Lipsky (1863–1937) Russian botanist

lisboae In honor of José Camillo Lisboa (c. 1822–1897) Indian physician and botanist

lisowskii In honor of Stanislaw Lisowski (1924–2002) Polish botanist

Litachne See *Lithachne*

litardiereana L. *-ana*, indicating connection. In honor of Rene Verriet de Litardière, French botanist (1888–1957)

Lithachne, Lithacne Gk *lithos*, stone; *achne*, scale. Paleas and lemmas thick and bony

Lithagrostis Gk *lithos*, stone; *agrostis*, grass. The cupule subtending the ultimate inflorescence units is indurated

lithobius Gk *lithos*, stone; *bios*, manner of living. Growing amongst rocks

lithophil-a, -um, -us Gk *lithos*, stone; *phileo*, love. Growing amongst rocks

lithuanica L. *-ica*, belonging to. From Lithuania

litigans L. *litigo*, dispute. In dispute in the sense of being very similar to a related species

litigiosum See *litigosa*

litigos-a, -um L. *litigium*, a dispute; *-osa*, abundance. Formerly confused with another species or placed in a different genus

litoral-e, -is See *littorale*, a widely used orthographic variant

litorosa L. lit(t)us, sea shore; *-osa*, abundant. Common on the sea-shores of some sub-Antarctic Islands

Littledalea In honor of St. George R. Littledale (c. 1851–1931) an English traveller to Tibet Autonomous Region, China

littoral-e, -is L. *lit(t)us*, sea shore; *-ale*, pertaining to. Species of sand dunes, salt marshes or river banks

littoreus L. *lit(t)us*, seashore; *-eus*, pertaining to. Seashore plants

litvinovii, litvinowii In honor of Dimitri Ivanovich Litvinov (1854–1929) Russian botanist

litwinowiana L. *-ana*, indicating connection. As for *litvinovii*

litwinowii See *litvinovii*

liukiuensis L. *-ensis*, denoting origin. From Liukiu, Taiwan

livid-a, -um, -us L. leaden. Spikelets grey or purple

liviensis L. *-ensis*, denoting origin. From Livia, Spain

lixin Vernacular name of the species in south eastern Xizang, China

llanganatensis L. *-ensis*, denoting origin. From Cordillera de los Llanganates, Ecuador

lloydianus L. *-anus*, indicating connection. In honor of James Lloyd (1810–1896) English-born French botanist

lloydii In honor of Frances Ernest Lloyd (1868–1947)

lobata L. *lobus*, lobe; *-ata*, possessing. Lemma lobed

lobelianum L. *-anum*, indicating connection. As for *lobelii*

lobelii In honor of Mathias de L'Obel (1538–1616) Flemish botanist

lodiculare L. *lodicula*, small blanket; *-are*, pertaining to. With conspicuous lodicules or mistakenly, the inflexed margins of the palea and lemma which became detached during dissection of the spikelet

Lodicularia L. *lodicula*, small blanket; *-aria*, pertaining to. Lodicules of upper flower conspicuous

lodunensis L. *-ensis*, denoting origin. From Loudun, now Vienna, Austria

loefflingiana See *loeflingiana*

loefgrenii In honor of Albert Löfgren (1854–1918) Swedish-born Brazilian botanist

loeflingian-a, -um L. *-ana*, indicating connection. In honor of Pehr Löfling (1729–1756) Swedish botanist and traveller

lofushanensis L. *-ensis*, denoting origin. From Luofu Shan, Guandong Province, China

loharduggae From Lohardugga, Bihar Province, India

loheri In honor of August Loher (?–1930) German-born Philippine plant collector

Lojaconoa In honor of Michele Lojacono-Pojero (1853–1919) Italian botanist

lokkomontana L. *mons*, mountain; *-ana*, indicating connection. From Lokkosan, a mountain in Settsu Province, now part of Hyogo and Osaka Prefectures, Japan

loliace-a, -um, -us L. *-acea*, resembling. Inflorescence resembles that of *Lolium*
loliiforme L. *forma*, appearance. See *loliacea*
lolioides Gk *-oides*, resembling. Inflorescence resembling that of *Lolium*
Loliolum L. *-olum*, diminutive but here used as a name-forming suffix. Resembling *Lolium*
Lolium Referred to by the Roman poet, Virgil, as a troublesome weed, possibly darnell (*Lolium temulentum*)
lolium Resembling *Lolium* with respect to the inflorescence
lomanensis L. *-ensis*, denoting origin. From Loman, Zaire
lomba Vernacular name for the species in Kikongo dialect, Zaire
Lombardochloa Gk *chloa*, grass. In honor of Atilio Lombardo Nolle (1902–1984) Uruguayan botanist
lommelii In honor of Lommel (fl. 1900) who collected in East Africa
londonoae In honor of Ximena Londoño (fl. 1990) Colombian botanist
long-a, -um L. long. – (1) Culms tall. *Leptochloa longa*, *Panicum longum* – (2) panicle contracted, long. *Helictotrichon longum*
longaevus L. long lived. Culms long lived
longearistat-a, -um, -us L. *longus*, long; *arista*, bristle. Lemmas or glumes long awned
longepedunculatum See *longipedunculata*
longeracemos-um, -us L. *longus*, long; *racemus*, raceme; *-osa*, abundance. With long inflorescence branches
longeradiata L. *longus*, long; *radius*, spoke of a wheel; *-ata*, possessing. Inflorescence branches long and whorled
longespicata See *longispicatus*
longianthera L. *longus*, long; Gk *antheros*, blooming. Flowering most of the year
longiarista L. *longus*, long; *arista*, bristle. Lemmas or glumes long awned
longiaristat-a, -um, -us L. *longus*, long; *arista*, bristle, *-ata*, indicating possession. Lemmas or glumes long awned
longiauriculata L. *longus*, long; *auris*, ear; *-ulus*, diminutive, *-ata*, possessing. Leaf-blades with long auricles
longiaurita L. *longus*, long; *auritus*, eared. Leaf-blades with long auricles
longiberbis L. *longus*, long; *barba*, beard. Callus or lemma invested with long hairs
longicaud-a, -um L. *longus*, long; *cauda*, tail. – (1) Lemma apex of lower floret or all florets long drawn out. *Cortaderia longicauda*, *Melinis longicauda*, *Panicum longicauda*, *Rhynchelytrum longicaudum* – (2) arms of triradiate awns drawn out. *Aristida longicauda*
longiceps L. *longus*, long; *-ceps*, pertaining to a head. Spikelets longer than those of related species
longicilius L. *longus*, long; *cilium*, hair. Leaf margins invested with long hairs
longicollis L. *longus*, long; *collum*, neck. Column of awn long
longicuspe L. *longus*, long; *cuspis*, point. Rhachis extending beyond the spikelet as a short stalk
longifimbriata L. *longus*, long; *fimbriae*, fringe; *-ata*, possessing. Bases of auricles bearing long hairs
longiflor-a, -um, -us L. *longus*, long; *flos*, flower. Spikelets longer than those of related species
longifoli-a, -um, -us L. *longus*, long; *folium*, leaf. Leaf-blades longer than those of related species
longiglum-a, -e, -is L. *longus*, long; *gluma*, husk. Spikelets with long glumes and or lemmas
longii In honor of David G. Long (1948–) who collected in Sikkim State, India
longiinternodus L. *longus*, long; *inter*, between; *nodus*, knot. The rhizome has long internodes
longijubatum L. *longus*, long; *juba*, mane; *-atum*, possessing. Panicles lax with abundant filiform branches
longilamina L. *longus*, long; *lamina*, sword blade. Leaf-blades long
longiligula L. *longus*, long; *ligula*, small tongue. Ligule, long
longiligulat-a, -um, -us L. *longus*, long; *ligula*, small tongue; *-ata*, possessing. Ligule long

L

longiloba L. *longus*, long; Gk *lobos*, lobe. Lemma deeply lobed

longiloreum L. *longus*, long; *loreum*, thong. Inflorescence whip-like

longinodis L. *longus*, long; *nodus*, knot. Nodes widely separated

longipalea L. *longus*, long. Palea longer than for related species

longipanicula L. *longus*, long; *panicula*, panicle. Panicle spreading

longipaniculata L. *longus*, long; *panicula*, panicle; *-ata*, possessing. Panicle longer than with related species

longipedicellat-a, -um L. *longus*, long; *pedicellus*, stalk; *-ata*, possessing. Spikelets with long pedicels

longipedunculata L. *longus*, long; *pedunculatus*, peduncule; *-ata*, possessing. Panicle borne on a long leafless stalk

longipes L. *longus*, long; *pes*, foot. – (1) Spikelets borne on long pedicels. *Cyrtococcum longipes*, *Eragrostis longipes*, *Loudetia longipes*, *Nematopoa longipes*, *Panicum longipes* – (2) the bases of culms lack leaves or the lower leaves lack blades. *Arundinaria longipes*

longipetiolat-a, -um L. *longus*, long; *petiolus*, little leg; *-ata*, possessing. The leaf-blade tapers gradually towards the sheath thereby generating a pseudopetiole

longipila L. *longus*, long; *pilus*, a hair. Plant with long hairs investing all or any of its parts

longipilosa L. *longus*, long; *pilus*, a hair; *-osa*, abundance. Oral setae *long*

longiplumosa L. *long*, long; *pluma*, small soft feather; *-osa*, abundance. Awn long and feather-like with hairs to one cm long

longiprophylla L. *longus*, long. Prophylls exceptionally long

longiramea L. *longus*, long; *ramus*, branch. Panicle with long branches

longiramosus L. *longus*, long; *ramus*, branch; *-osus*, abundance. Culm branches long

longiramum L. *longus*, long; *ramus*, branch. Spikelets borne on long pedicels

longiset-a, -um, -us L. *longus*, long; *seta*, bristle. – (1) Glumes or lemmas long awned. *Agrostis longiseta*, *Apera longiseta*, *Aristida longiseta*, *Brachypodium longisetum*, *Calamagrostis longiseta*, *Chaetochloa longiseta*, *Festuca longiseta*, *Oplismenus longisetus*, *Panicum longisetum*, *Pennisetum longisetum*, *Rhynchelytrum longisetum*, *Saccharum longisetum*, *Setaria longiseta*, *Tricholaena longiseta*, *Vulpia longiseta* – (2) arms of tripartite awns long. *Aristida longiseta*

longisetos-um, -us L. *longus*, long; *seta*, bristle; *-osa*, abundance. Lemma long-awned

longiset-um, -us See *longiseta*

longispatha L. *longus*, long; *spatha*, sheathing base and false petiole of a palm leaf. Inflorescence bracts long and leafy

longispica L. *longus*, long; *spica*, a point; hence, in particular, an ear or spike of grain. With long spikelets or inflorescence branches

longispicat-a, -us L. *longus*, long; *spica*, a point; hence, in particular, an ear or spike of grain; =*atus*, possessing. Inflorescence spicate and longer than that of related species

longispicul-a, -um L. *longus*, long; *spica*, a point; hence, in particular, an ear or spike of grain; *-ula*, diminutive. Spikelets long

longispiculat-a, -um L. *longus*, long; *spica*, a point; hence, in particular, an ear or spike of grain; *-ula*, diminutive; *-atum*, possessing. Spikelets long

longispiculum See *longispicula*

longispinus L. *longus*, long; *spina*, spine. Bristles of the subtending involucre longer than those of some other species

longissim-a, -um L. *longus*, long; *-issima*, most. – (1) Internodes very long. *Paspalum longissimum*, *Yushania longissima* – (2) subtending bristles very long. *Setaria longissima* – (3) spikelets very long. *Digitaria longissima*

longistolon L. *longus*, long; *stolo*, useless sucker. Plant with long stolons or rhizomes

longistylum L. *longus*, long; *stylum*, column. Style long

longiuscula L. *longius*, longer; *-ula*, tendency. Somewhat long

longivaginat-a, -um L. *longus*, long; *vagina*, sheath; *-ata*, possessing. Leaf-sheaths long

longivalvula L. *longus*, long; *valva*, leaf of a folding door; *-ula*, diminutive. Lemmas large

longum See *longa*

looseriana L. *-ana*, indicating connection. In honor of Gualterio Looser (1898–1982) Chilean botanist

Lophacme Gk *lophos*, crest; *akme*, highest point. Growing on ridge tops

Lophatherum Gk *lophos*, crest; *ather*, barb or spine. The sterile lemma is surmounted by a tuft of awns

Lophochlaena Gk *lophos*, crest; *chlaena*, cloak. The lemma apex is markedly erose

Lophochloa Gk *lophos*, crest; *chloa*, grass. Lemma apex shortly aristate

Lopholepis Gk *lophos*, crest; *lepis*, scale. Lemma apex shortly awned

Lophopogon Gk *lophos*, crest; *pogon*, beard. Upper glume awned and hairy at its apex

Lophopyrum Gk *lophos*, crest; *pyros*, wheat. Origin unclear, not given by the author but probably a reference to the upper midrib of the lemma bearing bristles or hairs

lophostachya Gk *lophos*, crest; *stachys*, spike as of an ear of wheat. The sterile upper florets are conspicuously three-awned

lophotrichus Gk *lophos*, crest; *thrix*, hair. The nine lemma awns are invested with hairs at the base causing it to resemble a crest

lopollensis L. *-ensis*, denoting origin. From the Lopollo District, Angola

lorentii In honor of J. August Lorent (1812–1884) American-born German botanist and traveller

lorentzian-a, -um L. *-ana*, indicating connection. In honor of Paul Günther Lorentz (1835–1881) German-born Urugayan botanist

Lorenzochloa Gk *chloa*, grass. In honor of Lorenzo Parodi; see *Parodiochloa*

loretensis L. *-ensis*, denoting origin. From Loreto Region, Peru

Loretia In honor of Henri Loret (1810–1888) French physician and botanist

loretii As for *Loretia*

loreum L. made of leather thongs. Leaf-blades leathery

loricata L. *lorica*, corselet; *-ata*, resembling. The rugose lower glume resembles the corselet of a Mediaeval soldier

losae In honor of Taurino Mariano Losa (1893–1966) Spanish botanist

Loudetia In honor of Loudet, German dentist at Karlsruhe

Loudetiopsis Gk *opsis*, resemblance. Similar to *Loudetia*

louisianae From Louisiana, USA

Louisiella L. *-ella*, diminutive but here used as a name-forming suffix. In honor of Jean Louis (1903–1944) Belgian botanist

lowanensis L. *-ensis*, denoting origin. From the Lowan, a district in south-eastern Australia recognized on account of its characteristic vegetation

loxensis L. *-ensis*, denoting origin. From Loja Province, Ecuador

Loxodera Gk *loxos*, slanting; *deire*, neck. The spikelets are obliquely placed on the rhachis

Loxostachys Gk *loxos*, slanting; *stachys*, spike as of an ear of wheat. Spikelets obliquely ovoid

lualabaensis L. *-ensis*, denoting origin. From Lualaba Region, Zaire

lubrica L. slippery. Growing on steep slopes with clay soils

Lucaea In honor of August Friedrich Theodor Lucae (1800–1840) German apothecary and botanist

luciae In honor of Lucy Kathleen Armitage Chippendall Crook (1913–) South African botanist

luciarum In honor of Lucy May Cranwell Smith (1907–1992) and Lucy Beatrice Moore (1906–1987) New Zealand botanists

lucid-a, -um L. clear. – **(1)** A new name required for the purposes of nomenclature thereby making clear the identity of the species. *Calamagrostis lucida* – **(2)** very different from another species of the same genus growing in a similar habitat. *Bracteola lucida*, *Festuca lucida*, *Panicum lucidum*, *Poa lucida*

lucidulum L. *lucidus*, clear; *-ulum*, diminutive. Leaf-blade somewhat transparent

lucidum See *lucida*

luconiae From Luzon, Philippines

lucorum L. *lucus*, a woodland. Growing in woodlands

ludens L. *ludo*, play. Applied to a species whose name was nomenclaturally invalid, thereby playing by the rules as required

ludianense L. *-ense*, denoting origin. From Loudian Xian, Guizhan Province, China

Ludolfia See *Ludolphia*

Ludolphia In honor of Michael Matthias Ludolph (1705–1756) German botanist

ludoviciana L. *-ana*, indicating connection. In honor of Ludovicius, that is, Louis, early deceased son of Durieu de Maisonneuve; see *duriaei*

ludwigii In honor of Carl Ferdinand Heinrich Ludwig (1784–1847). German-born South African pharmacist and merchant

luederitzianum L. *-anum*, indicating connection. In honor of Franz Adolf Edward Lüderitz (1834–1886) German merchant and explorer

luembensis L. *-ensis*, denoting origin. From the Luembe Valley, Zaire

luerssenii In honor of Christian Luerssen (1843–1916) German botanist

luetzelburgii In honor of Philipp Luetzelburg (1880–1948) German botanist

lugens L. *lugeo*, mourn. The spikelets are dull in color

lukwangulens-e, -is L. *-ense*, denoting origin. From Lukwangule Plateau, Tanzania

lumampao Vernacular name of the species in the Philippines

lunata L. *luna*, moon; *-ata*, possessing. – (1) Glumes and lemmas crescent-shaped. *Poa lunata*, *Raddiella lunata* – (2) auricles crescent-shaped. *Indosasa lunata*, *Pariana lunata*

lundellii In honor of Cyrus Longworth Lundell (1907–1994) United States botanical collector

luodianensis L. *-ensis*, denoting origin. From Luodian Xian, Guizhou Province, China

lupulina L. *lupulus*, hop plant; *-ina*, indicating resemblance. Panicle densely ovate resembling the inflorescence of the hop plant (*Humulus lupulina*)

luquensis L. *-ensis*, denoting origin. From Luqu County, Gansu Province, China

lurid-a, -um L. drab yellow. Lemmas or glumes brownish-yellow

lushuiensis L. *-ensis*, denoting origin. From Lushui Xian, Yunnan Province, China

lusitanic-a, -us From Lusitania, now Portugal

lustriale L. *lustrum*, bog; *-ale*, pertaining to. Bog dweller

lutchuensis L. *-ensis*, denoting origin. From Lutschu or Liukiu Island, Okinawa

lutensis L. *lutum*, mud; *-ensis*, denoting origin. Growing in mud

luteostriata L. *luteus*, golden; *striatus*, striped. Leaves striped when young

lutescens L. *lutesco*, become yellow. – (1) Panicles yellowish-green. *Agrostis lutescens*, *Arthratherum lutescens*, *Chaetochloa lutescens*, *Eragrostis lutescens*, *Melica lutescens*, *Panicum lutescens*, *Poa lutescens*, *Setaria lutescens* – (2) feathery awn yellow-green. *Aristida lutescens* – (3) anthers yellowish. *Schizostachyum lutescens*

lutetense L. *-ense*, denoting origin. From Lutete, Zaire

luticol-a L. *lutum*, mud; *-cola*, dweller. – (1) Growing on tidal flats. *Panicum luticola* – (2) growing beside lakes. *Paspalum luticola*

lutinflatum L. *lutum*, mud; *inflo*, inflate. Spikelets swollen and mud colored

lutos-a, -us L. *lutum*, mud; *-osa*, abundance. Growing in water or damp places

lutzii In honor of Adolpho Lutz (1855–1940) Brazilian medical researcher

luxurians L. *luxurio*, be abundant in growth. – (1) More robust than related species. *Euchlaena luxurians*, *Reana luxurians*, *Schizachyrium luxurians* – (2) growing abundantly. *Vilfa luxurians*

luzhiensis L. *-ensis*, denoting origin. From Liuzhi Xian, Guizhou Province, China

Luziola Modified from *Luzula* but reason for so doing uncertain

luzonens-e, -is L. *-ense*, denoting origin. From Luzon, Philippines. Also *luzoniense*
luzonicum L. *-icum*, belonging to. See *luzonense*
luzoniens-e, -is L. *-ense*, denoting origin. From Luzon, Philippines. Also *luzonense*
lyallii In honor of David Lyall (1817–1895) Scots-born botanist who collected in New Zealand
Lycochloa From the Lycus River, now Nahrel-Kelb, Lebanon
lycuroides Gk *-oides*, resembling. Similar to *Lycurus* with respect to the inflorescence
Lycurus Gk *lykos*, wolf; *oura*, tail. Panicles spike-like
Lygeum Gk *lygos*, willow twig. Culms widely used for weaving in North Africa
lynesii In honor of Hubert Lynes (1874–1942) British Naval Officer who collected in Africa
Lysurus See *Lycurus*

M

mabianensis L. *-ensis*, denoting origin. From Mabian, Sichuan Province, China
macala Origin uncertain, not given by author, but possibly from the Bengali vernacular
macalpinei In honor of Daniel McAlpine (1849–1932) Scots-born Australian mycologist
macbridei In honor of James Francis Macbride (1892–1976) United States botanist
maccannii In honor of Charles McCann (fl. 1930–1950) cleric and amateur botanist who collected in India
macclellandii In honor of John MacClelland (1805–1885) Public Health Officer who collected in India
macclounii In honor of John McClounie (fl. 1895) who collected in Malawi and Zambia
macclureana As for *Maclurochloa*
macclurei See *Maclurolyra*
macedoii In honor of Amaro Macedo (1914–) Brazilian botanist
macedonica L. *-ica*, belonging to. From Macedonia, formerly a region of south-eastern Europe, now divided between Greece, Bulgaria and the Republic of Macedonia
mac-er, -ra, -rum, -rus L. thin. Leaf-blades narrow
macgregorii In honor of William McGregor (1846–1919) Scots-born physician and British colonial administrator
macha From makha, the vernacular name of the species in western part of the Republic of Georgia
machrisianum In honor of Maurice A. Machris (fl. 1956–1977) United States philanthropist
macilent-a, -um L. thin. Culms slender
macivorii In honor of Ben McIvor (fl. 1964) Australian greenkeeper
mackayi In honor of A. E. Mackay (fl. 1851–1854) who collected in New Zealand
mackenzieana L. *-ana*, indicating connection. From the Mackenzie drainage basin, north-west Canada
mackenziei In honor of Kenneth K. Mackenzie (1877–1934) United States botanist
mackinlayi In honor of John McKinlay (1819–1872) Scots-born Australian explorer
mackliniae In honor of Jean Macklin (fl. 1955–1956) otherwise Mrs. Kingdon-Ward
maclaudii In honor of C. Maclaud (1895–?) who collected in West Africa
macleishii In honor of Ian McLeish, sometime agriculturalist at Royal Razat Farm, Salalah, Oman
macleodiae In honor of Miss Macleod who collected in Northern Nigeria
macloviana L. *-ana*, indicating connection. From Maclov, one of the Falkland Islands, United Kingdom Territory in the South Atlantic
maclurei See *Maclurolyra*
Maclurochloa Gk *chloa*, grass. As for *Maclurolyra*
Maclurolyra Similar to *Olyra* and in honor of Floyd Alonzo McClure (1897–1970) United States botanist with extensive experience of Chinese bamboos

macouniana L. *-ana*, indicating connection. As for *macounii*

macounii In honor of James Melville Macoun (1862–1920) Canadian botanist

macowanii In honor of Peter MacOwan (1830–1909) English-born South African educator and plant collector

macquariensis L. *-ensis*, denoting origin. From Macquarie Island in the Southern Ocean

M

macra See *macer*

macracteni-a, -um See *macractinia*

macractinia Gk *makros*, large; *ktenion*, small comb. The sterile lemma is ciliate with rigid hairs

macraei In honor of James Macrae (?–1830) who collected in Sri Lanka

macrandra Gk *makros*, large; *aner*, man. Anthers long

macranth-a, -um, -us Gk *makros*, large; *anthos*, flower. Spikelets large

macranthecium Gk *makros*, large; *anthos*, flower. Spikelets larger than those of related species

macranthela Gk *makros*, large; *anthele*, plume. Inflorescence large

macranther-a, -um, -us Gk *makros*, large; *antheros*, blooming. Panicle large

macranthos Gk *makros*, large; *anthos*, flower. Spikelets large

macranth-um, -us See *macrantha*

macrantoidea See *marantoidea*

macrarrhena Gk *makros*, large; *arrhen*, male. The anthers of the stalked male spikelets are larger than those of the sessile hermaphrodite spikelets

macrather-a, -us Gk *makros*, large; *ather*, spike or ear of wheat. Spikelets large

macroanthera Gk *makros*, large; *antheros*, blooming. Panicle large

macroblephar-a, -um Gk *makros*, large; *blepharon*, eye-lid. The glumes and sterile lemma bear copious long hairs

Macroblepharus Gk *makros*, large; *blepha-ron*, eye-lid. Lemma keels bear long cilia

Macrobriza Gk *makros*, large. Spikelets resemble those of Briza but are larger

macrocalyx Gk *makros*, large; *kalyx*, cup. Glumes more than half the length of the spikelet

macrocarp-a, -us Gk *makros*, large; *karpos*, fruit. – (1) Burr-forming spikelet clusters large. *Cenchrus macrocarpus* – (2) grain large. *Echinochloa macrocarpa*, *Setaria macrocarpa*

macrocarpon Gk *makros*, large; *karpos*, fruit. Anthoecium large

macrocarpus See *macrocarpa*

macrocephalus Gk *makros*, large; *kephale*, head. Burr-forming spikelet and associated involucre large

macrochaet-a, -um, -us Gk *makros*, large; *chaete*, bristle. Lemmas long-awned

Macrochaeta Gk *makros*, large; *chaete*, bristle. The spikelet clusters are surrounded by an involucre of long bristles

macrochlamys Gk *makros*, large; *chlamys*, cloak. Glumes long with respect to the length of the spikelet

Macrochloa Gk *macros*, large. Similar to *Briza* but with large spikelets

macrochloa Gk *macros*, large. Spikelets large compared with those of related species

macroclad-a, -um, -us Gk *makros*, large; *klados*, stem. – (1) Panicle branches long and slender. *Aristida macroclada*, *Bromus macrocladus*, *Poa macroclada* – (2) culms tall. *Panicum macrocladum*

macroculmis Gk *makros*, large; L. *culmus*, stalk. Culms large

macroglossa Gk *makros*, large; *glossa*, tongue. Ligule long

macrolemma Gk *makros*, large; *lemma*, husk. Lemmas large

macrolepis Gk *makros*, large; *lepis*, scale. Glumes and or lemmas large

Macronax Gk *makros*, large; L. *nax*, basket for catching fish. Used to make baskets for catching fish

macrophyll-a, -um, -us Gk *makros*, large; *phyllon*, leaf. Leaf-blades large

macropoda Gk *makros*, large; *pous*, foot. Pedicels long

macropodium Gk *makros*, large; *pous*, foot. Inflorescence borne on a long peduncle

macropogon Gk *makros*, large; *pogon*, beard. Leaf-blade densely hairy adjacent to ligule
macropus As for *macropoda*
macrorhinus Gk *makros*, large; *rhis*, nose. Lemma apex drawn out
macrosperm-a, -um, -us Gk *makros*, large; *sperma*, seed. Grains large
macrospicula Gk *makros*, large; L. *spica*, a point; hence, in particular, an ear or spike of grain; *-ula*, diminutive. Spikelets larger than those of related species
macrospiculata Gk *makros*, large; L. *spica*, a point; hence, in particular, an ear or spike of grain; *-ula*, diminutive; *-ata*, possessing. Spikelets very long compared with those of related species
macrospila Gk *makros*, large; L. *pilus*, a hair. Culms densely hairy at the base
macrostachy-a, -um, -us Gk *makros*, large; *stachys*, spike as of an ear of wheat. – (1) Inflorescence large. *Agrostis macrostachys*, *Arundinaria macrostachya*, *Avena macrostachya*, *Bambusa macrostachya*, *Bromus macrostachyus*, *Calotheca macrostachya*, *Cenchrus macrostachyus*, *Colanthelia macrostachya*, *Enteropogon macrostachyus*, *Ischaemum macrostachyum*, *Panicum macrostachyum*, *Pennisetum macrostachyum*, *Sclerodactylon macrostachyum*, *Sehima macrostachyum*, *Setaria macrostachya* – (2) spikelets with many florets. *Aeluropus macrostachyus*, *Chloris macrostachya*
macrostachyon Gk *makros*, large; *stachys*, spike as of an ear of wheat. Inflorescence large
macrostachys Gk *makros*, large; *stachys*, spike as of an ear of wheat. – (1) Spikelets large. *Andropogon macrostachys*, *Microbambus macrostachys* – (2) spikelets with many florets. *Bromus macrostachys*
macrostachy-um, -us See *Macrostachya*
macrothrix Gk *makros*, large; *thrix*, hair. Pedicels bearing long hairs
macrothyrsa Gk *makros*, large; *thyrsos*, ornamental wand. Inflorescence a large panicle
macrotis Gk *makros*, large; *ous*, ear. Auricles large
macrotrichum Gk *makros*, large; *thrix*, hair. Lower glume long-awned
macrour-a, -um, -us Gk *makros*, large; *oura*, tail. Inflorescence a spike or spike-like panicle
macrum See *macer*
macula L. *macula*, spot. Leaf-sheath with colored spots
maculat-a, -um, -us L. *macula*, spot; *-atum*, possessing. – (1) Culms and leaf-sheaths spotted. *Arthrostylidium maculatum*, *Sasa maculata*, *Sinobambusa maculata*, *Sucrea maculata*, *Yushania maculata* – (2) glumes with purple spots. *Iseilema maculatum*, *Notodanthonia maculata* – (3) apex of sterile lemma black-tipped. *Panicum maculatum*, *Paspalum maculatum* – (4) culms become spotted when mature. *Bambusa maculata*
maculos-a, -um L. *macula*, spot; *-osa*, abundance. Spikelets or culms spotted with reddish-brown or yellow patches
maculosoides Gk *-oides*, resembling. Similar to *Pleioblastus maculosa*
maculosum See *maculosa*
macusaniens-e, -is L. *-ense*, denoting origin. From Macusani, Puno Region, Peru
macutrensis L. *-ensis*, denoting origin. From Makutra, a mountain in the Ukraine
madagascariens-e, -is L. *-ense*, denoting origin. From Madagascar
madaropoda Gk *madaros*, bare; *pous*, foot. Stalks of upper racemes almost glabrous
madecassa From Madecassa, Madagascar
maderaspatan-a, -us L. *-ana*, indicating connection. From Maderaspata, that is Madras, India
maderense L. *-ense*, denoting origin. From Madeira, a Portuguese island in the North Atlantic Ocean
maderensis L. *-ensis*, denoting origin. From Madeira Islands
madida L. wet site. Growing in swamps
madiola L. *madidus*, wet site. Growing in swamps
madipirense L. *-ense*, denoting origin. From Madipir, East Africa
madorens L. *madeo*, be wet. Growing in damp sandy soil

madrensis L. *-ensis*, denoting origin. From Sierra Madre Mountains, Mexico
madritensis L. *-ensis*, denoting origin. From Madrid, Spain
maeotic-a, -um L. *-ica*, belonging to. From Maeotis Palus, now Sea of Azov
maeviae In honor of Maevia Noémi Correa (1914–2005) Argentinian botanist
magadanensis L. *-ensis*, denoting origin. See *magadanica*
magadanica L. *-ica*, belonging to. From Magadan Province, Russian Far East
magallanesiae In honor of Aurim Megallanes Nessi, Venezuelan botanist
Magastachya See *Megastachya*
magellanic-a, -um L. *-ica*, belonging to. – (1) From the Straits of Magellan. *Agropyron magellanicum, Agrostis magellanicus, Atropis magellanica, Catabrosa magellanica, Festuca magellanica, Glyceria magellanica, Hierochloe magellanica, Oplismenus magellanica, Torresia magellanica* – (2) in honor of José Carlos Reis de Magelhães collector of the holotype. *Merostachys magellanica*
magennisii From cultivar Magennis, a South African turf grass
magensiana L. *-ana*, indicating connection. In honor of Otto Magens (fl. 1958–1959) who collected in Chile
magentea L. *magenteus*, magenta. Anthers magenta-colored
magica Gk *magos*, Persian dream interpreter; *-ica*, belonging to. Here a reference to the occult, the species being used to make love charms
magn-a, -um, -us L. large. – (1) Plants large and generally with tall culms. *Bromus magnus, Chaetochloa magna, Digitaria magna, Eragrostis magna, Ischaemum magnum, Panicum magnum, Setaria magna, Syntherisma magna* – (2) caryopses large. *Avena magna*
magnicaespes L. *magnus*, large; *caespes*, clump of plants. Forming large tussocks
magnific-a, -us L. imposing. Culms tall compared with those of related species
magniflora L. *magnus*, large; *flos*, flower. Spikelets large
magnipoda L. *magna*, large; Gk *pous*, foot. Awn shorter than lemma base from which it develops
magnispicula L. *magnus*, large; *spica*, a point; hence, in particular, an ear or spike of grain; *-ula*, diminutive. Spikelets large compared with those of related species
magnolii In honor of Pierre Magnol (1638–1715) French botanist
magn-um, -us See *magna*
magohukuana L. *-ana*, indicating connection. In honor of T. Magohuku (fl. 1936) Japanese botanist
magrebensis L. *-ensis*, denoting origin. From the Maghreb region of northern Africa
maguirei In honor of Basset Maguire (1904–1991) United States botanist
maguireorum In honor of Basset Macguire (1904–1990) United States botanist and Basset Macguire, Jr. (fl. 1949) who together collected the type
mahafalense L. *-ense*, denoting origin. From Mahafaly Coast, Madagascar
mahendragiriensis L. *-ensis*, denoting origin. From Mahendragiri, Orissa State, India
maia The name of one of the Pleiades, a constellation associated with rain. A species of wet forests
maidenianus L. *-anus*, indicating connection. As for *maidenii*
maidenii In honor of Joseph Henry Maiden (1859–1925) English-born Australian botanist
Maillea In honor of Alphonse Maille (1813–1865) French botanist
mainborondroensis L. *-ensis*, denoting origin. From Mt Mainborondro, Madagascar
maipuriensis L. *-ensis*, denoting origin. From Maipuri Falls, Karaurieng River, Guyana
mairei In honor of – (1) Édouard-Ernest Maire (1848–1932) French cleric who collected in China. *Arundinaria mairei, Bromus mairei, Eragrostis mairei, Indocalamus mairei* – (2) René Charles Joseph Ernest Maire (1878–1949) French botanist. *Deschampsia mairei, Festuca mairei* – (3) R. P. Maire (fl. 1910–1921) who collected in China. *Poa mairei*

Mais A variant spelling of *mays*
maitlandii In honor of Thomas Douglas Maitland (1885–1978) Scots-born British colonial economic botanist
maiwa Hausa name for the species in northern Nigeria
maiz See *mays*
maizar From *zacatón maizor*, the vernacular name of the species in Mexico
Maizilla Vernacular name for the genus in Peru, literally "little maize"
majalcensis L. *-ensis*, denoting origin. From Cumbres de Majalca Parque Nacional, Chihuahua, Mexico
major L. larger. Larger in some respect than related species
majovskyi In honor of Jozef Májovský (1920–) Czech botanist
majungensis L. *-ensis*, denoting origin. From near Majunga, Madagascar
majus L. great. Culms tall
majuscul-a, -um, -us L. *majus*, greater; *-ula*, diminutive. Somewhat larger in some respect than related species
makinoi In honor of Tomitaro Makino (1861–1957) Japanese botanist
makoniensis L. *-ensis*, denoting origin. From Makoni, Zimbabwe
makuanensis L. *-ensis*, denoting origin. From Maguan Xian, Yunnan Province, China
makutrensis L. *-ensis*, denoting origin. As for *macutrensis*
malabaric-a, -um L. *-ica*, belonging to. From Malibar, India
malaca Gk *malakos*, soft to the touch. Plants softly hairy
malacanth-a, -um, -us Gk *malakos*, soft to the touch; *anthos*, flower. Lemmas softly hairy
malaccens-e, -is L. *-ense*, denoting origin. From Malacca, Malaysia
malacochaete Gk *malakos*, soft to the touch; *chaete*, bristle. Bristles of involucre densely and shortly ciliate
malacodes Gk *malakos*, soft to the touch; *-odes*, resembling. Surface of plants in whole or in part with texture of velvet
malacon Gk *malakos*, soft to the touch. Whole plant appressed-pubescent
malacophyll-a, -um, -us Gk *malakos*, soft to the touch; *phyllon*, leaf. Leaf-blades velvety-pilose
malacostachy-um, -us Gk *malakos*, soft to the touch; *stachys*, spike as of an ear of wheat. Peduncles of racemes plumose hairy
Malacurus Gk *malakos*, soft to the touch; *oura*, tail. Inflorescence a hairy spike
malalhuensis L. *-ensis*, denoting origin. From Malalhue, Argentina
malamalensis L. *-ensis*, denoting origin. From Cumbre de Malamala, Tucumán Province, Argentina
malampuzhaensis L. *-ensis*, denoting origin. From Malampuzha, India
malayana L. *-ana*, indicating connection. From Peninsula Malaysia and southern Thailand
maleschevica L. *-ica*, belonging to. From Malesheviska Planina, a mountain range on the Bulgarian-Serbian border
malesiae From Malesia, that is Malaysia
mali Vernacular name of the species in Yi, the language of south west Sichuan Province, China
malikoense L. *-ense*, denoting origin. From Maliko Gulchan Maui one of the Hawaiian Islands
maling Vernacular name of the species in Nepal
malingensis L. *-ensis*, denoting origin. Name derived from *Ma Ling Chuk*, the Cantonese name for the species
mallica L. *-ica*, belonging to. From the area occupied by the Malli in the time of Alexander the Great, now the Punjab, India
malmean-a, -um L. *-ana*, indicating connection. In honor of Gustaf Oskar Andersson Malme (1864–1937) Swedish botanist
malmei As for *malmeana*
malmgrenii In honor of Anders Johann Malmgren (1834–1897) Finnish zoologist and botanist
malmundariensis L. *-ensis*, denoting origin. From Malmundarium, now Malmédy, Belgium

malouinensis L. *-ensis*, denoting origin. From Malouin, now Falkland Islands, United Kingdom Territory in the South Atlantic

Maltea In honor of Malte Oskar Malte (1880–1933) Swedish-born Canadian botanist

Maltebrunia In honor of Konrad Malte Bruun (1775–1826) Danish-born French geographer

Malteburnia See *Maltebrunia*

maltei See *Maltea*

Malya In honor of Joseph Karl Maly (1797–1866) Bohemian-born Austrian botanist

malyschevii In honor of Leonid Ivanovich Malyschev (1931–) Russian botanist

malzevii In honor of Nikolai Ivanovich Malzer (fl. 1902–1910) Russian botanist

mamberamensis L. *-ensis*, denoting origin. From Mamberamo River, Papua, Indonesia

mamorae-um, -us Gk *-eum*, belonging to. From Mamora, North Africa

mampouw Vernacular name for the species in Sumatra, Indonesia

manabiense L. *-ense*, denoting origin. From Manabi Province, Ecuador

manacalensis L. *-ensis*, denoting origin. From Manacal, Venezuela

manatense L. *-ense*, denoting origin. From Manatee County, Florida, USA

manchuriensis L. *-ensis*, denoting origin. From Manchuria, now comprising the Provinces of Lianoning, Jilin and Heilongjiang in north-east China

mandalaiaensis L. *-ensis*, denoting origin. From Mandalay, Myanmar

mandarensis See *mandrarense*

mandavillei In honor of J. P. Mandaville Jr. who collected in the Middle East prior to 2000

Mandelorna An incomplete anagram of *Lenormandia*

mandiocanum L. *-anum*, indicating connection. From Mandioca, Brazil

mandioccanum See *mandiocanum*

mandonian-a, -us L. *-ana*, indicating connection. In honor of Gilbert Mandon (1799–1866) French botanist who collected in Bolivia

mandrarens-e, -is L. *-ense*, denoting origin. From the Mandrare River, Madagascar

mandshuric-a, -um, -us L. *-icum*, belonging to. From Manchuria, now comprising the Provinces of Lianoning, Jilin and Heilongjiang in north-east China

mangalorens-e, -is L. *-ense*, denoting origin. From Mangalore, Madras State, India

mangalorica L. *-ica*, belonging to. As for *mangalorense*

mangaluricum L. *-icum*, belonging to. As for *mangalorense*

manggong Vernacular name for the species in Java, Indonesia

mango The vernacular name for the species in Chile

manicat-a, -um L. *manica*, long sleeve; *-ata*, possessing. With long leaf-sheaths, of which the upper may enclose the base of the inflorescence

maniculata L. *manica*, handcuff; *-ula*, diminutive; *-ata*, possessing. The lower glume of the spikelet resembles handcuffs

manikensis L. *-ensis*, denoting origin. From the Manika plateau, Zaire

manillensis L. *-ensis*, denoting origin. From Manilla, Philippines

maniototo From the Maniototo Plain, South Island, New Zealand

manipurensis L. *-ensis*, denoting origin. From Manipur State, India

Manisuris Gk *manos*, necklace; *oura*, tail. The jointed racemes bear a fanciful resemblance to necklaces

manisuroides Gk *-oides*, resembling. Inflorescence resembles that of *Manisurus*

mannagettae In honor of Gunther Beck von Mannagetta und Lerchenau (1856–1931) Bohemian botanist

mannii In honor of Gustav Mann (1836–1916) German botanist and plant collector employed as gardener at Royal Botanic Gardens, Kew, England

manongarivens-e, -is L. *-ense*, denoting origin. From the Manongarivo Massif, Madagascar

manrakica L. *-ica*, belonging to. From Manrak Pass, Kazakstan

manschuricum L. *-ica*, belonging to. From Manchuria, now comprising the Provinces of Lianoning, Jilin and Heilongjiang in north-east China

manzanilloana L. *-ana*, indicating connection. From Manzanillo, Mexico

manzonzeensis L. *-ensis*, denoting origin. From Manzonza, Zaire

maokateiensis L. *-ensis*, denoting origin. From Maokatei, Sakhalin Island

maopingensis L. *-ensis*, denoting origin. From Maoping, Guandong Province, China

mapalense L. *-ense*, denoting origin. From Mapala forest, German East Africa

Mapira Meaning uncertain, origin not given by author

maranonensis L. *-ensis*, denoting origin. From Río Marañón, Peru

marantifolia L. *folium*, leaf. The leaf-blades resemble those of *Maranta*

marantoidea Gk *-oidea*, resembling. The leaf-blades resemble those of many Marantaceae

marathens-e, -is L. *-ense*, denoting origin. From Maratha Country, Bombay Presidency, India

marauense L. *-ense*, denoting origin. From Maraú, Bahia State, Brazil

marchica L. *-ica*, belonging to. From Mark of Brandenburg, Germany

marcida L. withered. The panicle branches droop at anthesis

Marcoduria From Marcodurum, now Düren, Germany

marcopetri In honor of Mark James Elder Coode (1937–) English botanist and Peter Francis Stevens (1944–) first collectors of the species

marcowiczii In honor of V. V. Marcowicz (Basil) (fl. 1926–1928) Russian botanist

mardinensis L. *-ensis*, denoting origin. From Mardin, south-eastern Anatolia, Turkey

margaritace-um, -us L. *margarita*, pearl; *-aceum*, resembling. The anthoecium is white and spherical resembling a pearl

margaritifer-a, -um L. *margarita*, pearl; *fero*, carry or bear. The glossy anthoecia resemble pearls

margelanica L. *-ica*, belonging to. From the Margelan District, Turkestan

marginalis L. *margo*, border; *-alis*, pertaining to. – (1) The leaf-blades have conspicuously thickened marginal nerves. *Aristida marginalis* – (2) leaf-blade margins undulate. *Cenchrus marginalis*

marginat-a, -um, -us L. *margo*, border; *-ata*, possessing. Leaf-blades or lemma with hairs in their margins or the margins otherwise conspicuous

marginellum Possibly a misspelling of *marginatum*

mariae (1) In honor of Mary Isobel Stuart McCallum (1899–1979) Rhodesian (Zimbabwean) nurse and amateur botanist. *Eragrostis mariae* – (2) from Maria, Madagascar. *Panicum mariae*

marianensis L. *-ensis*, denoting origin. From the Mariana Islands

mariesii In honor of Charles Maries (c. 1851–1902) English plant collector in China

marin-a, -um L. *mare*, sea; *-ina*, indicating possession. Growing by the sea

maritim-a, -um, -us L. belonging to the sea. Growing by the seaside

markgrafiae In honor of Ingeborg Markgraf-Dannenberg (1911–1996) Swiss botanist

markgrafii In honor of Friedrich Markgraf (1897–1987) German-born Swiss botanist

marliacea In honor of Joseph (Bory) Latour-Marliac (1830–1911) French botanist

marlothii In honor of Hermann Wilhelm Rudolf Marloth (1855–1931) German-born South African pharmacist, chemist and botanist

marmoratum L. *marmor*, marble; *-ata*, possessing. One of more structures has purple blotches

marmorea L. *marmor*, marble; *-ea*, indicating resemblance. Young culm-sheaths with purple-brown spots

maroccan-a, -us L. *-ana*, indicating connection. From Morocco

marojejyensis L. *-ensis*, denoting origin. From Marojejy Mts, Madagascar

M

marquisens-e, -is L. *-ense*, denoting origin. From the Marquesas Islands. part of French Polynesia, in the Pacific Ocean
marriettana L. *-ana*, indicating connection. From Marrietta, South Carolina, USA
marschalliana L. *-ana*, indicating connection. As for *biebersteiniana*
marshallense L. *-ense*, denoting origin. From the Republic of the Marshall Islands in the western Pacific Ocean
marshallii In honor of J. K. Marshall (fl. 1961) who collected in Peru
marshii In honor of Ernest George Marsh (1915–) United States botanist
marssonii In honor of Theodor Marsson (1816–1892) German apothecary and botanist
martha-gonzaleziae In honor of Martha González-Elizondo (1958–) Mexican botanist
martianus L. *-anus*, indicating connection. In honor of Karl Friedrich Philipp Martius (1794–1868) German botanist
martinezii In honor of Esteban Martínez Salas (fl. 1987–1992) Mexican botanist
martinianus L. *-anus*, indicating connection. As for *martinii*, as to Claude Martin
martinicens-e, -is L. *-ense*, denoting origin. From Martinique, West Indies
martinii In honor of – (1) Claude Martin (1731–1800) French-born British soldier who collected in India. *Andropogon martinii, Cymbopogon martinii, Gymnanthelia martinii* – (2) Dr. Martin, who collected in the Falkland Islands, United Kingdom Territory in the South Atlantic. *Deschampsia martinii* – (3) Ferdinando Martini (1841–?) Italian politician. *Afrotrichloris martinii*
martinovskyi In honor of Jan Otakar Martinovsky (1903–1980) Bohemian botanist
marungensis L. *-ensis*, denoting origin. From Marungu Plateaux, Zaire
marunguens-e, -is As for *marungensis*
maruyamana L. *-ana*, indicating connection. In honor of I. Maruyama (fl. 1979) Japanese botanist
masafueran-a, -us L. *-ana*, indicating connection. From Más Afuera, also called Alejandro Selkirk, an island in the Juan Fernández Archipelago, Chile
masambaensis L. *-ensis*, denoting origin. From Masamba, Zaire
masamunean-a, -us L. *-ana*, indicating connection. As for *masamunei*
masamunei In honor of Genkei Masamune (1899–?) Japanese botanist
mascatensis L. *-ensis*, denoting origin. From Mascate (Muscat), Oman
masenderana L. *-ana*, indicating connection. From East Masendran on the coast of the Caspian Sea
masirahensis L. *-ensis*, denoting origin. From Masirah Island, Persian Gulf
maskerensis L. *-ensis*, denoting origin. From Mt Masker, Morocco
massaicum L. *-icum*, belonging to. From the territory of the Massai (Maasai) in Kenya
massaiense L. *-ense*, denoting origin. From the Massai steppe, Uganda
massauensis L. *-ensis*, denoting origin. From Massawa in Eritraea
massauiensis See *massauensis*
Massia In honor of Joseph Massie (?–1794) English botanist who collected in Vietnam
massii As for *Massia*
massiliensis L. *-ensis*, denoting origin. From Massilia, Egypt
mastersii In honor of John W. Masters (c. 1792–1873) gardener, Calcutta Botanic Gardens
mastrucatum L. *mastruca*, rough sheep-skin garment; *-atum*, possessing. Lower glume rugose, its surface resembling that of a rough sheep-skin garment
mataniae From Matania, a mountain in Romania
matengoensis L. *-ensis*, denoting origin. From Matengo Hills, Tanzania
mathewsii In honor of Andrew Mathews (?–1841) who collected in Peru
matmat Besuki name of the species in Java, Indonesia
Matrella L. *mater*, mother; *-ella*, diminutive. Reason for choice of name is uncertain

matrella Resembling *Matrella*
matritensis L. *-ensis*, denoting origin. From Madrid, Spain
matsudae In honor of Magodji Matsuda (fl. 1931) Japanese botanist
matsudana L. *-ana*, indicating connection. In honor of E. Matsuda (fl. 1919) Japanese botanist
matsumoi In honor of Jutara Matsumo (1868–1946) Japanese botanist
matsumurae In honor of Jinzô Matsumura (1856–1928) Japanese botanist
matsumuri See *matsumurae*
matsushimensis L. *-ensis*, denoting origin. From Matsushima, Rikuzen Province, Japan
mattamuskeetense L. *-ense*, denoting origin. From Lake Mattamuskeet, North Carolina, USA
matteodanum L. *-anum*, indicating connection. In honor of S. Matteoda (fl. 1927) Italian botanist
mattheii In honor of Oscar Matthei Jensen (fl. 1964) Chilean agrostologist
matthewsii In honor of Henry John Matthews (1859–1909) New Zealand forester
mattogrossensis L. *-ensis*, denoting origin. From Mato Grosso, Brazil
Matudacalamus Gk *kalamos*, reed. Reed-like grass named in honor of Eizi Matuda (1894–1978) Japanese botanist
matudae See *Matudacalamus*
maudiae In honor of Maud Dunn (fl. 1909) wife of Stephen Troyte Dunn (1868–1938) English botanist who collected in China
mauiensis L. *-ensis*, denoting origin. From Maui, one of the Hawaiian Islands
maullinica L. *-ica*, belonging to. From river Maullin, Chile
mauritanic-a, -um, -us L. *-ica*, belonging to. From Mauritania
mauritian-a, -um, -us L. *-ana*, indicating connection. From Mauritius
mauritii (1) From Mauritius. *Eragrostis mauritii* – (2) In honor of Maurit. *Agrostis mauritii*
maurofernandeziana L. *-ana*, indicating connection. In honor of Mauro Fernández (fl. 1907) Costa Rican agriculturalist
mauryi In honor of Paul Jean Baptiste Maury (1858–1893) French botanist
maxim-a, -um, -us L. greatest. – (1) Culms very tall. *Agrostis maxima, Arundo maxima, Bambusa maxima, Briza maxima, Bromus maximus, Centotheca maxima, Diheteropogon maximus, Gigantochloa maxima, Glyceria maxima, Panicum maximum, Poa maxima, Saccharum maximum, Sasa maxima, Sasaella maxima, Thysanolaena maxima, Triticum maximum* – (2) anthoecia very large for genus. *Berriochloa maxima*
maximilianii, maximilianum In honor of Alexander Philipp Maximilian (1782–1867) Prince of Neuwiel who collected in Brazil
maximoviczii In honor of Karl Johann Maximovicz (1827–1891) Russian botanist
maxim-um, -us See *maxima*
maxonii In honor of William Ralph Maxon (1877–1948) United States botanist
maxwellii In honor of – (1) James F. Maxwell (1945–) United States botanist. *Eremochloa maxwellii* – (2) George Maxwell (1804–1880) who collected and dealt in Australian natural history specimens. *Poa maxwellii*
mayaensis, mayaënsis L. *-ensis*, denoting origin. From the Maya Mountains, Belize
mayanum L. *-anum*, indicating connection. In honor of the Mayan civilisation of northern Central America
mayarens-e, -is L. *-ense*, denoting origin. From Mayari, Provincia de Pinar del Río, Cuba
maydellii In honor of George Maydell (fl. 1867) German botanist
mayebarae As for *mayebarana*
mayebaran-a, -um, -us L. *-ana*, indicating connection. In honor of Kanjiro Mayebara (1890–?) Japanese botanist
mayeri In honor of Friedrich Mayer (1788–1828) German-born tutor in Italy
maynens-e, -is L. *-ense*, denoting origin. From Maynas, Peru
mayocoensis L. *-ensis*, denoting origin. From Mayoco, Zaire
maypurensis L. *-ensis*, denoting origin. From the cataract of "Saint Joseph Maypurensium" on the Orinoco River, Venezuela

Mays, mays A name of Caribbean origin for *Zea mays*
ma-yuen L. from the Ma-Yu Range in north-west Myanmar
mayumbense L. *-ense*, denoting origin. From Mayumbe forest, Gabon
mayumianus L. *-anus*, indicating connection. From Mayumi Province, Japan
Mayzea A contraction of *Zea mays*
mazelii In honor of Mazel, French nurseryman
mazettii, mazzettii In honor of Heinrich R. E. von Handel-Mazzetti (1882–1940) Austrian botanist
mazzettian-a, -um, -us L. *-ana*, indicating connection. In honor of Heinrich Handel-Mazzetti (1882–1940) Austrian botanist
mckiei In honor of Ernest Norman McKie (1882–1948) Australian cleric and amateur botanist
m'clellandii See *macclellandii*
meakensis L. *-ensis*, denoting origin. From Meakandake, Kushiro Province, Hokkaido, Japan
mearnsii In honor of Edward Alexander Mearns (1856–1916) United States botanist
meccana L. *-ana*, indicating connection. From Mecca, Saudi Arabia
medi-a, -um, -us L. middle. Characters midway between two or more other species
medica L. *-ica*, belonging to. From Media, north-western Iran
mediolanensis L. *-ensis*, denoting origin. From Mediola, now Milan, Italy
mediterranea Growing around the Mediterranean Sea
medius See *media*
medullosa L. *medulla*, bone-marrow; *-osa*, abundance. Culm internodes with an abundance of soft pith
Medusather Gk *ather*, barb or spine. Awns intertangled to form a dense mass reminiscent of the head of Medusa
meeboldii In honor of Alfred Karl Meebold (1863–1952) German botanist
megacarpum Gk *megas*, large; *karpos*, fruit. Florets large
Megalachne Gk *megas*, large; *achne*, scale. The glumes completely invest the spikelet
megalanth-a, -um Gk *megas*, large; *anthos*, flower. Spikelets large
megalogluma Gk *megas*, large; *gluma*, husk. Glumes larger than lemmas
megalophylla Gk *megas*, large; *phyllon*, leaf. Leaf-blades large
Megaloprotachne Gk *megas*, large; *proto*, before; *achne*, scale. The lower glume is much longer than the upper glume
megalosperma Gk *megas*, large; *sperma*, seed. Grains large
megalothyrsa Gk *megas*, large; *thyrsos*, ornamental wand. Inflorescence a large open panicle resembling the ornamented staff carried in procession by the devotees of Bacchus, in Roman mythology the God of Wine
megalura Gk *megas*, large; *oura*, tail. Inflorescence a narrow panicle and so somewhat resembling the tail of a large rat
megaphyll-a, -um Gk *megas*, large; *phyllon*, leaf. Leaf-blades large
megapotamia See *megapotamica*
megapotamic-a, -um, -us Gk *megas*, large; *potamos*, river; *-ica*, belonging to. – (1) From Rio Grande, southern Brazil. *Anatherum megapotamicum, Andropogon megapotamicus, Aristida megapotamica, Bouteloua megapotamica, Bromus megapotamicus, Deyeuxia megapotamica, Digitaria megapotamica, Elionurus megapotamicus, Eragrostis megapotamica, Panicum megapotamicum, Pappophorum megapotamica, Poa megapotamica* – (2) from Rio Grande do Sal. *Jarava megapotamica, Nasella megapotamica, Setaria megapotamica, Stipa megapotamica*
Megastachya Gk *megas*, large; *stachys*, spike as of an ear of wheat. Spikelets with more florets than those of related genera
megastachy-a, -um See *Megastachya*. – (1) Spikelets unusually large for the genus. *Brachiaria megastachya, Panicum megastachyum, Phyllostachys megastachya* – (2) spikelets with many florets. *Eragrostis megastachya, Poa megastachya*
megasthenes Gk *megas*, large; *sthenos*, strength. The central zone of the sterile lemma bears a patch of rigid hairs

megathyrsa See *Megathyrsus*
Megathyrsus Gk *megas*, large, *thyrsos*, wand or shaft. Inflorescence large with a conspicuous central axis
megiston Gk largest. Panicles large
meionectes Gk *meionektes*, one who has less. At first regarded as depauperate specimens of another species
mejlandii In honor of Yngvar Mejland (fl. 1937–1938) Norwegian botanist
mekiste Gk *mekistos*, tallest. Culms tall
mekongens-e, -is L. *-ense*, denoting origin. From Mékong, Laos
melaleuc-a, -um Gk *melas*, black; *leukos*, white. Glumes black and grains white
melananthum Gk *melas*, black; *anthos*, flower. Spikelets black
melanocarp-a, -us Gk *melas*, black; *karpos*, fruit. Mature spikelets black
Melanocenchris Gk *melas*, black. The glumes and lemma have scabrid purplish awns and the mature spikelets resemble the burr of *Cenchrus*
melanochila Gk *melas*, black; *chilos*, grass. The fertile floret is almost black from an early age
melanogonum Gk *melas*, black; *gony*, knee. Nodes dark-colored
melanosperm-a, -um Gk *melas*, black; *sperma*, seed. Lemma and palea darkly pigmented
melanotricha Gk *melas*, black; *thrix*, hair. Glumes and lower lemma invested in black hairs
melanotyl-a, -um Gk *melas*, black; *tylos*, knot. Spikelets borne on a black stipe
melanthera Gk *melas*, black; *anthera*, of flowers. Anthers dark
melanthes Gk *melas*, black; *anthos*, flower. Glumes dark-purple
melderisii In honor of Aleksandre Melderis (1909–1986) Latvian-born English botanist
Melica L. *mel*, honey; *-ica*, belonging to. Origin uncertain but three possible explanations are: a reference to the sweetness of its stem, an ancient Roman name for millet or an ancient Italian name for sorghum
melicacea L. *-acea*, resembling. Similar to *Melica*
melicari-a, -um L. *-aria*, pertaining to. Resembling *Melica*
melicoides Gk *-oides*, resembling. Similar to *Melica*
melicoideum Gk *-oideum*, resembling. Similar to *Melica*
melinacea L. *melinum*, kind of white color; *-acea*, resembling. Margins of leaf-blades white
melinacra See *melinacea*
melinioides Gk *-oides*, resembling. Similar to *Melinis*
Melinis, Melinum Gk *meline*. Ancient Greek name for a cereal, probably a millet
melinis Resembling *Melinis*
Melinum See *Melinis*
melitense L. *-ense*, denoting origin. From Melita, now Malta
Mellinis See *Melinis*
mellitum L. *mellitus*, pertaining to honey. Culms sweet to the taste
Melocalamus Gk *melon*, apple; *kalamos*, reed. Woody grass with fleshy, apple-like fruits
Melocanna Gk *melon*, apple; *kanna*, reed. Fruits fleshy and stems woody
melvillei In honor of George F. Melville (fl. 1937) who collected in Western Australia
membranace-a, -um, -us L. *membrana*, membrane; *-acea*, resembling. – (1) Inflorescence bracts or glumes papery. *Anthistiria membranacea, Ceresia membranacea, Dendrocalamus membranaceus, Eragrostis membranacea, Iseilema membranaceum, Paspalum membranaceum, Sorghum membranaceum, Vulpia membranacea* – (2) leaf-blades membranous. *Sasa membranacea*
membranifolia L. *membrana*, membrane; *folium*, leaf. Leaf-blades thin
membranigluma L. *membrana*, membrane; *glume*, husk. Glumes white and membranous
membranoidea L. *membrana*, membrane; Gk *-oidea*, resembling. Apex of culm-sheath papery

M

memphitica L. *-ica*, belonging to. From Memphis, Egypt

menachensis L. *-ensis*, denoting origin. From Menâcha, Arabia

mendocin-a, -um, -us L. *-ina*, indicating possession. From – (1) Mendoza, Chile. *Aristida mendocina, Distichlis mendocina, Elymus mendocinus, Munroa mendocina* – (2) Mendoza Province, Argentina. *Agropyron mendocinum, Chloris mendocina, Deschampsia mendocina, Diplachne mendocina, Ipnum mendocina, Trichloris mendocina*

mendocinensis L. *-ensis*, denoting origin. From Mendoza Province, Argentina

meneritana L. *-ana*, indicating connection. Locality not given in original description but name probably based on a place name in Sri Lanka

mengeanus L. *-anus*, indicating connection. Probably in honor of Franz Anton Menge (1808–1880) German botanist

menghaiensis L. *-ensis*, denoting origin. From Menghai, Yunnan Province, China

menhoferi In honor of Menhofer (fl. 1983) who collected in Bolivia

mensense L. *-ense*, denoting origin. From the Mensa region of Eritraea

menyharthii In honor of László Menyharth (1849–1897) Hungarian cleric who collected in Mozambique

Meoschium Meaning uncertain but probably an incomplete anagram of *Ischaemum*

mer-a, -um L. naked. The lower surface of the lemma is hairless

Merathrepta Gk *meros*, portion; *ather*, spike or ear of wheat; *hepta*, seven. Origin uncertain, not given by the author but may be a reference to the spikelets having seven florets

meratiana L. *-ana*, indicating connection. In honor of François Victor Mérat de Vaumartoise (1780–1851) French physician and botanist

meredisensis L. *meridies*, midday; *-ensis*, denoting origin. From the south but based on the French transliteration of the Latin, *meridies*

merguensis L. *-ensis*, denoting origin. From Mergui, Tennaserim, Myanmar

meridensis L. *-ensis*, denoting origin. From Mérida, Venezuela

meridional-e, -is L. *meridies*, midday; *-ale*, pertaining to. – (1) Occurring in places on the same meridian, as from North and South Carolina. *Dichanthelium meridionale, Panicum meridionale* – (2) growing on the same meridian as a related species. *Aristida meridionalis, Avena meridionalis*

meridisensis L. *meridies*, midday, *-ensis*, denoting origin. To the south, an allusion to the sun being in the southern sky at noon (in Europe)

Meringurus Gk *merinx*, bristle; *ourus*, tail. Each of the two glumes of the terminal spikelet is drawn out into a long awn

merinoi In honor of R. P. Baltasar Merino y Román (1845–1917) Spanish cleric and botanist

Merisachne Gk *merizo*, divided; *achne*, scale. Lemma deeply bifid

merkeri In honor of Merker Moritz (?–1908) who collected in East Africa

Merostachis See *Merostachys*

merostachyoides Gk *-oides*, resembling. Resembling *Merostachys*

Merostachys Gk *meros*, portion; *stachys*, spike as of an ear of wheat. The inflorescence is a one-sided spike

merrill-ana, -iana L. *-ana*, indicating connection. As for *merrillii*

merrillii In honor of Elmer Drew Merrill (1876–1956) United States botanist

mertensii In honor of Carl Heinrich Mertens (1796–1830) German-born physician and traveller

mertonensis L. *-ensis*, denoting origin. From Merton, England

meruensis L. *-ensis*, denoting origin. From Mt Meru, Tanzania

Merxmuellera In honor of Hermann Merxmüller (1920–1988) German botanist

mesathera Gk *mesos*, middle; *ather*, barb or spine. The awn arises from the middle of the lemma

mesocom-a, -um Gk *mesos*, middle; *koma*, hair of head. The lemma of the lower floret has tufts of hairs at its middle

mesopotamicus L. *-icus*, belonging to. From Mesopotamia, now largely Iraq

Mesosetum Gk *mesos*, middle; L. *seta*, bristle. Glumes bearing stiff hairs in the middle in contrast to *Holosetum* where they are hairy overall

messanensis L. *-ensis*, denoting origin. From Messana, now Messina, Italy

metabolon Gk *metaboulos*, changeful. Species variable

metake Japanese *me*, woman; *take*, bamboo. Culms slender

metallicola L. *metallicus*, belonging to metal; *-cola*, dweller. Growing in places where iron is mined

Metasasa Gk *meta-*, near. Similar to *Sasa*

metatoris L. *metator*, surveyor. Named in allusion to D. W. L. Henderson, surveyor in New South Wales, Australia

metcalfei In honor of Orrick Baylor Metcalfe (1879–1936) United States botanist

Metcalfia In honor of Charles Russell Metcalfe (1904–1991) English plant anatomist

metlesicsii In honor of H. Metlesics (fl. 1973) collector of the holotype

metuoensis L. *-ensis*, denoting origin. From Medong Xian, Xizang Autonomous Region, China

metzii In honor of Fr. Metz (1819–1885) who collected in India

mexican-a, -um, -us L. *-ana*, indicating connection. From Mexico

meyenian-a, -um L. *-ana*, indicating connection. In honor of Franz Julius Ferdinand Meyen (1804–1840) German physician, traveller and plant collector

meyenii As for *meyeniana*

meyeri In honor of – (1) Frederick Gustav Meyer (1917–) United States botanist. *Bromus meyeri* – (2) Frank N. Meyer (1875–1918) Dutch-born United States plant explorer. *Phyllostachys meyeri* – (3) ?Carl A. Meyer (1795–1855) Russian plant collector. *Poa meyeri*

meyerian-a, -um L. *-ana*, indicating connection. In honor of – (1) Ernst Heinrich Friedrich Meyer (1791–1858) German botanist who collected in South Africa. *Padia meyeriana, Panicum meyerianum, Eriochloa meyeriana, Oryza meyeriana* – (2) either Carl or Anton Andreevic Meyer (1795–1855) Russian botanists. *Chusquea meyeriana, Stipa meyeriana*

mezian-a, -um L. *-ana*, indicating connection. As for *Mezochloa*

mezii See *Mezochloa*

meziibrunneum In honor of Carl Mez (see *Mezochloa*) who described *Paspalum brunneum*

Mezochloa Gk *chloa*, grass. In honor of Carl Christian Mez (1866–1944) German botanist

mianningensis L. *-ensis*, denoting origin. From Mianning Xian, Sichuan Province, China

Mibora, Miboria Gk *mikros*, small; *bora*, food. The plant is diminutive providing little fodder or grain

Micagrostis Gk *mikros*, small; *agrostis*, grass. Dwarf annual

micans L. *mico*, tremble. The racemes are borne on slender pedicels and so readily shake in the breeze

micay Vernacular name for the species in the Colombia

michaelis L. of Michael. From the island of St. Michael, Azores, Portuguese islands in the Atlantic

michalkowii In honor of Michalkov

michauxian-a, -um L. *-ana*, indicating connection. As for *michauxii*

michauxii In honor of Andre Michaux (1746–1802) French botanist and traveller

Michelaria L. *-aria*, pertaining to. In honor of Pierre-Joseph Michel (1788–1854) Belgian gardener

michelii In honor of Pier Antonio Micheli (1679–1737) Italian botanist

michiganica L. *-ica*, belonging to. From Michigan State, USA

michinokuana L. *-ana*, indicating connection. From Michinoku, Japan

michisensis L. *-ensis*, denoting origin. From San Juan de Michis, Mexico

michnoi In honor of P. Michno (fl. 1924) Russian botanist

Micragrostis Gk *mikros*, small. Similar to *Agrostis* but small in some respect

Micraira Gk *mikros*, small. The spikelets and inflorescences are small and resemble those of Aira

micrandra Gk *mikros*, small; *aner*, man. Anthers small

micranth-a, -um, -us Gk *mikros*, small; *anthos*, flower. Spikelets small

micranther-a, -us Gk *mikros*, small; *antheros*, flowering. – (1) Inflorescence few-flowered. *Aeluropus micrantherus*, *Microlaena micranthera* – (2) sessile spikelets have small anthers. *Chrysopogon micrantherus*

micranth-um, -us See *micrantha*

micrather-a, -um, -us Gk *mikros*, small; *ather*, spike or ear of wheat. Inflorescence a small narrow panicle or composed of small racemes

microbachne Gk *mikros*, small; *bios*, manner of living; *achne*, scale. The lower glume is lacking or present only as a rim

Microbambus Gk *mikros*, small. Small herbaceous *Bambusa*-like grasses

Microbriza Gk *mikros*, small. Resembling *Briza* but with small spikelets

Microcalamus Gk *mikros*, small; *kalamos*, reed. A dwarf herbaceous bamboo

microcarp-a, -on, -um Gk *mikros*, small; *karpos*, fruit. Caryopsis very small

microcarpha Gk *mikros*, small; *karphos*, chip. The very small spikelets fall entire and so look like small chips of wood

microcarpon See *microcarpa*

microcephal-a, -um, -us Gk *mikros*, small; *kephale*, head. Inflorescence a short spike or contracted panicle

Microchlaena See *Microlaena*

Microchloa Gk *mikros*, small; *chloa*, grass. Some species of the genus are small plants

microclada Gk *mikros*, small; *klados*, branch. Culms slender

micrococc-a, -um Gk *mikros*, small; *kokkos*, kernel. Anthoecia very small

microdon Gk *mikros*, small; *odous*, tooth. The lemma is shortly tridentate

microfloscula Gk *mikros*, small; L. *flos*, flower; *-ula*, diminutive. Spikelets small

micrognostum Gk *mikros*, small; *gnostos*, known. Species poorly known

Microlaena, microlaena Gk *mikros*, small; *chlaina*, cloak. The subtending glumes are tiny

microlemma Gk *mikros*, small; *lemma*, husk. Upper lemma much reduced

microlepis Gk *micros*, small; *lepis*, scale. Glumes considerably shorter than the lemma of the first floret in spikelet

microphyll-a, -um, -us Gk *mikros*, small; *phyllon*, leaf. Leaf-blades short

micropoda Gk *mikros*, small; *pous*, foot. Pedicel very short

Micropogon Gk *mikros*, small; *pogon*, beard. Lemma shortly awned

microprolepis Gk *mikros*, small; *pro-*, before; *lepis*, scale. Lower glume much shorter than upper

microprotus Gk *mikros*, small; *protos*, first. Lower glume minute

micropyroides Gk *-oides*, resembling. Similar to *Micropyrum*

Micropyropsis Gk *opsis*, resemblance. Similar to *Micropyrum*

Micropyrum Gk *mikros*, small; *pyros*, wheat. Resembling small wheat plants

microseta Gk *mikros*, small; L. *seta*, bristle. Lemma shortly awned

microsperm-a, -um, -us Gk *mikros*, small; *sperma*, seed. Caryopses or spikelets small

microstachy-a, -um, -us Gk *mikros*, small; *stachys*, spike as of an ear of wheat. Inflorescences or spikelets small

microstachys Gk *mikros*, small; *stachys*, spike as of an ear of wheat. Inflorescences or spikelets small

microstachy-um, -us See *microstachya*

Microstegium Gk *mikros*, small; *stege*, cover. Glumes minute

microterus Gk *mikros*, small; *pteron*, wing or feather-like. Lower glume with two small wing-like keels

microtheca Gk *mikros*, small; *theke*, box. – (1) Apex of pedicels cup-shaped after spikelets have been shed. *Andropogon microtheca*, *Cymbopogon microtheca* – (2) glumes persistent forming a cup at the tip of the pedicel. *Glyceria microtheca*

Microthuareia Gk *mikros*, small. Resembling *Thuarea* but smaller

microthyrsum Gk *mikros*, small; *thyrsos*, ornamental wand. Panicle not well developed

microtis Gk *mikros*, small; *ous*, ear. Leaf-blades with small auricles

Micrurus Gk *mikros*, small; *oura*, tail. Inflorescence like a small tail

Miegia In honor of Achilles Mieg (1731–1799) Swiss botanist

migiurtina L. from Migiurtini, Somalia

migoi In honor of Hisao Migo (fl. 1937) Japanese botanist

mikamimonticola L. *mons*, mountain; *-cola*, dweller. From the Mikami Mountains, Japan

mikanii In honor of Johann Christian Mikan (1769–1814) Bohemian botanist

mikii In honor of Shigeru Miki (1905–1974) Japanese botanist

mikurensis L. *-ensis*, denoting origin. From Mikurajima Island, Idzu or Izu Islands, now part of Tokyo Prefecture, Japan

milanjian-a, -um, -us L. *-ana*, indicating connection. From Mt Milanje, Malawi

mildbraedian-a, -us L. *-ana*, indicating connection. In honor of Gottfried Wilhelm Johannes Mildbraed (1879–1954) German botanist

mildbraedii As for *mildbraediana*

Mildbraediochloa Gk *chloa*, grass. See *mildbraediana*

miliace-a, -um, -us L. *-acea*, resembling. Resembling *Milium*

miliare L. *-are*, connected to. Resembling *Milium*

miliaria L. *-aria*, pertaining to. Resembling millet

Miliarium L. *-arium*, connected to. Resembling *Milium*

Miliastrum L. *-aster*, incomplete resemblance. Able to be used as a millet

miliiform-e, -is L. *forma*, appearance. Similar to *Milium*

milioides Gk *-oides*, resembling. Similar to *Milium*

milioideum Gk *-oideum*, resembling. Spikelets resembling those of *Milium*

Milium, milium Name of Proso millet (*Panicum miliaceum*) in Roman times but name now applied to a different genus

M

milleana L. *-ana*, indicating connection. In honor of Louis *Mille* (1873–1954) Belgian-born Ecuadorean cleric and botanist

milleflorum L. *mille*, one thousand; *flos*, flower. Inflorecence with densely flowering branchlets

millegrana L. *mille*, one thousand; *granum*, grain. Inflorescence of many spikelets

milleri In honor of D. J. Miller (fl. 1984) rangeland and livestock specialist who studied the alpine grasses of the Kingdom of Bhutan

millettii In honor of Charles Millett (fl. 1825–1834) employee of the British East India Company and amateur botanist who collected in south-eastern China, Sri Lanka and southern India

Millium See *Milium*

milnei In honor of Edgar Wolston Bertram Handley Milne-Redhead (1906–1996) English botanist

milroyi In honor of Arthur John Wallace Milroy (1883–1936) sometime Conservator of Forests, Assam State, India

milsumii In honor of John Noel Milsum (1890–?) English-born Malayan agriculturalist

mimosa An abbreviation of *cará-mimosa*, the vernacular name for the species in Santa Catarina Province of Brazil

minaguchii In honor of K. Minaguchi (fl. 1929) Japanese botanist

minarovii In honor of Musea Minarovio, a Soviet cosmonaut from Daghestan, Russian Federation

minarum Of Minas Gerais Province, Brazil

mindanaens-e, -is L. *-ense*, denoting origin. From Mindanao, Philippines

mindorense L. *-ense*, denoting origin. From Mindoro, an island in the Philippines

mindoroensis As for *mindorense*

minim-a, -um, -us L. least. Regarded by the author as – (1) the smallest species in the genus. *Chaetostichium minimum, Dissanthelium minimum, Eragrostis minima, Gaimardia minima, Lepturus minimus, Oropetium minimum, Paspalum minimum, Stipidium minimum, Tripogon minimum, Zoysia minima* – (2) the smallest species in the genus in which originally described. *Mibora minima*

minimiflora L. *minimus*, least; *flos*, flower. Spikelets with only one or two florets

minim-um, -us See *minima*

minoensis L. *-ensis*, denoting origin. From Mino Province, now part of Gifu Prefecture, Japan

minomarsa From Mino Province, now part of Gifu Prefecture, Japan

minor L. lesser. Plants small in comparison with related species

minoriflor-a, -um L. *minor*, lesser; *flos*, flower. Spikelets smaller than those in a related species

minuartioides Gk *-oides*, resembling. Similar in habit to *Minuartia glomerata*

minus L. less. – (1) Smaller in stature than a related species. *Ctenium minus, Paspalum minus* – (2) differing in several respects from a related species. *Ischaemum minus*

minuscula L. *minus*, lesser; *-ula*, diminutive. Culms very small

minut-a, -um, -us L. very small. Smaller than usual in some respect

minutiflor-a, -um, -us L. *minutus*, very small; *flos*, flower. Spikelets or florets very small

minutissim-a, -us L. *minutus*, very small; *-issima*, most. Spikelets very small

minutul-a, -um L. *minutus*, very small; *-ula*, diminutive. Spikelets small

minut-um, -us See *minuta*

Miphragtes Anagram of *Phragmites*

Miquelia In honor of Friedrich Anton Wilhelm Miquel (1811–1871) Dutch botanist

mirabil-e, -is L. unusual. – (1) Plant with the habit of a rush rather than a grass. *Arundinaria mirabilis, Festuca mirabilis, Glaziophyton mirabile* – (2) leaf-blades wide for the genus. *Sporobolus mirabilis* – (3) differing markedly from most other members of genus. *Calamagrostis mirabilis, Panicum mirabile*

mirabunda L. full of wonder. Culms attractive in appearance

mirandana L. *-ana*, indicating connection. From Burgos à Miranda del Elro, Spain

mirandum L. strange. Quite unlike any other species in the genus

Miscanthidium Gk *-idium*, diminutive but here used as a name-forming suffix. Resembling *Miscanthus*

Miscanthus Gk *mischos*, pedicel; *anthos*, flower. Spikelets stalked

miser, -a, -um L. miserable. Small in comparison with related species

misionum L. from Misiones State, Argentina

missionum See *misionum*

mississippiense L. *-ense*, denoting origin. From the valley of the Mississippi River, USA

missuricum L. *-icum*, belonging to. From the vicinity of the Missouri River, USA

mistasypum Orthographic variant of *mystasipum*

mitchelliana L. *-ana*, indicating connection. As for *mitchellii*

mitchellii In honor of Thomas Livingstone Mitchell (1792–1855) Scots-born Australian surveyor and explorer

mit-e, -is Gk *mitos*, thread. – (1) Involucral bristles very slender. *Cenchrus mitis* – (2) oral setae very slender. *Bambusa mitis, Phyllostachys mitis* – (3) pedicels slender. *Panicum mite*

mitinokuensis L. *-ensis*, denoting origin. From Mitinoku, Japan

mitis See *mite*

mitophyllum Gk *mitos*, thread; *phyllon*, leaf. Leaf-blades narrow-linear

mitopus Gk *mitos*, thread; *pous*, foot. Pedicels thread-like

mitrushii In honor of I. Mitrush, Albanian botanist

Mitwabochloa Gk *chloa*, grass. From Mitwaba, Zaire
miurus Gk *mys*, mouse; *oura*, tail. Inflorescence a spike-like panicle
mixt-a, -um L. mixed. Sharing the characters of two or more other species, sometimes of hybrid origin
miyabei In honor of Kingo Miyabe (1860–1951) Japanese botanist
miyazawae In honor of Bungo Miyazawa (fl. 1932) Japanese botanist
mjobergii In honor of Eric G. Mjöberg (1882–1938) Swedish entomologist and explorer in Northern Australia
mlahiense L. *-ense*, denoting origin. From Mlahi, Tanzania
mnemateia Gk *mnema*, monument. Origin uncertain, not given by the author
Mnesithea In honor of Mnesitheos, 4th century B.C. Greek physician and writer on the subject of edible plants
Mniochloa Gk *mnion*, moss; *chloa*, grass. Vegetatively resembling the moss genus *Mnium*
moabitica L. *-ica*, belonging to. From Moab, Palestine
moandaensis L. *-ensis*, denoting origin. From Moanda, Zaire
mobukensis L. *-ensis*, denoting origin. From the Mobuku Valley, Uganda
mocquerysii In honor of Mocquerys (pre 1940) who collected in Venezuela
modatica L. *-ica*, belonging to. From Modat Province, Ethiopia
modensis L. *-ensis*, denoting origin. From Modan, Greece
moderabilis L. moderate. Culms of average height
modest-a, -um, -us L. moderate. Culms of average height
modic-a, -us L. moderate. Less robust than related species
moellendorfian-a, -us L. *-anus*, indicating connection. In honor of Otto Moellendorff (1848–1903) German diplomat who collected plants in Russia
moelleri In honor of Peter Möller (fl. 1887) who collected in Chile
Moenchia In honor of Konrad Moench (1744–1805) German botanist and pharmacist
moesiacus L. *-acus*, belonging to. From Moesia a region of the lower Danube straddling the boundary of present day Bulgaria and Serbia
moeszii In honor of Gustáv Moesz (1873–1946) Hungarian mycologist
mogaugensis L. *-ensis*, denoting origin. From the Mogaug forests, Myanmar
moggii In honor of Albert Oliver Dean Mogg (1886–1980) South African botanist
moharia From Mohari, India
mohavense L. *-ense*, denoting origin. From Mohave County, Arizona, USA
mohrii In honor of Charles Theodore Mohr (1824–1901) German-born, United States pharmacist and plant collector
mokaense L. *-ense*, denoting origin. From Moka, Fernando Po (now Bioko), Equatorial Guinea
mokensis L. *-ensis*, denoting origin. As for *mokaense*
mokogunensis L. *-ensis*, denoting origin. From Mukogun, Settsu Province, now part of Hyogo and Osaka Prefectures, Japan
mokuleiaense L. *-ense*, denoting origin. From Mokuleia on Maui, one of the Hawaiian Islands
moldavica L. *-ica*, belonging to. From Moldavia
molesta L. troublesome. The leaf-blades are rigid and spiny
molineri In honor of "Igantio Molineri" of Italy
Molineria In honor of Ignazio Bernardo Molineri (1741–1818) Italian botanist
Molineriella L. *-ella*, diminutive but here used as a name-forming suffix. Resembling *Molineria*
Molinia In honor of Juan Ignazio Molina (1740–1829) Chilean cleric and amateur botanist
molinioides Gk *-oides*, resembling. Resembling *Molinia*
Moliniopsis Gk *opsis*, appearance. Resembling *Molinia*

M

moll-e, -is L. soft. – (1) Softly hairy usually of leaf-blades. *Arthraxon molle, Bathratherum molle, Bromus mollis, Calamagrostis mollis, Enneapogon mollis, Eulalia mollis, Erianthus mollis, Festuca mollis, Holcus mollis, Ichnanthus mollis, Ischaemum molle, Leymus mollis, Melinis mollis, Panicum molle, Pennisetum molle, Poecilostachys mollis, Pollinia mollis, Stipa mollis, Trachypogon mollis* – (2) soft in aspect compared with the xeromorph facies of related species. *Plectrachne mollis*

mollendense L. *-ense*, denoting origin. From Mollendo, Peru

mollicom-a, -um, -us L. *molle*, soft; *coma*, head of hair. Covered in whole or in part with downy hairs

molliculmum L. *molle*, soft; *culmus*, stem. Culms softly hairy

molliculus L. *molle*, soft; *-ulus*, diminutive. Densely invested with short soft hairs

mollifolium L. *molle*, soft; *folium*, leaf. Leaf-blades densely and softly pilose

molliformis L. *forma*, appearance. Resembling *Bromus mollis*

mollior L. softer. Basal leaf-sheaths densely tomentose

mollipilum L. *molle*, soft; *pilus*, hair. Leaf-sheaths invested with soft hairs

mollis See *molle*

mollissim-a, -um, -us L. *molle*, soft; *-issima*, most. Plant in whole or in part densely covered with soft hairs

molokaiens-e, -is L. *-ense*, denoting origin. From Molokai, one of the Hawaiian Islands

molokaiensis L. *-ensis*, denoting origin. From Molokai, one of the Hawaiian Islands

moluccanus L. *-anus*, indicating connection. From the Moluccas, Indonesia

molybdea L. lead-colored. Spikelets lead-colored

mombasana, mombassana L. *-ana*, indicating connection. From Mombasa, Kenya

momosei In honor of Sizuo Momose (1906–1968) Japanese botanist

Monachather Gk *monarchos*, solitary; *ather*, barb or spine. Lemma one-awned in contrast to three-awned in *Danthonia*

Monachne, monachne Gk *monos*, one; *achne*, scale. The lower glume is very much reduced or absent

Monachyron, monachyron Gk *monos*, one; *achyron*, scale. The spikelet was interpreted by its author as having only one glume

monadelpha Gk *monos*, one; *adelphos*, close kinsman. Stamens united

monandr-a, -um, -us Gk *monos*, one; *aner*, man. Florets with one stamen

Monandraira Gk *monos*, one; *aner*, man. Resembling *Aira* but the florets with only one stamen

monandr-um, -us See *monandra*

monantha Gk *monos*, one; *anthos*, flower. Spikelets with a single floret

Monanthochloe, Monanthochloë Gk *monos*, one; *anthos*, flower; *chloa*, grass. The inflorescence has a single spikelet

Monathera Gk *monos*, one; *ather*, barb or spine. Racemes each of a solitary spikelet with an awned lemma

monatherus Gk *monos*, one; *ather*, barb or spine. Lemma only awned

Monelytrum Gk *monos*, one; *elytron*, cover. The lower glume of the spikelet is lacking

Monerma Gk *monos*, one; *erma*, support. The spikelets are subtended by only one glume

mongholica L. *-ica*, belonging to. From Mongolia

mongolic-a, -um See *mongholica*

mongolorum L. of the Mongols. From Mongolia

monguensis L. *-ensis*, denoting origin. From Mongui, Colombia

monianthum Gk *monos*, one; *anthos*, flower. Inflorescence a single spikelet

Monilia A misspelling of *Molinia*

moninens-e, -is L. *-ense*, denoting origin. From Monino, Angola

Monium Gk *monos*, one. Racemes reduced to a single pair of spikelets or spikelet

monocephala Gk *monos*, one; *kephale*, head. Inflorescence a single terminal cluster of racemes

Monocera Gk *monos*, one; *keras*, horn. The upper glume is long awned in contrast to the lower which is short awned or awnless

Monochaete Gk *monos*, one; *chaete*, bristle. Spikelets with a single bristle derived from the extended rhachilla
Monoclados See *Monocladus*
Monocladus, monocladus Gk *monos*, one; *klados*, a shoot. Culms with a solitary branch at each node
monococcum Gk *monos*, one; *kokkos*, kernal. Spikelets commonly producing a single grain
monococum See *monococcum*
Monocymbium Gk *monos*, one; *kymbe*, boat. The solitary racemes are each supported by boat-shaped spathes
monodactyl-a, -um Gk *monos*, one; *daktylos*, finger. Inflorescence a solitary raceme
Monodia Gk *monos*, one; *odous*, tooth. Lemma not toothed, a word play about the closely related genus *Triodia*
monodii In honor of – (1) Théodore Monod (1902–1950) French botanist. *Eragrostis monodii* – (2) Charles Monod de Froideville (1896–1978) Dutch botanist. *Digitaria monodii*
monogyna Gk *monos*, one; *gyne*, woman. Style single
monoica Gk *monos*, one; *oikos*, house. Florets hermophrodite
mononeurum Gk *monos*, one; *neuron*, nerve. Glumes one-nerved
monopholis Gk *monos*, solitary; *pholis*, scale of a snake. One glume fails to develop
monophylla Gk *monos*, one; *phyllon*, leaf. – (1) Mature culm producing a single leaf. *Sucrea monophylla* – (2) with branchlets terminating in a single leaf with a blade otherwise with leaf-sheaths only at lower nodes. *Yushania monophylla*
Monopogon Gk *monos*, one; *pogon*, beard. The lemma of the upper floret only is awned
monoracemum Gk *monos*, one. L. *racemus*, stalk of a bunch of grapes. Inflorescence of a single raceme
Monospatha Gk *monos*, one; *spatha*, sheathing base and false petiole of a palm leaf. Dichasium subtended by a broad spathe
Monostachya, -a, -os, -um, -us Gk *monos*, one; *stachys*, spike as of an ear of wheat. The inflorescence is a single spike or raceme
Monostemon Gk *monos*, one; *stemon*, thread. The florets possess a single stamen
monostigma Gk *monos*, one; *stigma*, column. Pistil with a single stigma
monothalamia Gk *monos*, one; *thalamos*, inner room. The single female floret is terminal and surrounded by five male florets
Monroa See *Munroa*
monspeliens-e, -is L. *mons*, mountain; *-ense*, denoting origin. From Montpellier, France
monspessulana L. *-ana*, indicating connection. From *mons Pessulanus*, the Latin name for Montpellier, France
monstruosa L. *monstrum*, monster; *-osa*, abundance. Inflorescence with spikelets replaced by bulbils
montalbanica L. *mons*, mountain; *-ica*, belonging to. Origin not given by the author but possibly named from the Europaean name for a Philippine mountain
montan-a, -um, -us L. *mons*, mountain; *-ana*, indicating connection. Growing on mountains
montanense L. *-ense*, denoting origin. From *Montana*, USA
montan-um, -us See *montana*
montevansi From Mount Evans, Colorado, USA
montevidensis L. *-ensis*, denoting origin. From Montevideo, Uruguay
montezumae In honor of Montezuma, last Aztec Emperor; species first described from Mexico
montianum L. *-anum*, indicating connection. In honor of Carlo del Monti, botanist at Florence
monticola L. *mons*, mountain; *-cola*, dweller. Growing on mountains
montigena L. *mons*, mountain; *gigno*, beget. Growing on mountains
montis-aurea L. *mons*, mountain; *aurea*, gold. From Mt d'Or, Auvergne, France
montis-celtici L. *mons*, mountain; *celticus*, of middle Gaul. From the mountains of the Auvergne, France

M

montis-wilhelmii From Mount Wilhelm, Papua New Guinea

montufarii In honor of Carlos Montúfar y Larrea (1780–1816) Ecuadoran naturalist and traveller

moomomiense L. *-ense*, denoting origin. From Moomomi, on Molokai, one of the Hawaiian Islands

mooneyi In honor of Herbert Francis Mooney (1897–1964) Irish-born Indian forester and plant collector

Moorea In honor of David Moore (1808–1879) Scots-born Irish botanist

moorei In honor of – (1) Charles Moore (1820–1905) Scots-born Australian botanist. *Chloris moorei* – (2) David Moresby Moore (1933–) English botanist. *Hierochloe moorei*

Moorochloa Gk *moros* (English) or *mooros* (Dutch), foolish; *chloa*, grass. Proposed in response to the failure of the Nomenclature Committee of the International Association of Plant Taxonomists to accept a proposal, by the author, for the conservation of a long-established generic name

mopane Growing in mopane, a type of woodland in Zimbabwe

morales-coelloi In honor of Morales y Coello, Cuban naval officer

moratii In honor of Phillipe Morat (1937–) French botanist

moreheadiana L. *-ana*, indicating connection. In honor of Boyd Dunlop Morehead (1843–1905) English-born Australian politician

morenensis L. *-ensis*, denoting origin. From Moreno Bluff, near Antofagasta, northern Chile

morichalense L. *-ense*, denoting origin. Growing around morichales, the Venezuelan name for palm swamps

morisiana L. *-ana*, indicating connection. As for *morisii*

morisii In honor of Giuseppe Giacinto Moris (1796–1869) Italian physician and botanist

morisonii In honor of Robert Morison (1620–1683) British botanist

moritzii In honor of Johann Wilhelm Karl Moritz (1797–1866) German botanist

mormonum From Utah, USA, the headquarters of the Church of Jesus Christ of the Latter-day Saints, whose members are commonly known as Mormons

morombense L. *-ense*, denoting origin. From Morombe, Madagascar

morotonensis L. *-ensis*, denoting origin. From Morotonomura, Ibaraki Prefecture, Japan

morrisii In honor of Patrick Francis Morris (1896–1974) Australian botanist

morrisonensis L. *-ensis*, denoting origin. From Nütaka Yara, known to the English as Mount Morrison, Taiwan

morronei In honor of Osvaldo Morrone (fl. 1993) Argentinian botanist

mortehanii In honor of Mortehan, who collected in Zaire

mortonian-a, -um In honor of Julius Sterling Morton (1832–1902) United States agricultural administrator

morulum L. *morus*, mulberry; *-ulum*, diminutive. Anthoecia black, like the fruits of mulberries

mosambicensis L. *-ensis*, denoting origin. From Mozambique

Mosdenia In recollection of Mosdene, the name of a farm near Naboomspruit, Waterberg, Transvaal, South Africa

mosquitiensis L. *-ensis*, denoting origin. From the Mosquitia region, sometimes called Costa de Miskitos, of Nicaragua and Honduras

mossambicensis See *mosambicensis*

mossamedens-e, -is L. *-ense*, denoting origin. From Mossamedes, Angola

mossii In honor of E. M. Moss (fl. 1946) Canadian botanist

mossulensis L. *-ensis*, denoting origin. From Mo(s)sul, Iraq

motembense L. *-ense*, denoting origin. From Sabana de Motembo, Cuba

motia Nimadi word meaning like a pearl, i.e. precious. The oil produced from the grass is much more valuable than that of related species

motidsukiana L. *-ana*, indicating connection. In honor of Seiti Motidsuki, Japanese botanist

Moulinsia In honor of Charles Robert Alexandre Moulins, also known as Desmoulins (1798–1875) French botanist

moupinensis L. *-ensis*, denoting origin. From Moupin, Sichuan Province, China

moyanii See *moyanoi*

moyanoi In honor of Carlos Moyano (fl. 1889) Argentinian botanist

Muantijamvella L. *-ella*, diminutive here used as a name-forming suffix. An allusion to Mwantijamva, a mediaeval East African empire

mucronat-a, -um, -us L. *mucro*, sharp point; *-ata*, possessing. – (1) With glumes or lemmas contracted into a short hard point or bifid and shortly awned from between the teeth. *Agropyron mucronatum, Agrostis mucronata, Arundinaria mucronata, Cleistogenes mucronata, Eleusine mucronata, Elytrigia × mucronata, Eriachne mucronata, Eutriana mucronata, Halopyrum mucronatum, Kengia mucronata, Leptochloa mucronata, Megastachya mucronata, Odyssea mucronata, Panicum mucronatum, Paspalidium mucronatum, Paspalum mucronatum, Podosaemum mucronatum, Schizostachyum mucronatum, Stipa mucronata, Trachys mucronata, Uniola mucronata* – (2) with inflorescences terminating in a sharp point. *Dactyloctenium mucronatum*

mucronulatum L. *mucronulus*, small hook. *-atus*, possessing. – (1) Lemma shortly aristate. *Panicum mucronulatum, Pappophorum mucronulatum* – (2) lower glume distinctly mucronate. *Dichanthium mucronulatum*

mucuchachensis L. *-ensis*, denoting origin. From Páramo de Mucuchíes, Venezuela

mueensis L. *-ensis*, denoting origin. From Mue, Zaire

muehlenbergian-a, -um, -us As for *Muhlenbergia*

muehlenbergii As for *Muhlenbergia*

Muehlenburga See *Muhlenbergia*

muelleri In honor of – (1) Ferdinand Jacob Heinrich Mueller (1825–1896) German-born Australian botanist. *Aristida muelleri, Diplachne muelleri, Eriachne muelleri, Festuca muelleri, Ichnanthus muelleri, Panicum muelleri, Paraneurachne muelleri, Stipa muelleri, Yakirra muelleri* – (2) Franz August Müller (1799–1877) German apothecary. *Agrostis muelleri, Trachypogon muelleri, Trichodium muelleri, Vilfa muelleri* – (3) Friedrich M. Müller (fl. 1853–1855) who collected in Mexico. *Schizachyrium muelleri*

muelleriana L. *-ana*, indicating connection. As for *muelleri* (1)

muenzneri In honor of Max Münzer (1908–) who collected in Malawi

muerensis L. *-ensis*, denoting origin. From Muera Plateau, Tanzania

muhavurensis L. *-ensis*, denoting origin. From Mount Muhavura in the Virunga Mountains on the border of Uganda and Rwanda

Muhlenbergia In honor of Gotthilf Heinrich Ernst Muhlenberg (1753–1815) a United States preacher, teacher and botanist

muhlenbergian-um L. *-anum*, indicating connection. As for *Muhlenbergia*

muhlenbergii As for *Muhlenbergia*

muhlenbergioides Gk *-oides*, resembling. Similar to *Muhlenbergia*

muiriana L. *-iana*, indicating connection. In honor of John Muir (1838–1914) United States naturalist

mukdenensis L. *-ensis*, denoting origin. From Mukden, now Shenyang, Liaoning Province, China

mukogunensis L. *-ensis*, denoting origin. From Mukogun, Japan

mukuku Origin uncertain, not given by author but possibly the vernacular name of the species in Zaire

mukuluens-e, -is L. *-ense*, denoting origin. From Mukulu, Zaire

mulalensis L. *-ensis*, denoting origin. From Mt Mulalo, Ecuador

muliensis L. *-ensis*, denoting origin. From Muli, Sichuan Province, China

mulleri In honor of C. H. Muller (fl. 1939) who collected in Venezuela and Mexico

Mullerochloa In honor of Lennox Andrew Graham Muller (1932–) Australian civil servant and amateur botanist

multibrachiatum L. *multus*, many; *brachium*, arm; *-atum*, possessing. Panicle much branched

multicaul-e, -is L. *multus*, many; *caulis*, stem. Culms much branched

multiciliata L. *multus*, many; *cilium*, hair; *-ata*, possessing. With several rows of hairs on the upper glume

multiculmis L. *multus*, many; *culmus*, stalk. Plants densely caespitose

multifida L. *multus*, many; *findo*, cleave. The second lemma is divided into five-seven awns

multiflor-a, -um, -us L. *multus*, many; *flos*, flower. – (1) Spikelets with more florets than those of related species. *Agropyron multiflorum, Arundinaria multiflora, Deschampsia multiflora, Eleusine multiflora, Elymus multiflorus, Greslania multiflora, Isachne multiflora, Lolium multiflorum, Poa multiflora, Spartina multiflora* – (2) inflorescence with many spikelets. *Andropogon multiflorus, Cenchrus multiflorus, Digitaria multiflora, Panicum multiflorum, Pennisetum multiflorum*

multifloscula L. *multus*, many; *flos*, flower; *-ula*, diminutive. Spikelet has many small florets

multifoli-a, -um L. *multus*, many; *folium*, leaf. Culms bearing many leaves widely separated

multinerva L. *multus*, many; *nervus*, nerve. Glumes or lemmas many-nerved

multinervi-a, -us L. *multus*, many; *nervus*, nerve. Glumes or lemmas many-nerved

multinervis L. *multus*, many; *nervus*, nerve. Leaf-blades many-nerved

multinervos-a, -um L. *multus*, many; *nervus*, nerve; *-osa*, abundance. The glumes or lemmas many-nerved

multinod-e, -is, -um L. *multus*, many; *nodus*, knot. Culms many-noded

multinodosum L. *multus*, many; *nodus*, knot; *-osum*, abundance. Culms with about 25 nodes

multinodum L. *multus*, many; *nodus*, knot. Culms many-noded

multiplex L. *multus*, many; *plexus*, network. – (1) Inflorescence much branched. *Andropogon multiplex, Arundo multiplex, Bambusa multiplex, Hyparrhenia multiplex* – (2) with more than the usual number of stalked spikelets. *Anthistiria multiplex*

multiradiata L. *multus*, many; *radius*, ray; *-ata*, possessing. Panicle branches numerous and arranged in semiverticels

multirame-a, -um L. *multus*, many; *ramus*, branch. Culms much branched

multiramosa L. *multus*, many; *ramus*, branch; *-osa*, abundance. Culms much branched

multiset-a, -um, -us L. *multus*, many; *seta*, bristle. Spikelets many-awned

multispica L. *multus*, many; *spica*, a point; hence, in particular, an ear or spike of grain. Inflorescence of many racemes

multispicat-a, -um, -us L. *multus*, many; *spica*, a point; hence, in particular, an ear or spike of grain; *-ata*, possessing. Inflorescence with many racemes

multispiculata L. *multus*, many; *spica*, a point; hence, in particular, an ear or spike of grain; *-ula*, diminutive; *-ata*, possessing. Panicle with many spikelets

multispiculis L. *multus*, many; *spica*, a point; hence, in particular, an ear or spike of grain; *-ula*, diminutive. Inflorescence with many spikelets

multnomae From Multnomah Falls, Oregon, USA

mundula L. *mundus*, elegant; *-ula*, diminutive. Attractive in appearance

mundum L. elegant. Attractive in appearance

munja Bengali *munj*, culm of sugar cane. Vernacular name for *Saccharum* on the Punjab plains, India

munozensis L. *-ensis*, denoting origin. From Cerro Muñoz, Department of Tafé, Argentina

Munroa In honor of William Munro (1818–1880) Scots-born British soldier and amateur botanist who collected extensively in India

munroan-a, -um L. *-ana*, indicating connection. As for *Munroa*

munroi See *Munroa*

munsuensis L. *-ensis*, denoting origin. From Munsu, South Korea

mupinensis See *moupinensis*

muralis L. *murus*, wall; *-alis*, pertaining to. Growing on walls

muramatsuana L. *-ana*, indicating connection. In honor of S. Muramatsu, Japanese botanist

murasabuana L. *-ana*, indicating connection. In honor of S. Murai, Japanese botanist

muratana L. *-ana*, indicating connection. In honor of Kazuye Murata

murayamae In honor of Y. Murayama (fl. 1928)

murcica L. *-ica*, belonging to. From Murcia Province, Spain

muricat-a, -um, -us L. *-ata*, possessing. – (1) Surface rough due to the presence of short hard points such as are present on the surface of *Murex* (gastropod) or otherwise terminating in a sharp point. *Aegilops muricata, Anatherum muricatum, Andropogon muricatus, Cyrtococcum muricatum, Echinochloa muricata, Eremochloa muricata, Oplismenus muricatus, Panicum muricatum, Rottboellia muricata, Trachys muricata, Triticum muricatum, Vetiveria muricata, Vilfa muricata* – (2) involucral bristles with many short barbs. *Cenchrus muricatus* – (3) leaf-apices sharp-pointed. *Sasa muricata*

muricatulus L. *muricatus*, rough; *-ulus*, diminutive. As for *muricatus* but surfaces less rough or pointed

muricat-um, -us See *muricata*

muricola L. *murus*, wall; *-cola*, dweller. Growing on walls

muriculata L. somewhat rough. Glumes conspicuously scabrid

murieliae In honor of Muriel Wilson, daughter of Ernest Henry Wilson (1876–1930) English botanist

murin-a, -um L. *murus*, wall; *-ina*, indicating possession. Growing on walls

muroian-a, -us L. *-ana*, indicating connection. In honor of Hiroshi Muroi (1914–) Japanese botanist

murphyi In honor of H. C. Murphy (fl. 1968) United States plant breeder

murriana L. *-ana*, indicating connection. In honor of Joseph Murr (1864–1932) Austrian botanist

musaefolium L. *folium*, leaf. Leaf-blades resembling those of *Musa*

musashiensis L. *-ensis*, denoting origin. From Musashi Province, now Tokyo Prefecture and parts of Saitama and Kanagawa Prefectures, Japan

muscarium L. *muscus*, moss; *-arium*, pertaining to. Growing amongst mosses

muscicola L. *muscus*, moss; *-cola*, dweller. Growing in moss-forests

muscosa L. *muscus*, moss; *-osa*, abundance. In habit resembling a moss

mustangensis L. *-ensis*, denoting origin. From Mustang District, Nepal

mustaphae From the Mustapha Valley, Algeria

Mustelia In honor of Nicolas-Alexandre Mustel (fl. 1772–1784) French horticulturalist

mustersii In honor of George Charworth Musters (fl. 1869) explorer in Patagonia

mutabil-e, -is L. changeable. – (1) Variable in accord with the season or habitat. *Agropyron mutabilis, Aristida mutabilis, Cenchrus mutabilis, Elymus mutabilis, Panicum mutabile, Paspalum mutabile* – (2) having sterile and fertile culms. *Calamagrostis mutabilis* – (3) the juvenile and adult plants resemble two other species. *Bambusa mutabilis*

mutic-a, -um, -us L. blunt. Lacking awns or lemmas and/or glumes truncate

mutilat-um, -us L. *mutilus*, maimed, especially of cattle which have lost one or both horns; *-atum*, possessing. One of more of the involucral bristles subtending the spikelet bent and so resembling a crumpled cattle horn

Mygalurus Gk *mygale*, field mouse; *oura*, tail. The spicate inflorescence resembles the tail of a field mouse

myojinensis L. *-ensis*, denoting origin. From Omyojinmura, Rikuchiu Province, Japan

myosotis Gk *myosotis*, madwort (*Asperugo procumbens*) one of the borages. Habit creeping and foliage asperous

myosuroides Gk *mys*, mouse; *ourus*, tail; *-oides*, resembling. Inflorescence resembling the tail of a mouse

myosurus Gk *mys*, mouse; *oura*, tail. Inflorescence branches short and narrow

Myriachaeta Gk *myrios*, countless; *chaete*, bristle. The inflorescence is stiff and many branched

myrianth-a, -um, -us Gk *myrios*, countless; *anthos*, flower. Inflorescence many spikelets

Myriocladus Gk *myrios*, countless; *klados*, branch. Branches many at each node

Myriostachya, -a, -um, -us Gk *myrios*, countless; *stachys*, spike as of an ear of wheat. Inflorescence of many racemes

Myriostachys See *Myriostachya*

myriostachyum See *Myriostachya*

myrthens-e, -is L. *-ense*, denoting origin. From Mirto, Sicily

mysorensis L. *-ensis*, denoting origin. From Mysore, India

mystasipum Gk *mystax*, hair on upper lip; *sipue*, case. Base of lemma subtending the grain invested with long hairs

myur-os, -um, -us Gk *mys*, mouse; *oura*, tail. Inflorescence a spike-like panicle

N

Nabelekia In honor of Frantisek Nabelek (1884–1965) Bohemian botanist

nabeshimana L. *-ana*, indicating connection. In honor of Y. Nabeshima (fl. 1932)

nachiczevanica L. *-ica*, belonging to. From Nakhichevan, Azerbaijan

nagalandiana L. *-ana*, indicating connection. From Nagaland, India

nagarum Of the Naga Hills, Assam State, India

nagasei In honor of Hideo Nagase (1918–) Japanese botanist

nagashima From Nagashima Island, Japan

nagensis L. *-ensis*, denoting origin. From Naga Hills, Assam State, India

nahuelhuapiensis L. *-ensis*, denoting origin. Growing on the shores of Lake Nahuel Huapí, Argentina

naibunensis L. *-ensis*, denoting origin. From Naibun, Taiwan

naigoensis L. *-ensis*, denoting origin. From Naigô, Niigata Prefecture, Japan

naiguatensis L. *-ensis*, denoting origin. From Pico de Naiguata, Miranda State, Venezuela

nairii In honor of N. Chandrasekharan Nair (1927–) Indian botanist

najad-a, -um L. *Naiad*, a river nymph. Growing in swamps and pools

nakaharae In honor of Gonji Nakahara (fl. 1907) Japanese botanist

nakaian-a, -um, -us L. *-ana*, indicating connection. – (1) In honor of Monoshin Nakai. *Panicum nakaianum*, *Phragmites nakaiana*, *Setaria nakaiana* – (2) See *nakaii*. *Miscanthus nakaianus*

nakaii In honor of Takenoshin Nakai (1882–1952) Japanese botanist

nakashimae In honor of Kadsuwo Nakashima (1904–1953) Japanese botanist

nakashimana L. *-ana*, indicating connection. As for *nakashima*

nakasiretokensis L. *-ensis*, denoting origin. From Nakasireto Peninsula, Sakhalin Island, Russian Far East

naltchikensis L. *-ensis*, denoting origin. From Nalczik, a district in the northern Caucasus, Russian Federation

naltozikensis L. *-ensis*, denoting origin. An orthographic variant of *nalchikensis*

namaquens-e, -is L. *-ense*, denoting origin. From Namaqualand, South Africa

namboodiriana L. *-ana*, indicating connection. In honor of A. N. Namboodiri (1930–) Indian botanist

nambuana L. *-ana*, indicating connection. From Nambu, Japan

nambuensis L. *-ensis*, denoting origin. See *nambuana*

namibensis L. *-ensis*, denoting origin. From Namibia

namuliensis L. *-ensis*, denoting origin. From Namuli Peaks, near Gurué, Mozambique

nan-a, -um, -us L. dwarf. Smaller than related species

nancaguense L. *-ense*, denoting origin. From Nancagua, Chile

nandadevica L. *-ica*, belonging to. From Nandevi National Park, Chamoli, Uttar Pradesh, India

nandaiensis L. *-ensis*, denoting origin. From Nandaisan, a mountain in Ibaraki Prefecture, Japan

nandanensis L. *-ensis*, denoting origin. From Nanda, Guangxi Province, China

nankoensis L. *-ensis*, denoting origin. – (1) From Nankotaisan, a mountain in Taiwan. *Poa nankoensis* – (2) from Nanko, Iwaki Province, Japan. *Sasa nankoensis*

nankotaizanensis L. *-ensis*, denoting origin. From Mt Nankotaisan, Taiwan

nannfeldtii In honor of Johan Axel Frithiof Nannfeldt (1904–1985) Norwegian botanist

nanningensis L. *-ensis*, denoting origin. From Nanning, Guangxi Province, China

nanpinensis L. *-ensis*, denoting origin. From Nan-pin Shi, Hau-pin, Fujian Province, China

nanpingensis See *nanpinensis*

nantaialpicola L. *alpes*, high mountain; *-cola*, dweller. From Nantaisan, Tochigi Prefecture, Japan

nan-um, -us See *nana*

nanunic-a, -us (1) L. *nanus*, dwarf; *unicus*, single. Branches solitary at the culm nodes. *Arundinaria nanunica, Indocalamus nanunicus* – (2) L. *-ica*, belonging to. From Nanun, Hunan Province, China. *Pseudosasa nanunica*

nanus See *nana*

napensis L. *-ensis*, denoting origin. From Napa County, California, USA

napocae From Napoca, Romania

napostaens-e, -is L. *-ense*, denoting origin. From Naposta, Argentina

naratavica See *karatavica*

narayanae In honor of N. Narayana, Indian botanist

nardifolia L. *folium*, leaf. Leaf-blades morphologically resemble those of *Nardus*

nardiformis L. *forma*, appearance. Resembling *Nardus* with respect to the inflorescence

nardoides Gk *-oides*, resembling. Inflorescence resembling that of *Nardus*

Narduretia Segregated from *Nardurus*

Narduroides, narduroides Gk *-oides*, resembling. Similar to *Nardurus*

Nardurus Gk *oura*, tail. The racemose inflorescence branches resemble the inflorescence of *Nardus*

nardus (1) Scented as with nard, see *Nardus*. *Andropogon nardus, Cymbopogon nardus* – (2) leaf-blades linear as with *Nardus*. *Agropyron nardus, Triticum nardus*

Nardus Gk *nardos*, spikenard. Derived from a Semite word for aromatic balm produced by spikenard (*Nardostachys jatamansi*) and by transfer of meaning to other species producing aromatic oils. How the name became transferred to *Nardus* is unclear because its species are unscented

Narenga Origin unclear, possibly a vernacular name for the species in Bengal

narenga Resembling *Narenga*

narihira See *narihiratake*

narihiratake Japanese *také*, bamboo. The stems are like those of *medake* (female bamboo) and the leaf-blades like those of *odake* (male bamboo). Hence, the plant, known in the vernacular as *narih iratake*, has both male and female characteristics as did Narihira, a character mentioned in "Kokinshu", a Japanese anthology first published in 904 C.E.

nascopieana L. *-ana*, indicating connection. Named for R. M. S. Nascopie which sailed regularly in Arctic waters

Nasella See *Nassella*

nashian-um, -us L. *-anum*, indicating connection. In honor of George Valentine Nash (1864–1921) United States botanist

Nassella L. *nassa*, wicker basket with a narrow *neck*, used for catching fish; *-ella*, diminutive. In lateral view the spikelets resemble such a fishing basket

Nastus Classical name used by Dioscorides for *Cenchrus frutescens*. Now applied to a quite different genus

natalens-e, -is L. *-ense*, denoting origin. From Natal, South Africa

natans L. *nato*, swim. Culms floating

N

nathalieae In honor of Nathalie. Origin uncertain, not given by author
nativitatis L. Pertaining to the Christian Festival of the Nativity. Endemic to Christmas Island, Indian Ocean
Natschia From *natsch*, the vernacular of the grass in alpine Switzerland
naucinodosa L. *naucum*, trifle; *nodus*, knot; *-osa*, abundance. Origin obscure, not given by author
naucinopilus L. *naucum*, trifle; *pilus*, hair. Plant somewhat hairy
Navicularia, navicularis L. *navis*, ship; *-ula*, diminutive; *-aria*, pertaining to. The spikelets are boat-shaped
Nazia Origin obscure, not given by author
ndemboensis L. *-ensis*, denoting origin. From Ndembo, Zaire
neaei In honor of Luis Née (Nea) (fl. 1789–1794) Spanish botanist who took part in the Malaspina Expedition
nealleyi In honor of Greenleaf Cilley Nealley (1846–1896) United States plant collector
neblinaensis L. *-ensis*, denoting origin. From Cerro de la Neblina, Venezuela
nebraskens-e, -is L. *-ense*, denoting origin. From Nebraska, USA
nebrodens-e, -is L. *-ense*, denoting origin. From Nebrodi Mountains, Sicily
nebulosa L. *nebula*, mist; *-osa*, abundance. Growing on high mountains
necopina L. unexpected. A segregate from another species
nedoluzhkoi In honor of Valeryi Aexeyevich Nedolushko (1953–2001) Russian plant ecologist and Botanical Gardens Administrator
Neeragrostis Gk *neos*, new. Resembling *Eragrostis* but dioecious
neesian-a, -um L. *-ana*, indicating connection. In honor of Christian Gottfried Daniel Nees von Esenbeck (1766–1858) German botanist
neesii As for *neesiana*
Neesiochloa Gk *chloa*, grass. See *neesiana*
neghellensis L. *-ensis*, denoting origin. From the vicinity of Neghelle in southern Ethiopia
neglect-a, -um, -us L. *negligo*, disregard. Often a segregate from another species with which it has been confused
Negria In honor of Giovanni Negri (1877–1960) Italian botanist
negrosense L. *-ense*, denoting origin. From Negros Islands, Philippines
nehruense L. *-ense*, denoting origin. In honor of Shri Jawaharlal Nehru (1889–1964) whose ashes are incorporated in Indian soil
neilreichiana L. *-ana*, indicating connection. In honor of August Neilreich (1803–1871) Austrian botanist
nekludowii In honor of Nekludow
nelsonii In honor of – (1) Edward William Nelson (1855–1934) United States explorer and naturalist. *Chusquea nelsonii, Eriochloa nelsonii, Paspalum nelsonii* – (2) Aven Nelson (1859–1952) United States botanist. *Stipa nelsonii*
Nemastachys Gk *nema*, thread; *stachys*, spike as of an ear of wheat. Racemes slender
nematanthus Gk *nema*, thread; *anthos*, flower. Without description but possibly a reference to thread-like peduncles
nematodes Gk *nema*, thread; *-odes*, indicating resemblance. Culms terete
nematophylla Gk *nema*, thread; *phyllon*, leaf. Leaf-blades filiform
Nematopoa Gk *nema*, thread; *poa*, grass. Leaf-blades filiform
nematostachya Gk *nema*, thread; *stachys*, spike as of an ear of wheat. Branches of panicle thread-like
nemopanthum Gk *nemos*, groove; *anthos*, flower. Originally collected in "Penitentiary Woods", North Carolina, USA
nemophyllus Gk *nema*, thread; *phyllon*, leaf. Leaf-blades narrow
nemoraliformis L. *forma*, appearance. Resembling *Poa nemoralis*
nemoralis L. *nemus*, wood; *-alis*, pertaining to. Woodland species
nemorivaga L. *nemus*, wood; *vagus*, wandering. Growing in open woodlands
nemoros-a, -um, -us L. *nemus*, wood; *-osa*, abundance. Woodland species

nemorum L. *nemus*, wood. Woodland species

neo Generic names and species epithets beginning with "neo" (L. *neos*, new) are commonly formed to distinguish a genus or species from that with which it was previously confused or to avoid the formation of a homonym. In the entries below, only those names are listed that are known not to strictly adhere to this format

neocaledonica L. *-ica*, belonging to. From New Caledonia

neogaea Gk *neos*, new; *ge*, earth. From the New World, that is, the Americas

Neohouzeaua Gk *neos*, new but here serving as a name-forming prefix. In honor of Jean Houzeau de Lehaie (1820–1888) Belgian botanist

Neohusnotia Gk *neos*, new but here serving as a name-forming prefix. In honor of Pierre Tranquille Husnot (1840–1929) French botanist

neomexican-a, -us L. *-ana*, indicating connection. From New Mexico, USA

neoyutakana L. *neos*, new; *-ana*, indicating connection. A replacement of an earlier name for the species, and in honor of Yutaka Hukuda (fl. 1937) Japanese botanist

nepalens-e, -is L. *-ense*, denoting origin. From Nepal

nepalica L. *-ica*, belonging to. From Nepal

nepheliphil-a, -um As for *nephelophila*

Nephelochloa Gk *nephele*, cloud; *chloa*, grass. Growing on mountain slopes

nephelochloides Gk *-oides*, resembling. Similar to *Nephelochloa*

nephelophil-a, -um Gk *nephele*, cloud; *phileo*, love. Growing in the clouds, that is a mountain species

nephroaurita Gk *nephron*, kidney; *aurita*, long-eared. With kidney-shaped auricles on the culm-sheaths

nepliana L. *-ana*, indicating connection. In honor of G. Nepli (fl. 1950) who collected in the Russian Far East

nereidaensis L. *-ensis*, denoting origin. From Río Nereidas, Colombia

nervalis L. *nervus*, nerve; *-alis*, pertaining to. Glumes conspicuously nerved

nervat-a, -um, -us L. *nervus*, nerve; *-ata*, possessing. – (1) Spikelets with conspicuously nerved lemmas or glumes. *Bouteloua nervata, Buchlomimus nervatus, Glyceria nervata, Gymnanthelia nervata, Isachne nervata, Panicularia nervata, Panicum nervatum, Poa nervata, Rehia nervatus* – (2) with conspicuously veined leaf-blades. *Andropogon nervatus, Cymbopogon nervatus*

nerviglum-e, -is L. *nervus*, nerve; *gluma*, husk. The glume(s) are conspicuously nerved

nervilemma L. *nervus*, nerve; Gk *lemma*, husk. Lemma nerves sharply defined

nervos-a, -um, -us L. *nervus*, nerve; *-osa*, abundance. Nerves, especially those of glumes, lemmas, culms or leaf-blades, conspicuous

nesiotes Gk islander. From St Lucia, one of the Leeward Islands

Nestlera In honor of Chrétien Geoffrey Nestler (1778–1832) French botanist

neumannii In honor of Roberto Neumann (fl. 1988) Argentinian agronomist and botanical collector

neumayerian-a, -um L. *-ana*, indicating connection. In honor of Joseph Neumayer (1791–1840) German botanist

neuquenensis L. *-ensis*, denoting origin. From Neuquén Territory, Argentina

Neurachne Gk *neuron*, nerve; *achne*, scale. Subtending glumes conspicuously nerved

neurachnoides Gk *-oides*, resembling. Similar to *Neurachne*

neuranthum Gk *neuron*, nerve; *anthos*, flower. Upper glume and lower lemma conspicuously seven-nerved

neuroelytrum Gk *neuron*, nerve; *elytron*, cover. Glumes conspicuously nerved

neuroglossa Gk *neuron*, nerve; *glossa*, tongue. Ligule conspicuously veined

Neurolepis Gk *neuron*, nerve; *lepis*, scale. Glumes and or lemmas conspicuously nerved

neurophylla Gk *neuron*, nerve; *phyllon*, leaf. Leaf-blade conspicuously veined

Neuropoa Gk *neuron*, nerve. Resembling *Poa* but the lemmas 9–11-nerved

neurosa Gk *neuron*, nerve; L. *-osa*, abundance. Spathes prominently nerved
neutralis In error for *australis*
nevadensis L. *-ensis*, denoting origin. – (1) From Nevada, USA. *Poa nevadensis* – (2) from Sierra Nevada, Spain. *Agrostis nevadensis, Festuca nevadensis*
nevenarae Origin uncertain, not given by author
nevinii In honor of James Cook Nevin (1835–1912/13) cleric and amateur botanist who collected in China and California
Nevroctola Gk *neuron*, nerve; *octo*, eight; *-ola*, diminutive but here used as a name-forming suffix. Lemma eight-nerved
Nevroloma Gk *neuron*, nerve; *loma*, border of a robe. Glumes with a single nerve and membranous margins
nevskiana L. *-ana*, indicating connection. As for *nevskii*
Nevskiella L. *-ella*, diminutive but here used as a name-forming suffix. See *nevskii*
nevskii In honor of Sergei Arseniyevich Nevsky (1908–1938) Russian agrostologist
newinii See *nevinii*
newmaniana L. *-ana*, indicating connection. In honor of L. M. Newman
newtonii In honor of Francisco Newton (1864–1909) Portuguese botanist who collected in Angola
Neyraudia Anagram of *Reynaudia*
neyrautii In honor of Jean Edmund Neyraut (1859–1942) French botanist
nhatrangensis L. *-ensis*, denoting origin. From Nha Trang, Vietnam
niamensis L. *-ensis*, denoting origin. From Niam-Niam, Zaire
niariensis L. *-ensis*, denoting origin. From Niari River, Congo
nicaraguense L. *-ense*, denoting origin. From Nicaragua
nicorae As for *Nicoraella*
Nicoraella L. *-ella*, diminutive but here serving as a name forming suffix. In honor of Elisa Gernaela Juana Raquel Nicora de Panza (1912–2001) Argentinian botanist
nidulans L. *nidular*, build a nest. Plant densely tufted and so resembling a bird's nest
nidularia L. *nidus*, nest; *-ulus*, diminutive; *-aria*, pertaining to. Resembling a small bird's nest in habit
niederleinii In honor of Gustav Niederlein (1858–1924) German botanist
nigamatake Japanese *niga*, bitter; *matake*, genuine bamboo. Origin uncertain, not given by author
nig-er, -ra, -rum L. black. Culms or spikelets dark-colored culms
nigerense L. *-ense*, denoting origin. From Republic of Niger, western Africa
nigerica L. *-ica*, belonging to. Growing on the banks of the *Niger*, a West African river
nigra See *niger*
nigrans L. *niger*, black; *-ans*, assuming the appearance of. Involucral bristles dark reddish-brown
nigrescens L. *nigresco*, become black. Spikelets black at maturity
nigricans L. *nigrico*, become blackish. Becoming black with maturity, usually of spikelets
nigriflora L. *niger*, black; *flos*, flower. Spikelets dark-colored
nigrinodis L. *niger*, black; *nodus*, knot. Culm nodes black
nigripes L. *niger*, black; *pes*, foot. Culm bases dark-colored
nigrirostr-e, -is L. *niger*, black; *rostrum*, beak. Fertile lemma has a dark tip
nigritan-a, -um, -us L. *-ana*, indicating connection. From the *Niger* region of Upper Guinea
nigritarum L. *niger*, black; *-arum*, pertaining to. Spikelets dark-colored
nigritella L. *niger*, black; *-ella*, diminutive. Panicle dark-purple
nigritian-a, -um L. *-ana*, indicating connection. From the *Niger* region of Upper Guinea
nigrivestis L. *niger*, black; *vestis*, clothes. Lemma-apices black
nigrociliata L. *niger*, black; *cilium*, hair; *-ata*, possessing. – (1) Internodes with dark hairs on the upper parts. *Gigantochloa nigrociliata, Oxytenanthera nigrociliata* – (2) lemmas invested with dark hairs. *Bambusa nigrociliata*

nigropedata L. *niger*, black; *pes*, food; *-ata*, possessing. Pedicels black

nigropurpurea L. *niger*, black; *purpurea*, purple. Dried plants dark-purple

nigrum See *niger*

niihauens-e, -is L. *-ense*, denoting origin. From Niihau, one of the Hawaiian Islands

niijimae In honor of Yoshinao Niijima (1871–?) Japanese botanist

niitakayamensis L. *-ensis*, denoting origin. From Niitaka Yama, Taiwan, known as Yu Shan to the Chinese (see *Yushania*) and as Mt Morrison to the English (see *morrisonensis*)

niitakensis L. *-ensis*, denoting origin. From Mt Niitake, Taiwan

nikitinii In honor of Sergei Nikolaevich Nikitin (1850–1909) Russian scientist

nikkoensis L. *-ensis*, denoting origin. From Nikko, Tochigi Prefecture, Japan

nikkomontana L. *mons*, mountain; *-ana*, indicating connection. From a mountain near Nikko, Tochigi Prefecture, Japan

nilagiric-a, -um, -us L. *-ica*, belonging to. From Nilgiri Hills, South India

niliac-a, -us L. belonging to the Nile. – (1) From an island in the Nile River, Egypt. *Crypsis niliaca* – (2) from the delta of river Nile, Egypt. *Aeluropus niliacus, Calotheca niliaca*

nilotic-um, -us L. *-icum*, belonging to. From areas close to the Nile River

nimbanum L. *-anum*, indicating connection. From Mt Nimba, Republic of Guinea

nimuana L. *-ana*, indicating connection. From Nimu, Tibet Autonomous Region, China

nindensis L. *-ensis*, denoting origin. From Ninda River, Angola

ningnanica L. *-ica*, belonging to. From Ningnan, Sichuan Province, China

ninoleense L. *-ense*, denoting origin. From Ninole, Hawaii

niphobia Gk *nipha*, snow; *bios*, mode of life. Growing near the snow-line

nipponensis L. *-ensis*, denoting origin. From Nippon, that is, Japan

nipponic-a, -um L. *-ica*, belonging to; *Nippon*, according to many nationals the Latin spelling most closely corresponding to the local pronunciation of the name of their country. From Japan

Nipponobambusa A *Bambusa*-like genus from Nippon, that is, Japan

Nipponocalamus Gk *kalamos*, reed. A reed-like genus from Nippon, that is, Japan

niquelandiae From Municipio Niquelândia, Brazil

nishigoensis L. *-ensis*, denoting origin. From Nishigô, Fukushima Prefecture, Japan

nishiyamensis L. *-ensis*, denoting origin. From Nishiyamamure, Rikuchui Province, Japan

nitens L. *niteo*, shine. Spikelets or lemmas glossy at maturity

nitid-a, -um, -us L. *niteo*, shine; *-ida*, condition. Spikelets, lemmas or leaf-sheaths shiny

nitidespiculata L. *niteo*, shine; *spica*, a point; hence, in particular, an ear or spike of grain; *-ula*, diminutive; *-ata*, possessing. Spikelets glossy

nitidul-a, -us L. *niteo*, shine; *-ula*, exhibiting tendency. – (1) Spikelets glossy. *Andropogon nitidulus, Aristida nitidula, Arthraxon nitidulus, Calamagrostis nitidula, Festuca nitidula* – (2) panicle branches glossy. *Koeleria nitidula*

nitid-um, -us See *nitida*

nival-e, -is L. *nix*, snow; *-ale*, pertaining to. Growing near to permanent snow line

nive-a, -us L. *nivea*, snowy. Rhachis densely invested with short white hairs

nivicola L. *nix*, snow; *-cola*, dweller. Alpine species

Nivieria In honor of Césaire Anthelme Alexis Niviere (1799–1879) French agriculturalist

nivifera L. *nix*, snow; *fero*, carry or bear. Able to tolerate burial in snow for several months of the year

nivosum L. *nix*, snow; *-osum*, abundance. Growing near the snow line on high mountains

niwahokori Japanese *niwa*, garden; *hokori*, dust. Vernacular name for the species in Japan

nlemfuensis L. *-ensis*, denoting origin. From Nlemfu, Zaire
nobilis L. tall. Culms tall for genus
nodatum L. *nodus*, knot; *-atum*, possessing. Lateral culms with numerous swollen nodes
nodibarbata L. *nodus*, knot; *barba*, beard; *-ata*, possessing. Nodes densely villous
nodiflorum L. *nodus*, knot; *flos*, flower. Culms with many short lateral branches each terminating in an inflorescence
nodiglumis L. *nodus*, knot; *gluma*, husk. Culms with reduced leaves and the nodes densely hairy
nodos-a, -um, -us L. *nodus*, knot; *-osa*, abundance. – (1) Culm nodes swollen. *Agropyron nodosum, Andropogon nodosus, Arthraxon nodosus, Arundinella nodosa, Dichanthium nodosum, Digitaria nodosa, Elytrigia nodosa, Hordeum nodosum, Microstegium nodosum, Ottochloa nodosum, Panicum nodosum, Poa nodosa, Triticum nodosum* – (2) culm bases swollen. *Phalaris nodosa* – (3) culms with more nodes than those of related species. *Stipa nodosa*
nodulibarbis L. *nodus*, knot; *-ula*, diminutive; *barba*, beard. Culm nodes invested with a collar of short hairs
nodulos-um, -us L. *nodus*, knot; *-ula*, diminutive; *-osum*, abundance. Apex of raceme joints cupuliform
noean-a, -us In honor of Friedrich Wilhelm Noë (?–1858) German apothecary and botanist
nogalensis L. *-ensis*, denoting origin. From Nogal, Ethiopa
nomokonovii In honor of Leontii Ivanovich Nomokonov (1902–?) Russian agronomist
nootkaensis See *nutkaensis*
norbergii In honor of Ingvar L. Norberg (fl. 1932–1945)
norfolkianum L. *-anum*, indicating connection. From Norfolk Island, an Australian Dependency, in the south-western Pacific
norica From Noricum, now largely included in Bavaria, Germany
Normanboria As for *Borinda*
normanii As for *Borinda*
norvegica L. *Norvegia*, Norway; *-ica*, belonging to. From Norway
nossibense L. *-ense*, denoting origin. From Nosy Bé, formerly Nossibé, an island close to the north-west coast of Madagascar
notabil-e, -is L. noteworthy. Culms tall and inflorescence large
notarisiana L. *-ana*, indicating connection. In honor of Giuseppe de Notaris (1805–1872) Italian botanist
notarisii As for *notarisiana*
notat-um, -us L. *nota*, distinguishing mark; *-ata*, possessing. Spikelets multi-colored
noterophil-a, -um Gk *noteros*, damp; *phileo*, love. Grows in damp sandy soil
Nothoholcus L. *nothum*, false. As defined by Linneus, *Holcus* included species now placed in *Sorghum*. Hence the name *Notholcus* was proposed for species currently included in *Holcus*
Notholcus See *Nothoholcus*
noth-um, -us L. false. Resembling two other species, that is combining the characters of both
Notochloe, Notochloë Gk *notos*, south; *chloe*, grass. Endemic to Australia
notochthon-a, -um Gk *notos*, south; *chthon*, the earth. From southern places such as Australia
notocoma Gk *noton*, the back; *kome*, head of hair. Keel of lemma hairy
Notodanthonia Gk *notos*, south. The spikelets of this largely New Zealand genus resemble those of *Danthonia*
notoensis L. *-ensis*, denoting origin. From Noto Province, now Ishikawa Prefecture, Japan
notolasia Gk *noton*, the back; *lasios*, shaggy. Subtending glumes densely hairy
Notonema Gk *noton*, the back; *nema*, thread. Lemma furnished with a dorsal awn
notopeninsulae L. *peninsula*, narrow strip of land jutting into the sea. From Noto Province, now Ishikawa Prefecture, Japan
notopogon Gk *noton*, the back; *pogon*, beard. Subtending glumes hairy
nototropus Gk *notos*, the south; *tropos*, direction. From southern localities

nottowayan-a, -us L. *-anus*, indicating connection. From valley of Nottoway River, Virginia, USA

nova-caledonica L. *-ica*, belonging to. See *novae-caledoniae*

novae-angliae From New England, USA

novae-caledoniae From New Caledonia

novae-guineae From New Guinea, now Papua New Guinea

novae-hollandiae From New Holland, now Australia

novae-zealandiae From New Zealand

novae-zelandiae From New Zealand

novae-zeylandiae From New Zealand

novakii In honor of František Antonín Novák (1892–1964) Czech botanist

novarae Commemorating the Austrian "Novara" Scientific Expedition (1857–1859)

novari In honor of Lázaro Juan Novara (1944–) Argentinian botanist

noveboracensis L. *-ensis*, denoting origin. From Noveboracum, that is New York, USA

novemnerve L. *novem*, nine; *nervus*, nerve. Both the upper glume and sterile lemma nine-nerved

novemnervia L. *novem*, nine; *nervus*, nerve. Glumes 9–11 nerved

novoagrariae L. *novus*, new; *agrarius*, belonging to the field. Origin unclear, not given by author. May be a reference to being an invader of cleared land

novocaledonica L. *-ica*, belonging to. From New Caledonia

novogaliciana L. *-ana*, indicating connection. From Novo Galicia, that is New Mexico, USA

novoguineensis L. *-ensis*, denoting origin. From New Guinea, now included in Papua New Guinea

novohibernica L. *novum*, new; *Hibernia*, Ireland; *-ica*, belonging to. From New Ireland, part of the Bismark Archipelago, Papua New Guinea

novozelandica L. *-ica*, belonging to. From New Zealand

novum L. new. A newly recognized species

Nowodworskia See *Nowodworskya*

Nowodworskya In honor of Johann Nowodworsky (?–1811) Bohemian Professor at Prague

nsoki Latinized form of *nsokia*, the vernacular name of the species in Zaire

nubic-a, -um, -us L. *-ica*, belonging to. From Nubia, either the Sudan or north-east Africa in general

nubicola L. *nubes*, cloud; *-cola*, dweller. Growing on high mountains

nubic-um, -us See *nubica*

nubigen-a, -um, -us L. *nubes*, cloud; *gigno*, bear. Growing on high mountains

nubila L. greyish-blue. Inflorescences purple

nud-a, -um, -us L. bare. – (1) Lemmas glabrous. *Andropogon nudus, Arthraxon nudus, Avena nuda, Bathratherum nudum, Digitaria nuda, Hordeum nudum, Poa nuda, Pollinia nuda* – (2) rachilla joints glabrous. *Microstegium nudum, Phragmites nudus* – (3) rachilla prolongation glabrous. *Deyeuxia nuda* – (4) leaf-sheaths glabrous. *Dendrocalamus nudus*

nudat-a, -um L. *nudus*, bare; *-ata*, possessing. – (1) Lemmas lacking woolly indumentum. *Poa nudata* – (2) spikelets lacking glumes. *Paspalum nudatum*

nudicaule L. *nudus*, bare; *caulis*, stem. Flowering culms leafless

nudiculme L. *nudus*, bare; *culmus*, stem. Culm leaves few, basal

nudideficiens L. *nudus*, bare; *deficio*, be missing. The grain is naked, that is shed without the palea and lemma but property of which the species is "deficient" is not given by the author

nudiflor-a, -um L. *nudus*, bare; *flos*, flower. – (1) Lemmas glabrous or nearly so. *Danthonia nudiflora, Poa nudiflora* – (2) callus glabrous. *Calamagrostis nudiflora, Deyeuxia nudiflora* – (3) upper floret lacking a palea. *Panicum nudiflorum*

nudiglume L. *nudus*, bare; *gluma*, husk. Glumes glabrous

nudipes L. *nudus*, bare; *pes*, foot. Pedicel glabrous

nudiramea L. *nudus*, bare; *ramus*, branch. Culm leaves deciduous

nudiramus L. *nudus*, bare; *ramus*, branch. Lower branches of the inflorescence lack spikelets

O

nud-um, -us See *nuda*
nulla L. *nullus*, nothing. Apex of the stipe lacks appendages
nullanulla Named for "Nulla Nulla" a property in New South Wales, Australia
nullarborensis L. *-ensis*, denoting origin. From Nullarbor Plain, Australia
numaeens-e, -is L. *-ense*, denoting origin. From Noumea, New Caledonia
numidian-a, -um L. *-ana*, indicating connection. From Numidia, now north-eastern Algeria
numidica L. *-ica*, belonging to. From Numidea, now north eastern Algeria
nummularium L. *nummus*, coin; *-arium*, indicating connection. Sterile lemma with raised rims giving them a coin-like appearance
nunobikiensis L. *-ensis*, denoting origin. From Mt Nunobikiyama, Mie Prefecture, Japan
nuriensis L. *-ensis*, denoting origin. From Nur, Spain
nuristanica L. *-ica*, belonging to. From Nuristan, Afghanistan
nuspicula L. *nuto*, nod; *spica*, a point; hence, in particular, an ear or spike of grain; *-ula*, diminutive. Inflorescence of nodding spikelets
nutabundum L. *nutans*, nodding; *abundum*, abundant. Inflorescence with copious nodding spikelets
nutans L. *nuto*, nod. Inflorescence branches slender, bending under the weight of the spikelets
nutkaensis L. *-ensis*, denoting origin. From Nootka Sound, Western Canada
nuttallian-a, -um L. *-ana*, indicating connection. In honor of Thomas Nuttall (1786–1859). United States naturalist
nuttallii As for *nuttalliana*
nyanzense L. *-ense*, denoting origin. From Bukoba near Lake Nyansa, Malawi
nyaradyana L. *-ana*, indicating connection. In honor of Erasmus Gyula Nyárády (1881–1966) Romanian botanist
nyaradyi As for *nyaradyana*
nyassae From Nyassaland, now Malawi
nyassan-a, -um L. *-ana*, indicating connection. From Nyassaland now Malawi
nyassense L. *-ense*, denoting origin. From Nyassaland, now Malawi
nyingchiensis L. *-ensis*, denoting origin. From Nyinchi, Tibet Autonomous Region, China
nymanii In honor of Karl Fredrik Nyman (1820–1893) Swedish botanist
nymphoides Gk *Nymphe*, a goddess presiding over localities including rocky places; *-oides*, resembling. Growing in the shelter of boulders
nyssana See *nyassana*

O

oahuaens-e, -is L. *-ense*, denoting origin. From Ohua, one of the Hawaiian Islands
oajacens-e, -is L. *-ense*, denoting origin. From Oajaca, Mexico
oaxacens-e, -is L. *-ense*, denoting origin. From Oaxaca State, Mexico
obclavata L. *ob-*, inversely; *clavis*, club; *-ata*, possessing. Club-shaped but with the swollen end towards the base
obconiciventris L. *obconicus*, conical with apex downwards; *venter*, belly. Upper glume broadly elliptical to obovate-elliptical
obensis L. *-ensis*, denoting origin. From the Obi River, western Siberia
obliqu-a, -um L. oblique. Base of leaf-blade is asymmetric
obliquiberbe L. *obliquus*, oblique; *berbe*, beard. Rhachis joints have an oblique band of hairs on one side
obliquifolia L. *obliquus*, oblique; *folium*, leaf. Leaf-blade asymmetric
obliquivalvis L. *obliquus*, oblique; *valvis*, leaf of a folding door. Lower glume winged towards apex and asymmetric when viewed from the side
obliquum See *obliqua*
oblita, oblitum L. *oblino*, cover. Lower glume half the length of the spikelet
obliter-a, -um L. weakly developed. Florets few per spikelet

oblong-a, -us L. *oblongus*, oblong. – (1) Leaf-blades oblong-lanceolate. *Yushania oblonga* – (2) inflorescences oblong. *Enneapogon oblongus*

oblongat-a, -um L. *oblongus*, oblong; *-ata*, possessing. Leaf-blades approaching elliptical in outline with the sides tending to be parallel

oblongo-ovata L. *oblongus*, oblong; *ovum*, egg; *-ata*, possessing. Spikelets narrow and somewhat egg-shaped

oblongula As for *oblongata*

oblongus See *oblonga*

obovatum L. *ob-*, contrary; *ovum*, egg; *-atum*, possessing. Spikelets egg-shaped with broad end distal to pedicel

obscur-a, -um, -us L. dark. – (1) Spikelets dark-colored. *Alopecurus obscurus, Andropogon obscurus, Aristida obscura, Cymbopogon obscurus, Nassella obscura, Paspalum obscurum* – (2) readily confused with other species. *Setaria obscura*

obseptum L. *obsaepio*, enclose. Inflorescence enclosed in leaf-sheath

obtect-a, -um, -us L. *obtego*, conceal. Segregated from a closely related species

obtorta L. *obtorqueo*, twist. Leaf-blade bent forward with a twist

obturbans L. *ob-*, contrary; *turbo*, spin. Spikelets elliptical-lanceolate in outline resembling an inverted spinning top

obtus-a, -um, -us L. blunt. – (1) Glumes or lemmas apically rounded. *Achneria obtusa, Andropogon obtusus, Aristida obtusa, Festuca obtusa, Glyceria obtusa, Helopus obtusus, Ortachne obtusa, Panicum obtusum, Pariana obtusa, Piptatherum obtusum, Stipa obtusa, Stipagrostis obtusa, Urachne obtusa* – (2) ligules truncate. *Oryzopsis obtusa*

obtusat-a, -um L. *obtusus*, blunt; *-ata*, possessing. – (1) Glumes truncate. *Paspalum obtusatum, Sphenopholis obtusata* – (2) lemmas truncate. *Arthrostylidium obtusatum, Arundinaria obtusata, Calamagrostis obtusata, Deyeuxia obtusata*

obtusiflor-a, -um, -us L. *obtusus*, blunt; *flos*, flower. Glume or lemma apices rounded

obtusifoli-a, -um, -us L. *obtusus*, blunt; *folium*, leaf. Apices of leaf-blades rounded

obtusiglume L. *obtusus*, blunt; *gluma*, husk. Apices of glumes and lemmas rounded

obtusissima L. *obtusus*, blunt; *-issima*, most. Glume apices rounded

obtusiusculum L. *obtusius*, blunter; *-ulum*, diminutive. Glumes short and rounded

obtus-um, -us See *obtusa*

obumbratum L. *obumbro*, overshadow. Overshadowed, that is growing in shady places

obvallat-a, -us L. *obvallo*, surround with a wall. Basal spikelets sterile forming a sheath around the fertile spikelet

obvipodus L. *obvius*, easily seen; Gk *pous*, foot. Spikelets conspicuously pedicellate

obvoluta L. *ob-*, on account of; *volutus*, rolled up. Lower glume wrapped around and exceeding in length the rest of the spikelet

occidental-e, -is L. *occident*, west; *-ale*, pertaining to. – (1) From the western states of the United States. *Dilophotriche occidentalis, Echinochloa occidentalis, Festuca occidentalis, Hierochloe occidentalis, Loliolum orientale, Panicum occidentale, Poa occidentalis, Stipa occidentalis* – (2) from Western Australia. *Brachiaria occidentalis, Danthonia occidentalis* – (3) from west coast of South America. *Pennisetum occidentale* – (4) from West Africa. *Coelachne occidentalis, Danthoniopsis occidentalis* – (5) from western Japan. *Sasa occidentalis* – (6) from western Europe. *Avena occidentalis, Avenula occidentalis*

occidentalialtaicus L. *occidentale*, of the west; *-icus*, belonging to. From the western Altai Mountains, Kazakhstan

occidentalis See *occidentale*

occitanica L. *-ica*, belonging to. From Occitana, now south-eastern France

occultum L. hidden. Sessile spikelet small, developing in the axil of the pedicel of the stalked spikelet, and thus easily overlooked

Ochlandra Gk *ochlos*, crowd; *aner*, man. Each floret has numerous anthers

ochotensis L. *-ensis*, denoting origin. From Ochotzh Province, eastern Siberia

ochroleuca Gk *ochros*, pale yellow; *leukos*, white. Spikelets pale yellow-white

ochrops Gk *ochros*, pale yellow; *ops*, eye. The involucre with its reflexed bristles and the yellow anthoecium combine to look like an eye

Ochthochloa Gk *ochthos*, hill; *chloa*, grass. Growing on hills

ocreata L. *ocrea*, metal armour for the lower leg; *-ata*, possessing. The setae around the orifice of the leaf-sheath give it the appearance of an ochrea, typical of the leaf-base of most Polygonaceae

octoflora L. *octo*, eight; *flos*, flower. The spikelets usually have eight florets

octonodum L. *octo*, eight; *nodus*, knot. Culms eight-noded

Odontelytrum Gk *odous*, tooth; *elytron*, cover. The spikelets are subtended by what appears to be a bract with a deeply dissected apex but which is an involucre of fused bristles

odorat-a, -um, -us L. *odora*, fragrance; *-ata*, possessing. – (1) Strongly scented with coumarin. *Anthoxanthum odoratum*, *Hierochloe odorata*, *Holcus odoratus*, *Torresia odorata* – (2) strongly scented with aromatic oils. *Amphilophis odorata*, *Andropogon odoratus*, *Bothriochloa odorata*, *Vetiveria odorata*

odoratissima L. *odora*, fragrance; *-issima*, most. Rhizomes rich in essential oils

odorat-um, -us See *odorata*

Odyssea In honor of the Odyssey, an epic Greek journey recounted by Homer. The single species included in the genus had been placed in several other genera before being segregated

Oedipachne Gk *oidos*, swelling; *achne*, scale. The lower glume is reduced to a crescent-shaped cushion

oedogonatum L. *-atum*, possessing. Nodes swollen thereby resembling the filamentous alga *Oedogonium*

oelandica L. *-ica*, belonging to. From Oeland, Sweden

oeningensis L. *-ensis*, denoting origin. From Oeningen, Germany

officinarum L. *officina*, drug shop; *-arum*, pertaining to. Used medicinally

offneri In honor of Jules Offner (1873–1957) French botanist

ogamiensis L. *-ensis*, denoting origin. From Ogami-mura, Ettyû Prefecture, Japan

ogiformis Gk *forma*, appearance. Resembling *ogi*, the vernacular name in Japan for a related species

ogowense L. *-ense*, denoting origin. From Région de l'Ogooué, Gabon

ohdana L. *-ana*, indicating connection. In honor of G. Ohda (fl. 1935) Japanese botanist

ohmiensis L. *-ensis*, denoting origin. From Ohmi or Omi Province, now Shiga Prefecture, Japan

ohminensis L. *-ensis*, denoting origin. From the Ohmine Ranges, Nara Prefecture, Japan

ohwiana L. *-ana*, indicating connection. As for *ohwii*

ohwii In honor of Jisaburo Ohwi (1905–1977) Japanese botanist

oiapocensis L. *-ensis*, denoting origin. From Oiapoque, Amapá, Brazil

oiwakensis L. *-ensis*, denoting origin. From Musha-Oiwake, Taiwan

okadana L. *-ana*, indicating connection. In honor of Yônosuke Okada, Japanese botanist

okamotoi In honor of S. Okamoto, Japanese botanist who collected in Taiwan

okuboi In honor of I. Okubo, Japanese botanist

okudana L. *-ana*, indicating connection. In honor of Seizô Okuda, Japanese educator

okuyezoensis L. *-ensis*, denoting origin. From Okuyezo, Sakhalin Island, Russian Far East

oldfieldii In honor of Augustus Frederick Oldfield (1820–1887) English-born Australian botanist

oldhamii In honor of Richard Oldham (1837–1864) English botanist

oleagina L. belonging to the olive tree. Spikelet a dull olive-green

oleosus L. oily. Culms sticky

olgae In honor of Olga Aleksandrovna Fedchenko (1845–1921) Russian botanist

olida L. *olea*, oil; *-ida*, tending to. Leaf-blades viscid

oliganth-a, -um, -us Gk *oligos*, few; *anthos*, flower. Inflorescence of few spikelets

oliganthos Gk *oligos*, few; *anthos*, flower. Spikelets few-flowered

oliganth-um, -us See *oligantha*

oligoadenotrichum Gk *oligos*, few; *aden*, gland; *thrix*, hair. Inflorescence branches bearing a few capitate hairs

oligobrachiat-a, -um Gk *oligos*, few; L. *brachium*, arm; *-ata*, possessing. The inflorescence has few branches

oligochaete Gk *oligos*, few; *chaete*, bristle. Bristles subtending spikelets few

oligoclada Gk *oligos*, few; *klados*, branch. Panicle branches paired rather than whorled

oligophylla Gk *oligos*, few; *phyllon*, leaf. Culms bearing few leaves

oligosanthes Gk *oligos*, few; *anthos*, flower. Inflorescence with few spikelets

oligospira Gk *oligos*, few; *spira*, spiral. Column of awn with few twists

Oligostachyum, -a, -um Gk *oligos*, few; *stachys*, spike as of an ear of wheat. Inflorescence with few branches

oligotrich-a, -um Gk *oligos*, few; *thrix*, hair. Plant in whole or part sparsely hairy

olivace-a, -um, -us L. *oliva*, olive; *-acea*, indicating resemblance. Spikelets or foliage olive-green

oliveri In honor of – (1) John William Oliver (1833–1907) Irish-born forester in India and Myanmar. *Thyrsostachys oliveri* – (2) Daniel Oliver (1830–1916) British botanist. *Agropyron oliveri*

olivetorum L. *olivetum*, olive-grove. Growing in olive groves

olivieri In honor of Guillaume Antoine Olivier (1756–1814) French biologist

Olmeca In honor of the Olmec Indians of Mexico

olmedoi In honor of Vincente Olmedo, Spanish botanist in Peru

olneyae In honor of Clara Olney, United States botanical collector

olympic-a, -um L. *-ica*, belonging to. – (1) From Mt Olympus, Greece. *Festuca olympica* – (2) from Mt Olympus, northwest Turkey. *Calamagrostis olympica*, *Pentatherum olympicum*

Olyra Name used by Theophrastus and Dioscorides for an unidentified species of grain

olyrachne Gk *achne*, chaff. Glumes resembling those of *Olyra*

olyraefolium L. *folium*, leaf. Plant with leaf-blades resembling those of *Olyra*

olyriformis L. *forma*, appearance. Resembling *Olyra*

olyroides Gk *-oides*, resembling. Similar to *Olyra*

omahekensis L. *-ensis*, denoting origin. From Omaheke. The sandy tree savannah of north-west South Africa known to the Herero people by that name

omega Final letter of Greek alphabet. From World's End View, Mbeya-Chunya, Tanzania

omeiensis L. *-ensis*, denoting origin. From Mt Omei, Sichuan Province, China

Omeiocalamus Gk *kalamos*, reed. Culms woody and from Mt Omei, Sichuan Province, China

omokoensis L. *-ensis*, denoting origin. From Omoko, Sikoku Province, Japan

oncothrix Gk *onkos*, barb; *thrix*, hair. Keels of the sterile lemmas stiffly hairy

onibensis L. *-ensis*, denoting origin. From bassin de l'Onibe, Madagascar

onibitoana L. *-ana*, indicating connection. From Valley of Onibito, Nagasaki Prefecture, Japan

Onoea In honor of Motoyoshi Ono (1837–1890) Japanese botanist

onoei See *Onoea*

ononbiense See *ouonbiense*

onslowense L. *-ense*, denoting origin. From Onslow County, North Carolina, USA

ontakensis L. *-ensis*, denoting origin. From Ontake-san, a mountain peak in Toyama Prefecture, Japan

ontariensis L. *-ensis*, denoting origin. From Ontario, Canada

onurus Gk *onos*, donkey; *oura*, tail. Inflorescence cylindrical

ooense L. *-ense*, denoting origin. From Puu Oo, Hawaii

ooh Vernacular name for the species in Bali, Indonesia

oostachys Gk *oon*, egg; *stachys*, spike as of an ear of wheat. Spikelets ovate

oostachyum Gk *oon*, egg; *stachys*, spike as of an ear of wheat. Inflorescence a solitary oblong spike

Ophiochloa Gk *chloa*, grass. Growing on soils derived from serpentine (ophiolite) rocks

ophioliticola Gk *ophios*, snake; *lithos*, stone; L. *-cola*, dweller. Growing on serpentine and amphibolite rocks

Ophismenus See *Oplismenus*

ophiticola Gk *ophis*, serpent; L. *-cola*, dweller. Growing on serpentinite rocks

ophitidis Gk *ophis*, snake. Growing on serpentine derived soils

Ophiura See *Ophiuros*

Ophiurinella L. *-ella*, diminutive but here used as a name-forming suffix. Resembling *Ophiurus*

ophiuroides Gk *-oides*, resembling. Resembling *Ophiuros*

Ophiuros Gk *ophis*, snake; *oura*, tail. The inflorescence is a cylindrical spike

Ophiurus See *Ophiuros*

ophryodes Gk *ophrys*, eyebrow; *-odes*, resembling. The upper glume and sterile lemma bear a band of glistening silky hairs which terminate below their apices and thereby resemble eyebrows

opienensis L. *-ensis*, denoting origin. From Opien (Ebian) Xian, Sichuan Province, China

Opitzia See *Opizia*

Opizia In honor of Philipp Maximilian Opiz (1787–1858) Bohemian botanist

oplismenoides Gk *-oides*, resembling. In habit similar to *Oplismenus*

Oplismenopsis Gk *opsis*, resemblance. The spikelets of *Oplismenopsis* differ only slightly from those of *Oplismenus*

Oplismenus Gk *hoplismenus*, bearing arms. Glumes and sterile lemmas awned

optimae In recognition of the important role played by the Organisation for Phyto-Taxonomic Investigation of the Mediterranean Area

opulenta L. wealthy. Spikelets with 8–10 florets, most of which are fertile

oranensis L. *-ensis*, denoting origin. From Oran, Algeria

orangense L. *-ense*, denoting origin. From Orange County, North Carolina, USA

oraria L. *ora*, coast; *-aria*, pertaining to. Growing on beaches or adjacent rocky cliffs

orba L. *orba*, orphan. The species is indigenous but was long assumed to be introduced

orbata L. *orbo*, deprive. Glumes reduced or absent

orbelica L. *-ica*, belonging to. From Orbelus the Classical name of a mountain on the border of Macedonia and Thrace, now Bulgaria

orbiculare L. *orbiculus*, round; *-are*, pertaining to. Spikelets circular in outline

orbiculat-a, -um L. *orbiculus*, round; *-ata*, possessing. – (1) Spikelets circular in outline. *Panicum orbiculatum*, *Paspalum orbiculatum* – (2) stem buds circular in outline. *Fargesia orbiculata*

orbignyana L. *-ana*, indicating connection. In honor of Alcide Dessalines d'Orbigny (1802–1857) French biologist

Orcuttia In honor of Charles Russell Orcutt (1864–1929) United States botanist

orcuttian-a, -us L. *-ana*, indicating connection. As for *Orcuttia*

ordubadense L. *-ense*, denoting origin. From Ordubad, Azerbaijan

oreades Gk *oreias*, belonging to mountains. Mountain species

oregonensis L. *-ensis*, denoting origin. From Oregon State, USA

oregonus From Oregon State, USA

Oreiostachys Gk *oreios*, mountain dweller; *stachys*, spike as of an ear of wheat. A mountain bamboo

orenuda L. *os*, mouth; *nuda*, base. Orifice of leaf-sheath lacking auricles

Oreobambos, Oreobambus Gk *oros*, mountain. Culms woody and growing on high mountains

oreoboloides Gk *-oides*, resembling. In habit resembling certain species of the sedge genus *Oreobolus*

Oreocalamus Gk *oros*, mountain; *kalamos*, reed. Reed-like mountain grasses

Oreochloa Gk *oros*, mountain; *chloa*, grass. Growing in alpine areas

oreodoxa Gk *oros*, mountain; *doxa*, splendour. The species adorning the high altitude pastures in which it grows

oreogena Gk *oros*, mountain; *genea*, birthplace. Mountain born, that is growing on mountains

oreophil-a, -um, -us Gk *oros*, mountain; *phileo*, love. – (1) Mountain species. *Agrostis oreophila, Chionochloa oreophila, Elymus oreophilus, Eragrostis oreophila, Festuca oreophila, Poa oreophila, Stipa oreophila, Tetrarrhena oreophila, Trikeraia oreophila, Trisetum oreophilum* – (2) growing only at high altitudes. *Rytidosperma oreophilum*

Oreopoa Gk *oreios*, mountain dweller; *poa*, grass. An alpine genus

oresbia Gk *oresbios*, mountain dweller. Alpine or subalpine species

oresigena Gk *oros*, mountain; *genea*, birthplace. Mountain species

orgyalis Gk *orgyia*, length of the outstretched arms; *-alis*, pertaining to. Culms about two metres tall

oricola L. *ora*, coast; *-cola*, dweller. Coastal species

oriental-e, -is L. *oriens*, the east; *-alis*, pertaining to. – (1) From the eastern Mediterranean through to Turkey. *Eremopyrum orientale, Chrysopogon orientalis, Hierochloe orientalis, Loliolum orientale, Nephelochloa orientalis, Pennisetum orientale, Rhaphis orientalis, Rhizocephalus orientalis, Secale orientale, Stipa orientalis, Triticum orientale* – (2) from Indo-China, now Cambodia, Laos and Vietnam. *Olyra orientalis* – (3) from Kamtchatka Peninsula, Russian Far East. *Glyceria orientalis*

orinocanum L. *-anum*, indicating connection. From Alto Orinoca, Venezuela

orinocense L. *-ense*, denoting origin. See *orinocanum*

orinosa Gk *oreinos*, mountain dweller. Growing at high altitudes

Orinus Gk *oreinos*, mountain dweller. Growing at high elevations in the Himalayas

orissae From Orissa State, India

Oriza See *Oryza*

orizabae From Pico de Orizaba, Mexico

orizabensis L. *-ensis*, denoting origin. – (1) From Orizaba Valley, Mexico. *Aristida orizabensis* – (2) from Pic d'Orizaba (Mount Orizaba), Mexico. *Festuca orizabensis, Poa orizabensis*

orizaeforme L. *forma*, appearance. Resembling *Oryza* in habit

ornans L. *orno*, adorn. Abundant in dry areas of the Kamtchatka Peninsula, Russian Far East

ornatum L. *orno*, adorn. Spikelets attractively colored

ornithocephala Gk *ornithos*, bird; *kephale*, head. See *Ornithocephalochloa*

Ornithocephalochloa Gk *ornithos*, bird; *kephale*, head; *chloa*, grass. The spikelets bear a fanciful resemblance to birds' heads

ornithopoda Gk *ornithos*, bird; *pous*, foot. Racemes often in threes and so disposed in inflorescence as to resemble a bird's foot

ornithopus Gk *ornithos*, bird; *pous*, foot. Inflorescence resembling a bird's foot

ornithorhyncha Gk *ornithos*, bird; *rhynchos*, beak. Spikelets shaped like the beak of a bird

Ornithospermum Gk *ornithos*, bird; *sperma*, seed. Grain grass with inflorescence resembling a bird's foot

oroana L. *-ana*, indicating connection. From El Oro, Bolivia

Oropetium Gk *oros*, mountain; *peta*, seek. Growing on high mountains

orophila Gk *oros*, mountain; *phileo*, love. Growing on high mountains

Orostachys Gk *oros*, mountain; *stachys*, spike as of an ear of wheat. Mountain species with a spicate panicle

Orrhopygium Gk *orros*, tail; *pygion*, bird. The tip of the lemma resembles a bird's tail
Ortachne See *Orthachne*
Orthachna See *Orthachne*
Orthachne Gk *orthos*, straight; *achne*, scale. Lemma narrow, not embracing palea
orthochaete Gk *orthos*, straight; *chaete*, bristle. Bristles of involucre rigid
orthochaetum Gk *orthos*, straight; *chaete*, bristle. Lemma awn straight or nearly so
Orthoclada Gk *orthos*, straight; *klados*, branch. The panicle arms are often reduced to racemes
orthoclada Gk *orthos*, straight; *klados*, branch. – (1) The panicle arms often reduced to racemes. *Eragrostis orthoclada* – (2) stolons and branched culms held stiffly erect. *Pogonarthria orthoclada*
orthonoton Gk *orthos*, straight; *noton*, back. Lemma keel not curved
orthophylla Gk *orthos*, straight; *phyllon*, leaf. Leaf-blades erect, stiff
Orthopogon Gk *orthos*, straight; *pogon*, beard. The glumes and sterile lemmas terminate in straight awns
Orthoraphium Gk *orthos*, straight; *rhaphis*, needle. Lemma terete with a rather stout loosely-twisted awn
orthos Gk *orthos*, straight. Culms erect
orthostachya Gk *orthos*, straight; *stachys*, spike as of an ear of wheat. Inflorescence a single raceme
Orthostachys, orthostachys Gk *orthos*, straight; *stachys*, spike as of an ear of wheat. The inflorescence is a spike
orthosticha Gk *orthos*, straight; *stichos*, row. Inflorescence an erect, contracted, narrow panicle
orurensis L. *-ensis*, denoting origin. From Oruro Department, Bolivia
Orysa See *Oryza*
Oryticum Hybrids between species of *Oryza* and *Triticum*
Oryza Classical Greek name for rice (*Oryza*), also spelled as *oryzon*
oryzetorum L. *oryzetum*, a rice field. Growing in rice fields or along their edges
oryzicola L. *-cola*, dweller. A weed of rice (*Oryza*) crops
Oryzidium Gk *-idium*, diminutive but here a name-forming suffix. Resembling *Oryza*
oryzinum L. *-inum*, indicating possession. A weed of rice (*Oryza*) crops
oryzoides Gk *-oides*, resembling. Resembling *Oryza*
Oryzopsis Gk *oryza*, rice; *opsis*, appearance. The grain resembles that of rice (*Oryza*)
osakae In honor of T. Osaka (fl. 1938) Japanese botanist
oscariana L. *-ana*, indicating connection. As for *Tovarochloa*
oschens-e, -is L. *-ense*, denoting origin. From Osh Kirgizya, Central Asia
oseana L. *-ana*, indicating connection. From Ozenuma, Oze Ushikubi and Ozegahara, Gunma Prefecture, Japan
oseroensis L. *-ensis*, denoting origin. From Osero now Ossero on the island of Cres, Yugoslavia
oshidensis L. *-ensis*, denoting origin. From Oshida, Rikuchu Province, now part of Iwate and Akita Prefectures, Japan
osikensis L. *-ensis*, denoting origin. From Oseka, Congo
osiridis In honor of Osiris, an Eygptian God. From Egypt
osmastonii In honor of Bertram Beresford Osmaston (1868–1961) English-born Indian forester
osoreyamensis L. *-ensis*, denoting origin. From Osoreyama, a mountain in Mutsu Province, Japan
osswaldii In honor of L. Osswald, schoolteacher at Nordhausen, Germany
osteniana L. *-ana*, indicating connection. In honor of Cornelius Osten (1863–1936) German-born Uruguayan botanist
Osterdamia In honor of Abrahamus Osterdam (fl. 1766) student of Carl Linnaeus (1707–1778)
ostrina L. purple. Anthers purple
otachyrioides Gk *-oides*, resembling. The stubby spikelets resemble those of *Otachyrium*

Otachyrium Gk *ous*, ear; *achyron*, chaff; *-ium*, indicating resemblance. At maturity the keels of the lower palea develop ear-like outgrowths

Otatea Latinized form of *otate* the Nahuatl word for several of the species

otaviensis L. *-ensis*, denoting origin. From Otavi, Angola

otayana L. *-ana*, indicating connection. In honor of Tasaku Otaya, Japanese botanist

oteroi In honor of Jorge Bamos de Otero (fl. 1935) Brazilian botanist

otisii In honor of Ira Clinton Otis (1861–1938) United States botanist

ottawensis L. *-ensis*, denoting origin. From Ottawa, Canada

Ottochloa In honor of Otto Stapf (1857–1933) Austrian-born English botanist

ottonis In honor of Friedrich Otto (1783–1856) German botanist and gardener

oubanguiensis L. *-ensis*, denoting origin. From Oubangui River, Central African Republic

ouonbiense L. *-ense*, denoting origin. From Ouonbi, Indo-China, now Cambodia, Laos and Vietnam

ourtana L. *-ana*, indicating connection. From l'Ourte now Ourthe, Belgium

ouwehandii In honor of Cornelius Dirk Ouwehand (1866–1951) Dutch-born Indonesian medical officer

oval-e, -is L. *ovum*, egg; *-ale*, pertaining to. Spikelets in outline the shape of an egg in longitudinal section

ovalifoli-a, -us L. *ovum*, egg; *-ale*, pertaining to; *folium*, leaf. Leaf-blades in outline the shape of an egg in longitudinal section

ovalis See *ovale*

ovat-a, -um, -us L. *ovum*, egg; *-ata*, possessing. Inflorescences, spikelets, or leaf-blades in outline the shape of an egg in longitudinal section

ovatoelliptica As for *ovata* together with *elliptica*. Leaf-blade ovate-elliptical

ovat-um, -us See *ovata*

ovczinnikovii In honor of Pavel Nikolaevich Ovchinnikov (1903–1975) Russian botanist

overi In honor of William Henry Over (1866–1956) English-born United States Museum Curator

ovin-a, -um (1) L. *ovis*, sheep; *-ina*, belonging to. A valuable species of sheep pastures. *Festuca ovina* – (2) L. *ovum*, egg; *-ina*, indicating resemblance. Spikelets egg-shaped. *Eragrostis ovina*, *Panicum ovinum*, *Poa ovina*, *Triscenia ovina*

oviniformis L. *forma*, appearance. The spikelets closely resemble those of *Festuca ovina*

ovuliferum L. *ovulum*, small egg; *fero*, carry or bear. Anthoecia egg-shaped

owarensis L. *-ensis*, denoting origin. See *owariensis*

owariensis L. *-ensis*, denoting origin. From Owari, Nigeria

owatarii In honor of Chûtarô Owatari (fl. 1892–1898) Japanese plant collector

oweniae In honor of Maria Louisa Owen (1825–1913) United States botanist

Oxyanthe, oxyanthum Gk *oxys*, sharp; *anthos*, flower. Spikelet oblong-lanceolate in outline

oxycephalum Gk *oxys*, sharp; *kephale*, head. Upper lemma terminating in a short mucro

Oxychloris Gk *oxys*, sharp. Like *Chloris* but the callus of the fertile floret long and needle-like

Oxydenia Gk *oxys*, sharp; *aden*, gland. The investing hairs exude an acid fluid

oxyglumis Gk *oxys*, sharp; L. *gluma*, husk. Glumes acute

oxylepis Gk *oxys*, sharp; *lepis*, scale. Glumes or lemmas tapering to a sharp point

oxyphloeus Gk *oxys*, sharp; *phleos*, an unidentified rush or reed. Lemma apex awned and so resembling the sharply tapering inflorescence bract of some rush (*Juncus*) species

oxyphyll-a, -um Gk *oxys*, sharp; *phyllon*, leaf. Tips of leaf-blades finely acuminate or pungent

Oxyrhachis Gk *oxys*, sharp; *rhachis*, backbone. The inflorescence axis is prolonged above the upper spikelet forming a narrow-lanceolate structure

Oxytenanthera Gk *oxytonos*, stretched to a point; *anthera*, bristle. Anthers long and narrow with connectives that extend into long apiculi

ozarkanus L. *-anus*, indicating connection. From the Ozark Mountains which straddle the border of Arkansas and Oklahoma States, USA

ozogonum Gk *oxos*, twig; *gony*, knee. Culms geniculate and prolifically branched from the lower nodes

P

paboan-um, -us L. *-anum*, indicating connection. In honor of Pabo (fl. 1848) who collected in European Russia

pabular-e, -is L. *pabulum*, fodder; *-are*, pertaining to. Considered a nutritious fodder for cattle

Pachea See *Pechea*

pachinensis L. *-ensis*, denoting origin. From Pachin, Taiwan

pachnodes Gk *pachys*, thick; L. *nodus*, knot. Nodes thick

pachyantha Gk *pachys*, thick; *anthos*, flower. Spikelets thick

pachyarthron Gk *pachys*, thick; *arthron*, joint. Rhachis internodes thickened

pachyclada Gk *pachys*, thick; *klados*, stem. Culms relative to their length thicker than those of related species

pachypholis Gk *pachys*, thick; *pholis*, scale. Glumes and lemmas thick

pachyphylla Gk *pachys*, thick; *phyllon*, leaf. Leaf-blades thick

pachypus Gk *pachys*, thick; *pous*, foot. Possibly a reference to a thickened pedicel

pachyrhizum, pachyrrhizum Gk *pachys*, thick; *rhiza*, root. Rhizome thick

pachystachya Gk *pachys*, thick; *stachys*, spike as of an ear of wheat. Racemes dense, somewhat stout

pachystachys Gk *pachys*, thick; *stachys*, spike as of an ear of wheat. Spike densely beset with spikelets and bristles

pacific-a, -um, -us L. of the Pacific Ocean. – (1) From the west coast of North America. *Bromus pacificus*, *Festuca pacifica* – (2) from the west coast of South America and the east coast of Australia. *Eriochloa pacifica* – (3) from islands in the Pacific. *Digitaria pacifica*, *Panicum pacificum*, *Zoysia pacifica* – (4) from east coast of the Russian Far East. ×*Leymotrigia pacifica*

pacuarense L. *-ense*, denoting origin. From Rio Pacuare, Costa Rica

Padia Vernacular name for cultivated rice in Malaya

pagophila Gk *pagos*, ice; rock; *phileo*, love. Growing at high altitudes in the Himalayas

pahangensis L. *-ensis*, denoting origin. From Pahang, Malaya

paianum L. *-anum*, indicating connection. In honor of Raghvendra Mukund Pai (1932–) Indian botanist

pairinii In honor of Datuk Pairin Kitingan (1940–) former Chief Minister of Sabah

paishanensis L. *-ensis*, denoting origin. From Paishan, Korea

palackyanum L. *-anum*, indicating connection. In honor of Johann Baptist Palacky (1830–1908) Bohemian botanist

palaeocolchicum Gk *palaios*, ancient; *-icum*, belonging to. From Colchis of the Ancients, that is from the Region south of the Caucasus and east of the Black Sea

Palaeoeriocoma Gk *palaios*, ancient. Fossil anthoecia resembling those of *Eriocoma*

Palaeophragmites Gk *palaios*, ancient. Fossils resembling *Phragmites*

Palaeopyrum Gk *palaios*, ancient; *pyros*, wheat. Fossils resembling *Triticum* or related genera

palaestin-a, -um From Palestine

palauense L. *-ense*, denoting origin. From Palau, Caroline Islands, Micronesia

palawanense L. *-ense*, denoting origin. From Palawan, one of the Philippine Islands

paleace-a, -um, -us L. *palea*, chaff; *-acea*, indicating resemblance. Glumes or lemmas papery

palenae From the Palena River, Chile

Paleoavena Gk *palaios*, ancient. Fossils resembling *Avena*
palibinii In honor of Ivan Vladimirovic Palibin (1872–1949) Russian botanist
palisotii In honor of Ambrose Marie Francis Joseph Palisot de Beauvois (1752–1820) French botanist and traveller
Pallasia In honor of Peter Simon Pallas (1741–1811) German physician and biologist
pallasii As for *Pallasia*
pallens L. pale. Inflorescences pale
pallescens L. *pallesco*, become pale. Losing color at maturity, especially spikelets
pallid-a, -um, -us L. pale-colored. Spikelets or grain pale-green
pallidefuscum See *pallidifusca*
pallidiflorus L. *pallidus*, pale-colored; *flos*, flower. Spikelets pale
pallidifolium L. *pallidus*, pale-colored; *folium*, leaf. Leaf-blades pale
pallidifusc-a, -um L. *pallidus*, pale-colored; *fusca*, dark. Anthoecia yellow-green with purple tips
pallidissimum L. *pallidus*, pale-colored; *-issimum*, most. Spikelets pale green
pallidiuscula L. *pallidius*, paler; *-ula*, diminutive. Spikelets somewhat pale
pallid-um, -us See *pallida*
palmaefolium See *palmifolia*
palmat-a, -um L. *palma*, hand; *-ata*, resembling. Leaves congested and their blades arranged like the fingers on a hand
palmerensis L. *-ensis*, denoting origin. From Palmer River, Alaska
palmeri In honor of Edward Palmer (1831–1911) United States botanist
palmifoli-a, -um L. *palma*, palm tree; *folium*, leaf. Leaf-blades longitudinally folded like those of certain palm trees
palmirensis L. *-ensis*, denoting origin. From Palmira, Ecuador
paludicola L. *palus*, swamp; *-cola*, dweller. Growing in swamps
paludificans L. *paludifer*, swamp making. Growing about swamps
paludigena L. *palus*, swamp; *gigno*, bring forth. Growing in swampy places
paludivagum L. *palus*, swamp; *vagum*, wandering. Runners much branched and often submerged
paludos-a, -um L. *palus*, swamp; *-osa*, abundance. Growing in swampy places
palustr-e, -is L. swampy place. Growing in swampy places
pamelae In honor of Pamela Nan Simon (1946–) South African born, Australian radiographer
pamiric-a, -um, -us L. *-ica*, belonging to. From the Pamir Mountains, on the border of Kyrgyzstan and Tajikistan
pamiroalaic-a, -um L. *-ica*, belonging to. From the Pamir Mountains and Alai Range, on the border of Kyrgyzstan and Tajikistan
pammelii In honor of Louis Hermann Pammel (1862–1931) United States botanist
pampagrandensis L. *-ensis*, denoting origin. From Pampagrande, Salta Province, Argentina
pampean-a, -um Quecha *pampas*. From the Pampas, that is grasslands of South America
pampinos-a, -um L. *pampinos*, vine leaf; *-osa*, abundance. Plants leafy as a consequence of freely branching from the base
pamplemoussense L. *-ense*, denoting origin. From Pamplemousses, Mauritius
panamens-e, -is L. *-ense*, denoting origin. From Panama
panasmitan-um, -us See *panormitana*
panchganiense L. *-ense*, denoting origin. From Panchangi, Bombay State, India
panciciana L. *-ana*, indicating connection. In honor of Josif Pancic (1814–1888) Serbian botanist
pand-a, -um L. bent. Pedicels flexuose
Paneion In honor of Pan, Greek God of shepherds and pastures, because many of the species are valuable fodder grasses
pangerangens-e, -is L. *-ense*, denoting origin. From Mt Pangerango, Java, Indonesia
Panicastrella Italian name for the *Setaria* section of *Panicum*
panice-a, -um, -us L. *-ea*, resembling. Resembling *Panicum*

paniciformis L. *forma*, appearance. Resembling *Panicum* with respect to the inflorescence

panicoides Gk *-oides*, resembling. Resembling *Panicum*, usually with respect to form of the inflorescence or spikelet

Panicularia L. *-ulus*, tending toward. *-aria*, pertaining to. Resembling *Panicum* but smaller

paniculat-a, -um, -us L. *paniculus*, panicle; *-ata*, indicating possession. Inflorescence a panicle

paniculifer-a, -um L. *paniculus*, panicle; *fero*, carry or bear. Inflorescence an open panicle

Paniculum L. *-ulum*, diminutive but here employed as a name-forming suffix

Panicum, panicum L. *panis*, bread; *-icum*, belonging to. In Roman times the vernacular name of the grass used for making bread

pankensis L. *-ensis*, denoting origin. From the banks of the Panke, a river in Hidaka Subprefecture, Hokkaido, Japan

pannonic-a, -us L. *-ica*, belonging to. From Pannonia, Hungary

pannosa L. *pannus*, piece of cloth; *-osa*, abundance. Leaf-blades densely covered with short hairs thereby resembling velvet

pannuceum L. *pannus*, piece of cloth; *-eum*, resembling. Upper glume and sterile lemma transversely wrinkled

panormitan-a, -um, -us L. *-ana*, indicating connection. From Panormus, now Palermo, Sicily

pans-a, -um L. extended. Inflorescence branches very long

Pantathera Gk *pante*, in every way; *ather*, *ather*, spike or ear of wheat. Grain similar to that of wheat (*Tricicum* species)

pantlingii In honor of Robert Pantling (1857–1910) British botanist, who later worked in Calcutta

pantrichum Gk *pas*, all; whole; *thrix*, hair. Plant softly papillose

panxianensis L. *-ensis*, denoting origin. From Panxian, Guizhan Province, China

paoliana L. *-ana*, indicating connection. In honor of Guido Paoli (1881–1947) Italian botanist who collected in Somaliland

paparistoi In honor of K. Paparisto (1914–1980) Albanian botanist

paphlagonica L. *-ica*, belonging to. From Paphlagonia, north-west Turkey

papilionacea L. *papilio*, butterfly; *-acea*, resembling. The broad lemmas resemble butterfly wings

papillatoides Gk *-oides*, resembling. Similar to *Bambusa papillatae*

papillipes L. *papilla*, nipple; *pes*, foot. Peduncles beset with spreading tubercle-based hairs

papillos-a, -um, -us L. *papilla*, nipple; *-osa*, abundance. With papillae on the glumes or lemmas

paposan-a, -us L. *-ana*, indicating connection. From Paposa, Chile

Pappagrostis Gk *pappos*, grandfather. The awn arises from within a crown of hairs at the apex of the bifid lemma. The spikelets otherwise resemble those of *Calamagrostis*

pappianum L. *-anum*, indicating connection. As for *pappii*

pappiferum Gk *pappos*, grandfather; L. *fero*, carry or bear. Lemma apices bearded, terminating in about thirteen slender awns

pappiform-e, -is Gk *pappos*, grandfather; *forme*, appearance. Upper part of the lemma hairy

pappii In honor of Agostino Pappi (fl. 1892–1934) sometime collector for the Herbarium at Massawa, Eritrea

pappinisseriense L. *-ense*, denoting origin. From Pappinisseri, India

pappophorea L. *-ea*, resembling. Similar to *Pappophorum*

pappophoroides Gk *-oides*, resembling. Resembling species of *Pappophorum* in the form of the lemma

Pappophorum, pappophorum Gk *pappos*, grandfather; *phero*, bear. Lemma invested in long hairs and so resembles an old man's beard

pappos-a, -um Gk *pappos*, grandfather. With long silky hairs. – (1) on the glumes. *Arundo papposa*, *Digitaria papposa*, *Panicum papposum* – (2) on the lemma. *Danthonia papposa* – (3) on the callus. *Agrostis papposa*, *Merxmuellera papposa* – (4) on the pedicel. *Stipa papposum* – (5) on the lemma awn. *Aristida papposa*

papuan-a, -um, -us L. *-ana*, indicating connection. From Papua, now Papua New Guinea

papyracea L. *papyrus*, paper; *-aceus*, indicating resemblance. Culm-sheaths with a papery margin

papyrascens L. *papyrasco*, become papery. Glumes papery

papyrifera L. *papyrus*, paper; *fero*, carry or bear. Useful for making paper

Parabambusa Gk *para*, near to. Similar to *Bambusa*

parabolicae L. *parabolicus*, comparison. Similar to *Lolium rigidum*

Paracolpodium Gk *para*, near to. Similar to *Colpodium*

paractaenoides Gk *-oides*, resembling. Resembling *Paractaenum*

Paractaenum, paractaenum Gk *paraktaomai*, get over and above. The axis of the inflorescence extends beyond the uppermost floret

Paractenium See *Paractaenum*

paradensis L. *-ensis*, denoting origin. From Parád, Hungary

paradisea Gk *paradeisos*, pleasure park. An oblique reference to the Garden of Eden, the species coming from the east, that is Aden in the Yemen. An alternate transliteration of Aden from the Arabic is Eden

paradisiac-a, -um L. *paradisus*, Christian Paradise. A dweller in Paradise, here taken from Valparaiso, the Chilean city which in Spanish means Valley of Paradise

paradox-a, -um, -us Gk *para*, irregular; *doxa*, opinion. Differing in some respect from related species

paraelegans Gk *para*, near to. Resembling *Sasa elegans*

Parafestuca Gk *para*, near to. Resembling *Festuca*

paraguaiensis L. *-ensis*, denoting origin. From Paraguay

paraguayana L. *-ana*, indicating connection. From Paraguay

paraguayens-e, -is L. *-ense*, denoting origin. From Paraguay

Parahyparrhenia Gk *para*, near to. Similar to *Hyparrhenia*

parallelum L. side by side. Leaf-blades held erect

paramilloensis L. *-ensis*, denoting origin. From Paramillo de Uspallata near Mendoza, Argentina

paramoensis L. *-ensis*, denoting origin. From Páram de las Cajas, Ecuador

paramushirensis L. *-ensis*, denoting origin. From Paramushir, Kuril'skye Ostrorava, Russian Federation

paranaens-e, -is L. *-ense*, denoting origin. – (1) From Rio Paraná, Argentina. *Axonopus paranaensis* – (2) from Paraná, Brazil. *Paspalum paranaense*

paranensis L. *-ensis*, denoting origin. From Paraná, Brazil

Paraneurachne Gk *para*, near to. Similar to *Neurachne*

paranjpyean-um, -us L. *-anum*, indicating connection. In honor of H. P. Paranjpye (fl. 1911) Indian botanist

parapaurochaetium Gk *para*, near to. Close to but not included in *Panicum* subgenus *Paurochaetium*

Parapholis Gk *para*, near to; *pholis*, scale of a snake. The glumes resemble the scales of a snake

paraquensis L. *-ensis*, denoting origin. From Parágue, Venezuela

Paratheria Gk *para*, near to; *ather*, barb or spine. The solitary spikelets are adpressed to the axis which projects beyond them as a bristle

paratunkensis L. *-ensis*, denoting origin. From Paratunka River, Kamchatka, Russian Far East

parceciliata, parciciliata L. *parcus*, frugal; *cilium*, hair; *-ata*, possessing. Spikelets and pedicels less hairy than those of related species

P

parciflor-a, -um L. *parcus*, frugal; *flos*, flower. – (1) Spikelets widely separated in inflorescence. *Panicum parciflorum* – (2) florets few per spikelet. *Festuca parciflora*

parcum L. frugal. Panicles with few spikelets

pardoana L. *-ana*, indicating connection. In honor of José Pardo y Sastron (1822–1909) Spanish botanist

pareisziana L. *-ana*, indicating connection. Probably in honor of Pareisz

P

Pariana L. *-ana*, indicating connection. Honoring the Paris, an Amerindian tribe living in Amazonia, Brazil

parilis L. like. Notwithstanding a lack of spikelets, assigned to a genus on the basis of its similar foliage

parinervium L. *par*, comparable; *nervus*, nerve. Leaf-blades filiform consisting of little more than the midrib

parishii In honor of – (1) Samuel Bonsall Parish (1838–1928) United States botanist. *Agropyron parishii, Aristida parishii, Puccinellia parishii, Stipa parishii* – (2) Charles Samuel Pollock Parish (1822–1897) cleric and Indian botanist. *Dendrocalamus parishii*

parisii In honor of Édouard Gabriel Paris (1827–1911) French soldier and botanist

parkeri In honor of Richard Neville Parker (1884–1958) British-born Indian Forester

parlatorei In honor of Filippo Parlatore (1816–1877) Italian botanist

parlatorii See *parlatorei*

parlatoris See *parlatorei*

parnassica L. *-ica*, belonging to. From Mt Parnassos, Greece

parnellii In honor of Richard Parnell (1810–1882) English-born physician and agrostologist

parodiana L. *-ana*, indicating connection. As for *Parodiochloa*

Parodiella L. *-ella*, diminutive but here employed as a name-forming suffix. See *Parodiochloa*

parodii As for *Parodiochloa*

Parodiochloa In honor of Lorenzo Raimondo Parodi (1895–1966) Argentinian agrostologist

Parodiolyra As for *Parodiochloa* but with *Olyra*

parontakensis Gk *para*, near to. Similar to *Sasa ontakensis*

parryi In honor of Charles Christopher Parry (1823–1890) United States botanist

parsana L. *-ana*, indicating connection. In honor of Ahmad Parsa (1907–) Iranian botanist

partimpatens L. *partim*, in part; *pateo*, lie open. Basal branches of the inflorescence divaricate

paruensis L. *-ensis*, denoting origin. From Rio Parú, Venezuela

parv-a, -um L. small. – (1) Of dwarf stature. *Danthoniopsis parva, Dichelachne parva, Hordeum parvum, Panicum parvum, Petrina parva* – (2) spikelets tiny. *Digitaria parva*

parviceps L. *parvus*, small; *caput*, head. Burrs small

parvicoccum L. *parvus*, small; Gk *kokkos*, kernel. Grains small

parviflor-a, -um, -us L. *parvus*, small; *flos*, flower. Spikelets small or with few florets

parvifoli-a, -um, -us L. *parvus*, small; *folium*, leaf. Leaf-blades small

parviglum-a, -e, -is L. *parvus*, small; *gluma*, husk. Glume or glumes short especially in comparison to lemmas

parvipaleata L. *parvus*, small; *-ata*, possessing. Florets with a small palea

parvipaniculat-a, -um L. *parvus*, small; *paniculus*, panicle; *-ata*, possessing. Panicle few-flowered

parviseta L. *parvus*, small; *seta*, bristle. Lemma shortly-awned

parvispic-a, -us L. *parvus*, small; *spica*, a point; hence, in particular, an ear or spike of grain. Inflorescence branches or spikelets small in comparison with those of related species

parvispicul-a, -um L. *parvus*, small; *spica*, a point; hence, in particular, an ear or spike of grain; *-ulum*, diminutive. Inflorescence or its components small

parvispiculata L. *parvus*, small; *spica*, spike or ear of grain; *-ula*, diminutive; *-ata*, possessing. Spikelets much smaller than those of related species

parvispicus See *parvispica*

parvissima L. *parvus*, small; *issima*, most. Culms very small

parvofolia L. *parvus*, small; *folium*, leaf. Leaf-blades small

Parvotrisetum L. *parvus*, small. Resembling *Trisetum* but small in some respect

parvul-a, -um L. *parvus*, small; *-ula*, diminutive. Dwarf in habit

parvum See *parva*

paryadrica L. *-ica*, belonging to. From Paryadres Ponti the Roman name for a series of mountain ranges in north-east Turkey

paschal-e, -is L. *pascha*, Passover; *-alis*, pertaining. From Easter Island

pascoeana L. *-ana*, indicating connection. In honor of Davis Pascoe (?–1989)

Pascopyrum L. *pascuum*, pasture; Gk *pyros*, wheat. Pasture species whose inflorescences resemble those of wheat

pascu-a, -us L. *pascuum*, pasture. Grassland species

Paspalanthium Gk *anthos*, flower. The spikelets resemble those of *Paspalum*

paspalidioides Gk *-oides*, resemblance. Inflorescence similar to that of *Paspalidium*

Paspalidium Gk *-idium*, a diminutive but here employed as a name-forming suffix

paspaliforme L. *forma*, appearance. Resembling *Paspalum*

paspalodes Gk *-odes*, resembling. Similar to *Paspalum*

paspaloides Gk *-oides*, resembling. Resembling *Paspalum*, usually with respect to the inflorescence

Paspalum, Paspalus Gk *paspalos*, a kind of millet mentioned by Classical authors

pasqualii In honor of Giuseppe Antonio Pasquale (1820–1893) Italian botanist

passa L. outstretched. Inflorescence branches curving

passourae From Campo de Passoura, near Kouran, French Guiana

pastoensis L. *-ensis*, denoting origin. From Pasto, Colombia

patagiata L. *patagium*, gold edging on a tunic; *-ata*, possessing. The sterile lemma bears a conspicuous transverse band of hairs midway along its length

patagonic-a, -um, -us L. *-ica*, belonging to. From Patagonia

patellaris L. *patella*, small dish; *-aris*, pertaining to. Small dishes are made from the culm internodes

Patellocalamus The type species of the genus is *Ampelocalamus patellaris*

patens L. *pateo*, lie open. Inflorescence or culm branches diverging at near right angles from the parent axis

patentiflora L. *pateo*, lie open; *flos*, flower. The spikelets are borne on flexuous spreading pedicels

patentifolium L. *pateo*, lie open; *folium*, leaf. Leaf-blades widely spreading

patentipilosa L. *pateo*, lie open; *pilum*, hair; *-osa*, abundance. The rhachis is invested in spreading hairs

patentissimum L. *pateo*, lie open; *-issimum*, most. Panicle branches spreading

patentivillosus L. *pateo*, lie open; *villi*, long weak hairs; *-osus*, abundance. Pedicels of the sterile florets and rachides of the inflorescence invested in long weak hairs

Patis Anagram of *Stipa*

patriae From the Rio Patria, Costa Rica

patris L. *pater*, father. In honor of Hyacinthe Julien Robert Vanderyst (1860–1934) Belgian cleric and botanist who wrote extensively on the grasses of Zaire

Patropyrum L. *pater*, father; Gk *pyros*, wheat. Species diploid and hence ancestral to cultivated wheat

pattersonii In honor of Harry Norton Patterson (1853–1919) United States botanist

patul-a, -um, -us L. outspread. Inflorescence an open panicle

patulifolia L. *patulus*, outspread; *folium*, leaf. Culm leaf-blades spreading

patul-um, -us See *patula*

patzkei In honor of Erwin Patzke (fl. 1964) German pharmacist and botanist

pauana L. *-ana*, indicating connection. In honor of Carlos Pau (1857–1937) Spanish botanist and pharmacist

pauciciliatum L. *paucus*, few; *cilium*, hair; *-atum*, possessing. Glabrous or the lower internodes puberulent

pauciflor-a, -um, -us L. *paucus*, few; *flos*, flower. With few florets per spikelet or few spikelets per inflorescence

paucifoli-a, -um L. *paucus*, few; *folium*, leaf. Culms bearing few leaves

paucinervis L. *paucus*, few; *nervus*, nerve. Lemma few-nerved

paucinod-e, -is L. *paucus*, few; *nodum*, knot. Culms one- or two-noded

paucipil-um, -us L. *paucus*, few; *pilus*, a single hair. Leaf-blades with few basal hairs

P

pauciracemosa L. *paucus*, few; *racemus*, stalk of a bunch of grapes; *-osa*, abundance. Inflorescence with few panicle branches

pauciramea L. *paucus*, few; *ramus*, branch. Inflorescence weakly branched

pauciseta L. *paucus*, few; *seta*, bristle. Bristles subtending the spikelets relatively few

paucisetosus L. *paucus*, few; *seta*, bristle; *-osa*, abundance. Rhachis with fewer hairs than related species

paucispicat-a, -um L. *paucus*, few; *spica*, a point; hence, in particular, an ear or spike of grain; *-ata*, possessing. Panicle with few branches

paucispiceus L. *paucus*, few; *spica*, a point; hence, in particular, an ear or spike of grain. *-eus*, resembling. Inflorescence with few branches

paucispicula L. *paucus*, few; *spica*, a point; hence, in particular, an ear or spike of grain; *-ula*, diminutive. Inflorescence of few spikelets

paucispina L. *paucus*, few; *spina*, spine. With few spines on the upper glume

paui As for *pauana*

paulsenii In honor of Ove Vilhelm Paulsen (1874–1947) Danish botanist and traveller

pauneroi In honor of Elena Paunero (1911–) Spanish botanist

pauper-a, -um L. poor. – (1) Culms diminutive. *Eragrostis paupera*, *Melinis paupera* – (2) bristles few in involucre. *Pennisetum pauperum*

paupercul-a, -us L. *pauper*, poor; *-ulus*, diminutive. Habit depauperate

pauperul-um, -us L. *pauper*, poor; *-ulus*, diminutive. Dwarf in habit

pauperum See *paupera*

pauzhetica L. *-ica*, belonging to. From Pauzhetka, Russian Far East

pavisii In honor of Pavis (fl. 1908) French botanist

pavlovii In honor of Nikolai Vasil'evich Pavlov (1893–1971) Russian botanist

pavonii In honor of José Antonio Pavón y Jiménez (1754–1844) Spanish botanist who collected in South America

pawlowskii In honor of Bogumil Pawlowski (1898–1971) Polish botanist

paytensis L. *-ensis*, denoting origin. From Payta, Peru

peacockii In honor of Robert William Peacock (c. 1869–1949) Australian agriculturalist

pearsonii In honor of Henry Harold Walsh Pearson (1870–1916) English-born South African botanist

Pechea In honor of Pierre André Pournet (1754–1818) French cleric and sometime Professor at Barcelona, Spain

peckii In honor of Morton Eaton Peck (1871–1951) United States botanist

peclardii In honor of Péclard who collected in Madagascar

pectinacea L. *pecten*, comb; *-acea*, indicating resemblance. The paleas are retained on the mature rhachilla giving it a comb-like appearance

Pectinaria L. *pecten*, comb; *-aria*, pertaining to. The margins of the lower glume bear a row of stiff hairs

pectinat-a, -um, -us L. *pecten*, comb; *-ata*, possessing. – (1) Inflorescence with spikes or racemes bearing a fanciful resemblance to a comb. *Agropyron pectinatum*, *Astrebla pectinata*, *Bromus pectinatus*, *Chloris pectinata*, *Muhlenbergia pectinata*, *Spartina pectinata*, *Triticum pectinatum* – (2) the margins of the sterile lemmas bear a series of stiff hairs. *Digitaria pectinata*, *Paspalum pectinatum* – (3) the mid-ribs of the glumes bear a series of stiff hairs. *Agrostis pectinata*, *Ischaemum pectinatum*, *Panicum pectinatum*, *Phalaris pectinata*

pectinell-a, -um, -us L. *pecten*, comb; *-ella*, diminutive. – (1) The inflorescence resembles a small comb. *Ctenopsis pectinella* – (2) the lemma bears a row of stiff hairs. *Panicum pectinellum, Sporobolus pectinella*
pectiniform-e, -is L. *pecten*, comb; *forma*, appearance. Inflorescence resembling a comb
pedalis L. *pes*, foot; *-alis*, pertaining to. Culms short
pedemontana From Pedemontanus, now Piedmont, Italy
pedersenii In honor of Troels Myndel Pedersen (1916–2000) Danish-born Argentinian botanist
pedicellar-e, -is L. *pes*, foot; *-ellus*, diminutive; *-are*, pertaining to. With at least some spikelets shortly stalked
pedicellat-a, -um, -us L. *pes*, foot; *-ellus*, diminutive; *-ata*, possessing. Spikelets shortly stalked
pedroensis L. *-ensis*, denoting origin. From Depto. San Pedro, Argentina
peduncular-e, -is L. *pedunculus*, small foot; *-are*, pertaining to. Inflorescence borne on a long leafless axis
pedunculat-a, -um, -us L. *pedunculus*, small foot; *-ata*, possessing. – (1) Inflorescence borne on a long leafless stalk. *Cenchrus pedunculata, Paspalum pedunculatum, Panicum pedunculatum* – (2) fascicles of spikelets borne on long peduncles. *Tragus pedunculatus*
peekelii In honor of Gerhard Peekel (1876–1949) German botanist
peguensis L. *-ensis*, denoting origin. From Pegu, Myanmar
peisonis From Peiso Lake, now Lake Neusiedler, Austria
pekinens-e, -is L. *-ense*, denoting origin. From Pekin, now Beijing, China
pekulnejensis L. *-ensis*, denoting origin. From Pekulnej, Siberia
peladoense L. *-ense*, denoting origin. From Cerro-Pelado, Paraguay
pelagica Gk *pelagos*, sea; *-ica*, belonging to. Growing on Pacific Islands
pelasgis In honor of the Pelasgoi, the oldest inhabitants of Greece
pelligera L. *pellis*, skin; *gero*, carry or bear. Spikelets densely hairy
pelliotii In honor of Paul Pelliot (1878–1945) French sinologist and explorer who, in company with Louis Vaillant, a medical doctor, collected c. 800 plant specimens from central Asia and China
pellit-a, -um, -us L. covered with skins. The fertile floret(s) is protected by hairy investing glumes or sterile florets
pellitoides Gk *-oides*, resembling. Similar to *Panicum pellitum*
pellit-um, -us See *pellita*
pellucid-a, -us L. translucent but not hyaline. Glumes transparent
pellytronis Gk *pellutra*, a sock or ankle bandage. The woolly leaf-sheath resembles a sock
peloponnesiaca L. a Peloponnesian. From Peloponnese, Greece
peltieri In honor of Peltier who collected in North Africa
Peltophora See *Peltophorus*
Peltophorus Gk *pelto*, shield; *phero*, bear. The lower glume in the sessile spikelet is leathery and transversely rugose, resembling a shield
pencanum L. *-anum*, indicating connection. From Penco, a valley near Chiguayante, Chile
pendul-a, -us L. *pendulus*, hanging down. Spikelets or inflorescence branches pendant
pendulin-a, -um, -us L. *pendeo*, hang down; *-ula*, tending towards; *-ina*, indicating resemblance. Panicle branches drooping
pendulosus L. *pendulus*, hanging downwards; *-osus*, abundance. Spikes distinctly drooping
pendulus See *pendula*
Penicellaria See *Penicillaria*
Penicillaria L. *penicillus*, a small brush; *-aria*, pertaining to. The anther tips bear a tuft of hairs
penicillat-a, -um L. *penicillus*, a small brush; *-ata*, possessing. – (1) With tufts of long hairs on the lemmas. *Arundo penicillata, Chaetochloa penicillata, Danthonia penicillata, Deschampsia penicillata, Mesosetum penicillatum* – (2) with tufts of hairs at the summit of the pedicel. *Schizachyrium penicillatum* – (3) with racemes arising in verticils. *Avena penicillatum, Panicum penicillatum, Paspalum penicillatum* – (4) with long involucral bristles. *Setaria penicillata*

penicilliger-a, -um L. *penicillus*, small brush; *gero*, carry or bear. Upper glume and lower lemma invested with long hairs

Peniculus L. little brush. The stiff hairs on the glumes and sterile lemmas give them the appearance of tiny brushes

peninsulae L. *paeninsula*, peninsula. From the Malay Peninsula

peninsulanum L. *-arum*, indicating connection. From peninsula India

P

peninsularis L. *-aris*, pertaining to. – (1) From peninsula India. *Agrostis peninsularis* – (2) from Baja California, Mexico. *Aristida peninsularis*

pennat-a, -um L. *penna*, feather; *-ata*, possessing. – (1) Awns villous. *Aristida pennata, Arthratherum pennatum, Stipa pennata, Stipagrostis pennata* – (2) pedicels villous. *Loudetia pennata, Trichopteryx pennata* – (3) peduncles villous. *Digitaria pennata, Panicum pennatum, Paspalum pennatum*

pennei In honor of Penne (pre 1908)

pennellii In honor of Francis Whittier Pennell (1886–1952) United States botanist

pennisetiformis L. *forma*, resemblance. Similar to *Pennisetum*

Pennisetum, pennisetum L. *penna*, feather; *seta*, bristle. The spikelets of most species of the genus are subtended by plumose bristles

pennsylvanic-a, -um L. *-ica*, belonging to. From Pennsylvania, USA. In the 17th and 18th centuries the spelling was Pensylvania

pensylvanic-a, -um See *pennsylvanica*

Pentacraspedon, Pentacrospedon Gk *penta*, five; *kraspedon*, fringe. The lemma apex is five toothed with each tooth bearing hairs

Pentameres See *Pentameris*

Pentameris Gk *penta*, five; *meros*, part. Lemma five-awned

Pentapogon Gk *penta*, five; *pogon*, beard. Lemma five-awned

pentapogonodes Gk *-odes*, resembling. Similar to *Pentapogon*

pentapolitana L. *-ana*, indicating connection. From Pentapolis, a region of Cyrenaica, Tripoli

Pentaraphis See *Pentarrhaphis*

Pentarhaphis See *Pentarrhaphis*

Pentarraphis See *Pentarrhaphis*

Pentarrhaphis Gk *penta*, five; *rhaphis*, needle. The glumes of the pair of spikelets forming the spike are reduced to awns. These together with the prolonged rhachis give the impression the spikelets are subtended by five awns

Pentaschistis Gk *penta*, five; *schistos*, split. Lemmas five-awned

Pentastachya, pentastachyum Gk *penta*, five; *stachys*, spike as of an ear of wheat. Spikelets or inflorescence branches develop in groups of five

Pentatherum Gk *penta*, five; *ather*, barb or spine. The lemmas of some species have five short awns

pentzii In honor of James Alexander Pentz (1896–1967) South African plant ecologist

penzesii In honor of Antal Pénzes (1895–1984) Hungarian botanist

pepeopaeense L. *-ense*, denoting origin. From Pepeopae on Molokai, one of the Hawaiian Islands

perakense L. *-ense*, denoting origin. From Perak, Malayasia

perangustatum L. *per*, very; *angustus*, narrow; *-atum*, possessing. Leaf-blades very narrow

perarta L. *per*, very; *arta*, confined. Inflorescence contracted

perbella L. *per*, very; *bella*, beautiful. Of attractive appearance

perberbis L. *per*, very; *berbe*, beard. The involucral bristles are numerous and more or less hide the spikelet

percivalianum L. *-anum*, indicating connection. In honor of John Percival (1863–1949) English agriculturalist

perconcinna L. *per*, very. The prefix has been employed to conserve a well established name which would otherwise be a later homonym to a relatively unknown species

perdensum L. *per*, very; *densum*, dense. Panicle branches densely floriferous

perdignus L. *per*, very; *dignus*, worthy. Worthy of recognition on several counts and especially the height of the culms

peregrin-a, -um L. foreigner. Country of origin not known with certainty

Pereilema Gk *per*, all round; *eilema*, covering. Spikelets subtended by a cluster of bristles

perennans L. *perenno*, persist for several years. Perennials

perenn-e, -is L. persisting for several years. Perennials

perexuguoseta L. *perexiguus*, very small; *seta*, bristle. Nodes shortly pubescent

perfecta L. complete. Pedicelled spikelet bisexual

perfoliatum L. *per*, through; *folium*, leaf; *-atum*, possessing. Leaf-blades amplexicaul

perforat-a, -um, -us L. *perforo*, bore into or through. – (1) Rhachis with a series of depressions in which the spikelets are partially enclosed. *Rottboellia perforata* – (2) lower glume with a small pit on its lower surface. *Andropogon perforatus* – (3) palea and lemma at maturity gape and expose the enclosed grain. *Panicum perforatum*

perfossus L. *per*, very; *fossa*, ditch. Lower glume with a conspicuous pit

pergracil-e, -is L. *per*, very; *gracile*, delicate. Culms slender

periantha See *eriantha*

Periballia Gk *peri-*, about; *ballo*, dance. The hygroscopic awn responding to changes in humidity is in constant movement

Peridictyon Gk *peri-*, near to; *dictyon*, net. The weathered basal leaf-sheaths are net-like

Perieilema See *Pereilema*

perinconspicua L. *per*, very. Prefix added to *inconspicua* thereby avoiding the formation of a homonym

perinvolucratus L. *per*, very; *involucrum*, sheath; *-atus*, possessing. Involucre better developed than in related species

peristerea Gk *-ea*, belonging to. From Mt Peristeri, Macedonia

peristypum Gk *peri-*, surrounding; *stypos*, stipe or stem. Leaf-blades amplexicaule

Perlaria French *perle*, bead; L. *-aria*, pertaining to. Meaning uncertain, origin not given by author, but possibly a reference to the swollen bead-like spikelets

perlax-a, -um L. *per*, very; *laxa*, weak. Culms decumbent and creeping

perligulat-a, -us L. *per*, very; *ligulus*, tongue; *-atus*, possessing. Ligule prominent

perlong-a, -um, -us L. *per*, very; *longa*, long. – (1) Leaf-blades long. *Fargesia perlonga* – (2) panicle on a long stalk. *Axonopus perlongus*, *Panicum perlongum*

permollis L. *per*, very; *mollis*, soft. Foliage densely covered with short soft hairs

pernambucens-e, -is L. *-ense*, denoting origin. From Pernambuco, Brazil

pernervosum L. *per*, very; *nervus*, nerve; *-osum*, abundance. Glumes and sterile lemma conspicuously nerved

perniciosa L. *per*, very; *noxius*, harmful; *-osa*, abundance. Callus very sharp and readily entangling in wool and clothing

Perobachne Gk *peros*, maimed; *achne*, scale. The lemmas are unawned unlike those of the related *Themeda*

peroninii In honor of A. Péronin (fl. 1872) who collected in Turkey

Perostis See *Perotis*

perotensis L. *-ensis*, denoting origin. Growing on the slopes of Volcán Cofre de Perote, Mexico

Perotis Gk *peros*, mutilated; *ous*, ear. The lemma is awnless

perplex-a, -um L. *per*, very; *plecto*, plait. Possesing the characters of two related species

perpusill-a, -um L. very small. Dwarf annual

Perrierbambus In honor of Joseph Marie Henri Alfred Perrier de la Bâthie (1872–1958) French botanist

perrieri See *Perrierbambus*

perrottetii In honor of Georges Samuel Perrottet (1793–1870) Swiss botanist

persarum L. *Persae*, the Persians. Of the Persians, now Iranians

perscabra L. *per*, very; *scaber*, rough. Plant scabrid

persic-a, -um, -us L. *-ica*, belonging to. From Persia, now Iran

persimilis L. *per*, very; *similis*, like. Closely related to another species

persistentia L. *persisto*, persist. The spikelets are not deciduous at maturity

personata L. *masked*, that *is*, resembling something else. Readily confused with another species

perspeciosum L. *per*, very; *speciosum*, showy. Inflorescence attractive

perspicinervium L. *per*, very; *spica*, a point; hence, in particular, an ear or spike of grain; *nervus*, nerve. Upper glume of spikelet conspicuously three-nerved

pertenu-e, -is L. *per*, very; *tenuis*, thin. Inflorescence a contracted panicle or spike

pertus-a, -um, -us L. with a pit. The lower glume has a deep pit in its lower surface

Perulifera L. *perula*, a small sac; *fero*, carry or bear. Lemmas and paleas of the hermaphrodite floret are leathery and form a sac about the grain

peruvian-a, -um, -us L. *-ana*, indicating connection. From Peru

pervariabilis L. *per*, very; *variabilis*, variable. Species variable

pes-avis L. *pes*, foot; *avis*, bird. The panicle resembles a bird's foot

peschkovae In honor of Galina A. Peshkova (1930–) Russian botanist

petelotii In honor of Paul Alfred Pételot (1885–?) French entomologist and plant collector in Indo-China, now Cambodia, Laos and Vietnam

peteri In honor of Gustav Albert Peter (1853–1937) German botanist

petersonii In honor of – (1) F. J. Peterson (fl. 1923) resident of Cuba. *Panicum petersonii* – (2) H. Peterson (fl. 1904) who collected in British Columbia. *Elymus petersonii* – (3) Paul M. Peterson (1954–) United States agrostologist. *Aristida petersonii, Festuca petersonii*

petilum L. slender. Culms very slender

petiolar-e, -is L. *petiolus*, little leg; *-are*, pertaining to. With a pseudopetiole between the leaf-blade and leaf-sheath

petiolat-a, -um, -us L. *petiolus*, little leg; *-ata*, possessing. Leaf-blades with a pseudopetiole

petitian-a, -us In honor of Antoine Petit (?–1843) French physician and zoologist who collected in Ethiopia

petiveri In honor of James Petiver (1663/4–1718) English pharmacist and naturalist

petrae-a, -um L. *petra*, rock; *-ea*, pertaining to. Growing in rocky places

petrens-e, -is L. *petra*, rock; *-ense*, denoting origin. Growing amongst exposed rocks

petriei In honor of Donald Petrie (1846–1925) Scots-born New Zealand educationalist and amateur botanist

Petriella L. *-ella*, diminutive but here used as a name-forming suffix. See *petriei*

Petrina Gk *petros*, rock; *-ina*, indicating possession. Growing in rock crevices

petrophila Gk *petros*, rock; *phileo*, love. Growing amongst rocks

petropolitanum L. *-anum*, indicating connection. From Mun. Petrópolis, Brazil

petros-a, -um L. *petra*, rock; *-osa*, full of. Growing in rocky places

petschorica L. *-ica*, belonging to. From Petschora Bay, Arctic Russia

Peyritschia In honor of Johann Joseph Peyritsch (1835–1889) Austrian botanist

pfisteri In honor of Augusto Pfister (fl. 1941–1943) who collected in Chile

pflanzii In honor of Karl Pflanz (1872–1925) German botanist

Phacellaria Gk *phakelos*, bundle; L. *-aria*, pertaining to. Racemes permanently adpressed and their rhachises bound together by interlocking hairs

phacellophora Gk *phakelos*, bundle; *phero*, bear. Branches arising in fascicles

Phacelurus Gk *phakelos*, bundle; *oura*, tail. Inflorescence subdigitate of more or less flattened racemes

phaeantha Gk *phaeos*, grey; *anthos*, flower. Spikelets dark-olive to black

Phaenanthoecium Gk *phaenestai*, becoming apparent; *anthos*, flower; *oikos*, house. Florets visible because lemmas exceed the glumes

Phaenosperma Gk *phaeinos*, shining; *sperma*, seed. The mature grain projects beyond the palea and lemma

phaenostachys Gk *phaeinos*, shining; *stachys*, spike as of an ear of wheat. Inflorescence a short shiny raceme resembling an ear of wheat

phaeocarp-a, -um Gk *phaeos*, grey; *karpos*, fruit. Anthoecia grey
phaeothrix, phaeotrix Gk *phaeos*, grey; *thrix*, hair. Spikelets invested with grey hairs
phaeotricha Gk *phaeos*, grey; *thrix*, hair. The glumes and sterile lemmas are invested with copious grey hairs
phaeotrix See *phaeothrix*
Phalarella L. *-ella*, diminutive but here used as a name-forming suffix. Resembling *Phalaris*
Phalaridantha Gk *phalaros*, coot; *anthos*, flower. At maturity the shiny white palea and lemma investing the grain resemble the white frontal-shield on the head of a coot (*Fulica atra*)
Phalaridium Gk *-idium*, diminutive but here used as a name-forming suffix. Resembling *Phalaris*
Phalaris Gk *phalaros*, coot. The phalaris of the Greeks was a grain enclosed in white scales thereby resembling the white frontal-shield on the head of a coot (*Fulica atra*)
phalaroides Gk *-oides*, resembling. Similar to *Phalaris*, usually with respect to the form of the inflorescence
Phalaroides Gk *-oides*, resembling. The spikelets resemble those of *Phalaris*
phalerata L. *phalerae*, a metal ornament worn on the breast; *-ata*, possessing. Leaf-blades, shield-like
Phalona See *Falona*, for which *Phalona* is a more conventional transliteration of the Greek from which the name is derived
phanerococca Gk *phaneros*, exposed; *kokkos*, kernel. The anthoecium is clearly visible at maturity
phaneroneuron Gk *phaneros*, exposed; *neuron*, nerve. Lemma when dry prominently nerved
Phanopyrum Gk *phanos*, bright; *pyros*, wheat. Anthoecium glossy
phar Vernacular name of the species in Lushai Hills, India
phariana L. *-ana*, indicating connection. From Phari, Tibet Autonomous Region, China
Pharus Gk *pharos*, sheet. In Jamaica the large leaves were used in former times to make clothes
Pheidochloa Gk *pheidos*, sparse; *chloa*, grass. Inflorescence weakly developed, with only three or four spikelets each with only two florets and two stamens
philadelphic-um, -us L. *-icum*, belonging to. From Philadelphia but not necessarily indigenous to that place, in that the specimens may have derived from a herbarium in that city
philippian-a, -um L. *-ana*, indicating connection. As for *philippii*
philippic-a, -um L. *-ica*, belonging to. From Philippine Islands
philippii In honor of Rudolf Amandus Philippi (1808–1904) German-born Chilean biologist and museum director
philippinensis L. *-ensis*, denoting origin. From the Philippines
philistaea L. from Philistea now mostly Israel
phillipsiana L. *-ana*, indicating connection. In honor of Mary Elizabeth Philipps (1917–1976) Australian botanist. The epithet is therefore misspelt
phillipsii In honor of Edwin Perez Phillips (1884–1967) South African botanist
Phippsia In honor of Constantine John Phipps (1744–1792) English-born Arctic explorer
phippsii In honor of James Bird Phipps (1934–) Canadian botanist
Phipsia See *Phippsia*
phleiforme L. *forma*, appearance. Inflorescence similar to *Phleum*
phleoides Gk *-oides*, resembling. Inflorescence a spike-like panicle. See *Phleum*
Phleum Gk *phleos*, a Classical Greek name for an unidentified marsh reed
phoenicia L. scarlet. Spikelets scarlet
phoenicoides Gk *-oides*, resembling. See *Phoenix*
phoenix A reference to the Phoenix, a mythical bird which sets itself alight and is then reborn from the ashes
phoiniclados Gk *phoinix*, purple-red; *klados*, stem. Culms and sheaths purple in color

pholiuroides Gk *-oides*, resembling. The inflorescence resembles that of *Pholiurus*
Pholiurus Gk *pholis*, scale of a snake; *oura*, tail. The inflorescence is a narrow cylindrical spike
phonoliticum L. *-icum*, belonging to. Growing on phonolite, a volcanic larva
phragmites Resembling *Phragmites* in habit
Phragmites, Phragmitis Gk *phragma*, a hedge; *-ites*, resembling. Name used by Dioscorides for a species whose stems were used for making hedges
Phragmitis See *Phragmites*
phragmitoides Gk *-oides*, resembling. Similar to *Phragmites* in habit
phryganodes Gk *phryganon*, dry stick; *-odes*, resemblance. Culms thin and leaf-blades short
phrygius From Phrygia, a region of western Turkey
phyllacantha Gk *phyllon*, leaf; akanthos, prickly plant. Leaves pungent
phyllanthum Gk *phyllon*, leaf; *anthos*, flower. Some panicle branches subtended by leafy bracts
phyllomacr-a, -um Gk *phyllon*, leaf; *makros*, large. Leaf-blades large
phyllophorachis Gk *phyllon*, leaf; *phero*, bear; *rhachis*, backbone. Origin not given by author
phyllopoda Gk *phyllon*, leaf; *pous*, foot. Pedicel of longer raceme produced into an ovate auricle
phyllopogon Gk *phyllon*, leaf; *pogon*, beard. Leaves densely pilose
Phyllorachis Gk *phyllon*, leaf; *rhachis*, backbone. The inflorescence is invested by a leafy bract
phyllorhachis Gk *phyllon*, leaf; *rhachis*, backbone. Rhachis winged
phylloryzoides Gk *phyllon*, leaf; *-oides*, resembling. Leaf-blades like those of *Oryza sativa*
Phyllostachys Gk *phyllon*, leaf; *stachys*, spike as of an ear of wheat. The lemmas of the spikelets have well developed blades
phyllotrichus Gk *phyllon*, leaf; *thrix*, hair. Leaf-blades long and thin
phymatonodosa Gk *phyma*, tumour; L. *nodus*, knot; L. *-osa*, abundance. Nodes swollen
piauiense L. *-ense*, denoting origin. From Piaui State, Brazil
picbaueri In honor of Richard Picbauer (1886–1955) Moravian botanist
piccae In honor of Pablo Picca (fl. 1999) Argentinian naturalist
pichinchae From Pichincha, Ecuador
pichleri In honor of Thomas Pichler (1828–1903)
pickeringii In honor of Charles Pickering (1805–1878) United States botanist
picoeuropeana L. *-ana*, indicating connection. From Picos de Europa
pict-a, -um, -us L. painted. – (1) Glumes and/or lemmas with conspicuous pigmented veins or margins. *Bothriochloa picta*, *Bromus pictus*, *Chusquea picta*, *Danthonia picta*, *Deyeuxia picta*, *Melica picta*, *Panicum pictum*, *Paspalum pictum*, *Rytidosperma picta*, *Suardia picta* – (2) culms with variegated leaves. *Phalaris picta*
pict-um, -us See *picta*
picturata L. *pictura*, painting; *-ata*, possessing. Spikelets green and variegated with purple
pictus See *picta*
piercei In honor of Edwin Pierce (fl. 1880s) who collected in Baluchistan
pierreana L. *-ana*, indicating connection. In honor of Pierre, who collected in Vietnam
pietrosii From Mt Pietrosii, Galicia, Spain
piettei In honor of Mme. Henri Fischer née Pietté
piettieri See *pittieri*
pignattii In honor of Alessandro Pignatti (1930–) Italian botanist
pignattiorum In honor of Alessandro Pignatti (1930–) and Erika Pignatti, Italian botanists
piifontii As for *fontqueri*
pilar-franceii In honor of Pilar Franco Rosseli (fl. 1993) Colombian botanist
pilata L. *pilus*, a hair; *-ata*, possessing. – (1) With long hairs on the leaf-blades or spikelets. *Panicum pilata*, *Stipa pilata* – (2) leaf-blades hair-like. *Poa pilata*

pilatii In honor of Albert Pilát (1903–1974) Czech mycologist
pilaxilis L. *pilus*, a hair; *axilis*, arm-pit. Lower axils of inflorescence densely hairy
pilcomayens-e, -is L. *-ense*, denoting origin. From Pilcomayo River, Gran Chaco, Paraguay
pilgeri As for *Pilgerochloa*
pilgerian-a, -um, -us L. *-ana*, indicating connection. As for *Pilgerochloa*
Pilgerochloa Gk *chloa*, grass. In honor of Robert Knuds Friedrich Pilger (1876–1953) German agrostologist, born on Helgoland before its transfer from British to German administration
pilifer-a, -um, -us L. *pilus*, a hair; *fero*, carry or bear. Hairy in some respect
piligens See *piligera*
piliger-a, -um L. *pilus*, a hair; *gero*, carry or bear. Hairy in some respect usually of the spikelet
pilipes L. *pilus*, a hair; *pes*, foot. – (1) Pedicels slender. *Panicum pilipes*, *Poa pilipes* – (2) peduncles densely pubescent. *Arthraxon pilipes*
pilisparsum L. *pilus*, a hair; *sparsum*, sparse. Inflorescence branches sparsely hairy
pilos-a, -um, -us L. *pilus*, a hair; *-osa*, abundance. The whole plant or any of its organs invested with long spreading hairs
pilosell-a, -us L. *pilus*, a hair; *-osa*, abundance. *-ella*, diminutive. Leaf-blades finely hirsute
pilosilemma L. *pilus*, a hair; *-osa*, abundance; Gk *lemma*, husk. Lemma densely hairy
pilosissim-a, -um L. *pilus*, a hair; *-osa*, abundance; *-issima*, abundantly. In whole or in part densely covered with long hairs
pilosiuscula L. *pilosius*, more hairy; *-ulus*, diminutive. Leaf-blades with abundant short hairs
pilosomarginatus L. *pilus*, a hair; *-osa*, abundance; *marginus*, edge; *-atus*, possessing. Margins of leaf-sheath hairy
pilosovaginatus L. *pilus*, a hair; *vagina*, sheath; *-ata*, possessing. Leaf-sheath densely hairy
pilosula L. *pilus*, a hair; *-osa*, abundance. *-ula*, diminutive
pilos-um, -us See *pilosa*
pilulifer-a, -um L. *pilula*, small pill; *fero*, carry or bear. Spikelets globose
pinalenoensis L. *-ensis*, denoting origin. From Pinaleno Mountains, Arizona, USA
pindic-a, -us L. *-ica*, belonging to. From Pindhes, Greece
pinegensis L. *-ensis*, denoting origin. From Pinega district, Archangelsk province, Russian Federation
pineti L. *pinetum*, pine-grove. Growing in pine (*Pinus*) woodlands
pinetorum L. *pinetum*, pine grove. Of, that is, growing in pine woods
Pinga Vernacular name for the type species in Manokwari, Papua, Indonesia
pingshanensis L. *-ensis*, denoting origin. From Pingshan, Sichuan Province, China
pinguipes L. *pinguis*, fatty; *pes*, foot. Pedicels clavate and glossy on the back
pinifoli-a, -um, -us L. *folium*, leaf. The leaf-blades resemble those of *Pinus*
pinnat-um, -us L. *pinna*, feather; *-atum*, possessing. Spikelets long sessile arising alternately from a central axis
piovanii, piovanoi In honor of Giovanni Piovano (fl. 1953) collector of the type
piperi In honor of Charles Vancouver Piper (1867–1926) United States agrostologist
Piptatherum, -um, -us Gk *pipto*, fall down; *ather*, barb or spine. The awns drop readily from their lemma
Piptochaetium Gk *pipto*, fall down; *chaete*, bristle. Awn is deciduous
Piptophyllum Gk *pipto*, fall down; *phyllon*, leaf. The leaf-blades disarticulate from their sheaths at maturity
piptopilum Gk *pipto*, fall down; L. *pilus*, a hair
Piptostachya, piptostachya Gk *pipto*, fall down; *stachys*, spike as of an ear of wheat. Spikelets deciduous
piptostachys See *Piptostachya*
Piresia In honor of Jonas Murça Pires (1917–) who collected in Brazil
Piresiella L. *-ella*, diminutive but here a name-forming suffix. Resembling *Piresia*
pirifer-a, -um L. *pyrus*, pear; *fero*, carry or bear. Spikelets pear-shaped

pirineosense L. *-ense*, denoting origin. From Pirineos, Brazil

pirinica L. *-ica*, belonging to. From Mt Pirin, Bulgaria

pirottae In honor of Pietro Romualdo Pirotta (1853–1936) Italian botanist

piscaporum L. *piscis*, fish; *capio*, capture. Culms used for making fishing rods

pishanic-a, -us L. *-ica*, belonging to. From Pishan, Xinjiang Uyghur Autonomous Region, China

P

pisidica L. *-ica*, belonging to. From Pisidia, Turkey

pisinn-a, -um L. little. Tiny in comparison with related species

pitardiana L. *-ana*, indicating connection. In honor of Charles-Joseph Marie Pitard (1873–1927) French botanist

pitensis L. *-ensis*, denoting origin. From Pita River, Ecuador

Pithecurus Gk *pithekos*, ape; *oura*, tail. Inflorescence a single narrow raceme reminiscent of the tail of an ape

pithogastrus Gk *pithos*, large earthenware wine-jar; *gaster*, belly. At maturity the lemma is conspicuously swollen

pithopus Gk *pithos*, large earthenware wine-jar; *pous*, foot. Lower internode of rhachilla assume the form of a cylindrical cup

pittieri In honor of Henry François de Fábrega Pittier (1857–1950) Swiss botanist and civil engineer

piurensis L. *-ensis*, denoting origin. From Piura Region, Peru

Plagiantha, plagianthum Gk *plagios*, placed sideways; *anthos*, flower. The rhachilla is almost at right angles to its pedicel

Plagiarthron Gk *plagios*, sloping; *arthron*, joint. Apices of disarticulating inflorescence branches sloping

Plagiochloa Gk *plagios*, placed sideways; *chloa*, grass. The spikelets are placed obliquely to the rhachis or central axis

Plagiolytrum Gk *plagios*, placed sideways; *elytron*, cover. The inflorescence is a one-sided spike

plagiopogon Gk *plagios*, placed sideways; *pogon*, beard. The column of the awn is unilaterally plumose

plagiopus Gk *plagios*, placed sideways; *pous*, foot. Successive spikelet clusters occur in different planes

Plagiosetum Gk *plagios*, placed sideways; L. *seta*, bristle. The spikelets are subtended by a pair of branches reduced to sterile bristles. In addition each spikelet is subtended by bristles

plana L. flat. Spikelets strongly compressed

planaltina Portuguese *planalto*, plateau. Growing on plateaux in southern Brazil

Planichloa L. *planus*, flat; Gk *chloa*, grass. Spikelets distinctly compressed

planiculm-e, -is L. *planus*, flat; *culmus*, stem. Culms distinctly compressed

planifoli-a, -um, -us L. *planus*, flat; *folium*, leaf. Leaf-blades flat

planipedicellatum L. *planus*, broad; *pedicellus*, short stalk; *-atus*, possessing. Pedicels oblanceolate

Planotia An anagram of *Platonia*

planotis Gk *planos*, flat; *ous*, ear. Upper lemma with two flat basal wings

plantagine-a, -um L. *-inea*, close resemblance. Resembling *Plantago* with respect to the inflorescence

Plantinia In honor of Christophe Plantin (c. 1514–1589) French publisher and naturalist

platatherus Gk *platys*, flat; *ather*, spike or ear of wheat. Inflorescence a compressed spike

platecaul-e, -is Gk *platys*, flat; *kaulos*, stem. Culms compressed

platens-e, -is L. *-ense*, denoting origin. From La Plata Province, Argentina, or places in Uruguay and Argentina close to the Rio de la Plate

platicaulis Gk *platys*, flat; *kaulos*, stem. Culms compressed

Platonia In honor of Carl Gottlieb Plato (fl. 1796) a Leipzig school-master

plattensis L. *-ensis*, denoting origin. From North Platte River, Nebraska, USA

platyacanthus Gk *platys*, flat; *acanthus*, spine. Involucral spines modified to acute scales

platyanth-a, -um Gk *platys*, flat; *anthos*, flower. Spikelets wider, relative to length, than those of related species

platycarph-a, -um Gk *platys*, flat; *karphe*, straw. The rhachis is broad with strongly compressed spikelets attached to one side

platycarpum Gk *platys*, flat; *karpos*, fruit. Spikelets dorsally compressed

platycaul-e, -is, -on, -os Gk *platys*, flat; *kaulos*, stem. Culms compressed

platychaeta Gk *platys*, broad; *chaete*, bristle. Awns flattened

platycoleum Gk *platys*, flat; *koleos*, sheath. Leaf-sheath strongly keeled

platyculmum Gk *platys*, flat; L. *culmus*, stalk. Culms flattened

platyglossa Gk *platys*, broad; *glossa*, tongue. Ligules broad truncate

platynot-a, -um Gk *platys*, flat; *noton*, back. Rhachis much flattened

platyphyll-a, -um, -us Gk *platys*, flat; *phyllon*, leaf. Leaf-blade broad

platypoda Gk *platys*, flat; *pous*, foot. The apex of the glume is expanded into an elliptical area from which the awn arises

platypus Gk *platys*, broad; *pous*, foot. Callus of sessile spikelet broad with few hairs

platyrhachis Gk *platys*, flat; *rhachis*, backbone. Rhachis scabrid

platyrrhachis See *platyrhachis*

platystachy-on, -s Gk *platys*, flat; *stachys*, spike as of an ear of wheat. With spikelets borne on a flat axis

platytaenia Gk *platys*, flat; *tainia*, band. Rhachis of raceme winged

Plazerium Origin uncertain, not given by author

plebeia L. *plebius*, common. Abundant and often widespread

plebeja See *plebeia*

plectostachy-a, -us Gk *plektos*, coil; *stachys*, spike as of an ear of wheat. Racemes of inflorescence curved

Plectrachne Gk *plektron*, spur; *achne*, scale. The lemma is stiff and three awned

plectrachnoides Gk *-oides*, resembling. Resembling *Plectrachne* with respect to spikelets

Pleiadelphia Gk *pleios*, several; *delphos*, brother. There are three or four pairs of homogamous spikelets at the base of the raceme in contrast to *Anadelphia* where there are none

pleianthemum Gk *pleios*, several; *anthemon*, flower. Spikelets four-flowered

pleianthum Gk *pleios*, several; *anthos*, flower. Panicle much branched and so bearing many spikelets

Pleioblastus Gk *pleios*, several; *blastos*, shoot or bud. The internodes are very short each with one bud but the overall impression is that of an internode with several buds

Pleiodon Gk *pleios*, several; *odous*, tooth. Lemma and rudiment of upper floret each three-awned

Pleioneura Gk *pleios*, several; *neuron*, nerve. Lemma with several nerves

pleiophyll-a, -um Gk *pleios*, several; *phyllon*, leaf. Culm with several leaves

pleiostachya Gk *pleios*, several; *stachys*, spike as of an ear of wheat. – (1) Inflorescence of several spikes. *Saugetia pleiostachya* – (2) of several racemes. *Pollinia pleiostachya*

plenum L. *plenus*, bulky. Culms tall and densely tufted

Pleopogon Gk *pleos*, several; *pogon*, beard. Spikelets with several awns because both glumes and lemmas are awned

pleostachyum Gk *pleios*, several; *stachys*, spike as of an ear of wheat. The inflorescence has several racemes

plesiantha Gk *plesios*, neighbour; *anthos*, flower. Spikelets crowded in inflorescence

Pleuraphis Gk *pleura*, several; *rhaphis*, needle. The glumes and lemmas each have two or more awns

Pleurhaphis See *Pleuraphis*

pleurigluma L. *pleura*, several; *gluma*, husk. The proximal lemmas of the spikelet are sterile

pleuriracemosum L. *pleura*, several; *racemosum*, racemed. Inflorescence of several racemes

Pleuroplitis Gk *pleuros*, rib; *hoplitis*, armed soldier. Midrib of lower glume extended as an awn

Pleuropogon, pleuropogon Gk *pleuron*, rib; *pogon*, beard. Lemma nerves densely bearded

plexipes L. *plecto*, interweave; *pes*, foot. With intertwining rhizomes

plica-polonica L. *plico*, fold; *polonia*, Poland; *-ica*, belonging to. A note on the type specimen indicates the author wished to honor the Polish nation

P

plicat-a, -um L. *plico*, fold. – (1) Leaf-blade with a single longitudinal fold. *Glyceria plicata, Poa plicata* – (2) leaf-blade with several longitudinal folds. *Panicum plicatum, Setaria plicata*

plicatil-e, -is L. *plicata*, folded; *-ile*, property. Surface of leaf-blade undulate

plicatulum L. *plicatus*, folded; *-ulum*, tending towards. The sterile lemma is transversely wrinkled

plicatum See *plicata*

pliniana L. *-ana*, indicating connection. As for *plinii*

plinii In honor of Gaius Plinius Secundus (23–79 C.E.) Roman admiral and natural historian

Plinthanthesis Gk *plinthos*, plinth; *anthos*, flower. Inflorescence a raceme

plonkae In honor of François Plonka (fl. 1988) French botanist

Plotia In honor of Robert Plot (1640–1696) English naturalist

plowmanii In honor of Timothy C. Plowman (1944–1989) who collected in Brazil

plukenetii In honor of Leonard Plukenet (1642–1706) English physician and botanist

plumbe-a, -us L. leaden. Spikelets dark-grey

plumiger, -a, -um L. *pluma*, feather; *gero*, carry or bear. – (1) Pedicels of sterile florets with long hairs. *Andropogon plumiger, Schizachyrium plumigerum* – (2) with a villous awn. *Stipa plumigera*

plumos-a, -um, -us L. feathery. – (1) With long hairs giving on the pedicels a feathery appearance. *Agrostis plumosa, Andropogon plumosus, Arthratherum plumosum, Leptochloa plumosa, Panicum plumosum, Ptiloneilema plumosum, Santia plumosa, Sorghum plumosum, Trachypogon plumosus* – (2) with the awn or one of its members densely hairy. *Aristida plumosa, Stipa plumosa, Stipagrostis plumosa* – (3) with the upper floret reduced to a feathery axis. *Achaeta plumosa* – (4) with lemmas densely hairy. *Eragrostis plumosa, Poa plumosa*

pluriflora L. *plus*, several; *flos*, flower. Florets several per spikelet

plurifolia L. *plus*, several; *folium*, leaf. Culms many-leaved

plurigluma L. *plus*, several; *gluma*, husk. Some of the lower florets are sterile, their lemmas thereby resembling glumes

plurinervata L. *plus*, several; *nervus*, nerve; *-ata*, possessing. With more nerves in the glumes than for related species

plurinervis L. *plus*, several; *nervus*, nerve. Lower lemma and upper glume with several nerves

plurinodis L. *plus*, several; *nodus*, knot. Culms several-noded

plurisetosa L. *plus*, several; *seta*, bristle. Lower culm leaf-sheaths with stiff abundant hairs

Poa, poa Gk *poa*, herb or grass. In Classical Greek, a word applied to grasses and other herbs useful for fodder. Other dialect spellings have also been employed, e.g., Ionic *poe* and *poie*; Doric, *poia*

poacea L. *-acea*, resembling. Similar to *Poa*

Poacites Gk *poa*, grass; *-ites*, resemblance. Grass-like fossils

poaeflorum See *poiflorum*

poaeform-e, -is See *poiforme*

poaemorph-a, -um See *poimorpha*

poaeoides Gk *-oides*, resembling. – (1) The inflorescence resembles that of *Poa*. *Brachiaria poaeoides, Koeleria poaeoides* – (2) resembling *Poa* in habit or spikelets. *Calamagrostis poaeoides, Eragrostis poaeoides, Melica poaeoides, Panicum poaeoides, Sporobolus poaeoides, Uralepis poaeoides*

Poagris Gk *poa*, grass; *agrios*, living in the fields. Meaning obscure but possibly a reference to not being cultivated

Poagrostis Combining the characters of *Poa* and *Agrostis*

Poarion Gk *-ion*, indicating condition. Employed as a suffix to *Poa* to form a new generic name

Pobeguinea In honor of Charles Henri Oliver Pobéguin (1856–1951) French colonial administrator and amateur botanist

pobeguinii See *Pobeguinea*

poculiformis L. *poculum*, goblet; *forma*, appearance. Palea goblet-like

pocutica L. *-ica*, belonging to. From Pokutia the region north of the river Doriester in the vicinity of Chernovtsy, Ukraine

podachne Gk *pous*, foot; *achne*, scale. Basal culm-leaves lacking a well-formed blade

Podagrostis Gk *pous*, foot. Rhachilla prolonged as a hairy rudiment about half the length of the palea

Podionapus Gk *podion*, stalk; *a-*, not; *pous*, foot. Lower spikelets of inflorescence sessile, upper stalked

podolica L. *-ica*, belonging to. From Podolicus now Podielen, Galicia, Poland

podophora See *Podophorus*

Podophorus Gk *pous*, foot; *phero*, bear. Distal sterile floret borne on a long rhachilla extension

Podopogon Gk *pous*, foot; *pogon*, beard. Spikelets with a bearded callus

Podosaemon, Podosaemum See *Podosemum*

Podosemum Gk *pous*, foot; *haima*, blood. Pedicels purple

podotrich-a, -us Gk *pous*, foot; *thrix*, hair. Peduncles shortly hairy at their apices

podperae In honor of Josef Podpera (1878–1954) Bohemian botanist

poecilanth-a, -um Gk *poikilos*, variable; *anthos*, flower. Spikelets with variable numbers of sterile and fertile florets

Poecilostachys Gk *poikilos*, variable; *stachys*, spike as of an ear of wheat. The spikelets are variable in color or some other respect

poecilotrich-a, -us Gk *poikilos*, variable; *thrix*, hair. Raceme hairy, whitish proximally fulvous distally

poeppigiana L. *-ana*, indicating connection. In honor of Eduard Friedrich Poeppig (1798–1868) German botanist

poggeana L. *-ana*, indicating connection. In honor of Karl Pogge (fl. 1882–1907) who collected in Zaire and S.W. Africa

Pogochloa Gk *pogon*, beard; *chloa*, grass. Lemma nerves densely villous

Pogonachne Gk *pogon*, beard; *achne*, scale. The upper glume bears a conspicuous tuft of hairs

pogonanthus Gk *pogon*, beard; *anthos*, flower. Spikelets densely hirsute

Pogonarthria Gk *pogon*, beard; *arthron*, joint. Rhachilla internodes fringed with hairs

pogonathera Gk *pogon*, beard; *ather*, barb or spine. Awn plumose

Pogonatherum Gk *pogon*, beard; *ather*, barb or spine. The upper glumes and upper lemmas are hair-like

Pogoneura See *Pogononeura*

pogonia Gk *pogon*, beard. Nodes densely hairy

Pogonochloa Gk *pogon*, beard; *chloa*, grass. Lemmas are awned

Pogononeura Gk *pogon*, beard; *neuron*, nerve. Lemma nerves invested with long hairs

Pogonopsis Gk *pogon*, beard; *opsis*, appearance. Spikelets subtended by an involucre of hairs

pogonoptil-a, -um Gk *pogon*, beard; *ptilon*, feather. One branch of the trifid awn is bearded

pogonostachyum Gk *pogon*, beard; *stachys*, spike as of an ear of wheat. Callus with a low spreading beard

pohlean-a, -us In honor of Richard Pohle (1869–1926) Latvian-born German botanist

pohlian-um, -us L. *-anum*, indicating connection. – (1) In honor of Johann Emanuel Pohl (1782–1834) Bohemian botanist. *Andropogon pohlianus* – (2) as for *Pohlidium*. *Sorghastrum pohlianum*

Pohlidium Gk *-idium*, diminutive but here used as a name-forming suffix. In honor of Richard Walter Pohl (1916–1993) United States agrostologist

pohlii As for *Pohlidium*

Poidium Gk *-idium*, diminutive. A genus with species in some way resembling *Poa* but smaller

poidium A species resembling in some way one of the species of *Poidium*

poiflorum L. *flos*, flower. Spikelets with hairy lemmas resembling those of *Poa*

poiform-e, -is L. *forma*, appearance. Resembling *Poa* in some respect

poilanei In honor of Eugene Poilane (1888–1964) French botanist

poimorph-a, -um L. *morphe*, shape. Resembling one or more *Poa* species

poiophyllus Gk *poa*, grass; *phyllon*, leaf. Apices of leaf-blades boat-shaped as in *Poa*

poiphagorum Gk *poa*, grass; *phagos*, a glutton. Grass of the gluttons, that is from the yak pastures of the Himalayas

poiretian-a, -um L. *-ana*, indicating connection. In honor of Jean Louis Marie Poiret (1755–1834) French cleric and encyclopedist

P

poiretii As for *poiretiana*

polesica L. *-ica*, belonging to. From Polesia, now Poles'ye a marshy plain across the border of Belorussuja (Belarus) and Ukraine

Polevansia In honor of Illtyd Buller Pole-Evans (1879–1968) Welsh-born South African botanist

polevansii As for *Polevansia*

poliophyllum Gk *polios*, gray; *phyllon*, leaf. Leaf-blades glaucous or grey-green

polita L. *polio*, polish. Foliage glabrous

politii In honor of Louis P. Politi (1916–1972) Venezuelan botanist

pollinensis L. *-ensis*, denoting origin. From Mount Pollino, Italy

Pollinia In honor of Ciro Pollini (1782–1833) Italian botanist and physician

polliniaefolius L. *folium*, leaf. Leaf-blades resembling those of *Pollinia*

Polliniastrum Gk *-astrum*, incomplete resemblance but here used as a name-forming suffix. Similar to *Pollinia*

Pollinidium Gk *-idium*, diminutive but here used as a name-forming suffix. Resembling *Pollinia*

pollinioides Gk *-oides*, resembling. Resembling *Pollinia*

Polliniopsis Gk *opsis*, appearance. Resembling *Pollinia*

polliniopsis Gk *opsis*, resemblance. Similar to *Pollinia*

pollockii In honor of Norman Arthur Robert Pollock (1874–1951) Australian agriculturalist

polo A creek in North Queensland, Australia

polonic-a, -um L. *-ica*, belonging to. From Polonia, now Poland

poluninii In honor of – (1) Oleg Vladimir Polunin (1914–1985) English botanist. *Festuca poluninii* – (2) Nicholas Vladimir Polunin (1909–1997) English botanist. *Calamagrostis poluninii*

polyanth-a, -us Gk *polys*, many; *anthos*, flower. Spikelets with more than the expected number of florets

Polyantherix Gk *polys*, several; *anthos*, flower; *thrix*, hair. The glumes and lemmas are long-awned

polyanthes Gk *polys*, many; *anthos*, flower. Inflorescence richly endowed with spikelets

polyanthus See *polyantha*

polyather-a, -us Gk *polys*, several; *ather*, spike or ear of wheat. Inflorescence of spike-like racemes bearing many spikelets

polybotrya Gk *polys*, many; *botrys*, bunch of grapes. The panicle bears many racemes

polybotryoides Gk *-oides*, resembling. Similar to *Digitaria polybotryoides*

polybracteatus Gk *polys*, several; L. *bracteus*, bract; *-atus*, possessing. With several sterile spikelets in each cluster of spikelets

polycarpha Gk *polys*, many; *karphos*, dry stalk. Lower spikelets on each spike replaced by short deciduous branchlets

polycaulis Gk *polys*, many; *kaulos*, stem. Culms much branched at the base

polycaulon Gk *polys*, several; *kaulos*, stem. Culms densely caespitose. Culms numerous, arising in bunches from rhizome

polychaet-a, -um Gk *polys*, several; *chaete*, bristle. – (1) Foliage invested in long erect hairs. *Paspalum polychaetum* – (2) ultimate spikelet of inflorescence with several awns or awn-like structures. *Anadelphia polychaeta*

polychroa Gk *polys*, several; *chroia*, color. Spikelets mostly purple

polyclad-a, -os, -um Gk *polys*, several; *klados*, branch. Plants with much branched culms often with the branches in fascicles

polycolea Gk *polys*, several; *koleos*, sheath. Lower leaf-sheaths very lax

polycomum Gk *polys*, several; *kome*, head of hair. Spikelets pubescent

polydactyl-a, -on Gk *polys*, several; *daktylon*, finger. The panicle has several finger-like branches

polygam-a, -um Gk *polys*, several; *gamos*, marriage. – (1) Florets numerous per inflorescence. *Cinnagrostis polygama, Dactylis polygama, Festuca polygama, Panicum polygamum, Poa polygama* – (2) florets of different sex in the same inflorescence. *Gouinia polygama*

polygonatum Gk *polys*, several; *gony*, knee; L. *-atum*, possessing. Culms with hairy nodes, thereby resembling some species of *Polygonum*

polygonoides Gk *-oides*, resembling. Habit creeping as with some *Polygonum* species

polymorph-a, -um, -us Gk *polys*, many; *morphe*, shape. Producing spikelets of two kinds or otherwise variable

Polyneura, -a, -on, -os Gk *polys*, several; *neuron*, nerve. The glumes and lemmas are many-nerved

polynoda Gk *polys*, several; L. *nodus*, knot. Culms several-noded

polynodon Gk *polys*, several; L. *nodus*, knot. Culms several-noded

Polyodon Gk *polys*, several; *odous*, tooth. Lemmas five- or seven-toothed

polyphyll-a, -um, -us Gk *polys*, several; *phyllon*, leaf. Culms many-leaved in comparison with related species

polypodioides Gk *-oides*, resembling. The leaves on the middle to upper part of the culms have leaf-blades whose arrangement resembles that of the pinnules of *Polypodium*

Polypogon Gk *polys*, several; *pogon*, beard. Inflorescence bristly

polypogon Resembling *Polypogon*

Polypogonagrostis Hybrids between species of *Polypogon* and *Agrostis*

polypogonoides Gk *-oides*, resembling. Resembling *Polypogon* with respect to the inflorescence

Polyraphis Gk *polys*, several; *rhaphis*, needle. The lemma is many-awned

polyrhizum Gk *polys*, several; *rhiza*, root. Rooting freely from the lower nodes

Polyschistis Gk *polys*, several; *schizo*, split. The upper glume and the lemmas are divided into several lobes each of which is awned

polysetus Gk *polys*, several; L. *seta*, bristle

polysperma Gk *polys*, many; *sperma*, seed. Spikelets with many florets

polystachi-on, -os See *polystachya*

polystachy-a, -um, -us Gk *polys*, many; *stachys*, spike as of an ear of wheat. – (1) Plants with many branched culms or inflorescences. *Arundinaria polystachya, Axonopus polystachyus, Beckera polystachya, Chloris polystachya, Cynodon polystachyus, Deyeuxia polystachya, Diplachne polystachya, Echinochloa polystachya, Echinolaena polystachya, Eriochloa polystachya, Festuca polystachya, Heteropogon polystachyus, Ischaemum polystachyum, Leptochloa polystachya, Manisuris polystachya, Panicum polystachyum, Paspalum polystachyum, Pseudechinolaena polystachya, Spartina polystachya, Stylagrostis polystachya* – (2) with many close-set culms. *Anthistiria polystachya, Echinochloa polystachya, Eleusine polystachya, Eriochloa polystachya, Gymnothrix polystachya, Hymenachne polystachyum*

polystachy-on, -os See *polystachya*

polystachy-um, -us See *polystachya*

polystichus Gk *poly-*, many; *stichos*, row. Spikelets in several rows

Polytoca Gk *polys*, many; *tokas*, offspring. There are the three types of spikelets in the one inflorescence

Polytrias Gk *polys*, several; *treis*, three. The spikelets are arranged in triads

polytricha Gk *polys*, several; *thrix*, hair. – (1) Auricles bearing abundant long hairs on their margins. *Yushania polytricha* – (2) axils of the panicle branches hairy. *Eragrostis polytricha*

Pomereulla See *Pommereulla*

Pommereulla In honor of Madame Du-gage née Pommereul, French botanist especially interested in grasses

Pommeureuilla See *Pommereulla*

pompale L. showy. Spikelets tinged with purple

ponapensis L. *-ensis*, denoting origin. From Ponape (Pohnpei), an island in the Federated States of Micronesia

Ponceletia In honor of Polycarpe Poncelet (fl. 1755–1800) French biologist

ponderos-a, -us L. of great weight. Inflorescence densely congested

ponojensis L. *-ensis*, denoting origin. From Ponoj (Ponoy), Kola Peninsula, Russian Federation

pontanal-e, -is Portuguese *pontanal*, marsh; L. *-alis*, pertaining to. Growing in swampy places

pontarlieri In honor of Nicolas Charles Pontarlier (1812–1889) French botanist

pontic-a, -um, -us L. *-ica*, belonging to. From Pontus in ancient times, a Province of Asia Minor, now Turkey

Pooideites Gk *-ites*, similar to. Form genus for grass-like fossils

poophagorum See *poiphagorum*

popinensis L. *popina*, eating place; *-ensis*, denoting origin. First collected near a cafe

poplawskiae In honor of Henrietta Ippolitovna Poplavskja (1885–1956) who collected in the Transbaikal region, Russian Federation

popovii In honor of Mikhail Grégorievic Popov (1893–1955) Russian botanist

poppelwellii In honor of Dugald Louis Poppelwell (1863–1939) New Zealand amateur botanist

Poranthera Gk *poros*, pore, *anthera*, anther. Anthers opening by pores rather than slits

porcat-a, -us L. *porca*, ridge between two furrows; *-ata*, possessing. Ridged as of culms or glumes

porcii In honor of Florian Porcius (1816–1906) Romanian botanist

porifera L. *porus*, pore; *fero*, carry or bear. Surface of lower glume honey-comb pitted

porosa L. *porus*, pore; *-osa*, abundance. Bearing pit-like glands

porphyrantha Gk *porphyra*, purple dye; *anthos*, flower. Spikelets purple

porphyrea Gk purple-red. Culm buds purple-red

porphyroclados Gk *porphyra*, purple; *klados*, branch. Culms reddish-brown

porphyrocoma Gk *porphyra*, purple dye; *coma*, head of hair. Spikelets invested with purple hairs

porphyrrhizos Gk *porphyra*, purple dye; *rhiza*, root. Lower nodes brown-purple

porranth-a, -um Gk *porro*, far off; *anthos*, flower. Possibly a reference to the spikelets being widely separated in the inflorescence

porrect-a, -us L. stretched outwards and forwards. Inflorescence of stiffly spreading pedunculate racemes bare of spikelets for a considerable distance from the base

Porroteranthe Gk *porrotero*, furthest off; *anthos*, flower. From Australia

porsildii In honor of Alf Erling Porsild (1901–1977) Danish-born Canadian botanist

Porteresia In honor of Roland Portères (1906–1974) French ethnobotanist

porteri In honor of Thomas Conrad Porter (1822–1901) United States botanist

porterianum L. *-anum*, indicating connection. As for *porteri*

portoi In honor of Paulo Campos Porto (1889–?) Brazilian plant collector

portoricens-e, -is L. *-ense*, denoting origin. From Puerto Rico

pospischilii In honor of Pospischil who collected in East Africa

potamium Gk *potamos*, river; L. *-ium*, characteristic of. From Amazonian Brazil

Potamochloa Gk *potamos*, river; *chloa*, grass. A floating grass

Potamophila, potamophila Gk *potamos*, river; *philos*, love. Grows on river banks and shingle beds

potaninii In honor of Grigorij Nikolajevic Potanin (1835–1920) Russian botanist

potaroensis L. *-ensis*, denoting origin. From Potaro Gorge, Guyana

potosiana L. *-ana*, indicating connection. From Potosi Department, Bolivia

potosiensis L. *-ensis*, denoting origin. From San Luis Potosi State, Mexico

pourretii In honor of Pierre André Pourret de Figeae (1754–1818) French cleric and botanist

pouzolzii In honor of Pierre Casimir Marie de Pouzolz (1785–1858) French botanist

pradan-a, -um L. *-ana*, indicating connection. From Mesa de Prada, Cuba

praealt-a, -um L. *prae-*, very; *altus*, tall. Taller than related species

praecaespitos-a, -um, -us L. *prae-*, very; *caespes*, turf; *-osa*, abundance. Forming a dense turf

praecapillata L. *prae*, very; *capillis*, hair; *-ata*, possessing. Inflorescence branches filiform

praecipua L. special. Attractive in appearance

praeclusa L. *praecludo*, close. Origin not given by the author, but may refer to the overlapping leaf-sheath margins

praecocioides Gk *-oides*, resembling. Similar to *Avena praecoqua* in being early maturing

praecoci-um, -us L. developing early. – (1) The branching of the autumnal phase develops early in the season before the first panicle is expanded. *Panicum praecocius* – (2) early maturing. *Hordeum praecocium*

praecoqua L. early ripening. Grain early maturing

praecox L. early. Flowering early in the spring

praegnans L. pregnant. Spikelets turgid

praegravis L. very heavy. High yielding cereal

praelongum L. *prae-*, very; *longum*, long. Inflorescence effuse

praemorsa L. *praemordeo*, bite off. – (1) Leaf-blades much shorter than those of related species. *Eulalia praemorsa* – (2) Glume apices truncate, erose. *Phalaris praemorsa*

praerupt-a, -us L. *praerumpo*, break off. Inflorescence readily disarticulating

praestans L. *praesto*, stand out. Culms taller than those of related species

praestantissima L. *praesto*, stand out; *-issima*, most. Clearly distinguished from related species

praeteritus L. *praetereo*, escape notice. Species previously overlooked

praetermissa L. *praetermitto*, make no mention of. Ignored by previous writers

praetervis-a, -um, -us L. *praeter*, beyond; *visum*, seen. Similar to but beyond the range of variability of another species

praetutiana L. a Praetutian, that is a resident of Picenum, an ancient district comprising the present-day Abruzzi and Southern Marche, Italy; *-ana*, indicating connection. From Picinum

praeusta L. *praeuro*, burn at the tip. Awns black at their tips

prahliana L. *-ana*, indicating connection. In honor of Peter Prahl (1843–1911) German physician and botanist

prainii In honor of David Prain (1857–1944) Scots-born physician and botanist, sometime Director, Royal Botanic Gardens, Kew

prasina Gk *prason*, leek; *-ina*, indicating resemblance. Young shoots reminiscent of leeks

pratens-e, -is L. *pratum*, a meadow; *-ense*, denoting origin. Meadow species

pratensiformis L. *forma*, appearance. Resembling *Poa pratensis*

pratensis See *pratense*

pratericola L. *pratum*, meadow; *-cola*, dweller. Meadow grass

praticola L. *pratum*, meadow; *-cola*, dweller. Of meadows, that is growing in open places

pratorum L. *pratum*, meadow. Of meadows, that is growing in open places

precatoria L. *precatorius*, one who prays. The nodding spikelets resemble the bowed heads of worshippers

prehensilis L. *prehendo*, seize; *-ilis*, property. The leaf tips are coiled enabling the plant to climb

Preissia, preissia In honor of Johann August Ludwig Preiss (1811–1883) German botanist

preissiana L. *-ana*, indicating connection. As for *Preissia*

prenticeanum L. *-anum*, indicating connection. In honor of Charles Brightly Prentice (1820–1894) Queensland physician and amateur botanist

presliana L. *-ana*, indicating connection. In honor of Karel Borivoj Presl (1794–1852) Bohemian botanist

preslii As for *presliana*

press-um, -us L. compressed. Culms and leaf-sheaths compressed

P

prestoei In honor of Henry Prestoe (1842–1923) sometime Director of Botanic Gardens, Trinidad

pretoriensis L. *-ensis*, denoting origin. From Pretoria, South Africa

pricei In honor of Morgan Phillips Price (1885–1973) who collected in north-west Mongolia

prichardii In honor of Hesketh Vernon Hesketh Prichard (1876–1922) who collected in Argentina

P

prieurii Dedicated to F. R. Leprieur (1799–1869) French naval officer and amateur botanist

prilipkoana L. *-ana*, indicating connection. In honor of Leonid Ivanovich Prilipko (1907–1983) Russian botanist

primae In honor of V. M. Prima (fl. 1971) who collected along the upper reaches of Shon-Den River, Caucasus, Russian Federation

primaeva L. *primus*, first; *aevum*, age. The oldest species in the genus

princeps L. most distinguished. – (1) Attractive as in appearance. *Andropogon princeps*, *Cymbopogon princeps*, *Rhiniachne princeps*, *Rhytachne princeps* – (2) agriculturally significant. *Hordeum princeps*

pringlei As for *Pringleochloa*

Pringleochloa Gk *chloa*, grass. In honor of Cyrus Guernsey Pringle (1838–1911) United States botanist

Prionachne Gk *prion*, saw; *achne*, scale. The glumes have serrated keels

Prionanthium Gk *prion*, saw; *anthos*, flower. The glumes have well developed tooth-like projections arising from their nerves

prionitis Gk *prion*, saw; *-itis*, similar to. The leaf-blades are rigid and the margins furnished with short hairs

prionodes Gk *prion*, saw; *-odes*, resembling. Lower glume has stiff hairs on the marginal and submarginal nerves

probatovae In honor of N. S. Probatova (1939–) Russian botanist

proboscideum Gk *proboscis*, means of providing food and so by transference of meaning a beak; *-eum*, belonging to. Glumes and sterile lemma apically attenuated

procer-a, -um, -us L. tall. Culms tall

procerior L. taller. Culms taller than most other species of genus

procerrim-a, -um L. tallest. Tallest of a group of related species

procer-um, -us See *procera*

procumbens L. *procumbo*, fall down. Culms creeping

procurrens L. *procurro*, project. The inflorescence projects beyond the leaf-sheath

prodigiosa L. *prodiguosa*, strange or prodigious. – (1) Leaf-margins white. *Sasa prodigiosa* – (2) awn very large. *Aristida prodigiosa*

product-a, -us L. *produco*, extend. Rhachilla drawn out

Programinis Fossil grass spikelets preserved in amber from Myanmar

projectum L. *proicio*, stretch out. Panicle interrupted

prokudinii In honor of Jurij Nikolajevic Prokudin (1911–) Russian botanist

prolifer-a, -um L. *proles*, offspring; *fero*, carry or bear. Producing runners

prolificum L. *proles*, offspring; *facio*, make. Inflorescence with many flowers and so capable of producing an abundance of seed

prolixior L. more than usually spreading abroad. Somewhat rampant

prolixus L. widely extended. The panicle is strongly exserted

prolutum L. *proluo*, wash; *lutus*, swamp. Growing near water

prominens L. *promineo*, jut out. Nodes conspicuous

prona L. inclined forward. Flowering culms procumbent

propinqu-a, -um, -us L. near to. Similar to another species

prorepens L. *prorepo*, creep forward. Rhizomatous creeper

Prosphysis Gk *prosphysis*, adherence. The caryopsis adheres to the glumes

Prosphytochloa Gk *prosphytuo*, grow upon; *chloa*, grass. A climber with leaf tendrils

prostrat-a, -um L. *prosterno*, throw to the ground. Culms creeping or wiry so as to fall on the ground

protens-a, -um L. *protendo*, stretch out. – (1) Inflorescence a long spike. *Hemarthria protensa, Manisuris protensa, Rottboellia protensa* – (2) culms creeping with long internodes. *Panicum protensum*

protractum L. *protraho*, reveal. Distinguished from a related species

protrusus L. *protrudo*, push out. Origin uncertain, not given by the author

provincialis L. *provincia*, province; *-alis*, pertaining to. Relating to a province, in particular to Provincia, now Provence, France

proxim-a, -um, -us L. near to. Readily confused with another species

prudhommei In honor of J. Prudhomme

pruinifer-a, -um L. *pruina*, hoar frost; *fero*, carry or bear. The leaf-blades are bluish-green

pruinos-a, -um, -us L. *pruina*, hoar frost; *-osa*, abundance. Leaf-blades are covered with short white hairs

prunifera L. *prunum*, plum; *fero*, carry or bear. Fruit about the size of a plum

pruriens L. *prurio*, itch. Densely hairy causing itching to sensitive skins

przewalskii See *przewalskyi*

przewalskyi In honor of Nikolai Michailowicz Przewalsky (also Przhevalsky or Przewalski) (1839–1888) Russian geographer and explorer

Psamma Gk *psammos*, sand. Growing in sandy habitats

Psammagrostis Gk *psammos*, sand; *agrostis*, grass. Growing on sand hills

Psammochloa Gk *psammos*, sand; *chloa*, grass. From high altitude sand dunes in Mongolia

Psammophila Gk *psammos*, sand; *phileo*, love. Dune species

psammophil-a, -um, -us Gk *psammos*, sand; *phileo*, love. Growing in sandy places

Psammopyrum Gk *psammos*, sand; *pyros*, wheat. Wheat-like grass preferring sandy habitats

Psathyrostachys Gk *psathyros*, brittle; *stachys*, spike as of an ear of wheat. The rhachis is fragile and readily breaks into segments when the spikes are mature

Psatyrostachys See *Psathyrostachys*

pseud, pseudo Generic names and species epithets beginning with "pseud" or "pseudo" (Gk *pseudos*, false) are commonly formed to distinguish a genus or species from that with which it was previously confused or to avoid the formation of a homonym. In the entries below, only those names are listed that are known not to strictly adhere to this format

Pseudachne Gk *pseudos*, false; *achne*, scale. The side-lobes of the glumes are very small

pseudanceps Gk *pseudos*, false. Readily confused with *Panicum anceps*

pseudaristata Gk *pseudos*, false; L. *arista*, bristle; *-ata*, possessing. Apices of upper glume and lower lemma attenuate

Pseudelymus Hybrids between species of *Pseudoregneria* and *Elymus*

pseudisachne Gk *pseudos*, false. Spikelets resembling those of *Isachne*

Pseudobromus Gk *pseudos*, false. Spikelets resemble those of *Bromus* but are not laterally compressed

pseudobtusa Gk *pseudos*, false. Intermediate between *Eragrostis obtusa* and *Eragrostis echinochiloidea*

pseudobulbosa Gk *pseudos*, false. Culms slightly thickened at base

Pseudocoix Gk *pseudos*, false. The inflated glossy glumes resemble the cupule of *Coix*

pseudodurva Gk *pseudos*, false. Possibly a reference to the species resembling *Cynodon dactylon* which is known in India as durva grass

pseudoligulata Gk *pseudos*, false; *ligulus*, small tongue; *-ata*, possessing. The collar of the leaf-blade is readily confused with the ligule

pseudopetiolata Gk *pseudos*, false; L. *petiolus*, little leg; *-ata*, possessing. Leaf with a petiole-like constriction between the blade and sheath

Pseudophragmites Gk *pseudos*, false. Fossil genus resembling *Phragmites*

pseudopubescens, pseudo-pubescens Gk *pseudos*, false; *pubescens*, hairy. The leaf-blades sometimes lack hairs down the centre

pseudoracemosum Gk *pseudos*, false; *racemus*, stalk of a bunch of grapes; *-osum*, abundance. Primary inflorescence axes raceme-like

pseudosetaria Gk *pseudos*, false. Panicle elongated resembling that of *Setaria*

Pseudostachyum Gk *pseudos*, false; *stachys*, spike as of an ear of wheat. Glumes mostly bear bulbils and not florets

Psilantha, -a, -um Gk *psilos*, bare; *anthos*, flower. Florets glabrous

psilantherum Gk *psilos*, bare; *antheros*, blooming. Lemma awns smooth

Psilathera Gk *psilos*, bare; *ather*, barb or spine. Lemma awns smooth

psilobasis Gk *psilos*, bare; *basis*, base. Culms glabrous

psilocaulum Gk *psilos*, bare; L. *caulis*, stem. Culms glabrous

Psilochloa Gk *psilos*, bare; *chloa*, grass. The upper glume and lemmas lack apical appendages

Psilolemma Gk *psilos*, bare; *lemma*, husk. Lemmas glabrous

psilolepis Gk *psilos*, bare; *lepis*, scale. Lemmas glabrous

psilophylla Gk *psilos*, bare; *phyllon*, leaf. Leaf-blade glabrous

psilopodium Gk *psilos*, bare; *pous*, foot. Pedicels glabrous

Psilopogon Gk *psilos*, bare; *pogon*, beard. Awn glabrous

psilosanth-a, -um Gk *psilos*, bare; *anthos*, flower. Lemma glabrous

Psilostachys, -s, -um Gk *psilos*, bare; *stachys*, spike as of an ear of wheat. Inflorescence glabrous

Psilurus Gk *psilos*, bare; *oura*, tail. The inflorescence is bare like the tail of a rat

psittacorum L. *psittacus*, parrot. From *Arroyos de la papagallow* (Valley of the Parrots) near Mendoza, Argentina

psychrophila Gk *psychros*, cold; *phileo*, love. Alpine species

psylantha See *psilantha*

pteridigodium See *Pterigodium*

Pterigodium, pterygodium Gk *pteryx*, wing; L. *-odium*, resemblance. Palea of lower floret winged at maturity

Pterium Gk *pteron*, wing or feather-like; *-ium*, resembling. Three sterile florets invest the fertile floret, as the wings of a bird cover its body

Pterochlaena Gk *pteron*, wing or feather-like; *chlaena*, cloak. The upper glume is winged on the margins

Pterochloris Gk *pteron*, wing or feather-like. The lemma of the lower floret is winged at the apex, otherwise resembling *Chloris*

pteropechys Gk *pteron*, wing or feather-like; *pechys*, fore-arm. Raceme internodes with densely hairy margins and in shape resembling the radius of a human forearm

pteropholis Gk *pteron*, wing or feather-like; *pholis*, scale of a snake. Glume keels winged

Pteropodium Gk *pteron*, wing or feather-like; *pous*, foot. Callus densely hairy, resembling down feathers

pterostachys Gk *pteron*, wing or feather-like; *stachys*, spike as of an ear of wheat. The clusters of sterile lemmas projecting from the florets render the spicate inflorescence a wing-like appearance

pterygodium See *Pterigodium*

Pterygostachyum Gk *pterygion*, little wing; *stachys*, spike as of an ear of wheat. The inflorescence branches are flattened

Ptilagrostis Gk *ptilon*, fluff. Spikelets with a single floret as in *Agrostis* but lemma awns feathery-pilose

Ptiloneilema Gk *ptilon*, fluff; *eilema*, covering. Glumes invested with long hairs

Ptilonema See *Ptiloneilema*

pubens L. downy. Leaf-sheaths downy

puberul-a, -um L. *pubes*, hair of adulthood; *-ula*, diminutive. Plant covered in whole or in part with short hairs

pubescens L. *pubesco*, become hairy. Plant whole or in part hairy

pubiannula L. hair of adulthood; *annulus*, ring. Nodes densely hairy

pubicalyx L. *pubes*, hair of adulthood; Gk *kalyx*, cup. Glumes hairy

pubicaulis L. *pubes*, hair of adulthood; *caulis*, stem. With hairy culms

pubiculmis L. *pubes*, hair of adulthood; *culmus*, stalk. Leaf-sheaths densely hairy

pubiflor-a, -um, -us L. *pubes*, hair of adulthood; *flos*, flower. With some or all parts of the inflorescence or spikelets densely hairy

pubifoli-a, -um L. *pubes*, hair of adulthood; *folium*, leaf. Leaf-blades hairy

pubigera L. *pubes*, hair of adulthood; *gero*, carry or bear. Plant pubescent in all parts

pubiglum-a, -e, -is L. *pubes*, hair of adulthood; *gluma*, husk. Glumes densely hairy

pubinervis L. *pubes*, hair of adulthood; *nervis*, nerve. – (1) The glumes and lemmas are hairy, especially on the nerves. *Festuca pubinervis* – (2) main nerve of leaf-blade hairy beneath. *Gigantochloa pubinervis*

pubinod-e, -is L. *pubes*, hair of adulthood; *nodus*, knot. Nodes hairy

pubipetiolata L. *pubes*, hair of adulthood; *petiolus*, little leg; *-ata*, possessing. Petiole of leaf hairy

pubispicula L. *pubes*, hair of adulthood; *spica*, a point; hence, in particular, an ear or spike of grain; *-ula*, diminutive. Glumes and sterile lemmas pubescent

pubivagina L. *pubes*, hair of adulthood; *vagina*, sheath. Leaf-sheaths hairy

pubivaginat-um, -us L. *pubes*, hair of adulthood; *vagina*, sheath; *-atum*, possessing. Leaf-sheath hairy

Puccinellia In honor of Benedetto Puccinelli (1808–1850) Italian botanist

puccinellii As for *Puccinellia*

Pucciphippsia Hybrids between species of *Puccinellia* and *Phippsia*

puchiparensis L. *-ensis*, denoting origin. From Puchipar near Madras, India

pudica L. modest. On account of the nodding spikelets

puelches Chilean, a native of the eastern side of the Andes. Growing east of the Andes

Puelia In honor of Timothée Puel (1812–1890) French physician and amateur botanist

puelii As for *Puelia*

puellarum L. *puella*, girl. Of little girls who in East Asia make necklaces from the cupules

pugae In honor of Frid. Puga who collected in Chile

pugionifoli-a, -um L. *pugio*, dagger; *folium*, leaf. Leaf-blade apices pungent

pulanensis L. *-ensis*, denoting origin. From Pulan, Tibet Autonomous Region (Xizang), China

pulchell-a, -um, -us L. pretty. Attractive in some respect, usually the inflorescence

pulcherrim-a, -um L. most beautiful. The most beautiful of several related species

pulchr-a, -um L. beautiful. Attractive in appearance, usually with respect to the inflorescence

Puliculum L. *pulex*, flea; *-ulum*, diminutive. The spikelets bear a fanciful resemblance to fleas

pullei In honor of August Adriaan Pulle (1878–1955) Dutch botanist

pullulans L. *pullulo*, sprout out. Culms develop from buds that break through the bases of the leaf-sheaths

pulvinat-a, -us L. *pulvinus*, cushion; *-ata*, possessing. – (1) Glumes convex. *Atropis pulvinata* – (2) habit cushion-like. *Aciachne pulvinata*, *Calamagrostis pulvinata*, *Sporobolus pulvinatus*

pulviniformis L. *pulvinus*, cushion; *formis*, appearance. In habit cushion-shaped

pulvinorum L. *pulvinus*, cushion. Of cushions, that is growing amongst cushion plants

pumil-a, -um, -us L. *dwarf*, low growing. Habit typically depauperate

pumilio L. a dwarf. Plants small compared with those of related species

pumil-um, -us See *pumila*

pumpellianus L. *-anus*, indicating connection. In honor of Raphael Pumpelly (1837–1923) United States geologist

punctat-a, -um, -us L. *pungo*, prick; *-ata*, possessing. – (1) Glumes spotted with color. *Anthistiria punctata*, *Oryza punctata*, *Panicum punctatum*, *Paspalidium punctatum*, *Paspalum punctatum*, *Poa punctata*, *Polytoca punctata*, *Saccharum punctatum*, *Sclerachne punctata* – (2) lower glume very reduced and colored. *Agrostis punctata*, *Eriochloa punctata*, *Helopus punctatus*, *Milium punctatum*, *Monachne punctata*, *Oedipachne punctata* – (3) glume pitted. *Andropogon punctatus*

P

punctiglandulosa L. *punctus*, point; *glans*, gland; *-ulus*, diminutive; *-osa*, abundance. Lemma nerves with abundant small glands

punctoria L. *pungo*, prick; *-oria*, indicating capability. Tips of leaf-blades pungent

punctulat-a, -um L. *punctum*, point; *-ulus*, diminutive; *-ata*, possessing. Young culm surfaces marked with small purple blotches

punensis L. *-ensis*, denoting origin. From the Puna or Altiplano region of north-west Argentina

pungens L. *pungo*, prick. – (1) Leaf-blades sharp-pointed. *Aeluropus pungens, Agropyron pungens, Agrostis pungens, Ammochloa pungens, Aristida pungens, Avena pungens, Avenula pungens, Cortaderia pungens, Elytrigia pungens, Micraira pungens, Oryzopsis pungens, Panicum pungens, Pentaschistis pungens, Phragmites pungens, Plectrachne pungens, Poa pungens, Sacciolepis pungens, Stipagrostis pungens, Triodia pungens, Triraphis pungens, Triticum pungens, Vulpia pungens, Zoysia pungens* – (2) stipes sharp-pointed. *Anthephora pungens* – (3) involucral bristles sharp-pointed. *Cenchrus pungens, Pennisetum pungens* – (4) branches very spiny. *Bambusa pungens* – (5) callus pungent. *Andropogon pungens*

pungipes L. *pungo*, prick; *pes*, foot. Spikelet contracted at the base into an acute callus

punicea L. *puniceus*, red. Panicle branches red

purandharensis L. *-ensis*, denoting origin. From Purandhar, near Bombay, India

purdieana L. *-ana*, indicating connection. In honor of William Purdie (c. 1817–1857) Scots-born plant collector in West Indies and South America

purgans L. *purgo*, cleanse. Scours the gut if eaten

purpuraristatus L. *purpureus*, purple; *arista*, bristle; *-ata*, possessing. Glumes are purple in color

purpurascens L. *purpurasco*, become purple. Inflorescences or foliage reddish-purple

purpurat-a, -um L. *purpureus*, purple; *-ata*, possessing. – (1) Inflorescence or spikelets purple. *Arthrostylidium purpuratum, Aulonemia purpurata, Eriochrysis purpurata, Piptochaetium purpuratum, Saccharum purpuratum* – (2) leaf-blades purple-red. *Phyllostachys purpurata*

purpure-a, -um, -us L. purple to reddish. Spikelets purple

purpurellus L. *purpurea*, purple to reddish; *-ellus*, diminutive. Spikelets pale purple

purpureoargentea L. *purpurea*, purple to reddish; *argentea*, silvery. Spikelets purple-silvery

purpureomaculata L. *purpureus*, purple; *macula*, spot; *-ata*, indicating possession. Culm internodes purple-spotted

purpureopedicellata L. *purpurea*, purple to reddish; *pes*, foot; *-ella*, diminutive; *-ata*, possessing. Pedicels reddish-purple

purpureoserice-um, -us L. *purpurea*, purple to reddish; *sericeum*, silky. Inflorescence invested with long purple to reddish hairs

purpurescens A misspelling of *purpurascens*

purpure-um, -us See *purpurea*

purpusiana L. *-ana*, indicating connection. As for *purpusii*

purpusii In honor of Carl Albert Purpus (1853–1941) German-born United States botanist

purshii In honor of Friedrich Traugott Purs(c)h (1774–1826) German-born United States botanist

purushothamanii In honor of K. G. Purushothaman (fl. 1970) Indian botanist

puser Local name for species in Abra Province, Philippines

pushpangadanii In honor of P. Pushpangadan, Indian botanist

pusill-a, -um, -us L. very small. Plants of small stature

pycnanth-a, -um, -us Gk *pyknos*, thick; *anthos*, flower. Spikelets relatively broad

pycnocephalus Gk *pyknos*, thick; *kephale*, head. Panicle obovate-oblong

pycnostachy-a, -um, -us Gk *pyknos*, thick; *stachys*, spike as of an ear of wheat. – (1) Spike cylindric, very dense and stout. *Cenchrus pycnostachyus*, *Pennisetum pycnostachyum* – (2) inflorescence a contracted panicle. *Aristida pycnostachya*, *Stipa pycnostachya*

pycnostachys Gk *pycnos*, thick; *stachys*, spike as of an ear of wheat. See *pycnostachya* (2)

pycnothrix Gk *pyknos*, thick; *thrix*, hair. Glumes with short thick hairs on their keels

pycnotricha Gk *pyknos*, thick; *thrix*, hair. Leaf-blades densely hairy

pygmae-a, -um L. dwarf. Culms shorter than those of many other species in the genus

pynaertii In honor of Édouard Christophe Pynaert Geert (1845–1900) Belgian botanist

pyramidal-e, -is L. *pyramis*, pyramid; *-ale*, pertaining to. Panicle pyramid-shaped

pyramidat-a, -um, -us L. like a pyramid. Panicle pyramid-shaped

pyrenaica L. *-ica*, belonging to. From Pyrenaei Montes, that is the Pyrenees

pyrifera L. *pyrus*, pear; *fero*, carry or bear. Spikelets pear-shaped

pyriform-e, -is L. *pyrus*, pear; *forma*, appearance. Spikelets pear-shaped

pyrogea Gk *pyr*, fire; *ge*, earth. From Patagonia, otherwise known as Land of Fire

pyrophila Gk *pyr*, fire; *philos*, friend. Regenerates well and flowers after fire

Pyrrhanthera Gk *pyrrhos*, flame-colored; *antheros*, flowering. Inflorescence red

pyrularium L. *pyrus*, pear; *-ulus*, diminutive; *-arium*, pertaining to. Spikelets resemble small pears

Q

qiaojiaensis L. *-ensis*, denoting origin. From Qiaojia Xian, Yunnan Province, China

qinghaic-a, -us L. *-ica*, belonging to. From Qinghai, Guinon Xian, China

qingyuanensis L. *-ensis*, denoting origin. From Qingyuan, Zheijiang Province, China

qinlingensis L. *-ensis*, denoting origin. From Qin Ling Mountains, Shaanxi Province, China

Qiongzhuea A compound of *zhu*, a general term for bamboo in Chinese and Mt Qiang-Lai supposed habitat of at least one species of the genus

quadrangula L. *quadrus*, square; *angulus*, corner. Branches quadrangular in transverse section

quadrangularis L. *quatuor*, four; *angulus*, angle; *-aris*, pertaining to. Culms square in cross-section

quadrat-a, -us L. *quadrus*, square; *-ata*, indicating possession. – (1) Transverse veins of leaf-blade conspicuously mark the surface into small squares. *Indocalamus quadratus* – (2) panicle branches arranged at right angles. *Poa quadratus*

quadridens L. *quatuor*, four; *dens*, tooth. Lemma four-toothed

quadridentat-a, -um L. *quatuor*, four; *dens*, tooth; *-ata*, possessing. – (1) Lower glume four-toothed. *Festuca quadridentata* – (2) upper glume four-toothed. *Muhlenbergia quadridentata*, *Podosaemum quadridentatum*

quadridentulus L. *quatuor*, four; *dens*, tooth; *-ulus*, diminutive. The lemma is bilobed and each lobe two-toothed

quadrifari-a, -um L. in four parts. – (1) Spikelets borne in clusters of four. *Stipa quadrifaria* – (2) in two rows of pairs. *Panicum quadrifarium*, *Paspalum quadrifarium*

quadrifida L. *quatuor*, four; *findo*, divide. Lemma apex terminating in four awns

quadriflora L. *quatuor*, four; *flos*, flower. Spikelets mostly with four florets

quadriglume L. *quatuor*, four; *gluma*, husk. Spikelets with two sterile lemmas in addition to the two glumes

quadrinerv-e, -is L. *quatuor*, four; *nervus*, nerve. Lateral nerves of the leaf-blade mostly four

quadriseta L. *quatuor*, four; *seta*, bristle. Lemma four-awned

quadrivalvis L. *quatuor*, four; *valva*, leaf of a folding door. The four sessile male spikelets form an involucre below the hermaphrodite spikelet

quarinii In honor of Camilo Luis Quarín (1943–) Argentinian agrostologist
quarrei In honor of Paul Quarre (1904–1980) Belgian botanist
quartinian-a, -um, -us L. *-ana*, indicating connection. In honor of Richard Quartin-Dillon (?–1841) French botanist who collected in Ethiopia
queenslandic-a, -um L. *-ica*, belonging to. From Queensland, Australia
queko Vernacular name in Colombia for the flute made from the internode of *Aulonemia queko*
quelpaertensis L. *-ensis*, denoting origin. From Quelpeart, now Cheju do, a Korean island
quercetopinetorum Of, that is growing in mixed oak-pine (*Quercus-Pinus*) forests
quercetorum L. *quercetum*, oak grove. Growing in oak-woods
queriana L. *-ana*, indicating connection. In honor of Pio Font i Quer (1888–1964) Spanish botanist
queribunda L. complaining. Origin uncertain, but may refer to the difficulty in distinguishing it from related species
quetameense L. *-ense*, denoting origin. From Quetame, Colombia
quexo See *queko*
quila Vernacular name in Chile for several reed-like grasses
quilioi In honor of A. L. M. le Couriault du Quilio (1815–?) French naval officer
quillinga Origin uncertain, not given by author but possibly the vernacular name of the species in Chile
quilonens-e, -is L. *-ense*, denoting origin. From Quilon, now Kollam, Kerala State, India
quingchengsanensis L. *-ensis*, denoting origin. From Quin Cheng Shan, Sichuan Province, China
quinghaiensis L. *-ensis*, denoting origin. From Quinghai, China
quinhonensis L. *-ensis*, denoting origin. From Qui-nhon, Annam Province, Vietnam
quinqueciliata L. *quinque*, five; *cilium*, eyelid; *-ata*, possessing. All nerves of the lemma are ciliate
quinquefida L. *quinque*, five; *findo*, divide. Lemma five-toothed
quinquenervata L. *quinque*, five; *nervus*, nerve; *-ata*, possessing. Lemma five-nerved
quinquenerv-e, -ia, -is L. *quinque*, five; *nervus*, nerve. Lemma five-nerved
quinqueplumis L. *quinque*, five; *pluma*, feather. Each spikelet cluster has two stalked spikelets each with a pair of plumose setae and the awn of the stalked spikelet is hairy
quinqueset-a, -um, -us L. *quinque*, five; *seta*, bristle. Lemma five-awned
quinquesetica L. *quinque*, five; *seta*, bristle; *-ica*, belonging to. Spikelets with four sterile and one fertile lemma, all awned
quinqueset-um, -us See *quinqueseta*
quinquevalvis L. *quinque*, five; *valva*, leaf of a folding door. The spikelet has five scales comprising the glumes, sterile lemma, fertile lemma and palea
quintasii In honor of Francisco Joachim Dias Quintas (fl. 1893) Portugese civil servant and amateur botanist in Mozambique
Quiongzhuea See *Qiongzhuea*
quiriegoensis L. *-ensis*, denoting origin. From Municipio de Quiriego, Mexico
quirihuense L. *-ense*, denoting origin. From hacia Quirihue, Chile
quitens-e, -is L. *-ense*, denoting origin. – (1) From Quito, Ecuador. *Chusquea quitensis, Eragrostis quitensis, Muhlenbergia quitensis, Paspalum quitense, Poa quitensis* – (2) mistakenly from Quito. *Calamagrostis quitensis*

R

Rabdochloa Gk *rhabdos*, rod; *chloa*, grass. Racemes fastigate
racemiflor-a, -um L. *racemus*, stalk of a cluster of grapes; *flos*, flower. The spikelets are borne in racemes
racemigera L. *racemus*, stalk of a cluster of grapes; *gero*, carry or bear. Inflorescence composed of racemes

Racemobambos Similar to *Bambusa*, with a racemose inflorescence

racemos-a, -um, -us L. *racemus*, stalk of a cluster of grapes; *-osa*, abundance. The spikelets are borne in racemes or contracted panicles

racemulosum L. *racemus*, stalk of a bunch of grapes; *-ulus*, diminutive; *-osum*, abundance. Inflorescence of several shortly stalked racemes

Raddia In honor of Guiseppe Raddi (1770–1829) Italian botanist

raddian-a, -um L. *-ana*, indicating connection. As for *Raddia*

Raddiella L. *-ella*, diminutive but here used as a name-forming suffix. See *Raddia*

radiat-a, -um, -us L. *radius*, spoke of a wheel; *-ata*, possessing. – (1) With racemes arranged in fascicles along a central axis. *Agrostis radiata*, *Arundinaria radiata*, *Atractantha radiata*, *Aulonemia radiata*, *Chloris radiata*, *Digitaria radiata*, *Panicum radiatum*, *Paspalum radiatum* – (2) with culms radiating from a caespitose base. *Paspalidium radiatum*

radicans L. *radico*, take root. Putting forth aerial roots from lower nodes

radiciflora L. *radix*, root; *flos*, flower. Culms dimorphic with the fertile arising separately from the rhizome

radicos-a, -um L. *radix*, root; *-osa*, abundance. Plant with well developed roots or rhizomes

radonensis L. *-ensis*, denoting origin. From Radon Creek, Northern Territory, Australia

radula L. scraper. Rough to the touch. Leaf-blades or other parts asperous

radulans L. *rado*, scrape; *-ula*, tendency or action. Leaf-blades scabrid, that is rasp-like

raduliformis L. *radula*, scraper; *forma*, appearance. Leaf-blades rough to the touch

raegneri See *roegneri*

rafinesqueanum L. *-anum*, indicating connection. In honor of Constantin Samuel Rafinesque-Small (1783–1840) Turkish-born United States botanist and traveller

ragamowski In honor of Ragamowsky

ragonesei In honor of Arturo E. Ragonese (fl. 1934–1946) who collected in Argentina

rahmeri In honor of Carlos F. Rahmer (1858–1917) German-born Chilean taxidermist

raiateensis L. *-ensis*, denoting origin. From Raiatea, French Polynesia

raizadae In honor of Mukat Behari Raizada (1907–) Indian botanist

rajbhandarii In honor of Keshab R. Rajbhandari (fl. 1988–2002) Nepalese botanist

ramboi In honor of P. Balduino Rambo (1905–1961) Brazilian cleric and amateur botanist

ramifera L. *ramus*, branch; *fero*, carry or bear. Culms erect, woody and branched

ramiparum L. *ramus*, branch; *parum*, a little. Inflorescence weakly branched

ramisetum L. *ramus*, branch; *seta*, bristle. Inflorescence branches slender

ramnagarensis L. *-ensis*, denoting origin. From Ramnagar, India

ramonae A contraction of *páramo* at the limits of which the species grows in Venezuela

ramondii In honor of Ramond

ramos-a, -um, -us L. *ramus*, branch; *-osa*, abundance. Inflorescences or culms much branched

Ramosia In honor of Maximo Ramos (1882–1932) Philippine plant collector

ramosissim-a, -um, -us L. *ramus*, branch; *-osa*, abundance; *-issima*, most. – (1) Inflorescence an open, much branched panicle. *Agrostis ramosissima*, *Cenchrus ramosissimus*, *Ischaemum ramosissimum*, *Olyra ramosissima*, *Panicum ramosissimum*, *Parodiolyra ramosissima*, *Paspalum ramosissimum*, *Triplopogon ramosissimus* – (2) culms much branched. *Arundinaria ramosissima*, *Aulonemia ramosissima*, *Chusquea ramosissima*, *Muhlenbergia ramosissima*, *Pennisetum ramosissimum*, *Pleioblastus ramosissimus*, *Poa ramosissima*, *Sasa ramosissima*, *Sehima ramosissima*, *Stipa ramosissima*, *Triraphis ramosissima*, *Vilfa ramosissima*

ramos-um, -us See *ramosa*

Rampholepis Gk *rhamphos*, the crooked-beak of a bird of prey; *lepis*, scale. The spikelets are gaping and gibbous in outline

ramular-e, -is L. *ramulus*, small branch; *-are*, pertaining to. Inflorescence much branched

ramulos-a, -us L. *ramulus*, small branch; *-osa*, abundance. – (1) Culms branching. *Agrostis ramulosa, Sporobolus ramulosus, Vilfa ramulosa* – (2) inflorescences branching. *Setaria ramulosa*

R

rangacharianum L. *-anum*, indicating connection. As for *rangacharii*

rangacharii In honor of Kadami Ranga Achariyar (1868–1934) Indian botanist

rangei In honor of – (1) Paul Range (1879–1952) German geologist who collected plants in South Africa. *Sporobolus rangei* – (2) Max Range, German physician who collected in S.W. Africa. *Melinis rangei, Merxmuellera rangei, Pennisetum rangei*

rangkulensis L. *-ensis*, denoting origin. From Rangkul, a lake in Tajikistan

rankingii In honor of Robert Archibald Ranking (1843–1912) British-born Australian magistrate

raoulii In honor of Édouard Fiacre Louis Raoul (1815–1852) French naval surgeon

rapensis L. *-ensis*, denoting origin. From Rapa Island, in southeast Polynesia

Raphis See *Rhaphis*

rar-a, -um L. far apart. Spikelets far apart in panicle

Raram Meaning obscure, origin not given by the author

rariflor-a, -um, -us L. *rarus*, far apart; *flos*, flower. – (1) With florets well separated on the rachilla. *Hierochloe rariflora, Ortachne rariflora, Orthoclada rariflora, Thaumastochloa rariflora* – (2) with spikelets well separated in inflorescence. *Oplismenus rariflorus, Panicum rariflorum, Setaria rariflora* – (3) panicle with few spikelets. *Muhlenbergia rariflora*

raripilum L. *rarus*, far apart; *pilus*, a hair. Sparsely hairy

rarisetum L. *rarus*, far apart; *seta*, bristle. Leaf-blades with a few long, scattered hairs

raroflorens L. *rarus*, far apart; *floreo*, flower. Plants rarely flower

rarum See *rara*

Raspailia In honor of François Vincent Raspail (1791–1878) French physician and botanist

Raspalia See *Raspailia*

Rattraya In honor of James McFarlane Rattray (1907–1974) agronomist in Zimbabwe

Ratzeburgia In honor of Julius Theodor Christian Ratzeburg (1801–1871) German forester, botanist and zoologist

rauhii In honor of Werner Hermann Heinrich Rauh (1913–2000) German botanist

raunkiaeri In honor of Christen Christiansen Raunkiaer (1860–1938) Danish botanist

rautanenii In honor of Martin Rautanen (1845–1926) Finnish missionary who collected in Amboland, S.W. Africa

ravenelii In honor of Henry William Ravenel (1814–1887) United States botanist, plant collector and mycologist

ravennae From the valley of Ravenna, Italy

ravianus L. *-anus*, indicating connection. In honor of Nanoo Ravi (1938–) Indian botanist

rawitscheri In honor of Felix Rawitscher (or Rawitcher) (1890–1957) German-born Brazilian botanist

raynaliana L. *-ana*, indicating connection. In honor of Aline Marie Roques Raynal (1933–) French botanist

readeri In honor of Felix Maximilian Reader otherwise von Reyder (1850–1911) German-born Australian pharmacist and botanist

Reana Named for Reana del Royale, Province of Udine, Italy, where the author of the name resided

Rebentischia In honor of Johann Friedrich Rebentisch (1772–1810) Prussian botanist

Reboulea In honor of Eugenè de Reboul (1781–1851) French-born Italian botanist

rechingeri In honor of Karl Heinz Rechinger (1906–1998) Austrian botanist

reclinat-a, -um L. *reclino*, lean back. Culms weakly procumbent

recognitum L. *recognesco*, recognize. Long recognized before formally described

rect-a, -um, -us L. upright. Panicle branches erect or spike-like
recticlada L. *rectus*, upright; Gk *klados*, stem. Culms upright
rectirhachis L. *rectus*, upright; Gk *rhachis*, backbone. Pedicels of racemes erect
rectocuneatus L. *rectus*, straight. Leaf-blades cuneate without basal nodes
rect-um, -us See *recta*
recurvat-a, -us L. reflexed. Spikelets with spreading or reflexed awns
redacta L. *reduco*, reduce. Lateral branches of awn very reduced
Redfieldia In honor of John Howard Redfield (1815–1895) United States amateur botanist
redheadii In honor of Edgar Wolston Bertram Handsley Milne-Redhead (1906–1996) English botanist
redivivum L. reviving from a dried state. Able to withstand drought
redolens L. *redoleo*, give off an odor. Inflorescences smell of coumarin
redondense L. *-ense*, denoting origin. From Fazenda Capão Redondo, Paraná, Brazil
redowskii In honor of Ivan Redowski (1774–1807) Russian botanist
reduncum L. bent backwards. Racemes of inflorescence recurved
redundans L. *redundo*, be abundant. Locally abundant
reederi In honor of John Raymond Reeder (1914–) United States botanist
Reederochloa See *reederorum*
reederorum In honor of John Raymond (1914–) and Charlotte Gooding (1916–) Reeder, United States botanists
reedii In honor of A. C. Reed, railroad manager in Cuba
reflex-a, -um, -us L. bent sharply backwards. – (1) Panicle branches reflexed. *Anthistiria reflexa*, *Bouteloua reflexa*, *Cymbopogon reflexus*, *Deyeuxia reflexa*, *Digitaria reflexa*, *Eragrostis reflexa*, *Festuca reflexa*, *Olmeca reflexa*, *Poa reflexa*, *Sporobolus reflexa* – (2) spikelets bent in the middle. *Arthrostylidium reflexum* – (3) spikelets reflexed. *Pentaschistis reflexa* – (4) leaf-blades reflexed. *Trichopteryx reflexa*
reflexiaristat-a, -um L. *reflexa*, bent sharply backwards; *arista*, bristle; *-ata*, possessing. Awn bent backwards
reflex-um, -us See *reflexa*
refract-a, -um, -us L. curved back abruptly. Inflorescence branches or awns reflexed at maturity
regelian-a, -us In honor of Eduard August Regel (1815–1892) German-born Russian botanist
regelii As for *regeliana*
regis L. *rex*, king. From Laguna del Rey, that is "Lagoon of the King", Coahuila State, Mexico
regnellii In honor of Anders Frederick Regnell (1807–1884) Swedish physician and botanist
regnii In honor of Karl Heinz Rechinger (1906–1998) Austrian botanist, using in reverse the last four letters of his surname
regular-e, -is L. regular. Typical for the genus
Rehia In honor of Richard Eric Holttum (1895–1990) English botanist
rehmannii In honor of Antoni Rehmann (1840–1917) Polish botanist and geographer who collected in South Africa
reholttumianus L. *-anus*, indicating connection. As for *Rehia*
reimannii In honor of Karl Reimann (1843–1904) German engineer
Reimaria In honor of Juan Alberto Enrique Reimar (1729–1814) German physician and biologist
reimarioides Gk *-oides*, resembling. Similar to *Reimaria*
Reimarochloa Gk *chloa*, grass. See *Reimaria*
Reimbolea In honor of Reimbole who collected in Sicily
reinwardtii In honor of Caspar Georg Carl Reinwardt (1773–1854) German-born Dutch botanist
Reitzia In honor of Raulino Reitz (1919–1990) Brazilian botanist
rejuvenescens L. *rejuvensco*, able to rejuvenate. Recovering readily after fire
Relchela An anagram of *Lechlera*
remissa L. *remitto*, drive back. Culms retrorsely scabrid

remot-a, -um, -us L. distant. – (1) Spikelets widely separated. *Brachiaria remota, Digitaria remota, Glyceria remota, Orthopogon remotus, Panicum remotum, Poa remota* – (2) from an isolated locality. *Danthonia remota*

remotiflor-a,-us L. *remotus*, distant; *flos*, flower. Spikelets with widely separated florets

remotigluma L. *remotus*, distant; *gluma*, husk. Lower glume is minute and remote from the upper

remot-um, -us See *remota*

R

remyi In honor of Esprit Alexandre Remy (1826–1893) French botanist

rendlei As for *Rendlia*

Rendlia In honor of Alfred Barton Rendle (1865–1938) English botanist

renggeri In honor of Johann Rudolf Rengger (1795–1832) Swiss botanist who collected in Paraguay

reniformis L. *renes*, kidney; *forma*, appearance. Lemmas kidney-shaped

renvoizei In honor of Stephen Andrew Renvoize (1944–) English agrostologist

repandum L. bent backwards. Axis of inflorescences winged with the margins reflexed

repatrix L. *repo*, crawl; *thrix*, hair. Rhizome well developed

repens L. *repo*, crawl. Rhizome well developed

repentellum L. *repens*, creeping; *-ellum*, diminutive. Related to *Panicum repens* but much more slender

reptans L. *repo*, crawl. Culms ascending from a creeping rhizome or runner

reptatum L. *repto*, creep. Culms procumbent and root at the nodes

requienii In honor of Esprit Requien (1788–1851) French botanist

rescissum L. *rescindo*, cut off. The apices of the glumes and sterile lemma are somewhat erose

respiciens L. *respico*, look backwards. The barbs on the involucral bristles are directed away from their apices

restingae Portuguese *restinga*, a sandy spit. Growing in restinga forests, so-called because they grow on sandy soils subject to flooding

restingense L. *-ense*, denoting origin. As for *restingae*

restioide-a, -um Gk *-oidea*, resemblance. Similar to *Restio*

restionaceus L. *-aceus*, resembling. Similar to *Restio*

reticulat-a, -um L. *reticulum*, net; *-ata*, possessing. – (1) Glumes net-veined. *Brachiaria reticulata, Panicum reticulatum, Paspalum reticulatum, Thrasya reticulata* – (2) leaf-blades net-veined. *Phyllostachys reticulata, Sinoarundinaria reticulata*

reticulinerve L. *reticulum*, net; *nervum*, nerve. Venation of lower glume reticulate

retiglume L. *retis*, net; *gluma*, husk. Upper glume and lower lemma with reticulate venation

retinorrhoea Gk *retine*, resin; *rhoia*, flux. Plants sticky

retorta L. *retorqueo*, bend back. Awn hygroscopic

retrofacta See *retrofracta*

retroflex-a, -us L. *retro*, backwards; *flexus*, bend. Inflorescence branches or leaf-blades reflexed

retrofract-a, -um L. *retro*, backwards; *fractus*, broken. – (1) With retrorse hairs on the rhachis. *Agropyron retrofractum* – (2) with retrorse hairs on the glumes. *Agrostis retrofracta*

retropila L. *retro*, backwards; *pilus*, a hair. Internodes densely retrorse-ciliate

retrorsa L. turned backwards. – (1) Hairs on upper surface of palea retrorsely disposed. *Olyra retrorsa* – (2) with retrorse hairs on the leaf-sheaths. *Merostachys retrorsa*

Rettbergia In honor of Elmann Rettberg, German botanist

retus-a, -um L. *retundo*, blunt. Apices of lemmas rounded or notched sometimes with a small mucro

retzii In honor of Anders Jahan Retzius (1742–1821) Swedish botanist

reuteri In honor of Guillaume Reuter (1808–1872) Swiss botanist

reuteriana L. *-ana*, indicating connection. As for *reuteri*

reverchonii In honor of Julien Reverchon (1837–1905) French-born United States plant collector

reverdattoi In honor of Viktor Vladimirovich Reverdatto (1891–1969) Russian botanist
reversipilum L. *reverto*, turn back; *pilus*, a hair. Hairs on the leaf-sheath retrorse
reversum L. *reverto*, turn back. Spikes initially erect then reversed
reygeri In honor of Gottfried Reyger (1704–1788) Prussian botanist
Reynandia See *Reynaudia*
Reynaudia, reynaudia In honor of A. A. M. Reynaud (1804–?) French Naval Surgeon and plant collector
reynaudiana L. *-ana*, indicating connection. As for *Reynaudia*
reynaudioides Gk *-oides*, resembling. Resembling *Reynaudia*
reynoldensis L. *-ensis*, denoting origin. From Reynolds Creek, southeast Queensland, Australia
Rhabdochloa See *Rabdochloa*
Rhachidospermum Gk *rhachis*, backbone; *sperma*, seed. The spikelets are embedded in depressions of the spongy material of the thickened rhachilla
rhachitrich-a, -um, -us Gk *rhachis*, backbone; *thrix*, hair. Rhachis hirsute
rhadina Gk *rhadinos*, delicate. Habit tufted, leaf-blades filiform
rhaetica L. *-ica*, belonging to. From Rhaetia, a Roman Province now included in the Austrian Tyrol, Bavaria and northern Italy
Rhampholepis Gk *rhamphos*, the curved beak of a bird of prey; *lepis*, scale. Long axis of spikelet curved
Rhaphis Gk needle. The fertile spikelet bears a needle-like callus
rheedii In honor of Heinrich van Rheede tot Droakenstein (1637–1692) Governor of Dutch possessions in Malabar, India
rhenana L. *-ana*, indicating connection. From Rhenanus, now Rhine River, Western Europe
Rheochloa Gk *chloa*, grass. First collected in the Parc Nacional das Emas, so named for the flightless bird, *Rhea americana*
rhigiophyllum Gk *rhigos*, frost; *phyllon*, leaf. Leaf-sheath invested with dense white hairs appearing as if frosted
rhignon Gk shrivelled with old age. Origin uncertain, not given by author but possibly a reference to a rugose lemma
Rhiniachne Gk *rhine*, file; *achne*, scale. The lower glume is leathery with transverse ribs
rhiniochloa Gk *rhine*, file; *chloa*, grass. Leaf-blades scabrid
Rhipidocladum Gk *rhipis*, fan; *klados*, branch. Branch complement fan-like
rhizantha Gk *rhiza*, a root; *anthos*, flower. Flowering culms discrete and arising directly from the rhizome
Rhizocephalus Gk *rhiza*, a root; *kephale*, head. The capitate inflorescence is borne near the base of the culms
rhizogonum Gk *rhiza*, a root; *gony*, knee. Runners root at the nodes
rhizomat-a, -um, -us Gk *rhizoma*, a root. Rhizomes well developed
rhizomatis Gk *rhizoma*, a root. Plants rhizomatous
rhizomatosum Gk *rhizoma*, root; L. *-osum*, abundance. Conspicuously rhizomatous
rhizomat-um, -us See *rhizomata*
rhizomophora Gk *rhizoma*, a root; *phero*, bear. Plant with well developed scaley rhizomes
rhizophor-a, -um, -us Gk *rhiza*, a root; *phero*, bear. Rooting at the lower nodes
rhodesian-a, -um L. *-ana*, indicating connection. From Rhodesia, now Zimbabwe
rhodopea From Mount Rhodopea, Bulgaria
rhodopedum Gk *rhodon*, rose-colored; *pedon*, ground. From Rioda Terra Vermelha, Santa Catarina Province, Brazil
rhomboidea Gk *rhombos*, rhombus; *-oidea*, resemblance. Spikelets rhomboid in outline
Rhombolytrum Gk *rhombos*, rhombus; *elytrum*, cover. The glumes are rhombus-shaped
rhyncantha Gk *rhynchos*, snout; *anthos*, flower. Origin uncertain, not given by author
Rhynchelythrum See *Rhynchelytrum*
rhynchelytroides Gk *-oides*, resembling. Similar to *Rhynchelytrum*
Rhynchelytrum Gk *rhynchos*, snout; *elytron*, cover. The upper glume and sterile lemma are shortly beaked

R

rhynchophorus Gk *rhynchos*, snout; *phero*, bear. Lower glume of pedicelled spikelet acuminate

Rhynchoryza See *Rynchoryza*

Rhytachne, rhytachne Gk *rhytis*, a wrinkle; *achne*, scale. The lower glume is transversely rugose

rhytachnoides Gk *-oides*, resembling. See *Rhytachne*

Rhytidachne Gk *rhytis*, a wrinkle; *achne*, scale. See *Rhytachne*

R

riabuschinskii In honor of Th. P. Riabuschinskij (fl. 1908) Russian botanist who collected in Kamchatka, Russian Far East

riauensis L. *-ensis*, denoting origin. From Riau Province, Sumatra, Indonesia

ribbentropii Named for Joachim Ribbentrop (1893–1946) Nazi politician

riccerii In honor of Carlo Ricceri (1933–) Italian botanist

richardii In honor of – (1) Achille Richard (1794–1852) French botanist. *Arundo richardii* – (2) Jean Michel Claude Richard (1784–1868) French botanist and Garden's Curator at Réunion Island. *Panicum richardii, Pseudostreptogyne richardii* – (3) Louis Claude Marie Richard (1754–1821) French botanist and plant collector in Antilles and South America. *Lachnagrostis richardii, Paspalum richardii, Pennisetum richardii, Vilfa richardii*

Richardsiella L. *-ella*, diminutive but here used as a name-forming suffix. In honor of Mary Alice Eleanor Richards (1895–1977) English botanist resident in Zambia

richardsonii, richardsonis (1) In honor of John Richardson (1787–1865) English physician, naturalist and Arctic explorer. *Agropyron richardsonii, Agrostis richardsonii, Bromus richardsonii, Festuca richardsonii, Hordeum richardsonii, Stipa richardsonii, Zerna richardsonii, Muhlenbergia richardsonis, Vilfa richardsonis* – (2) in honor of Arnold Edwin Victor Richardson (1883–1949) Australian agricultural scientist. *Danthonia richardsonii* – (3) in honor of Arthur Johnstone Richardson (fl. 1898) British Army Officer who collected in Nigeria. *Digitaria richardsonii*

richteriana L. *-ana*, indicating connection. In honor of Herman Eberhard Friedrich Richter (1808–1876) German botanist

ridleyi In honor of Henry Nicholas Ridley (1855–1956) English-born Malayan botanist

Riedelia In honor of Ludwig Riedel (1790–1861) German plant collector and traveler

riedeliana L. *-ana*, indicating connection. As for *Riedelia*

riedelii As for *Riedelia*

rifana L. *-ana*, indicating connection. From the Rif district of Morocco

rigens L. *rigeo*, be stiff. – (1) Leaf-blades cylindrical or inrolled when dry. *Isachne rigens, Panicum rigens, Sporobolus rigens, Vilfa rigens* – (2) culms erect. *Poa rigens* – (3) racemes ascending. *Paspalum rigens*

rigescens L. *rigesco*, become rigid. Leaf-blades stiff and erect

rigid-a, -um, -us L. stiff. Culms, spikelets or inflorescence branches held stiffly erect

rigidifoli-a, -um, -us L. *rigidus*, stiff; *folium*, leaf. Leaf-blades stiff and often with a rigid tip

rigidior L. more rigid. Culms to a metre tall

rigidiseta L. *rigidus*, stiff; *seta*, bristle. Lemma awn rigid

rigidissima L. *rigidus*, stiff; *-issimus*, most. Leaf-blades rolled, rigid

rigidiuscula L. *rigidius*, stiffer; *-ula*, diminutive. Leaf-blades tending to be held erect

rigidul-a, -um L. *rigidus*, stiff; *-ula*, diminutive. Plant with stiffly erect inflorescence branches or leaf-blades

rigid-um, -us See *rigida*

rigoi In honor of Gregorio Rigo (1841–1922) Italian botanist

riguorum L. *riguus*, a well-watered place. Of well watered places

rigurosa L. *rigeo*, be stiff; *-osa*, abundance. Leaf-blades sub-pungent

riloensis L. *-ensis*, denoting origin. From Mt Rila, Bulgaria

rimpaui In honor of Wilhelm Rimpau (1842–?) German plant breeder

ringoetii In honor of A. Ringoet (fl. 1889) who collected in Zaire

rinihuensis L. *-ensis*, denoting origin. From river Rinihue, Patagonia

riobrancensis L. *-ensis*, denoting origin. From land drained by the Rio Branco, Territory of Roraima, Brazil

riograndensis L. *-ensis*, denoting origin. From Rio Grande de Sul, Brazil

rioplatensis L. *-ensis*, denoting origin. From Rio Plate, Uruguay

riosaltensis L. *-ensis*, denoting origin. From Paradão do Rio Salto, Minas Gerais State, Brazil

ripari-a, -um, -us L. *ripa*, river bank; *-aria*, pertaining to. Growing on river banks

riparioides L. *ripa*, river bank; Gk *-oides*, resembling. Growing in water or along river banks

ripari-um, -us See *riparia*

riphaea A Riphaean, that is an inhabitant of the *Rhiphaei* (or *Riphaei*) *montes*, classical name for the area around the source of the Don River in southeast Russia

Ripidium Gk *rhipis*, fan; *-idium*, diminutive. Inflorescence a large panicle

ritcheyi See *ritchiei*

ritchiei In honor of David Ritchie (1809–1866) physician and plant collector in India

rivae In honor of Domenico Riva (1856?–1895) Italian physician who collected in Somalia

rival-e, -is L. pertaining to brooks. Growing along river banks

rivas-martinezii In honor of Salvador Rivas-Martínez (1935–) Spanish botanist

rivular-e, -is L. *rivulus*, river; *-are*, pertaining to. Growing adjacent to rivers

rivulorum L. *rivulus*, river. Of the rivers, that is in communities associated with river banks

roanokense L. *-ense*, denoting origin. From Roanoke Island, North Carolina, USA

robecchii In honor of Luigi Robecchi-Bricchetti (1855–1926) Italian botanist who worked in Somalia

robertianus L. *-anus*, indicating connection. In honor of Robert Brown (1773–1858) Scots born English botanist

robertii In honor of Robert Thorbjörn Porsild (1898–1977) Danish-born Canadian botanist

robinsoniana L. *-ana*, indicating connection. In honor of – (1) Benjamin Lincoln Robinson (1864–1935) United States botanist. *Koeleria robinsoniana* – (2) the literary character Robinson Crusoe, who was marooned on the island of Robinson Crusoe (also known as Más a Tierra) in the Juan Fernandez Archipelago, from where the species was first collected. *Phalaris robinsoniana*, *Phalaridantha robinsoniana*

robinsonii (1) In honor of Charles Budd Robinson (1871–1913) Canadian-born United States botanist murdered while collecting on Amboina. *Digitaria robinsonii* – (2) in honor of Frederick Robinson (fl. 1911–1923) English botanist. ×*Agropogon robinsonii* – (3) origin uncertain. *Agrostis* × *robinsonii*

roblensis L. *-ensis*, denoting origin. From Rancho El Roble near El Derramandero, Mexico

roborowskyi In honor of Vsevolod Ivanovi Roborowsky (1856–1910) Russian botanist

robust-a, -um L. robust. Culms tall, or leaf-blades or spikelets large

robustifolia L. *robustus*, robust; *folium*, leaf. Foliage coarse

robustiramea L. *robustus*, robust; *ramus*, branch. Lateral branches well developed

robustissimus L. most robust. Culms very tall for the genus

robustum See *robusta*

robynsii As for *Robynsiochloa*

Robynsiochloa Gk *chloa*, grass. Named in honor of Frans Hubert Edouard Arthur Walter Robyns (1901–1986) Belgian botanist

rocanum L. *-anum*, indicating connection. In honor of Modesta Roca, Cuban cleric and friend of Brother, Frère or Hermano Léon (also known as Joseph Sylvestre Sauget-Bargier). See *Saugetia* for details of the latter

rochelianus L. *-anus*, indicating connection. As for *rochelii*

rochelii In honor of Anton Rochel (1770–1847) Austrian horticulturalist

rockii In honor of Joseph Francis Charles Rock (1884–1962) Austrian-born United States botanist

rodetii In honor of Commandant Rodet, French military governor of the district in Algeria where this hybrid grew abundantly
rodnensis L. *-ensis*, denoting origin. From Rodnei Muntii, a mountain range in Romania
rodriguezii In honor of Rodriguez
rodwayi In honor of – (1) Frederick Arthur Rodway (1880–1956) Australian physician and plant collector. *Danthonia rodwayi*, *Deyeuxia rodwayi* – (2) Leonard Rodway (1853–1936) English-born Australian dental surgeon and amateur botanist. *Poa rodwayi*
roegneri As for *Roegneria*
Roegneria In honor of Roegner of Orcanda (fl. 1844) Crimea
roemeri In honor of – (1) Hans L. Roemer, Canadian ecologist. *Festuca roemeri* – (2) Lucien Sophie Albert Marie von Roemer (1873–?) Dutch physician in Indonesia. *Setaria roemeri*
Roemeria In honor of Johann Jacob Roemer (1763–1819) Swiss physician and naturalist
roemeriana L. *-ana*, indicating connection. As for *Roemeria*
rogeri In honor of Roger who collected in Senegal
rogersii In honor of – (1) Frederick Arundel Rogers (1876–1944) English cleric and botanist who collected widely in Africa and Iran. *Eragrostis rogersii* – (2) Charles Gilbert Rogers (1864–1937) English-born Indian forester. *Schizostachyum rogersii*
rohlfsii In honor of Gerhard Rohlfs (1831–1896) German traveller and collector in Africa
rohmooana L. *-ana*, indicating connection. In honor of Rohmoo (fl. 1910) a Lepcha plant collector probably born in Sikkim State, India
roigii In honor of Fidel A. Roig (fl. 1990) Argentinian agriculturalist
rojasii In honor of Teodoro Rojas (1877–1954) Paraguayan botanist
rolloana L. *-ana*, indicating connection. In honor of James Rollo (fl. 1889) who collected in India
rollotii In honor of Maurice A. Rollot who collected in Republic of Columbia
romae In honor of Angel Maria Romo (1955–) Spanish botanist
romeroi-zarcoi In honor of Carlos Romero-Zarco (1953–) Spanish botanist
rondoniensis L. *-ensis*, denoting origin. From Rôndonia, Brazil
ronnigeri In honor of Karl Ronniger (1871–1954) Austrian botanist
ropalotrich-a, -um Gk *rhopalos*, club; *thrix*, hair. Glumes and lower lemma bearing club-shaped hairs
roraimensis L. *-ensis*, denoting origin. From Mt Roraima, Guyana
rosacea L. *rosea*, pink; *-acea*, indicating resemblance. Spikelets pink
roscida L. covered with a dew-like exudation. Stems and leaves covered with resin
rose-a, -um, -us L. pink. Spatheoles and/or spikelets pink
rosei In honor of Joseph Nelson Rose (1862–1928) who collected widely in the Americas
rosengurttii In honor of Bernado Rosengurtt (1916–) who collected in South America
rosenkrantzii In honor of A. Rosenkrantz (fl. 1926–1953) Danish geologist
roseotomentosum L. *rosea*, pink; *tomentum*, stuffing material of a pillow; *-osa*, abundance. Indumentum rose-pink when fresh
rosettae In memory of Rosette Cugnac, daughter of A. de Cugnac
rosettanum L. *-anum*, indicating connection. From Rosetta, otherwise Rashid, Egypt
rose-um, -us See *rosea*
Roshevitsia In honor of Romain Julievic Roshevitz (1882–1949) Russian agrostologist
roshevitsian-a, -um, -us L. *-ana*, indicating connection. As for *Roshevitsia*
roshevitsii As for *Roshevitsia*
roshevitzii As for *Roshevitsia*
rossbergiana L. *-ana*, indicating connection. In honor of William Rossberg (?–1940) United States botanist
rossiae In honor of Edith A. Ross (fl. 1885–1895) United States amateur botanist

Rostraria L. *rostrum*, beak; *-aria* pertaining to. The lemma is beaked

rostrat-a, -us L. *rostrum*, beak; *-ata*, possessing. – (1) Lower glume with two short beaks. *Andropogon rostratus*, *Elionurus rostratus* – (2) lemma with a rostrate apex. *Gigantochloa rostrata*

rotae In honor of Lorenzo Rota (1819–1855) Italian physician and botanist

Rotbolla See *Rottboelia*

Rotbollia See *Rottboelia*

Rothia In honor of Albrecht Wilhelm Roth (1757–1834) German physician and botanist

rothmaleri In honor of Werner Hugo Paul Rothmaler (1908–1967) German botanist

rothrockii In honor of Joseph Trimble Rothrock (1839–1922) United States physician and botanist

rotifer L. *rota*, wheel; *fero*, carry or bear. Lower nodes of inflorescence bear dense whorls of branches

Rottboelia, Rottboella, Rottboellia, rottboellia In honor of Christen Friis Rottboell (1727–1797) Danish botanist

rottboellioides Gk *-oides*, resembling. Resembling *Rottboellia* with respect to the inflorescence

rottleri In honor of Johan Peter Rottler (1749–1836) French-born Indian cleric and botanist

rotundat-a, -us L. *rotundus*, round; *-ata*, possessing. Spikelets subrotund

rotundiflora L. *rotundus*, round; *flos*, flower. Spikelets spherical

rotundissima L. *rotundus*, round; *-issima*, most. Culms terete

rotundum L. round. Spikelets subrotund and very turgid

Rouxia In honor of Nisius Roux (1854–1923) French botanist

rouxii In honor of Honoré Roux (1812–1892) French botanist

rovellii In honor of Renato Rovelli (1806–1880)

rovumense L. *-ense*, denoting origin. From Rovuma now Mozambique

rowlandii In honor of John William Rowland (1852–1925) who collected in Nigeria

roxburghian-a, -um, -us L. *-ana*, indicating connection. As for *roxburghii*

roxburghii In honor of John Roxburgh (fl. 1770s–1820s) sometime Overseer, Botanic Garden, Calcutta

Roylea In honor of John Forbes Royle (1798–1858) English botanist, sometime resident in India

roylean-a, -um, -us L. *-ana*, indicating connection. As for *Roylea*

roylei See *Roylea*

ruahensis L. *-ensis*, denoting origin. From Ruaha National Park, Tanzania

rubella L. *ruber*, red; *-ella*, diminutive. Spikelets pale-red

rubens L. *rubeo*, be red. Spikelets reddish-purple

rubicund-a, -us L. red. Stems and sheaths at first reddish

rubida L. reddish. Inflorescence branches and spikelets reddish

rubiginos-a, -um L. rusty red. Inflorescences reddish

rubra L. red. – (1) Foliage red. *Agrostis rubra*, *Arundo rubra*, *Chionochloa rubra* – (2) spike-lets red. *Briza rubra*, *Festuca rubra*

rubroligula L. *ruber*, red; *ligula*, small tongue. Ligule red

rubromarginata L. *ruber*, red; *marginis*, edge; *-ata*, possessing. Ligule and oral setae red

rubrotinctum L. *ruber*, red; *tingo*, color. Plant reddish

rud-e, -is L. uncultivated. Species whose relatives are often cultivated

ruderalis L. growing wild near human habitation

rudgei In honor of Edward Rudge (1763–1846) English magistrate and amateur botanist

rudimentifer L. *rudimentum*, beginning; *fero*, carry or bear. The rudiments of a second floret are sometimes developed by the spikelet

rudis See *rude*

rudiuscula L. *rudius*, wilder; *-ula*, diminutive

rueppelianum, rueppellianus L. *-anum*, indicating connection. As for *ruppellii*

ruf-a, -um, -us L. reddish. – (1) Inflorescence purple to red. *Andropogon rufus, Anthaenantia rufa, Aulaxanthus rufus, Briza rufa, Chascolytrum rufum, Cymbopogon rufus, Hyparrhenia rufa, Monium rufum, Sorghum rufum, Trachypogon rufus* – (2) culm-sheaths purple to red. *Fargesia rufa*

rufescens L. *rufesco*, grow reddish. Spikelets or inflorescence reddish-brown

ruficom-a, -um L. *rufus*, reddish; *coma*, hair of the head. Glumes and sterile lemmas with abundant reddish hairs

rufinflatum L. *rufus*, reddish; *inflo*, inflate. Mature spikelets swollen and reddish

rufipil-um, -us L. *rufus*, reddish; *pilus*, a hair. Spikelets surrounded by mauve-colored long hairs

rufipogon L. *rufus*, reddish; Gk *pogon*, beard. Awns reddish-brown

rufispicum L. *rufus*, red; *spica*, a point; hence, in particular, an ear or spike of grain. Inflorescence invested with red hairs

rufobarbatum L. *rufus*, red; *barba*, beard; *-ata*, possessing. Awn with reddish hairs

ruf-um, -us See *rufa*

rugelii In honor of Ferdinand Ignatius Xavier Rugel (1806–1878) German-born United States botanist, physician and apothecary

rugi Vernacular name of the species in Southern Chile

rugoloana L. *-ana*, indicating connection. In honor of Zulma E. Rúgolo de Agrasar (1940–) Argentinian botanist

rugos-a, -um L. *ruga*, wrinkle; *-osa*, abundance. Usually with sculptured glumes

rugosiglumis L. *ruga*, wrinkle; *-osa*, abundance; *gluma*, husk. Lemma margins transversely rugose

rugulos-a, -um L. *ruga*, wrinkle; *-ula*, diminutive; *-osa*, abundance. Fertile lemma conspicuously wrinkled

ruiz-lealii In honor of Adrian Ruiz Leal (fl. 1933–1942) who collected in Argentina

rukwae From Rukwa, Tanzania

rumphiana L. *-ana*, indicating connection. In honor of Georg Eberhard Rumphius (Rumpf) (1628–1702) German-born Dutch naturalist

Runcina In honor of Runcina, a Roman Goddess invoked to prevent the growth of weeds and so promote the harvest

runemarkii In honor of Hans Runemark (1927–) Swedish botanist

runssoroensis L. *-ensis*, denoting origin. From Ruwenzori, East Africa

runyonii In honor of Robert Runyon (1881–1968) United States botanist

rupestr-e, -is L. *rupes*, rock; *-estre*, place of growth. Growing amongst rocks

Rupestrina L. *rupes*, rock; *-estre*, place of growth; *-ina*, indicating possession. Growing in rocky places

rupestris See *rupestre*

rupicaprina L. *rupes*, rock; *caper*, he-goat; *-ina*, indicating possession. A species of high mountain goat pastures

rupicola L. *rupes*, rock; *-cola*, dweller. Growing on rocky slopes

rupincola See *rupicola*

rupium L. *rupes*, rock; *-ium*, characteristic of. Growing amongst rocks

ruppelian-a, -us, ruppelliana L. *-anum*, indicating connection. As for *ruppellii*

ruppellii In honor of (Wilhelm Peter) Eduard (Simon) Rüppell (1794–1884) traveller in North Africa

ruprechtii In honor of Franz Josef Iwanowitsch Ruprecht (1814–1870) German-born Russian botanist

rura L. *rus*, countryside. Growing wild

ruschii In honor of Ernest Julius Rusch (1867–1957) or Ernst Franz Theodor Rusch (1897–1964). German-born/South African-born South African business men and plant collectors

ruscifoli-a, -um L. *folium*, leaf. Leaf-blades ovate, resembling the cladodes of *Ruscus*

ruscinonensis L. *-ensis*, denoting origin. From Ruscinon, now Roussillen, France

ruspolian-um, -us L. *-anum*, indicating connection. As for *ruspolii*

ruspolii In honor of Eugenio Ruspoli (1866–1893) Italian nobleman who travelled in Somalia

russellii In honor of R. Scott-Russell (fl. 1939)

ruthenic-a, -us L. *-ica*, belonging to. From Ruthenia, now mainly Moldavia and the Ukraine

rutila L. red. Spikelets purplish-red

rutilans L. *rutilo*, make reddish. Culms and leaf-sheaths are reddish-orange when young

ruwensorensis L. *-ensis*, denoting origin. From Mt Ruwenzori, one of the peaks in a range of that name on the border of Zaïre and Uganda

ruwenzoriensis See *ruwensoreensis*

ruziziensis L. *-ensis*, denoting origin. From Ruzizi Plains, Burundi

rydbergii In honor of Per Axel Rydberg (1860–1931) Swedish-born United States botanist, included in anticipation of finding a species

Rynchoryza Gk *rhynchos*, beak. Resembling *Oryza* with the fertile lemma tapering into a long awn

Rytachne See *Rhytachne*

Rytidosperma Gk *rhytis*, wrinkle; *sperma*, seed. Wrinkled larvae mistaken for caryopses

Rytilix Gk *rhytis*, wrinkle; *kalyx*, cup. The grain is enfolded by the lower glume of the sessile spikelet, whose surface is pitted

ryukyuensis L. *-ensis*, denoting origin. From Ryukyu, Japan

rzedowskiana L. *-ana*, indicating connection. In honor of Jerzy Rzedowski (1926–1969) Polish-born Mexican botanist

S

sabalanica L. *-ica*, belonging to. From Sabalan, Kuhha-ye Mountains, Iran

sabarimalayana L. *-ana*, indicating connection. From Sabarimala, Kerala State, India

sabauda From Sabauda now mostly included in Savoie, France

sabeana L. *-ana*, indicating connection. From Sana Saba county, Texas, USA

sabiense L. *-ense*, denoting origin. From Lower Sabi, Zimbabwe

sabineae In honor of Sabine Lüdtke née Bleissner (1943–) who collected in southern Africa

sabinei In honor of Edward Sabine (1788–1883) English astronomer and Arctic explorer

Sabsab Origin not given by the author, but probably the Senegalese name for a local cereal grass

sabuli L. *sabulum*, coarse sand. Growing in damp sand

sabulicola L. *sabulum*, coarse sand; *-cola*, dweller. Growing in damp, sandy places

sabulorum L. *sabulum*, coarse sand. Of coarse sands, that is a beach species

sabulos-a, -us L. *subulum*, coarse sand; *-osa*, abundance. Growing in sandy soils

sacandros L. *sakos*, shield; *aner*, male. The dense weft of hairs on the upper leaf-surface immediately above the ligule resemble the male pubes

sacatilla From Sacatilla, Mexico

Saccarum See *Saccharum*

saccatus L. *saccus*, sac; *-atus*, possessing. Upper leaf-sheaths inflated

saccharat-a, -um, -us L. *saccharum*, sugar; *-ata*, possessing. Culm-juice sweet

Saccharifera, saccharifera L. *saccharum*, sugar; *fero*, carry or bear. Sugar producing

sacchariflor-a, -us L. *flos*, flower. Inflorescence resembles that of *Saccharum*

saccharoides Gk *-oides*, resembling. – (1) Resembling *Saccharum* in its production of sugar. *Amphilophis saccharoides, Andropogon saccharoides, Erianthus saccharoides, Holcus saccharoides, Sorghum saccharoides* – (2) resembling *Saccharum* in spikelet or inflorescence form. *Arundo saccharoides, Gynerium saccharoides, Panicum saccharoides, Paspalum saccharoides*

saccharoideum Gk *-oideum*, resembling. Resembling *Saccharum* in some respect

Saccharum L. *saccharum*, sugar. Some species are cultivated for their sugar content

Sacciolepis, Saccolepis Gk *sakkion*, small sack; *lepis*, scale. Upper glume inflated

sacculata L. *saccus*, sac; *-ulus*, diminutive; *-ata*, possessing. The base of the upper glume is expanded into a small sac

sachalinens-e, -is L. *-ense*, denoting origin. From Sakhalin Island, Russian Far East

Sacharum See *Saccharum*

S

sacrariocola L. *sacrarium*, place for sacred objects; *-cola*, dweller. Origin uncertain, not given by author but possibly from near a wayside shrine

sacrosancta L. *sacrum*, sacred; *sanctum*, holy place. Origin not given by author, probably collected from the vicinity of a temple

sadae Origin not given by author but apparently in honor of Sada

sadaoi In honor of Sadao Suzuki, Japanese botanist

sadinii Very likely a misspelling of *sabinei*

sadleriana L. *-ana*, indicating connection. In honor of Josef Sadler (1791–1849) Hungarian physician and botanist

sadoensis L. *-ensis*, denoting origin. From Sado, a Japanese island

sagittat-a, -um L. *sagitta*, arrow; *-ata*, possessing. – (1) Leaf-blades resemble an arrow-head. *Phyllorachis sagittata* – (2) inflorescences resemble arrow-heads. *Arundo sagittata, Gynerium sagittatum, Saccharum sagittatum*

sagittatinea L. *sagittus*, shaped like an arrow-head; *-inea*, close resemblance. Culms used for making arrows

sagittatum See *sagittata*

sagittifoli-a, -um, -us L. *sagittus*, shaped like an arrow-head; *folium*, leaf. With leaf-blades resembling an arrow-head

sagraeana, sagrana L. *-ana*, indicating connection. In honor of Ramón de la Sagra (1798–1871) Spanish naturalist, sometime resident of Cuba

sahelica L. *-ica*, belonging to. From the Sahel, a region of North Africa

saigonense L. *-ense*, denoting origin. From Saigon, now Ho Chi Min City, Vietnam

saikanica L. *-ica*, belonging to. From Saikan, or Saykhan, Kazakhstan

saitoana L. *-ana*, indicating connection. In honor of Saito Chiken (fl. 1932) Japanese botanist

sajanensis L. *-ensis*, denoting origin. From the Sajan Mountains, Irkutsk Province, Siberia

sakaigunensis L. *-ensis*, denoting origin. From Sakaigun, Yetsizenn Province, Japan

sakaii In honor of Tadatosi Sakai, Japanese botanist

salamanca From Salamanca, Spain

salarkhanii In honor of Mohammed Salar Khan (1924–2002) Indian-born Bangladeshi botanist

salaziensis L. *-ensis*, denoting origin. From Salazes, La Réunion, Mascarenes

salifex L. *sal*, salt; *-fex*, maker. The ash is a source of salt

salin-a, -us (1) From Salina Pass, Utah, USA. *Agrostis salina, Elymus salinus, Poa salina* – (2) growing in saline soils. *Calamagrostis salina, Festuca salina, Triodia salina*

salinaria L. *salina*, saline; *-aria*, pertaining to. Growing in salt marshes

salinus See *salina*

sallacustris L. *sal*, salt; *lacus*, lake; *-estris*, indicating place of growth. Growing along shore lines of salt lakes

sallentiana In honor of Angel Sallent y Gotés (1857–1934) Spanish philologist and botanist

salmantic-a, -um From Salmantica, now Salamanca, Spain

Salmasia In honor of Claudus Salmasium, otherwise Claude de Saumaise (1588–1658) Belgian botanist

sals-a, -us L. saline. Growing in salty soils

salsuginosus L. *salsugo*, saltiness; *-osa*, abundance. Growing in salt marshes

salsus See *salsa*

saltana L. *-ana*, indicating connection. As for *saltense*

saltens-e, -is L. *-ense*, denoting origin. From Salta Province, Argentina

saltuensis L. *saltus*, forest pasture or woodland; *-ensis*, denoting origin. Growing in woodland

salzmanniana See *salzmannii*

salzmannii In honor of Philipp Salzmann (1781–1851) German-born physician, naturalist and traveller

samaniana L. *-ana*, indicating connection. From Samani, Hokkaido Province, Japan

sambiensis L. *-ensis*, denoting origin. From Sambi, Bizenn (Buzen?) Province, Japan

sambiranens-e, -is L. *-ense*, denoting origin. From Sambirano, Madagascar

sampaioana L. *-ana*, indicating connection. In honor of Alberto José de Sampaio (1881–1946) Brazilian botanist

sampsonii In honor of Hugh Charles Sampson (1878–1953) who collected in Northern Nigeria

sanct-a, -um, -us L. sacred. First collected from Mt Athos in Greece, where mountain tops were historically held to be sacred

sanctaecruziensis L. *-ensis*, denoting origin. From Rio Santa Cruz, Argentina

sanctae-luciae From St. Luzia, Brazil

sanctae-martae From Sierra Nevada de Santa Marta, Colombia

sanctae-marthae (1) From Province of St. Marthe, Venezuela. *Panicum sanctae-marthae* – (2) an alternative spelling of *sanctae-martae. Festuca sancta-martae*

sanct-um, -us See *sancta*

sandaensis L. *-ensis*, denoting origin. From Sanda, Zaire

sandangorgiana L. *-ana*, indicating connection. From Sandankio, Hiroshima Prefecture, Japan

sandbergii In honor of John Herman Sandberg (1848–1917) Swedish-born, United States physician and amateur botanist

sandiens-e, -is L. *-ense*, denoting origin. From Sandía, Peru

sandinensis L. *-ensis*, denoting origin. From Sandino, Cuba

sandorii In honor of Josef Sándor, Hungarian botanist

sandvicens-is, -is L. *-ensis*, denoting origin. From the Sandwich, now Hawaiian Islands

sanguinal-e, -is L. *sanguineus*, dull-red; *-alis*, pertaining to. Foliage or inflorescence purplish

Sanguinaria L. pertaining to the blood. Applied by Pliny to a plant used for staunching blood and currently a vernacular name for *Digitaria sanguinalis* in Italy

sanguine-a, -um, -us L. *sanguineus*, dull-red. Inflorescence dull-red in color

Sanguinella L. *-ella*, diminutive but here used as a name-forming suffix together with *sanguinalis*. A vernacular name in Italy for *Digitaria sanguinalis*

sanguine-um, -us See *sanguinea*

sanguinolentum L. *sanguineus*, dull-red; *-olentum*, markedly developed. Culm bases dull-red

sangvinale See *sanguinale*

sanionis In honor of Karl Gustav Sanio (1832–1891) German botanist

sanlorenzanus L. *-anus*, indicating connection. From Seranía de San Lorenzo, Santa Cruz Department, Bolivia

sanluisensis L. *-ensis*, denoting origin. From San Luis Province, Cordoba, Argentina

sanmingensis L. *-ensis*, denoting origin. From Sanming, Fijian, Japan

santacrucense See *santacruzense*

santacruzensis L. *-ensis*, denoting origin. From Santa Cruz

santanensis Japanese *san*, three; L. *-ensis*, denoting origin. From the three Tan Districts, Tanzen, Tango and Tamba of Japan

santapaui In honor of Hermenegild Santapau (1903–1970) Spanish-born Indian cleric and botanist

Santia In honor of Georgio Santi (1746–1822) Italian botanist

santyvesii As for *Yvesia*

sape Vernacular name of the species in Brazil

sapinii In honor of Adolphe Sapin (1869–1914) who collected in Zaire

saposhnikovii In honor of Vasili Vasilievic Sapozhnikov (1861–1924) Russian botanist

sara Bengali vernacular name for the species in reference to its fleshy stems

sarcocarpa Gk *sarx*, flesh; *karpos*, fruit. Fruit fleshy

sardo-a, -us Sardous, now Sardinia

sareptana L. *-ana*, indicating connection. From Sarepta, now Krasnoarmeysk, Russian Federation

saresberiensis L. *-ensis*, denoting origin. Latinized form of Salisbury, now Harare, Zimbabwe

Sarga Meaning obscure, origin not given by author but possibly an allusion to being intermediate between *Agrostis* and *Stipa*

sarmentos-a, -um, -us L. *sarmentum*, small branch; *-osa*, abundance. Culms much branched

Sarocalamus Gk *saron*, broom; *kalamos*, reed. The type species used for sweeping and with its erect branching habit, the plants resemble brooms

sarracenorum L. *Saracenus*, Saracen. Of the Saracens, that is from southern Spain

Sarsa A misspelling of *Sarga*

Sartidia An anagram of *Aristida*

sartorii In honor of Joseph Sartori (1809–1880) German apothecary and botanist

saruwagetica L. *-ica*, belonging to. From the Saruwaged Mountains of Papua New Guinea

S

sarymsactensis L. *-ensis*, denoting origin. From Sarymsacty Pass in the southern Altai Mountains, Kazakhstan

Sasa Vernacular name in Japan for several species of small bamboo

Sasaella L. *-ella*, diminutive but here used as a name-forming suffix. Related to *Sasa*

sasaelloides Gk *-oides*, resembling. Similar to *Sasaella*

sasagaminensis L. *-ensis*, denoting origin. From Mt Sasagamine, Tamba Province, now part of Kyoto and Hyogo Prefectures, Japan

sasakiana L. *-ana*, indicating connection. As for *sasakii*

sasakii In honor of Shun-ichi Sasaki (1888–1960) Japanese botanist

Sasamorpha L. *morpha*, appearance. Resembling *Sasa*

sat A contraction of *kai-sat*, the vernacular name for the species in Annam

satarensis L. *-ensis*, denoting origin. From Satara, Keshyaturda, India

sativ-a, -um L. cultivated. Crop species

sattosasa Japanese *sasa*, a dwarf bamboo. From Satto, Sakhalin Island, Russian Far East

Saugetia In honor of Joseph Sylvestre Sauget-Barbier, also known as Brother, Frère or Hermano León (1871–1955) French-born cleric and Cuban botanist

saugetii As for *Saugetia*

saundersii In honor of William Saunders (1822–1900) Scots born United States horticulturalist

saurae In honor of Fulgencio Saura (fl. 1948) Argentinian cytologist

sauric-a, -us L. *-ica*, belonging to. – (1) From Sauria, in Classical times a town in Akarania, Greece. *Festuca saurica* – (2) from the Saur-Tarbagatai Ranges, Kazakstan. *Elymus sauricus, Stipa saurica*

sauvagei In honor of Charles Phillippe Felix Sauvage (1909–1980) French botanist

savannarum Through Spanish from *zavana*, Caribbean for a tree-less plain. Species of grasslands

Savastana In honor of Francesco Eulalio Savastano (1657–1717) Italian cleric and botanist

savignonii In honor of Francesco Savignone (1818–?) Italian physician and botanist

savii In honor of Gaetano Savi (1769–1844) Italian botanist

savulescui In honor of Trajan Savulescu (1889–1963) Romanian botanist

sawadae In honor of Taketarô Sawada (1899–1938) Japanese botanist

saxatile, saxatilis L. *saxum*, rock; *-atile*, place of growth. Dwelling among rocks

saxicola L. *saxum*, rock; *-cola*, dweller. Growing on or amongst boulders

saxifraga L. *saxum*, rock; *frango*, shatter. Growing amongst rocks or in habit resembling *Saxifraga*

saximontana L. *saxum*, rock; *mons*, mountain; *-ana*, indicating connection. Growing amongst rocks on mountains

sayanuka In honor of Sayanuka, Japanese botanist

sayapensis L. *-ensis*, denoting origin. From Laguna Sayape, Argentina

sayekiensis L. *-ensis*, denoting origin. From Sayekigun, Hiroshima Prefecture, Japan

scab-er, -ra, -rum, -rus L. rough or gritty to the touch. Plants with rough leaf-blades, spikelets or stems

scaberrim-a, -us L. *scaber*, rough; *-rima*, most. Leaf-blades very scabrous

scaberula L. *scaber*, rough; *-ula*, diminutive. Somewhat scabrous, usually referring to the lemma

scabra See *scaber*

scabrat-a, -us L. *scaber*, rough; *-ata*, possessing. Plant totally or in part scabrid

scabrella L. *scaber*, rough; *-ella*, diminutive. Leaf-blades somewhat scabrous

scabrescens L. *scabresco*, becoming rough. Leaf-blades rough

scabriculmis L. *scaber*, rough; *culmus*, stem. Culms rough

scabrid-a, -um, -us L. *scaber*, rough; *-ida*, becoming. Plants in part or whole rough to the touch

scabridulum L. *scabrida*, rough; *-ulum*, diminutive. Leaf-blades somewhat scabrid

scabrid-um, -us See *scabrida*

scabriflor-a, -um L. *scaber*, rough; *flos*, flower. Spikelets with scabrous glumes and/or lemmas

scabrifoli-a, -um L. *scaber*, rough; *folium*, leaf. Leaf-blades rough

scabriglumis L. *scaber*, rough; *gluma*, husk. Glumes scabrous

scabrimarginatus L. *scaber*, rough; *marginus*, edge; *-atus*, possessing. Margin of leaf-blade rough

scabriolus L. *scaber*, rough; *-olus*, diminutive. Somewhat rough to the touch

scabrior L. rougher. Leaf-sheaths more or less hispidulus

scabristemmed Origin uncertain, not given by author but probably a reference to the scabrid culms

scabriuscul-a, -um L. *scabrius*, rougher; *-ula*, diminutive. Somewhat scabrous usually of leaf-blades or leaf-sheaths

scabrivaginata L. *scaber*, rough; *vagina*, sheath; *-ata*, possessing. Leaf-sheath scabrid

scabrivalvis L. *scaber*, rough; *valva*, leaf of a folding door. The lemmas and/or glumes are densely hispid

scabrosa L. *scaber*, rough; *-osa*, abundance. Leaf-blades distinctly scabrous

scabrum See *scaber*

scaettae In honor of Helios Francesco Antonio Scaetta (1894–1941)

scalar-e, -is L. *scala*, ladder; *-aris*, pertaining to. Lemma ornamented with longitudinal striations

scalarum L. *scala*, ladder. Leaf-blades short and held at right angles to culms thereby resembling a ladder with a central axis

scandens L. *scando*, climb up. Of scrambling habit

scandica L. *-ica*, belonging to. From Scandia, now Scandinavia

scandinavicum L. *-icum*, belonging to. From Scandinavia

scaposum L. *scapus*, stalk; *-osa*, abundance. Inflorescence a spicate panicle borne on a long leafless axis

scarios-a, -us L. of thin and membranous texture, but not green. In general of glumes or lemmas

schaackianum In honor of George B. Van Schaack (fl. 1945) United States soldier and amateur botanist

schaeferi In honor of Fritz Schaefer (?–1911) medical practitioner and plant collector in South Africa

schaenfeldia See *Schoenefeldia*

Schaffnera In honor of Wilhelm Darmstadt Schaffner (?–1802) who collected in Mexico

Schaffnerella L. *-ella*, diminutive but here used as a name-forming suffix. Similar to *Schaffnera*

schaffneri See *Schaffnera*

schafkatii In honor of Schafkat

schangulensis L. *-ensis*, denoting origin. From Schangula, Ethiopia

schantzii In honor of Schantz who collected in Zaire

schebehliensis L. *-ensis*, denoting origin. From Wabi-Shabali, Ethiopia

Schedololium Hybrids between species of *Schedonurus* and *Lolium*

Schedonnardus Gk *schedon*, near to. Resembles *Nardus* with respect to the inflorescence

Schedonorus, Schoenodorus Gk *schedon*, near to; *oura*, tail. Lower glume shortly awned

scheelei In honor of Georg Heinrich Scheele (1808–1864) German cleric and botanist

scheelii See *scheelei*

schelkownikowii In honor of A. B. Schelkovnikov (fl. 1926) Russian botanist

schellian-a, -um L. *-ana*, indicating connection. In honor of – (1) Ernst Schelle (1864–1945) German botanist. *Helictotrichon schellianum* – (2) Julian Schell (c. 1850–1881) Russian botanist. *Avena schelliana*, *Avenula schelliana*

S

Schellingia In honor of Friedrich William Joseph Schelling (1806–1854) German philosopher

schenckii In honor of Johann Heinrich Rudolf Schenck (1860–1924) German botanist

Schenckochloa Gk *chloa*, grass. See *schenckii*

schereri In honor of Oliver Joseph Scherer (1906–) United States geologist

scheuchzeri In honor of Johann Scheuchzer (1684–1738) Swiss physician and botanist

scheuchzeriformis L. *forma*, appearance. Resembling *Festuca scheuzeri*

schiedean-a, -um, -us L. *-ana*, indicating connection. In honor of Christian Julius Wilhelm Schiede (1798–1836) German-born Mexican botanist

schiemanniana In honor of E. Schiemann (fl. 1921) German cereal breeder

schiffneri In honor of Victor Felix Schiffner (1862–1944)

schimperi In honor of Georg Heinrich Wilhelm Schimper (1804–1878) German plant collector in Near East and northeastern Africa

schimperian-a, -um, -us L. *-ana*, indicating connection. As for *schimperi*

schinzii In honor of Hans Schinz (1858–1941) Swiss traveller and botanist

schirensis L. *-ensis*, denoting origin. From Schire Highlands, Ethiopia

Schirostachyum See *Schizostachyum*

Schisachyrum See *Schizachyrium*

schischkinii In honor of Boris Konstantinovich Shishkin (1886–1963) Russian botanist

schismoides Gk *-oides*, resembling. Similar to *Schismus*

Schismus Gk *schismos*, a splitting. The lemma apex is bidentate

schistacea L. *-acea*, indicting resemblance. Growing on soils derived from schists

Schistachne Gk *schistos*, divided; *achne*, scale. The lemma is bifid

schisticola L. *-cola*, dweller. Growing on schist

Schizachne Gk *schizo*, split; *achne*, scale. Lemma apex bifid

Schizachyrium Gk *schizo*, split; *achyron*, chaff. The upper lemma is deeply bilobed

schizantha Gk *schizo*, split; *anthos*, flower. The male and female flowers occur on different plants

Schizopogon Gk *schizo*, split; *pogon*, beard. Internodes of the inflorescence plumose and apically bifid

schizostachyoides Gk *-oides*, resembling. Resembling *Schizostachyum*

Schizostachyum Gk *schizo*, split; *stachys*, spike as of an ear of wheat. Spikelets widely separated on axis

schlagintweitii In honor of one or both brothers, Hermann Alfred Rudolf (1826–1882) and Robert (1833–1885) Schlagintweit, German botanists and the first Europeans to visit Tibet

schlanstedtensis L. *-ensis*, denoting origin. From Schlanstedt, Germany

schlechteri In honor of Friedrich Richard Rudolf Schlechter (1872–1925) German botanist and traveller

schleicheri In honor of Johann Christoph Schleicher (1768–1834) Swiss botanist

Schlerochloa See *Sclerochloa*

Schleropelta Gk *skleros*, hard; *pelte*, shield. The glumes are leathery

schlickumii In honor of Julius Schlickum (1804–1884) German apothecary

schliebenii In honor of Hans Joachim Schlieben (1902–1975) German plant collector in Tanzania

schlumbergeri In honor of F. Schlumberger (?–1893)

Schmidetia Orthographic variant of *Smidetia*

schmidian-a, -us L. *-ana*, indicating connection. As for *schmidii*

schmidii In honor of Ludwig Bernhard Ehregott Schmid (1788–1859) German-born Indian missionary and plant collector

Schmidtia In honor of – (1) Johann Anton Schmidt (1823–1905) German botanist. *Schmidtia pappophoroides* – (2) Franz Wilibald Schmidt (1763–1796) Bohemian botanist. *Schmidtia subtilis*

schmidtianum L. *-anum*, indicating connection. In honor of Karl Schmidt (fl. 1848) who collected in the Crimea

schmidtii As for *Schmidtia* (1)
Schmistachne See *Schistachne*
schmitzii In honor of Albert Schmitz (pre 1879) who collected in Mexico
schmutzii In honor of E. Schmutz (fl. 1971) who collected in West Flores, Indonesia
schneideri In honor of Camillo Karl Schneider (1876–1951) German botanist
Schnizleinia In honor of Adalbert Carl Friedrich Hellwig Conrad Schnizlein (1813–1868) German botanist
schoenanthus Plants with the habit of *Schoenanthus*
Schoenefeldia In honor of Melchior Schoenefeld (fl. 1619) German botanist
schoenfelderi In honor of Eberhard Bruno Willie Schoenfelder (1892–1969) South African farm manager and plant collector
schoenites Gk *-ites*, closely connected. The spikelets superficially resemble those of *Schoenus*
Schoenodorus See *Schedonorus*
schoenoides Gk *-oides*, resembling. Similar to *Schoenus*
Schoenus Now a genus of *Cyperaceae*, but Carl Linnaeus (1707–1778) included therein a species of *Crypsis*
schomburgkii In honor of Robert Hermann Schomburgk (1808–1865) German botanist
schottii In honor of Heinrich Wilhelm Schott (1794–1865) Austrian botanist
schraderi In honor of Heinrich Adolph Schrader (1767–1836) German botanist
schraderiana L. *-ana*, indicating connection. As for *schraderi*
schreberi In honor of Johan Christian Daniel Schreber (1739–1810) German botanist
schrenkian-a, -um, -us L. *-ana*, indicating connection. In honor of Alexander Gustav Schrenk (1816–1876) Russian botanist
schroederi In honor of J. Schroeder (fl. 1920–1922) who collected in Uruguay
schroeteriana L. *-ana*, indicating connection. In honor of Carl Schröter (1855–1939) Swiss botanist
schuetzeana L. *-ana*, indicating connection. In honor of Schuetze
schugnanic-a, -um, -us L. *-ica*, belonging to. From Schugnan (Shugnan) Province, Turkestan region of Central Asia
Schultesia, Schultezia In honor of Josef August Schultes (1773–1831) Austrian botanist
schultesii (1) As for *Schultesia*. *Agrostis schultesii*, *Panicum schultesii*, *Poa schultesii* – (2) in honor of Richard Evans Schultes (1915–) who collected in Colombia. *Axonopus schultesii*, *Paspalum schultesii*
Schultezia See *Schultesia*
schultzei In honor of Leonard Sigismund Schultze (1872–1955) German botanist and traveller
schultziana L. *-ana*, indicating connection. As for *schultzii*
schultzii In honor of Frederick Schultz (fl. 1869) who collected in northern Australia
schumannian-a, -um L. *-ana*, indicating connection. In honor of Karl Schumann (1851–1904) German botanist
schurii In honor of Philipp Johann Ferdinand Schur (1799–1878) German botanist and chemist
schwabii In honor of Samuel Heinrich Schwabe (1789–1875) German astronomer and botanist
schwackeanum L. *-anum*, indicating connection. In honor of Carl August Wilhelm Schwacke (1846–1904) German-born Brazilian botanist
schweinfurthiana L. *-ana*, indicating connection. As for *schweinfurthii*
schweinfurthii In honor of Georg August Schweinfurth (1836–1925) German botanist and anthropologist
schweinitzii In honor of Ludwig David von Schweinitz (1780–1834) United States botanist
sciaphil-a, -um Gk *skia*, shade; *phileo*, love. Growing in the shade
scindens L. *scindo*, cut. Apices of glumes and sterile lemma erose
scindic-um, -us L. *-icum*, belonging to. From Scinde, now Sind, Province of Pakistan
scintillans L. *scintillo*, sparkle. Hairs on inflorescence silvery, glistening

scirpe-a, -um L. *scirpus*, name of a rush; *-ea*, resembling. In habit resembling *Scirpus*

scirpifolia L. *scirpus*, a rush; *folium*, leaf. Leaf-blades rush-like

Scirpobambus Resembling *Bambusa* with respect to its woody culms and *Scirpus* in possessing cylindrical spikelets

scirpoid-ea, -es, -eum L. *scirpus*, a rush; *-oidea*, resembling. As for *scirpea*

scitul-a, -um L. pretty. The spikelets are colored thereby making the inflorescence attractive

S

sciurea L. *sciurea*, squirrel; *-ea*, resembling. The inflorescences resemble the tail of a squirrel

sciuroidea Gk *skiouros*, squirrel; *-oides*, resembling. Inflorescence resembles a squirrel's tail

sciuroides See *sciuroidea*

sciurotis Gk *skiouros*, squirrel. The inflorescence resembles a squirrel's tail

sciurotoides Gk *-oides*, resembling. Resembling *Panicum sciurotis*

sciurus L. *sciurus*, squirrel. Culms covered with snow-white woolly indument

Sclerachne Gk *skleros*, hard; *achne*, scale. The glumes are indurated

sclerachne Spikelets resembling those of *Sclerachne*

Sclerandrium Gk *skleros*, hard; *aner*, man. The glumes of the pedicellate male spikelets are indurated

sclerantha Gk *skleros*, hard; *anthos*, flower. Lemmas cartilagenous

scleranthoides Gk *-oides*, resembling. Similar to *Scleranthus*

sclerioides Gk *-oides*, resembling. Inflorescence similar to that of *Scleria*

sclerocalamos Gk *skleros*, hard; *kalamos*, reed. Culms reed-like

sclerochlaena Gk *skleros*, hard; *chlaena*, cloak. Lemmas and sometimes glumes cartilaginous

Sclerochloa, sclerochloa Gk *skleros*, hard; *chloa*, grass. The glumes are indurate

scleroclad-a, -um Gk *skleros*, hard; *klados*, branch. Culm moderately stout

Sclerodactylon Gk *skleros*, hard; *daktylon*, finger. The inflorescence comprises two or three one-sided densely crowded one-sided spikes

sclerodes Gk *skleros*, hard; *-odes*, resembling. Leaf-blades rigid

Sclerodeyeuxia Gk *skleros*, hard. Resembling *Deyeuxia* but lemma cartilaginous

Sclerolaena Gk *skleros*, hard; *chlaena*, cloak. The lemma is cartilagenous

sclerophyll-a, -um, -us Gk *skleros*, hard; *phyllon*, leaf. Leaf-blades indurate or leathery

Scleropoa Gk *skleros*, hard; *poa*, grass. The spikelets resemble those of *Poa* but have leathery glumes and lemmas

Scleropogon Gk *skleros*, hard; *pogon*, beard. The upper florets of the spikelet are reduced to a bunch of long awns

scleropoides Gk *-oides*, resembling. Similar to *Scleropoa*

Sclerostachya Gk *skleros*, hard; *stachys*, spike as of an ear of wheat. Subtending glumes leathery in texture

Scolochloa Gk *skolos*, spine; *chloa*, grass. The lemma apex has one-three short cusps. The name has been applied to two genera sharing this characteristic

scopari-a, -um, -us L. *scopa*, twig or *scopae*, several twigs or a broom; *-aria*, pertaining to. – (1) Inflorescences condensed resembling a broom. *Andropogon scoparius, Agrostis scoparia, Aristida scoparia, Axonopus scoparius, Bromus scoparius, Dichanthelium scoparium, Enneapogon scoparius, Festuca scoparia, Panicum scoparium, Paspalum scoparius, Paspalus scoparius, Pennisetum scoparium, Schizachyrium scoparium, Stipagrostis scoparia, Thysanachne scoparia* – (2) culms fasciculate. *Muhlenbergia scoparia, Poa scoparia* – (3) inflorescences with sterile shoots resembling brooms. *Distichlis scoparia*

scoparioide L. *scopae*, broom; Gk *-oides*, resembling. Resembling *Panicum scoparium*

scopari-um, -us See *scoparia*

scopelophila Gk *skopelos*, lookout place; *phileo*, love. Growing on rocky outcrops

scopolianum L. *-anum*, indicating connection. In honor of Giovanni Antonio Scopoli (1723–1788) Tirol-born physician and botanist

scopula L. a small broom. There is a brush-like row of cilia along each keel of the exserted palea

scopuliferum L. *scopa*, branch; *-ula*, diminutive; *fero*, carry or bear. Plant a tuft of small leafy branches

scopulorum L. *scopulus*, cliff. Growing amongst rocks or at the bases of cliffs

scorpioides Gk *-oides*, resembling. The inflorescence bears a fanciful resemblance to a scorpion

scortechinii In honor of Benedetto Scortechini (1845–1886) Italian cleric and botanist

scotantha Gk *skotos*, darkness; *anthos*, flower. Inflorescence not fully exserted

scotelliana L. *-ana*, indicating connection. In honor of George Francis Scott Elliott (1862–1934) British administrator in West Africa

scotica L. *-ica*, belonging to. From Scotia, now Scotland

scottii As for *scottelliana*

scott-thomsonii In honor of John Scott-Thomson (1882–1943) New Zealand chemist and amateur botanist

scouleri In honor of John Scouler (1804–1871) Scots-born physician and naturalist

scoutii From Scout Canyon, near Lewellen, Nebraska, USA

scribneri As for *Scribneria*

Scribneria In honor of Frank Lamson Scribner (1851–1938) United States agrostologist

scribnerian-a, -um L. *-ana*, indicating connection. As for *Scribneria*

scriptori-a, -us L. belonging to writing. Origin uncertain, but may be a reference to the plants being used to make paper

scrobiculat-um, -us L. *scrobis*, ditch; *-ulus*, diminutive; *-atus*, possessing. Glumes or lemmas furrowed

Scrotochloa L. *scrotum*, scrotum; Gk *chloa*, grass. Lemmas urn-shaped with connate margins

Scutachne Gk *skytos*, leather; *achne*, scale. The upper glume and sterile lemma are leathery in texture

scyphofera Gk *skyphos*, cup; L. *fero*, carry or bear. There is a trumpet-shaped appendage at the apex of the peduncle

scythic-a, -um L. *-ica*, belonging to. From Scythia, in Classical times the name for the plains north and west of the Black Sea

scytophylla Gk *skytos*, leather; *phyllon*, leaf. Leaf-blades leathery

sczerbakovii In honor of B. V. Sczerbakov, Kazakhstan biologist

searsii In honor of Ernest R. Sears (1910–1991) United States plant breeder and geneticist

seatonii In honor of Henny Eliason Seaton (1869–1893) United States botanist

sebastinei In honor of Kunju Mathew Sebastine (1918–1967) Indian botanist

Secale L. *seco*, cut. Latin name of a cereal, possibly rye

Secalidium Gk *-idium*, diminutive but here a name-forming suffix. Resembling *Secale*

secalin-um, -us L. *-inum*, indicating possession. Growing in fields of rye (*Secale*)

Secalotricum Hybrids between species of *Secale* and *Triticum*

secans L. *seco*, cut. The margins of the leaf-blades are sharp and capable of cutting

secernenda L. *secerno*, set apart. Readily distinguished from related species

sechellens-e, -is L. *-ense*, denoting origin. From the Seychelles

secund-a, -um, -us L. bent to one side. – (1) Branches restricted to one side of inflorescence. *Agropyron secundum*, *Andropogon secundus*, *Chloachne secunda*, *Deyeuxia secunda*, *Eragrostiella secunda*, *Eragrostis secunda*, *Heteropogon secundus*, *Melica secunda*, *Melinis secunda*, *Oplismenus secundus*, *Perobachne secunda*, *Trachypogon secundus*, *Triticum secundum* – (2) leaf-blades twisted to one side of culm. *Panicum secundum*

secundat-a, -um L. *secundus*, bent to one side; *-atus*, possessing. – (1) Inflorescence a curved fleshy axis. *Ischaemum secundatum*, *Stenotaphrum secundatum* – (2) panicle branches secund. *Triodia secunda*

secundiflor-a, -um L. *secundus*, turned to one side; *flos*, flower. Spikelets restricted to one side of inflorescence branches

secundispiculus L. *secundus*, turned to one side; *spica*, a point; hence, in particular, an ear or spike of grain; *-ulus*, diminutive. Spikelets turned to one side on panicle branches

secund-um, -us See *secunda*

sedan Burmese smoking pipe. Culms used for making pipes

sedenens-e, -is L. *-ense*, denoting origin. From Sedena that is the mountains of de la Seyna, France

S

seelyae In honor of M. K. Seely (fl. 1991) South African plant ecologist

seemenianus L. *-anus*, indicating connection. In honor of Karl Otto Seemen (1838–1910) German botanist

segaenensis L. *-ensis*, denoting origin. From Segaen, India

segawana L. *-ana*, indicating connection. In honor of Segawa Kikuji, Japanese botanist

segetalis L. *seges*, cornfield; *-alis*, pertaining to. Growing amongst cultivated cereals

segetum L. *seges*, cornfield; *-etum*, place of growth. Growing amongst cultivated cereals

Sehima Arabic *saehim*. The vernacular name in Egypt of the type species of the genus

sehima Resembling *Sehima*

seidlii In honor of Wenzel Berno Seidl (1773–1842) Bohemian botanist

seineri In honor of Franz Seiner (1874–c. 1940) German botanist who collected in Angola

sejuncta L. separated. Spikelets widely separated along inflorescence branches

sekimotoi In honor of H. Sekimoto (fl. 1931) Japanese botanist

seleri In honor of Caecilie Seler (1855–1933) and Georg Eduard Seler (1849–1922) who collected in Central and South America

selloan-a, -us L. *-ana*, indicating connection. In honor of Friedrich Sellow (1789–1831) German botanist

sellovii, sellowii As for *selloana*

sellowiana As for *selloana*

sellowii As for *selloana*

Sellulocalamus L. *sella*, chair; *-ula*, diminutive; Gk *kalamos*, reed. Origin unclear

Semeiostachys Gk *semeion*, flag; *stachys*, spike as of an ear of wheat. Spikes erect or slightly droopy

semenovii In honor of Peter Petrowitsch von Semenow-Tiam-Shansky (1827–1914) Russian traveller in Central Asia

semialat-a, -um, -us L. *semi-*, half; *ala*, wing; *-ata*, possessing. – (1) Upper glumes winged. *Alloteropsis semialata*, *Axonopus semialatus*, *Oplismenus semialatus*, *Panicum semialatum*, *Urochloa semialata* – (2) paleas auricled at the base. *Coridochloa semialata*

semiannularis L. *semi-*, half; *annulus*, a ring; *-aris*, pertaining to. The lemma bears a half ring of hairs

Semiarundinaria L. *semi-*, half. Resembling *Arundinaria*

semibarbata L. *semi-*, half; *barba*, beard; *-ata*, possessing. Awn column hairy but bristle asperous

semiberb-e, -is L. *semi-*, half; *berbe*, beard. Pedicels ciliate only along outer edge

semiciliata L. *semi-*, half; *cilium*, hair; *-ata*, possessing. Glumes incompletely hairy, when compared with *Eriachne ciliata*, with which it was previously confused

semidecumbens L. *semi-*, half; *decumbo*, lie down. Culm bases resting on the ground

semienensis L. *-ensis*, denoting origin. From Semien, Ethiopia

semiglabrum L. *semi-*, half; *glaber*, smooth. Internodes and pedicels glabrous on the back

seminud-a, -um L. *semi-*, half; *nuda*, bare. Only part of the plant bearing hairs

semiorbiculata L. *semi-*, half; *orbiculus*, circular; *-ata*, possessing. Possibly a reference to the club-shaped rhizomes

semiovata L. *semi-*, half; *ovata*, ovate. Leaf-blades ovate-lanceolate

semisagittat-um, -us L. *semi-*, half; *sagittatum*, arrow-like. Leaf-blades narrow-cordate and separated from their sheaths by a pseudopetiole

semispirale L. *semi-*, half; *spira*, spiral; *-ale*, pertaining to. The spikelets bear a long awn which spirals once around the raceme and then flattens out

semisterilis L. *semi-*, half; *sterilis*, sterile. Terminal spikelets sterile

semitect-um, -us L. *semi-*, half; *tectum*, roof. – (1) The racemes are solitary and partially covered by the spathe. *Andropogon semitectus* – (2) glumes almost as long as the spikelet. *Panicum semitectum* – (3) blades of upper culm-leaves overtopping panicle. *Dissanthelium semitectum*

semiteres L. *semi-*, half; *teres*, narrow cylindric. The inflorescence at maturity separates into single-seeded segments which serve as dispersal units

semitons-a, -um L. *semi-*, half; *tonsa*, shaven. Upper subtending glume less pilose than lower and more pilose than sterile lemma

semiundulat-a, -um L. *semi-*, half; *undulatus*, wavy. Inflorescence branches flexuous

semiverticillat-a, -um, -us L. *semi-*, half; *verticillus*, whorl; *-ata*, possessing. Having inflorescences with secondary branching tending towards verticillate

semperiana In honor of Juan *Semper* (fl. 1944–1945) who collected in Argentina

sempervirens L. *semper*, always; *virens*, green. Perennial species

semplei In honor of A. T. Semple (fl. 1955) who collected in Mexico

senanensis L. *-ensis*, denoting origin. From Senano Island, Japan

sendaica L. *-ica*, belonging to. From Sendai Hill, Rikuzen Province, Japan

sendulskyae In honor of Tatiana Skvortzov Sendulsky (1922–) Russian parentage but born in Harbin, Heilongjiang Province, China, who became a Brazilian botanist

senegalensis L. *-ensis*, denoting origin. From Senegal, now Senagambia

senescens L. *senesco*, grow old. Leaf-blades with indumentum of white hairs

senex L. old man. From the Old Man Range, Central Otago District, South Island, New Zealand

Senisetum L. *seni*, six apiece; *seta*, bristle. Lemma six-awned

Senites L. *seni*, six apiece; Gk *-ites*, closely connected. The basal floret of the spikelet is female and both the succeeding florets are male, each with three stamens

sennarensis L. *-ensis*, denoting origin. From Sennar, Sudan

Sennenia In honor of Gustav Senn (1875–1945) Swiss botanist or Frère Sennen otherwise Etienne Marcelin Grenier-Blanc (1864–1943)

sennii In honor of Lorenzo Senni (1879–1954) Italian botanist

seorsa L. apart from. Segregated from a similar species

sepang From Sepang, Bali, Indonesia

separatum L. separate. Spikelets remote and solitary and so inflorescences unlike those of related species

sepium L. cuttle-bone. Spikelets the shape of a cuttle-bone

septentrional-e, -is L. northern. Northern in distribution

serana L. *-ana*, indicating connection. From Seran (Ceram) Island in the Moluccan Islands, Indonesia

seravschanic-um, -us L. *-icum*, belonging to. From Seravschan (Zeravshan), Tajikistan)

seredinii In honor of R. M. Seredin (1912–)

seretii In honor of Félix Seret (1875–1910) who collected in Zaire

sergievskajae In honor of C. V. Sergievskaja (1926–) Russian botanist

seriata L. *series*, row; *-ata*, possessing. Culms arise in more or less close succession from extravaginal innovations

sericans L. *sericeus*, silken; *-ans*, assuming the appearance of. Spikelets silky

sericantha Gk *serikos*, silken; *anthos*, flower. Glumes and lemmas long, hairy

sericat-a, -um, -us L. *sericus*, silken; *-atus*, possessing. – (1) Rhachis conspicuously hairy. *Andropogon sericatus* – (2) leaf-sheath densely hairy. *Paspalum sericatum*

serice-a, -um, -us L. *sericus*, silken; *-ea*, indicating resemblance. Densely invested in part or totally with long hairs

Sericrostis Origin uncertain, not given by author but possibly a contraction of *serikos* (Gk silken) plus *Agrostis*

S

Sericura Gk *Seres*, Indian tribe from whom silk was bought; *oura*, tail. Now applied to two genera. One has an inflorescence with pedicels invested with long silky hairs; the other has the spikelet clusters subtended by long, often flexuose bristles

Serigrostis See *Sericrostis*

serik Vernacular name of the species in Sumatra, Indonesia

serotin-a, -um L. late. Flowering late in the season

serpens L. *serpo*, creep. Plants conspicuously rhizomatous or scandent

serpentin-a, -um L. *serpens*, snake; *-ina*, indicating resemblance. – (1) From the Javanese vernacular, snake-bamboo, on account of the culms hardly raised above the ground. *Schizostachyum serpentinum* – (2) lemmas transversed with dark lines. *Paspalum serpentinum* – (3) growing on soils derived from serpentine rocks. *Roegneria serpentina*, *Trisetum serpentinum*

serpentini L. of serpentine. Growing on serpentine soils

serpentum From the Serpentine River, Western Australia

serraefolium L. *serra*, saw; *folium*, leaf. Margin of leaf-blades rough from the tuberuclar hair bases

serrafalcoides Gk *-oides*, resembling. Racemes resembling the spikelets of *Serrafalcus*

Serrafalcus In honor of Domenico Lo Faso Pietrasanta Duca di Serrafalco (1783–1869) Italian archeologist

serrana L. *serra*, saw; *-ana*, indicating connection. From a mountain ridge of the Cerra de la Ánimas, Brazil

serranoi In honor of A. Serrano (prior to 1886) who collected in South America

serrat-a, -um, -us L. *serra*, saw; *-ata*, possessing. – (1) Margins of leaf-blades or pedicels with short stiff hairs or hair-bases. *Andropogon serratus*, *Brachiaria serratus*, *Holcus serratus*, *Lepeocercis serrata*, *Panicum serratum*, *Sorghum serratum* – (2) keels of lemmas toothed. *Dactylis serrata*

serratifolia L. *serra*, saw; *-ata*, possessing; *folium*, leaf. Leaf-margin finely denticulate

serratiglumis L. *serra*, saw; *-ata*, possessing; *gluma*, husk. Glumes with serrated keels

serrat-um, -us See *serrata*

serrifolia L. *serra*, saw; *folium*, leaf. Leaf-blade margins spinulosly toothed

serrulat-a, -um, -us L. *serra*, saw; *-ula*, diminutive; *-ata*, possessing. – (1) Margin of leaf-blade bearing short, stiff hairs. *Andropogon serrulatus*, *Chrysopogon serrulatus*, *Chusquea serrulatus*, *Zoysia serrulata* – (2) margin of glume bearing short stiff hairs. *Arthraxon serrulatus*, *Bathratherum serrulatum*

Sesleria In honor of Lionardo Sesler (?–1785) Venetian physician and botanist

sesleriaeformis L. *forma*, appearance. Resembling *Sesleria*

Sesleriella L. *-ella*, diminutive but here employed as a name-forming suffix. Resembling *Sesleria*

seslerioides Gk *-oides*, resembling. Similar to *Sesleria* usually with respect to habit or inflorescence

sesquiflor-a, -um L. *sesqui*, one and a half; *flos*, flower. Spikelets with one fertile floret and a second sterile or male floret

sesquiglume An error for *subsesquiglume*

sesquimetralis L. *sesqui*, one and a half; *-alis*, pertaining to. Culms about 1.5 m tall

sesquiterti-a, -um L. *sesqui*, one and a half; *tertia*, bearing the ratio of four to three. Only three of the four florets fertile

sessiliflorus L. *sessilis*, sessile; *flos*, flower. Origin uncertain, not given by the author

sessilis L. sessile. – (1) Spikelets sessile. *Pseudozoysia sessilis* – (2) leaves sessile. *Racemobambos sessilis*

sessilispic-a, -us L. *sessilus*, sessile; *spica*, a point; hence, in particular, an ear or spike of grain. Spikelets sessile or very shortly stalked

setace-a, -um, -us L. *seta*, bristle; *-acea*, indicating resemblance. – (1) With bristle-like leaf-blades. *Agrostis setacea, Aristida setacea, Deschampsia setacea, Ehrharta setacea, Merxmuellera setacea, Microchloa setacea, Panicum setaceum, Rottboellia setacea, Stipa setacea, Tricholaena setacea* – (2) with long bristles in the inflorescence. *Pennisetum setaceum, Phalaris setacea* – (3) upper florets of spikelet abortive and forming bristles. *Bromus setaceus*

Setaria L. *seta*, bristle; *-aria*, possessing. The spikelets are subtended by one or more persistent bristles

setarioides Gk *-oides*, resembling. Inflorescence spicate resembling that of *Setaria*

Setariopsis Gk *opsis*, resemblance. Resembling *Setaria* with respect to the spikelets being subtended by bristles

setari-um, -us L. *seta*, bristle; *-aria*, possessing. The glumes and sterile lemmas terminate in bristles

Setiacis L. *seta*, bristle; Gk *akis*, pointed object. Upper glume bears a tuft of apical hairs

seticulmis L. *seta*, bristle; *culmus*, stem. Culms filiform

setifer L. *seta*, bristle; *fero*, carry or bear. Rhachis sparsely setose

setifera L. *seta*, bristle; *fero*, carry or bear. Lodicules with long deciduous bristles

setifoli-a, -um, -us L. *seta*, bristle; *folium*, leaf. Leaf-blades bristle-like

setiformis L. *seta*, bristle; *forma*, appearance. Leaf-sheaths with long oral setae

setiger L. *seta*, bristle; *gero*, carry or bear. Glumes shortly aristate

setiger-a, -um, -us L. *seta*, bristle; *gero*, carry or bear. – (1) With hairs or awns on the glumes or lemmas. *Arundinaria setigera, Aulonemia setigera, Brachiaria setigera, Cyrtococcum setigerum, Digitaria setigera, Panicum setigerum, Stipa setigera, Urochloa setigera* – (2) with spikelets subtended by bristles. *Cenchrus setigerus* – (3) with well developed oral setae. *Sasa setigera*

setiglum-e, -is L. *seta*, bristle; *gluma*, husk. Glumes and/or sterile lemmas terminating in bristles

setinsigne L. *seta*, bristle; *insigne*, outstanding. Upper lemma of floret with a well developed awn

setivalva L. *seta*, bristle; *valva*, leaf of a folding door. The sterile lemmas bear long stiff hairs on their margins in addition to shorter silky hairs

setoides L. *seta*, bristle; Gk *-oides*, resembling. Spikelet surrounded by an involucre of bristles

setos-a, -um, -us L. *seta*, bristle; *-osa*, abundance. – (1) Glumes and/or lemmas awned or attenuated. *Agrostis setosa, Andropogon setosus, Arundinella setosa, Chaetochloa setosa, Chamaeraphis setosa, Digitaria setosa, Fargesia setosa, Holcus setosus, Panicum setosum, Pariana setosa, Pennisetum setosum, Pleopogon setosum, Podosaemum setosum, Sorghastrum setosum, Syntherisma setosa* – (2) spikelets subtended by bristles. *Cenchrus setosus, Panicum setosum, Setaria setosa* – (3) leaf-sheath with bristle-like hairs. *Rottboellia setosa*

Setosa L. *seta*, bristle; *-osa*, abundance. Each raceme subtended by a stout bristle

settsuensis L. *-ensis*, denoting origin. From Settsu Province, now part of Hyogo and Osaka Prefectures, Japan

setulifer-a, -um L. *seta*, bristle; *-ula*, diminutive; *fero*, carry or bear. Lemma apex apiculate

setulosa L. *seta*, bristle; *-ula*, diminutive; *-osa*, abundance. – (1) Glumes shortly hairy. *Urochondra setulosa* – (2) leaves shortly hairy. *Setaria setulosa* – (3) lemma terminating in a short bristle. *Vilfa setulosa*

sevangensis L. *-ensis*, denoting origin. From Lake Sevang, now Lake Goktcha, Armenia

sewerzowii In honor of Nicolai Alexyevich Severzoff (1827–1885) Russian botanist

seyrigii In honor of André Seyrig who collected in Madagascar

shaanxiense L. *-ense*, denoting origin. From Shaanxi Province, China

shallote From Shallote, North Carolina, USA

shandongensis L. *-ensis*, denoting origin. From Shandong Province, China

shansiensis L. *-ensis*, denoting origin. From Shansa, China

shapoensis L. *-ensis*, denoting origin. From Sha Po Ling, Hainan Province, China

sharonensis L. *-ensis*, denoting origin. From Plain of Sharon, Palestine

sharpii In honor of Aaron John Sharp (1904–1997) United States botanist who collected in Mexico

shastense L. *-ense*, denoting origin. From Mt Shasta, California, USA

shatilowiana In honor of Shatilow

shawanensis L. *-ensis*, denoting origin. From Shawan, Xinjiang Uyghur Autonomous Region, China

shawii In honor of W. B. Kennedy Shaw (1901–?) English-born forester in the Sudan

shearii In honor of Cornelius Lott Shear (1865–1956) United States mycologist

sheldonii In honor of C. S. Sheldon (fl. 1882) United States botanist

shelkovnikovii In honor of A. Shelkovnikov, the collector

shensiana L. *-ana*, indicating connection. From Shensi, China

shepherdii In honor of A. H. Shepherd who extended hospitality in Mexico to Dr. Palmer, plant collector

Shibataea In honor of Keita Shibata (1877–1949) Japanese biochemist and botanist

shibataeaoides, shibataeoides Gk *-oides*, resembling. Similar to *Shibataea*

shibutamensis L. *-ensis*, denoting origin. From Shibutamimura, Ribuchiu Province, Japan

shigaensis L. *-ensis*, denoting origin. From Shiga, Shiga Prefecture, Japan

shikotanensis L. *-ensis*, denoting origin. From Shikotan, one of the Kurile Islands

shimabarensis L. *-ensis*, denoting origin. From Shimabara Peninsula, Nagasaki Prefecture, Japan

shimadae In honor of – (1) T. Shimada, Japanese botanist. *Andropogon shimadae* – (2) S. Shimada (fl. 1911) Japanese botanist. *Bambusa shimadae*

shimadan-a, -us L. *-ana*, indicating connection. In honor of Yaichi Shimada (?–1971 or 1972) Japanese botanist

shimidzuana L. *-ana*, indicating connection. In honor of Tôtarô Shimidzu (1932–) Japanese botanist

shinanoana L. *-ana*, indicating connection. From Shinano Province, now Nagano Prefecture, Japan

shinyangense L. *-ense*, denoting origin. From Shinyanga, Tanzania

shiobarensis L. *-ensis*, denoting origin. From Shiobara, Tochigi Prefecture, Japan

shirensis L. *-ensis*, denoting origin. From Mt Chiré, Ethiopia

shirleyanum L. *-anum*, indicating connection. In honor of John Francis Shirley (1849–1922) English-born Australian educator and botanist

shiwotae In honor of Kenzo Shiwota (fl. 1934) Japanese botanist

shiwotana L. *-ana*, indicating connection. As for *shiwotae*

shoshoneana L. *-ana*, indicating connection. In reference to the Shoshone people whose ancestral lands encompass the known geographical distribution of the species

shouliangiae See *cheniae*

shrevei In honor of Forrest Shreve (1878–1950) United States botanist

shrirangii In honor of Shrirang Ramachandra Yadav (1954–) Indian botanist

shuka Vernacular name for several grasses in Argentina

shumushuensis L. *-ensis*, denoting origin. From Sumushu, one of the Kurile Islands

siamens-e, -is L. *-ense*, denoting origin. From Siam, now Thailand

Sibertia See *Libertia*

sibilans L. *sibilo*, whistle. Culms used for making whistles

sibinicus L. *-icus*, belonging to. From the Sibin depression in the eastern Kalba Mountains of Kazakhstan

sibiric-a, -um, -us L. *-ica*, belonging to. From Siberia

sibthorpii In honor of John Sibthorp (1758–1796) English botanist

siccaneum L. *siccus*, dry; *-an*, indicating connection; *-eum*, pertaining to. Growing in sandy places

sicc-um, -us L. dry. Growing on dry grassy plains
sichotensis L. *-ensis*, denoting origin. From Sichote-Alinj Mountains, Siberia
sichuanensis L. *-ensis*, denoting origin. From Sichuan Province, China
sichuanicus L. *-icus*, belonging to. From Sichuan Province, China
siciliensis L. *-ensis*, denoting origin. From Sicilia, now Sicily
sicul-a, -um, -us L. a Sicilian. From Sicily
siderograpta Gk *sideros*, iron; *graptos*, painted. The sterile lemma is densely pubescent with brown hairs between the lateral nerves
sieberi In honor of Franz Wilhelm Sieber (1789–1844) Bohemian botanist and traveller
sieberian-a, -um L. *-ana*, indicating connection. As for *sieberi*
sieboldii In honor of Philipp Franz van Siebold (1796–1866) German-born physician in Dutch service who collected in Japan
Sieglingia In honor of Johann Blasius Siegling (1760–1835) German mathematician at Erfurt
sierrae From Sierra Nevada Mountains, California, USA
signata L. *signo*, mark out. Conspicuous species
sikangensis L. *-ensis*, denoting origin. From the former Sikang Province, now part of the Tibet Autonomous Region and Sichuan Province, China
sikkimens-e, -is L. *-ense*, denoting origin. From Sikkim State, India
sikokian-a, -um L. *-ana*, indicating connection. From Shikoku Prefecture, Japan
Silentvalleya From Silent Valley dam site, Kerala State, India
silicatum English, silica; L. *-atum*, possessing. The culms possess an abundance of silica
sillingeri In honor of Pavel Sillinger (1905–1938) Czech botanist
silvatic-a, -us L. *silva*, wood; *-ica*, belonging to. Growing in woodlands
silvean-a, -us L. *-ana*, indicating connection. In honor of William Arents Silveus (1875–1953) United States botanist and attorney
silverstonei In honor of P. A. Silverstone-Sopkin (fl. 1982–1988)
silvestris L. *silva*, wood; *-estris*, place of growth. Plants of woodlands
simaoensis L. *-ensis*, denoting origin. From Simao, Yunnan Province, China
simbense L. *-ense*, denoting origin. From Simba, Kenya
simensis L. *-ensis*, denoting origin. From Siemen Province, Ethiopia
simeonis In honor of Simeon Delmas, French cleric in the Marquesas
similaris L. resembling. Readily confused with another species
simil-e, -is L. like. Readily confused with one or more other species
simillimus L. very similar. Readily confused with another species
simlensis L. *-ensis*, denoting origin. From Simla, Kashmir, India
simonensis As for *simoniana*
simoniana In honor of Bryan Kenneth Simon (1943–) South African born Australian botanist
simonii (1) In honor of Eugène L. Simon (1838–1924) French diplomat who sent plants to the Simon-Louis brothers' nursery at Metz in Alsace, France. *Arundinaria simonii*, *Bambusa simonii*, *Pleioblastus simonii* – (2) as for *simoniana*. *Brachyachne simonii*
simonkaii In honor of Lajos Simonkai (1837–1910) Hungarian botanist
simonsonii In honor of Simonson
simplex L. simple. Ultimate unit of inflorescence a spikelet rather than a pseudopspikelet
Simplicia L. *simplex*, simple. Inflorescence unbranched
simpliciflora L. *simplex*, simple; *flos*, flower. – (1) Inflorescence a raceme. *Aristida simpliciflora* – (2) inflorescence of few florets. *Eragrostis simpliciflora*, *Megastachya simpliciflora*
simplicissim-a, -us L. *simplex*, simple; *-issima*, most. Culms unbranched

S

simpliciuscul-a, -um L. *simplicius*, simpler; *-ula*, diminutive
simpsonii In honor of Joseph Herman Simpson (1841–1918) United States amateur botanist
simulans L. *simulo*, to assume the appearance of something. Closely resembling another species
sinaic-a, -us L. *-ica*, belonging to. From the *Sinai*
Sinarundinaria L. *Sina*, China. From China and resembling *Arundinaria*
sinattenuata L. *Sina*, China. Resembling *Poa attenuata* and from China

S

sincoranum L. *-anum*, indicating connection. From Serra do Sincora, Brazil
sinelatior L. *Sina*, China; *elatior*, taller. From China, and resembling or separated from *Deyeuxia elatior*
sinens-e, -is L. *-ense*, denoting origin. From Sina, now China
Singlingia See *Sieglingia*
singuaensis L. *-ensis*, denoting origin. From Markt Singua, Republic of Cameroon
singular-e, -is L. *singulus*, solitary; *-are*, pertaining to. – (1) Inflorescence consisting of a single raceme. *Centrochloa singulare, Digitaria singularis, Paspalum singulare* – (2) peduncle terminating in a single spikelet. *Festuca singularis*
sinic-a, -us L. *Sina*, China; *-ica*, belonging to. From China
Sinoarundinaria L. *Sina*, China. An *Arundinaria*-like genus from China
Sinobambusa L. *Sina*, China. A *Bambusa*-like genus from China
Sinocalamus L. *Sina*, China; *kalamos*, reed. Tall woody genus from China
Sinochasea L. *Sina*, China. In honor of Mary Agnes Merrill Chase (1869–1963) United States agrostologist who gave much assistance to the author Keng during his visit to Washington studying Chinese grasses
sinoflexuosus L. *sina*, China; *flecto*, bend; *-osa*, abundance. A Chinese grass with a strongly bent awn
sinoglauca L. *Sina*, China. Resembling *Poa glauca* and from China
sinomongholica L. *-ica*, belonging to. From Inner Mongolia, an autonomous region of China
sinomutica L. *Sina*, China. Resembling *Festuca mutica* and coming from China
sinospinosa L. *Sina*, China; *spina*, spine; *-osa*, abundance. A spiny species from China
sintenisii In honor of Paul Ernst Emil Sintenis (1847–1907) German apothecary and plant collector
sinuat-a, -um, -us L. *sinus*, curve; *-ata*, possessing. – (1) Lemma apex sinuate. *Agropyron sinuatum, Elytrigia sinuata* – (2) epidermal cell walls of lemma sinuous. *Amphibromus sinuatus*
sipapoense L. *-ense*, denoting origin. From Sipapo, Venezuela
siphonoglossa Gk *siphon*, a hollow body; *glossa*, tongue. Ligule forming a tube about the culm
sipitangensis L. *-ensis*, denoting origin. From Sipitang, Sabah, Malaysia
sipyle-a, -us From Mt Sipylo, now Manissa Dagh, Turkey
Sirochloa Gk *sira*, chord; *chloa*, grass. Culms wire-like
siroyamensis L. *-ensis*, denoting origin. From Siroyamamura, a mountain in Japan
sisca Vernacular name of the species in Spain
sitanioides Gk *-oides*, resembling. Similar in habitat to *Sitanion*
Sitanion Gk *sitos*, grain of either wheat or barley; *-ion*, indicating occurrence. Inflorescences similar to those of certain species of wheat and barley
sitchensis L. *-ensis*, denoting origin. From Sitcha Island, Alaska
Sitopsis Gk *sitos*, an edible grain; *opsis*, resemblance. Similar to cultivated *Triticum*
Sitordeum Intergeneric hybrids between species of *Sitanion* and *Hordeum*
Sitospelos Gk *sitos*, wheat; *pelos*, mud. Possibly used for making mud bricks
sivagiriana L. *-ana*, indicating connection. From Sivagiri Hills, India
sivarajanii In honor of V. V. Sivarajan (1944–1995) Indian botanist
sjuzevii In honor of Paul W. Sjuzew, Russian botanist
skorpilii In honor of H. and K. Skorpil (fl. 1892) Bulgarian geographers

skottsbergii In honor of Carl Johan Fredrik Skottsberg (1880–1963) Swedish botanist

skrjabinii In honor of S. Skrjabin (fl. 1967) Russian botanist

skvortzovii In honor of – (1) Boris Vassilievich Skvortzov (1890–1980) Russian-born Manchurian botanist. *Merostachys skvortzovii* – (2) Alexei Konstantinovich Skvortzov (1920–) Russian botanist. *Festuca skvortzovii, Poa skvortzovii*

sloanei In honor of Hans Sloane (1660–1753) English physician and botanist who collected in West Indies

smaragdina L. *smaragdus*, precious stone of green color but probably not applied to emerald; *-ina*, indicating resemblance. Foliage emerald-green in color

Smidetia See Franz Wilibald Schmidt under entry for *Schmidtia*

smidetia See Franz Wilibald Schmidt under entry for *Schmidtia*

smilacifolia L. *folium*, leaf. Leaf-blades similar to those of *Smilax*

smirnovii, smirnowii In honor of Valentin Ivanovich Smirnow (1879–1942) Russian botanist

smithian-a,-us L. *-ana*, indicating connection. In honor of Christen Smith (1785–1816) Norwegian botanist

smithii In honor of – (1) Charles Eastwick Smith (1820–1900) United States engineer and amateur botanist. *Avena smithii, Melica smithii* – (2) Jared Gage Smith (1866–1925) United States botanist. *Agropyron smithii* – (3) Albert Charles Smith (1906–1997) United States botanist. *Stipa smithii* – (4) Lyman Bradford Smith (1904–1999) United States botanist. *Reitzia smithii* – (5) James Edward Smith (1759–1829) English botanist. *Festuca smithii, Poa smithii* – (6) Philip Morgans Smith (1941–2004) English-born Scottish botanist and educator. *Panicum smithii* – (7) Christen Smith (1785–1816) Norwegian botanist. *Aeluropus smithii, Dactylis smithii* – (8) Jeremy Michael Bayliss Smith (1945–) English born Australian ecologist. *Deyeuxia smithii* – (9) David Nelson Smith (1945–1991) United States botanist. *Chusquea smithii*

smitinandiana L. *-ana*, indicating connection. As for *Temochloa*

smutsii In honor of Jan Christiaan Smuts (1870–1950) South African statesman and amateur botanist

sneidernii In honor of Kjell von Sneidern (fl. 1910) who collected in Colombia

Snowdenia In honor of Joseph Davenport Snowden (1886–1973) English-born Ugandan economic botanist

snowdenii As for *Snowdenia*

sobolevskiana L. *-ana*, indicating connection. In honor of Kira Arkadyevna Sobolevskaja (1911–) Russian botanist

sobolifer-a,-um L. *soboles*, offshoot; *fero*, carry or bear. – (1) Strongly rhizomatous. *Achnatherum soboliferum, Agrostis sobolifera, Muhlenbergia sobolifera, Podosaemum soboliferum* – (2) having culms that push through the vegetation. *Paspalum soboliferum*

socotranum L. *-ana*, indicating connection. From Socotra, a Yemeni island in the Gulf of Aden

soczawae In honor of Victor Borisovich Soczawa (1905–) Russian botanist and plant geographer

Soderstromia In honor of Thomas Robert Soderstrom (1936–1987) United States agrostologist

soderstromiana L. *-ana*, indicating connection. As for *Soderstromia*

soderstromii As for *Soderstromia*

sodiroan-a,-um L. *-ana*, indicating connection. In honor of Luigi Sodiro (1836–1909) Italian-born Ecuadorean cleric and amateur botanist

Soejatmia In honor of Soejatmi Dransfield (1939–) Indonesian-born English botanist

soerensenii In honor of Thorvald Sørensen (1902–1973) Danish forester

sogdian-a,-um From Sogdiana, a district in Central Asia between the Jaxartes and Oxus Rivers, also known as Turkestan

Sohnsia In honor of Ernest Reeves Sohns (1917–) United States botanist

sokotranum See *socotranum*

solandri In honor of Daniel Carl Solander (1736–1782) Swedish-born English botanist

solearis L. sandle-shaped. The apex of the palea is folded forward to form a shallow pocket

Solenachne Gk *solen*, channel; *achne*, scale. Lower glume two-keeled

Solenophyllum Gk *solen*, channel; *phyllon*, leaf. The subulate leaf-blades on drying roll inwards to form a groove

solid-a, -us L. solid. – (1) Culms solid. *Fargesia solida, Gelidocalamus solidus, Indocalamus solidus, Monocladus solidus* – (2) panicles contracted. *Poa solida*

solitaria L. solitary. Inflorescene unbranched

solomonensis L. *-ensis*, denoting origin. From Solomon Islands, in Pacific Ocean

solut-a, -us L. completely separate. Clearly distinct from related species

somae In honor of T. Soma (fl. 1914) Japanese botanist

somalens-e, -is L. *-ense*, denoting origin. From Somalia

somdevae In honor of Som Deva (fl. 1991) Indian botanist

sommieri As for *sommieranum*

sommierianum L. *-anum*, indicating connection. In honor of Carlo Pietro Stefano Sommier (1848–1922) Italian botanist

sondongensis L. *-ensis*, denoting origin. From Son-Dong, Ha Bac Province, Vietnam

songorica See *soongarica*

sonorum From Sonora, Mexico

soongarica L. *-ica*, belonging to. From Soongaria, now Sungaria (Dzungaria, Zungaria), north-western China

soratana L. *-ana*, indicating connection. From Sorata, Bolivia

soratensis L. *-ensis*, denoting origin. From Sorata, Bolivia

sordid-a, -um L. dirty. Spikelets dark-green

sorgerae In honor of Friederike Sorger (1914–2001) Austrian merchant and plant collector

Sorghastrum L. *-astrum*, somewhat resembling. Similar to *Sorghum* in appearance

sorghi Of, that is resembling *Sorghum*

sorghoide-a, -um Gk *-oidea*, resembling. Resembling *Sorghum*

sorghoides Gk *-oides*, resembling. Resembling *Sorghum*

sorghoideum See *sorghoidea*

Sorghum Italian, *sorgho*. Vernacular name for the genus

sorghum In habit resembling *Sorghum halepense*

Sorgum See *Sorghum* but quite unrelated

sorianoi In honor of Alberto Soriano (fl. 1960) Argentinian agriculturalist

sororia L. *soror*, sister. Readily confused with related species

sorrentini From Sorrentino, Sicily

sorstitialis L. *solstitium*, solstice (longest day of the year); *-alis*, pertaining to. Meaning uncertain and possibly a misspelling of *solstitialis*, and if so, may be a reference to the flowering season

sorzogonensis L. *-ensis*, denoting origin. From Sorzogon on Luzon Island, Philippines

sosnovskyi, sosnowskyi In honor of Dimitrii Ivanovich Sosnowsky (1885–1952) Russian botanist

soukupii In honor of Jaroslev Soukup (1903–1989) Czech missionary and ethnobotanist who collected in Peru

southwoodii From Southwood National Park, Queensland, Australia

sovieticum L. *-icum*, belonging to. From the Soviet Union, now replaced by many self-governing Republics in Europe and Asia

soyensis L. *-ensis*, denoting origin. From Sôya, Kitami Province, Hokkaido, Japan

sozanensis L. *-ensis*, denoting origin. From Sozan, Taihoku, Taiwan

spadice-a, -us L. *spadix*, chestnut-brown; *-ea*, pertaining to. The spikelets and/or inflorescence branches are chestnut-brown

spania Gk *spanos*, scarce. Known only from Waitaki Valley, New Zealand

spanianth-a, -us Gk *spanios*, rare; *anthos*, flower. Spikelets distant in the inflorescence

spanospicula Gk *spanos*, scarce; L. *spica*, spike; *-ula*, diminutive. Spikelets fewer than in related species

spanostachya Gk *spanios*, rare; *stachys*, spike as of an ear of wheat. Inflorescence with few branches

sparmannii In honor of Anders Sparmann (1748–1820) Swedish botanist and traveller

sparshottiorum In honor of Kym Margaret Sparshott (1970–) Australian botanist and Peter Edward Sparshott (1969–) Australian naturalist

sparsicomum L. *sparsus*, few or scattered; *como*, hair of head. Spikelets few and distant on the panicle

sparsiflor-a, -um L. *sparsus*, few; *flos*, flower. – (1) Florets few per spikelet. *Arundinaria sparsiflora, Sinarundinaria sparsiflora* – (2) spikelets few per inflorescence. *Panicum sparsiflorum, Triplasis sparsiflora, Trisetum sparsiflora*

sparsifructus L. *sparsus*, few; *fructus*, fruit. The spikelets are often solitary rather than paired

sparta Gk *spartos*, broom (*Spartium junceum*) used for making rope. See *Spartum*

spartea Gk *spartos*, broom (*Spartium junceum*) used for making rope. See *Spartum*

spartellum Gk *spartos*, broom; L. *-ellum*, diminutive. The rigid terete culms resemble the terete, almost leafless stems of Spanish Broom (*Spartium junceum*)

Sparteum See *Spartum*

Spartina Gk *spartos*, broom (*Spartium junceum*); *-ina*, indicating resemblance. The tough leaves may be used for making cordage as is spartos or broom (*Spartium junceum*)

spartinae Resembling *Spartina*

spartinoides Gk *-oides*, resembling. Superficially similar to *Spartina*

Spartochloa Gk *spartos*, broom (*Spartium junceum*); *chloa*, grass. Culm leaves reduced thereby resembling broom (*Spartium junceum*)

Spartum, spartum Gk *spartos*, broom (*Spartium junceum*) used for making rope. Name transferred to the grass because it too is a source of fibre

spathace-a, -um, -us L. *spatha*, sheathing base and false petiole of a palm leaf; *-acea*, indicating resemblance. – (1) Inflorescence pedunculate and so scarcely exserted from the uppermost leaf-sheath. *Arthraxon spathaceus, Arundinaria spathacea, Chloris spathacea, Cryptochloris spathacea, Fargesia spathacea, Paspalum spathaceum, Tetrapogon spathaceus* – (2) inflorescence bracts spathe-like. *Chusquea spathacea*

spathellosum L. *spatha*, sheathing base and false petiole of a palm leaf; *-ella*, diminutive; *-osum*, abundance. Lemma of male floret large with a broad margin

Spathia L. *spatha*, sheathing base and false petiole of a palm leaf. Racemes protected by inflated sheathing leaf-base

spathiflor-a, -um, -us L. *spatha*, sheathing base and false petiole of a palm leaf; *flos*, flower. Racemes enclosed in sheathing leaf-bases

speciana L. *species*, beauty; *-ana*, indicating connection. Origin uncertain, not given by the author

specios-a, -us L. *species*, beauty; *-osa*, abundance. Showy in some respect, in particular the inflorescence

speciosissimum L. *speciosus*, showy; *-issimum*, most. Inflorescences very beautiful

speciosus See *speciosa*

spectabil-e, -is L. showy. Attractive or outstanding in some respect such as height

spegazzinii In honor of Carlo Luigi Spegazzini (1858–1926) Italian-born Argentinian botanist

speirostachya Gk *speira*, anything twisted; *stachys*, spike as of an ear of wheat. Florets disposed spirally along rachilla

spellenbergii In honor of Richard William Spellenberg (1940–) United States botanist

Spelta, spelta Old Saxon for a species of wheat formerly widely cultivated in southern Europe

speltaeform-e, -is See *speltiforme*

speltiform-e, -is L. *forma*, appearance. Resembling *Triticum spelta*

speltoides Gk *-oides*, resembling. Similar to *Spelta*

speluncarum L. *spelunca*, cave. Of caves, growing in moist shady caverns

spencei In honor of James F. Spence (fl. 1812) who collected in Venezuela

sperguliifolium L. *folium*, leaf. Leaf-blades resembling those of *Spergula*

Spermachiton Gk *sperma*, seed; *chiton*, tunic. Seed readily separated from pericarp

Spermatochiton See *Spermachiton*

sphacelat-a, -um L. speckled with brown or black. The apices of the anthoecia are purple or black

S

sphacioticus Gk *sphakos* or *sphagnos*, a fragrant moss; *-icus*, belonging to. Growing in alpine sphagnum bogs

Sphaerella Gk *sphaera*, ball; *-ella*, diminutive. Inflorescence spherical

Sphaerium Gk *sphaera*, ball. The pistillate portion of the inflorescence is enclosed in a hard, bead-like structure

Sphaerobambos Gk *sphaera*, ball. Fruit spherical otherwise resembling *Bambusa*

sphaerocarp-a, -on, -um Gk *sphaera*, ball; *karpos*, fruit. Spikelets almost circular in outline

Sphaerocaryum Gk *sphaera*, ball; *karyon*, nut. The grain is hard and resembles a ball

sphaerocephal-a, -us Gk *sphaera*, ball; *kephale*, head. Inflorescence a contracted globose panicle

sphaerococcum Gk *sphaera*, ball; *kokkos*, grain. Grain spherical

sphagnicola L. *-cola*, dweller. Growing in *Sphagnum* bogs

Spheneria Gk *sphen*, wedge; L. *-aria*, pertaining to. The spikelets are top-shaped

Sphenopholis Gk *sphen*, wedge; *pholis*, scale as of snake. The upper glume is wedge-shaped and indurated

Sphenopus Gk *sphen*, wedge; *pous*, foot. Pedicels wedge-shaped towards the tip

sphondylodes Gk *sphondylos*, vertebra; *-odes*, resembling. The spikelets resemble vertebrae

spicaeformis See *spiciforma*

spicat-a, -um, -us L. *spica*, a point; hence, in particular, an ear or spike of grain; *-ata*, possessing. Inflorescence a spike or spicate panicle

spicaventi L. *spica*, a point; hence, in particular, an ear or spike of grain; *ventus*, wind. – (1) Panicle branches long and thin enabling spikelets to wave in the breeze. *Apera spicaventi*, *Stipa spicaventi* – (2) as used by Linneus, a translation into Latin of the Swedish vernacular name, *vindhren*. *Agrostis spicaventi*

spiciform-a, -e, -is L. *spica*, a point; hence, in particular, an ear or spike of grain; *forma*, appearance. Inflorescence a condensed spike-like panicle

spiciger-a, -um L. *spica*, a point; hence, in particular, an ear or spike of grain; *gero*, carry or bear. Inflorescence a spicate panicle

spiculosa L. *spica*, a point; hence, in particular, an ear or spike of grain; *-ula*, diminutive; *-osa*, well-developed. Inflorescence spike-like as with *Triticum*

spinescens L. *spinesco*, become thorny. Inflorescence branches terminally pungent

spinifera L. *spina*, thorn; *fero*, carry or bear. The glumes terminate in an apical spine

Spinifex L. *spina*, thorn; *facio*, make. The leaf-blades of some species have sharply tipped apices

spinifex L. *spina*, thorn; *facio*, make. Involucral bristles spiny

spinos-a, -um L. *spina*, thorn; *-osa*, abundance. – (1) Inflorescence branches terminating in spines. *Cladoraphis spinosa* – (2) roots or stems bearing thorns. *Arthrostylidium spinosum*, *Arundo spinosa*, *Bambusa spinosa*, *Guadua spinosa*, *Ischurochloa spinosa* – (3) leaf-blades pungent. *Eragrostis spinosa*, *Festuca spinosa*, *Poa spinosa*

spinosissima L. *spinosus*, spiny; *-issima*, most. With abundant thorns derived from lateral shoots whose growth has been arrested

spinosum See *spinosa*

spiralis L. *spira*, anything coiled; *-alis*, pertaining to. – (1) Leaf-blades spirally coiled. *Chionochloa spiralis* – (2) panicle branches spirally arranged. *Echinochloa spiralis* – (3) spikelets spirally arranged on the rhachis. *Cynosurus spiralis*

spirathera Gk *speira*, anything twisted; *ather*, barb or spine. Lemma awn large and column spirally twisted when dry

spiridonovii In honor of Maxim Demitrievic Spiridonov (1878–1939) Russian botanist

spirifera L. *spira*, coil; *fero*, carry or bear. With circinate hairs on the upper glume and sterile lemma

Spirochloe, Spirochloë Gk *speiran*, twist; *chloa*, grass. At maturity the panicle becomes elongated and spiral

spirostylis Gk *speira*, coil; *stylos*, column. Style bent

Spirotheros Gk *speira*, anything twisted; *ather*, barb or spine. Lemma with a very long spirally twisted awn

spissifolium L. *spissus*, dense; *folium*, leaf. Culms freely branching from the lower nodes thereby generating a dense mass of foliage at the base of the plant

spissum L. compact. Leaf-blades erect, appressed to stem

splendens L. shining. – (1) Culms glossy. *Cortaderia splendens* – (2) lemmas glossy. *Achnatherum splendens, Koeleria splendens, Paspalum splendens, Stipa splendens*

splendid-a, -um L. splendid. Culms tall

Spodiopogon Gk *spodios*, grey; *pogon*, beard. The spikelets and pedicels are invested with long grey hairs

spongiosum L. spongey. Spikelets somewhat turgid

spontane-a, -um L. naturally growing wild. Growing in the wild but closely related to species known in cultivation

Sporabolus See *Sporobolus*

Sporichloe A misspelling of *Spirochloe*

Sporobolus Gk *sporos*, seed; *ballo*, throw. At maturity the seeds are squeezed out of the fruits

sprengelii In honor of Kurt Sprengel (1766–1833) German botanist

spretum L. *sperno*, separate. Treated as a distinct taxon from a species complex

spruceana L. *-ana*, indicating connection. In honor of Richard Spruce (1817–1893) English botanist in South America

sprucei As for *spruceana*

spuria L. spurious. The lateral awns of the lemma are insignificant or absent, the spikelets then resembling those of *Stipa*

squamulat-um, -us L. *squama*, scale; *-ula*, diminutive; *-atum*, possessing. – (1) Spikelet clusters subtended by scale-like spatheoles. *Andropogon squamulatus* – (2) paleas, lemmas and glumes well developed. *Paspalum squamulatum, Pennisetum squamulatum, Pentastachya squamulatum*

squarros-a, -um, -us L. spreading at right angles from a common axis. – (1) Inflorescence arms held at right angles to common axis. *Aegilops squarrosa, Andropogon squarrosus, Astrebla squarrosa, Boissiera squarrosa, Bromus squarrosus, Cleistogenes squarrosa, Diplachne squarrosa, Eragrostis squarrosa, Eriachne squarrosa, Kengia squarrosa, Leptochloa squarrosa, Molinia squarrosa, Pogonarthria squarrosa, Pseudoraphis squarrosa, Spinifex squarrosa, Stipa squarrosa, Triticum squarrosum, Urelytrum squarrosum* – (2) lemma square in outline. *Munroa squarrosa* – (3) leaf-blades held at right angles to culm axis. *Panicum squarrosa*

sreenarayanae In honor of Sree Narayana Guru, Indian Saint and Social Reformer who founded a number of educational establishments in Kerala, India

srilankensis L. *-ensis*, denoting origin. From Sri Lanka

stachydanthus Gk *stachys*, spike as of an ear of wheat; *anthos*, flower. Panicle spiciform

stachyodes Gk *stachys*, spike as of an ear of wheat; *-odes*, resembling. Panicle spiciform

stagnalis L. *stagnum*, standing water; *-alis*, pertaining to. Growing along the margins of still water

stagnatile L. *stagnum*, pool; *-atile*, place of growth. Growing in pools

stagnin-a, -um, -us L. *stagnum*, pool; *-ina*, indicating possession. Aquatic species

staintonii In honor of John David Adam Stainton (1921–) English botanist

standleyi In honor of Paul Carpenter Standley (1884–1963) United States botanist

S

Stapfia In honor of Otto Stapf (1857–1933) Austrian-born English botanist
stapfian-a, -um, -us L. *-ana*, indicating connection. As for *Stapfia*
stapfii As for *Stapfia*
Stapfiola L. *-ola*, diminutive but here employed as a name-forming suffix. As for *Stapfia*
Stapfochloa Gk *chloa*, grass. As for *Stapfia*
staroplaninica L. *-ica*, belonging to. From the Stara Planin Mountains, Serbia
starosselskyi In honor of V. Starosselsky (fl. 1921–1922) Russian botanist
stassewitschii In honor of Stassewitsch
stauntonii In honor of George Leonard Staunton (1737–1801) Irish-born physician and British civil servant who collected in China
stebbinsianum L. *-anum*, indicating connection. In honor of George Ledyard Stebbins (1906–2000) United States botanist
stebbinsii As for *stebbinsianum*
stebeckii In honor of Stephan G. Beck (1944–) Bolivian botanist
stebleri In honor of Friedrich Gottlieb Stebler (1852–1935) Swiss botanist
steenisii In honor of Cornelis Gijsbert Gerrit Jan van Steenis (1901–1986) Dutch botanist
stefaninii In honor of Giuseppe Stephanini (fl. 1882–1938) Italian botanist
Stegosia Gk *stegos*, roof. Used for roofing material in Indo-China, now Cambodia, Laos and Vietnam
steinbachii In honor of J. Steinbach (fl. 1929) who collected in Bolivia
steinbergii In honor of E. Steinberg (fl. 1931–1934) Russian plant collector
Steinchisma Gk *steinos*, narrow; *chasma*, yawning hollow. The lower floret is gaping
Steirachne Gk *steira*, forepart of ship's keel; *achne*, scale. In outline the lemma of the fertile floret has the shape of the prow of a ship
stejnegeri In honor of Leonard Hess Stejner (fl. 1882–1897)
Stelephuros Gk *stelephouros*, hare's foot plantain. Plants softly hairy like the hare's foot plantain (*Plantago lagopus*)
stellaris L. *stella*, star; *-aris*, pertaining to. From the Star Mountains, Papua New Guinea
stellat-a, -um L. star-shaped. Hairs at the base of the spikelet spreading stellately at maturity
stelleri In honor of Georg Wilhelm Steller (1709–1746) German traveller who collected in Siberia
Stematospermum See *Stemmatospermum*
Stemmatosperma See *Stemmatospermum*
Stemmatospermum Gk *stemma*, wreath made of wool or wool itself; *sperma*, seed. Possibly a reference to the pinnate-plumose stigmas
stenachyr-a, -um Gk *stenos*, narrow; *achyron*, chaff. Spikelets with narrow glumes and lemmas
stenanth-a, -um Gk *stenos*, narrow; *anthos*, flower. Spikelets narrow
stenoauritus Gk *stenos*, narrow; L. *auritis*, long-eared. Auricles linear
Stenobromus Gk *stenos*, narrow. Spikelets similar to but narrower than those of most *Bromus* species
stenocarpa Gk *stenos*, narrow; *karpos*, fruit. Inflorescence segments subtended by cylindrical cupules
Stenochloa Gk *stenos*, narrow; *chloa*, grass. Inflorescence a narrow panicle
stenoclad-a, -um Gk *stenos*, narrow; *klados*, branch. Panicle branches thin
stenodes Gk *stenos*, narrow. Culms slender and wiry
stenodoides Gk *-oides*, resembling. Similar to *Panicum stenodes*
Stenofestuca Gk *stenos*, narrow. Spikelets resembling those of *Festuca* but with narrow glumes
stenolemma Gk *stenos*, arrow; *lemma*, scale. Lemmas narrow-lanceolate
stenophyll-a, -um Gk *stenos*, narrow; *phyllon*, leaf. Leaf-blades filiform
stenoptera Gk *stenos*, narrow; *pteron*, wing or feather-like. Margins of glumes and sterile lemmas white-winged
stenorrhachis Gk *stenos*, narrow; *rhachis*, backbone. Culms slender

stenosoma Gk *stenos*, narrow; *soma*, body. Culms slender

stenostachy-a, -um, -us Gk *stenos*, narrow; *stachys*, spike as of an ear of wheat. – (1) Inflorescence a narrow spike or spike-like panicle. *Agropyron stenostachyum, Aristida stenostachya, Bambusa stenostachya, Digitaria stenostachya, Panicum stenostachyum, Penicillaria stenostachya, Pennisetum stenostachyum, Roegneria stenostachya, Sporobolus stenostachyus, Triodia stenostachya* – (2) raceme narrow. *Paspalum stenostachyum*

Stenostachys, stenostachys See *stenostachya*. – (1) Spikelets narrow. *Eragrostis stenostachys* – (2) inflorescence narrow. *Hordeum stenostachys*

Stenostachyum See *stenostachya*

Stenotaphrium See *Stenotaphrum*

stenotaphrodes Gk *-odes*, resembling. Similar to *Stenotaphrum*

Stenotaphron See *Stenotaphrum*

Stenotaphrum Gk *stenos*, slender; *taphros*, ditch. The spikelets are sunken in small pits on the surface of the rhachis

stenothyrs-a, -um, -us Gk *stenos*, narrow; *thyrsos*, ornamental wand. Panicle contracted

stentiana L. *-ana*, indicating connection. In honor of Sydney Margaret Stent (1875–1942) South African botanist

Stephanachne Gk *stephanos*, crown; *achne*, scale. The lemma has a crown of long hairs around the two lobes

stepparia English *steppe* from Russian *stip*, grassland; *-aria*, pertaining to. From grasslands of Argentina

stepposa English *steppe* from Russian *stip*, grassland; *-osa*, abundant. A common steppe species

Stereochlaena Gk *stereos*, rigid; *chlaena*, cloak. The fertile lemma is rigid

stereophylla Gk *stereos*, rigid; *phyllon*, leaf. Leaf-blades stiff

sterilis L. sterile. – (1) The spikelets fall soon after attaining maturity. *Avena sterilis, Bromus sterilis, Koeleria sterilis* – (2) the lower florets are sterile. *Eragrostis sterilis*

Steudelella L. *-ella*, diminutive but here employed as a name-forming suffix. In honor of Ernest Gottlieb von Steudel (1783–1856) German botanist and physician

steudelian-a, -um L. *-ana*, indicating connection. As for *Steudelella*

steudelii As for *Steudelella*

stevenii In honor of Christian von Steven (1781–1863) Finnish-born Russian botanist

stevensianum L. *-anum*, indicating connection. In honor of Frank Lincoln Stevens (1871–1934) United States botanist

stevensii In honor of Warren Douglas Stevens (1944–) United States botanist

stewartiana In honor of Ralph Randles Stewart (1890–1993) United States missionary and botanist at Rawalpindi, Pakistan

stewartii As for *stewartiana*

steyermarkii As for *Steyermarkochloa*

Steyermarkochloa Gk *chloa*, grass. In honor of Julian Alfred Steyermark (1909–1988) United States botanist who worked principally in Venezuela

Stiburus Gk *stibi*, powdered antimony used for eye paint; *oura*, tail. Spikelets dark-purple and inflorescence a dense spicate panicle

stickhania In honor of Eve and James Stickha, United States rock collectors

stigmatisat-a, -um L. *stigma*, mark; *-atum*, possessing. Subtending glumes with red spots

stigmos-a, -um L. *stigma*, mark; *-osa*, abundance. Leaf-sheath with very small glands

stillmanii In honor of Jacob Davis Babcock Stillman (1819–1888) United States physician and amateur botanist

stillmannii See *stillmanii*

Stilpnophleum Gk *stilphnos*, glistening. Glumes membranous and often pigmented, otherwise resembling *Phleum*

Stipa Gk *stype*, coarse part of hemp or other plant fibre. Used for making rope

stipacea L. *-acea*, indicating resemblence. Similar to *Stipa*

stipaeculmis L. *culmus*, stem. With the habit of *Stipa*

stipaeformis L. *forma*, appearance. Resembling *Stipa* in some respect

Stipagrostis Compounded of the generic names *Stipa* and *Agrostis*. The type species has feathery awns like those of *Stipa pennata*

stipatum L. *stipes*, stalk; *-atum*, possessing. Spikelets distinctly stalked

Stipavena A contraction of *Stipa* and *Avena* the genus sharing characters of both

Stipidium Gk *-idium*, resemblance. Fossil fruits resembling those of *Stipa*

stipiflorum L. *stipes*, stalk; *flos*, flower. Glumes separated by a small stipe

stipifoli-a, -um L. *folium*, leaf. The leaf-blades resemble those of *Stipa*

stipiformis L. *forma*, appearance. Resembling *Stipa* in some respect

stipitat-a, -um L. *stipes*, stalk; *-ata*, possessing. – (1) Pedicels long. *Aristida stipitata* – (2) rachilla internodes long. *Panicum stipitatum* – (3) stigmas shortly stalked. *Ischaemum stipitatum*

stipoide-a, -um Gk *-oidea*, resembling. The spikelet has a long sharp callus and resembles *Stipa*

stipoides Gk *-oides*, resembling. With spikelets resembling those of *Stipa*

Stiporyzopsis Hybrids between species of *Stipa* and *Oryzopsis*

stiriaca L. *stiria*, icicle. Growing on high mountains

stjohnii In honor of Harold St John (1892–1991) United States botanist especially concerned with the Pacific region

stocksii In honor of John Ellerton Stocks (1822–1854) English-born physician and plant collector in India

stokesii In honor of A. M. Stokes (fl. 1922) who collected in the Marquesas

stoliczkae In honor of Ferdinand Stoliczka (1838–1874) Austrian/Czech palaeontologist who collected in Kashmir

stolonifer, -a, -um L. *stolo*, shoot; *fero*, carry or bear. Plant with well developed underground stems

stolziana L. *-ana*, indicating connection. As for *stolzii*

stolzii In honor of Adolph Ferdinand Stolz (1871–1917) German missionary in Nyssaland, now Malawi

stracheyi In honor of Richard Strachey (1817–1908) English-born Indian Army officer and plant collector

stragulus L. creeping. Plants with freely branching stolons

stramine-a, -um, -us L. straw-yellow. – (1) Spike-lets and attendant structures straw-colored. *Bromus stramineus*, *Chusquea straminea*, *Jouvea straminea*, *Panicum stramineum*, *Paspalum stramineum*, *Pennisetum stramineum* – (2) glumes and leaves straw-coloured. *Muhlenbergia straminea* – (3) leaves and glume apices straw-coloured. *Deschampsia straminea*

strangulata L. *strangulo*, torment. The inflorescences are burr-like and catch in clothes

streblochaeta See *Streblochaete*

Streblochaete Gk *streblos*, twisted; *chaete*, bristle. The lemmas bear long, twisted awns

strephioides Gk *-oides*, resembling. The foliage resembles that of *Strephium*

Strephium Gk *strepho*, twist. The leaf-blades exhibit sleep-movements

strephoides Gk *-oides*, resembling. Superficially similar to *Strephium*

Streptachne Gk *streptos*, flexible; *achne*, scale. The lemma terminates in a hygroscopic awn in both genera so named

Streptia Gk *streptos*, flexible. The style is twisted into three long tortuous stigmas

streptobotrys Gk *streptos*, flexible; *botrys*, bunch of grapes. Spikelets widely separated in spike

Streptochaeta Gk *streptos*, flexible; *chaete*, bristle. The awn is twisted

Streptogyna, Streptogyne Gk *streptos*, flexible; *gyne*, woman. The style is twisted and divided into three long tortuous stigmas

Streptolophus Gk *streptos*, flexible; *lophos*, crest. The spikelets are subtended by a basket-like involucre formed by the recurved spinous tips of a dwarf branch system

Streptostachis See *Streptostachys*

Streptostachys, streptostachys Gk *streptos*, flexible; *stachys*, spike as of an ear of wheat. Spikelets appressed to secondary branches of inflorescence and so, with their recurved awns, resemble spikes of wheat

striat-a, -um, -us L. *stria*, furrow; *-ata*, possessing. – (1) Glumes and lemmas streaked. *Andropogon striatus, Bromus striatus, Cleistogenes striata, Coelachne striata, Glyceria striata, Kengia striata, Panicum striatum, Pollinia striata, Rottboellia striata, Sacciolepis striata* – (2) leaf-sheath streaked. *Holcus striatus, Sinobambusa striata, Trochera striata*

striatulum L. *stria*, furrow; *-ata*, possessing; *-ulum*, diminutive. Glumes weakly grooved

striat-um, -us See *striata*

stribrnyi In honor of Václav Stríbrny (1853–1927) Bohemian-born Bulgarian botanist

strict-a, -um, -us L. erect. – (1) Inflorescence branches erect. *Amphipogon strictus, Andropogon strictus, Aristida stricta, Arundinella stricta, Arundo stricta, Atropis stricta, Avenella stricta, Bambusa stricta, Briza stricta, Calamagrostis stricta, Calotheca stricta, Crypsis stricta, Danthonia stricta, Dendrocalamus strictus, Deschampsia stricta, Deyeuxia stricta, Digitaria stricta, Distichlis stricta, Eleusine stricta, Epicampes strictus, Eremopogon strictus, Ferrocalamus strictus, Festuca stricta, Garnotia stricta, Glyceria stricta, Isachne stricta, Merxmuellera stricta, Nardus stricta, Oplismenus strictus, Panicum strictum, Puccinellia stricta, Roegneria stricta, Saccharum strictum, Sporobolus strictus, Stipa stricta, Tridens strictus* – (2) culms erect. *Cenchrus strictus, Ehrharta stricta, Ferrocalamus strictus*

strictiflor-a, -um L. *strictus*, erect; *flos*, flower. Panicles narrow

strictifolium L. *strictus*, erect; *folium*, leaf. Leaf-blades erect or nearly so

strictior L. more erect. Panicle branches more appressed than in related taxa

strictiramea L. *strictus*, erect; *ramus*, branch. Inflorescence branches erect

strictissimum L. *strictus*, erect; *-issimum*, indicating to a high degree. Panicle branches very closely appressed to central axis

strictula L. *strictus*, erect; *-ula*, tending to. Panicles erect, somewhat constricted

strict-um, -us See *stricta*

stridula L. rustling. The leaves make a crackling sound when trodden on

strigatus L. *stringo*, press together. Spikelets appressed to the axis of the inflorescence

strigos-a, -um L. covered with short, bristle-like hairs. One or more organs covered with bristle-like hairs

stripitans Origin unclear, not given by the author but may be a misspelling of present participle of L. *strepito*, rustle. Rustling is an allusion to sounds fallen leaves make when trodden upon

Strombodurus Gk *strombos*, a spiral shell; *oura*, tail. Meaning uncertain, origin not given by author; manuscript name only

strumosum L. *struma*, a scrofulous tumour; *-osum*, abundance. Fertile lemma marked with swellings

stuartiana L. *-ana*, indicating connection. In honor of Charles Stuart (1802–1877) English-born Australian plant collector

stuckertii In honor of Teodoro Juan Vicente Stuckert (1852–1932) Swiss-born Argentinian botanist

stuebelii In honor of Moritz Alphons Stübel (1835–1904) German botanist-traveller who collected in South America

stuhlmannii In honor of Franz Ludwig Stuhlmann (1863–1927) German Army Officer and administrator in East Africa where he collected plants

Stupa See *Stipa*

stupos-a, -um L. *tow*, that is fibre for rope-making. Leaf-bases breaking into fibres

Sturmia In honor of Jakol Sturm (1771–1848) German natural history illustrator

stygia L. *-ia*, indicating connection. From the Styx Valley, Peloponnisos, Greece

Stylagrostis Gk *stylos*, mast of a ship. Resembling *Agrostis* but lemma long awned

Stypa See *Stipa*

Styppeiochloa Gk *stuppion*, coarse flax or hemp; *chloa*, grass. The plant base is tough and fibrous

stypticus L. with the power to contract living tissue. Used in Angola to stop bleeding

Suardia In honor of Paulus Suardus (fl. 1528) Venetian physician who wrote a text on medicinal herbs

suaveolens L. sweet-scented. Usually of species with sweet-scented inflorescences

suavis (1) L. *suavis*, agreeable. Habit graceful. *Poa suavis* – (2) L. *suavis*, sweet. Possibly a reference to the culms tasting sweet if chewed. *Indosasa suavis*

S

sub Many epithets beginning with "*sub-*" (L. *sub-*, approaching) are intended to distinguish a genus or species from that with which it was previously confused, or to avoid the formation of a homonym. In the entries below, only those names are listed that are known not to adhere to this rule

subacaul-e, -is L. *sub-*, approaching; *a-*, without; *caulis*, stem. Dwarf plants with short culms

subacrochaeta L. *sub-*, approaching; Gk *akros*, sharp; *chaete*, bristle. Lemma very shortly aristate

subaequiglum-a, -is L. *sub-*, approaching; *aequis*, equal; *gluma*, husk. Glumes similar

subalpinum L. *sub-*, approaching. Growing on the lower slopes of high mountains

subandina L. *sub-*, approaching; *-ina*, indicating possession. From the foothills of the Andes

subaphylla L. *sub-*, almost; Gk *a-*, without; Gk *phyllon*, leaf. Leaf-blades poorly developed

subaristat-a, -um L. *sub-*, approaching; *arista*, bristle. Lemma sharply acute

subarticulata L. *sub-*, almost; *articulus*, joint; *-ata*, possessing. Florets tardilly articulating

subatra L. *sub-*, approaching; *ater*, dark. Spikelets dark-colored

subbiflora L. *sub-*, approaching; *bis*, two; *flos*, flower. Spikelets often with two florets

subbulbos-um, -us L. *sub-*, approaching; *bulbus*, bulb; *-osus*, abundance. Bases of culms somewhat swollen

subcaerulea L. *sub-*, approaching; *caerulea*, blue. Leaf-blades whitish-green

subcalva L. *sub-*, approaching; *calva*, a bald scalp. The upper glume and sterile lemma are basally glabrous and distally hairy

subcordatifolius L. *sub-*, approaching; *cordus*, heart; *-ata*, possessing; *folium*, leaf. Leaf-blade cordate at the base

subcordatum L. *sub-*, approaching; *corda*, heart; *-atum*, possessing. Leaf-blade subcordate

subeglume L. *sub-*, almost; *e-*, without; *gluma*, husk. – (1) Lower glume missing and upper glume much reduced. *Panicum subeglume*, *Pennisetum subeglume* – (2) both glumes much reduced. *Agropyron subeglume*

subenervis L. *sub-*, approaching; *-e*, without; *nerva*, nerve. Glumes and/or lemmas weakly nerved

suberostratum L. *sub-*, approaching; *e-*, without; *rostrum*, beak. Sterile lemma with or without a terminal bristle

subesetosa L. *sub-*, approaching; *e-*, without; seta, bristle; *-osa*, abundance. Bristles in inflorescence very few

subfastigiat-a, -um L. *sub-*, approaching; *fastigio*, sharpen to a point. Panicle branches held erect and produced in twos or threes from the same node

subflexuosa L. *sub-*, approaching; *flexuosa*, bent. Culms with a tendency to bend

subglabr-a, -um L. *sub-*, approaching; *glaber*, without hairs. Having few hairs on the glumes and/or sterile lemmas

subglabratum L. *sub-*, approaching; *glaber*, smooth; *-atum*, possessing. Leaf-blades hairy but glumes glabrous

subglabriflora L. *sub-*, approaching; *glaber*, smooth; *flos*, flower. Spikelets almost glabrous

subglabrum See *subglabra*

subglandulosa L. *sub-*, approaching; *glans*, gland; *-ulus*, diminutive; *-osa*, abundance. Pedicels of the inflorescence branches bear inconspicuous glands

subglobosum L. *sub-*, approaching; *globa*, sphere; *-osa*, abundance. Spikelets almost globose

subinclusum L. *sub-*, almost; *includo*, include. Panicle partly enclosed in subtending leaf-sheath

subjunceum L. *sub-*, approaching. Culms resembling those of certain *Juncus* species

sublaevigata L. *sub-*, approaching; *laevigata*, smooth and polished. Culms with stiff hairs when young but becoming smooth and glabrous with age

sublima L. lofty. Alpine species

sublimis L. *sublimo*, raise up. Culms tall

submutic-a, -us L. *sub-*, almost; *mutica*, blunt. Apices of glumes or lemmas rounded

subnudum L. *sub-*, approaching; *nudus*, bare. Lemma almost glabrous at the base

subpectinat-a, -um L. *sub-*, approaching; *pecten*, comb; *-ata*, possessing. With setose fimbrae on the margins of the leaf-sheath

subquadripar-a, -um L. *sub-*, almost; *quadri*, four; *pario*, bear. Inflorescence mostly of four racemes

subreflexa L. *sub-*, almost; *reflecto*, bend back. Lower branches of panicle drooping

subrostrat-a, -um L. *sub-*, almost; *rostrum*, beak. Upper glumes and sterile lemmas less conspicuously beaked than in other species of the genus

subsericans L. *sub-*, approaching; *sericus*, silken; *-ans*, assuming the appearance of. A putative hybrid between *Themeda arundinacea* and *T. villosa*

subsesquiglume L. *sub-*, approaching; *sesqui*, one half more; *gluma*, scale. Spikelets appearing to have one long and one short glume

subsessilis L. *sub-*, approaching; *sessilis*, sessile. Pseudopetiole very short

subspicat-a, -um L. *sub-*, approaching; *spica*, spike; *-ata*, possessing. Inflorescence a spike-like panicle

subsulcata L. *sub-*, spproaching; *sulcus*, furrow; *-ata*, possessing. The palea of the sterile floret is somewhat sulcate

subtil-e, -is L. delicate. Leaf-blades linear

subtiliracemosum L. *subtilis*, delicate; *racemus*, stalk of a cluster of grapes; *-osum*, abundance. Inflorescence has many thread-like branches

subtilissimum L. very subtle. – (1) Distinguished from related species only by careful comparison. *Tripogon subtilissimum* – (2) culms delicate. *Panicum subtilissimum*

subtiramulosum L. *subtilis*, delicate; *ramus*, branch; *-ula*, diminutive; *-osum*, abundance. Inflorescence branches are delicate and thread-like

subtriflora L. *sub-*, approaching; *tres*, three; *flos*, flower. Spikelets mostly of five florets of which two or three are often male or sterile

subtrivialis L. *sub-*, approaching. Similar to *Poa trivialis*

subulat-a, -um L. *subulus*, a fine point; *-ata*, possessing. – (1) Glumes, lemmas or calluses sharply tapered. *Andropogon subulatus, Chusquea subulata, Elymandra subulata, Festuca subulata, Loliolum subulatum, Melica subulata, Oryza subulata, Phleum subulatum, Rytidosperma subulata, Stenotaphrum subulatum, Stipa subulata* – (2) leaf-blades sharply tapered. *Agrostis subulata, Brachypodium subulatum, Danthonia subulata, Panicum subulatum*

subuliflora L. *subulus*, a fine sharp point; *flos*, flower. Lemmas drawn out into long non-hygroscopic awns

subulifolia L. *subulus*, fine sharp point; *folium*, a leaf. Leaf-blades narrow, tapering

subunifoveolatus L. *sub-*, approaching; *unus*, one; *fovea*, pit; *-olus*, diminutive; *-atus*, possessing. Lower glume mostly with a single abaxial depression

subverticillata L. *sub-*, approaching; *verticillus*, whorl. – (1) Lateral branches tending to form whorls in the inflorescence. *Echinochloa subverticillata, Festuca subverticillata, Poa subverticillata* – (2) apical leaves of culm subverticillate. *Sasa subverticillata*

subvestita L. *sub-*, approaching; *vestita*, clothing. Lemmas softly hairy at the base

subxerophilum L. *sub-*, approaching; Gk *xerophilum*, desert lover. From semi-desert regions

succinct-a, -um L. compact. Inflorescence branches held erect
succulentus L. *succus*, juice; *-ulentus*, well developed. Foliage a useful fodder
sucosum L. *succus*, juice; *-osa*, abundance. Culms fleshy
Sucrea In honor of Dimitri Sucre Benjamin (c. 1945–) a Panamanian-born Brazilian botanist
sudanens-e, -is L. *-ense*, denoting origin. From the Sudan
sudans L. *sudo*, exude. Plant sticky
sudavica L. *-ica*, belonging to. From Suduva, Lithuania
Suddia Arabic *sudd*. A major component of the floating islands which obstruct navigation on the White Nile
sudetica L. *-ica*, belonging to. From Sudeten Mountains on the border of Czech Republic and Poland
sudhanshui As for *jainiana*
sudicola L. *sudis*, crag; *-cola*, dweller. Inhabitating steep mountain slopes
suecic-a, -um L. *-ica*, belonging to. From Suecia, that is Sweden
suffrutescens L. *suffrutesco*, become woody. Culms somewhat woody
suffultiformis L. *forma*, appearance. Similar in appearance to *Axonopus suffultus*
suffult-um, -us L. *suffulcio*, support from beneath. Rhizome raised off the soil by prop roots
suffusca L. brownish. Spikelets brown
sugawarae In honor of Shigezo Sugawara (fl. 1937) Japanese botanist
sugimotoi In honor of Junichi Sugimoto (1901–?) Japanese botanist
suijiangensis L. *-ensis*, denoting origin. From Suijiang, Yunnan Province, China
suishaense L. *-ense*, denoting origin. From Suisha, Taiwan
suizanensis L. *-ensis*, denoting origin. From Suizan, Taiwan
suka Vernacular name for the species in Chile
sukatschewii In honor of Vladimir Nikolajevic Sukatschew (1880–1967)
suksdorfii In honor of Wilhelm Nikolaus Suksdorf (1850–1932) German-born United States botanist
sulcat-a, -um, -us L. *sulcus*, furrow; *-ata*, possessing. – (1) Internodes grooved. *Chaetochloa sulcata, Deyeuxia sulcata, Festuca sulcata, Heteropholis sulcatus, Ischaemum sulcatum, Oligostachyum sulcatum, Peltophorus sulcatus, Setaria sulcata* – (2) glume or glumes grooved. *Capillipedium sulcatum, Schizachyrium sulcatum, Sehima sulcatum* – (3) palea grooved. *Chusquea sulcata* – (4) fertile lemmas grooved. *Axonopus sulcatus, Panicum sulcatum* – (5) rhachis grooved. *Rottboellia sulcata*
sulcigluma L. *sulcus*, furrow; *gluma*, husk. Sterile lemma grooved
sulphurea L. pale-yellow. Culms golden-yellow
sumapana L. *-ana*, indicating connection. From Páramo de Sumapaz, Colombia
sumatran-a, -um, -us L. *-ana*, indicating connection. From Sumatra, Indonesia
sumatrense L. *-ense*, denoting origin. See *sumatrana*
sumichrasti In honor of Adrian Luis Jean Francois Sumichrast (1829–1882) Mexican plant collector
summilusitana L. *summa*, highest place; *Lusitana*, Lusitania. From the highest mountains of Lusitania, now mainly Portugal
sumneviczii In honor of Georgij Prokopievic Sumnevicz (1909–1947) Russian botanist
sundaic-a, -us L. *-ica*, belonging to. From Java (Indonesia) and near to the Sunda Strait
sundararajii In honor of Daniel Sundararaj (1919–) Indian botanist
suniana In honor of B. S. Sun, Chinese agrostologist
superat-a, -um L. *supero*, overtop. Inflorescence projecting conspicuously at anthesis
superba L. extra. Plants large in some respect and especially with tall culms or long leaf-blades
superbiens L. *superbio*, be splendid. At maturity the inflorescence is amethyst colored
superciliat-um, -us L. *supercilium*, eyebrow. Margins of the lower glume are densely ciliate

supernum L. upper half. Leaf-blades pilose only on inner surface
superpendens L. *super*, above; *pendeo*, hang. Spikelets pendulous in upper part of spikelet
supervacu-a, -um L. redundant. There is an extra second sterile lemma between the lower sterile lemma and fertile lemma
supin-a, -um L. prostrate. Strongly rhizomatous
suprapilosa L. *super*, above; *pilum*, a hair; *-osa*, abundance. Upper leaf surface hairy
suraboja L. from Suraboja, Java, Indonesia
surculosa L. *surculus*, young twig; *-osa*, abundance. Inflorescence much branched
surgens L. *surgo*, raise. Culms long, ascending
surinamens-e, -is L. *-ense*, denoting origin. From Surinam
suringarii In honor of Willem Frederick Reinier Suringar (1832–1898) Dutch botanist
surrect-a, -um L. *surgo* (*subrigo*), raise aloft. Stems initially procumbent
suruana L. *-ana*, indicating connection. From Suru-Tal, Kashmir
surugensis L. *-ensis*, denoting origin. From Suruga Province, now Shizuoka Prefecture, Japan
suwekoana L. *-ana*, indicating connection. In honor of Suwe-ko Makino (?–1928) wife of T. Makino
suzukaensis L. *-ensis*, denoting origin. From Suzukayama, a mountain in Mie Prefecture, Japan
suzukii In honor of Sadao Suzuki (fl. 1930s) Japanese botanist
swainsonii In honor of William Swainson (1789–1855) British naturalist and explorer
Swallenia In honor of Jason Richard Swallen (1903–1991) United States botanist
swalleniana L. *-ana*, indicating connection. As for *Swallenia*
swallenii As for *Swallenia*
Swallenochloa Gk *chloa*, grass. See *Swallenia*
swartbergensis L. *-ensis*, denoting origin. From Swartberg, South Africa
swartzian-a, -um L. *-ana*, indicating connection. In honor of Olof Peter Swartz (1760–1818) Swedish botanist
swartzii As for *swartziana*
swazilandensis L. *-ensis*, denoting origin. From Swaziland
swynnertonii In honor of Charles Francis Massey Swynnerton (1877–1938) Indian-born African farmer and biologist
sykesii In honor of William Russell Sykes (1927–) English-born New Zealand botanist
Syllepis Gk *syllephis*, a putting together. The spicate inflorescence is condensed in contrast to the open panicles of related genera
sylvanum L. *-anum*, indicating connection. In honor of René Sylva, Hawaiian biologist
sylvatic-a, -um, -us L. *silva*, wood; *-ica*, belonging to. Growing in woodlands
sylvestris L. *silva*, wood; *-estris*, place of growth. Plants of woodlands
sylviae In honor of Sylvia Mabel Phillips (1945–) English botanist
sylvicola L. *silva*, wood; *-cola*, dweller. Woodland species
Symbasiandra Gk *syn*, together with; *basis*, pedestal; *andros*, male. The spikelets are borne in triads, the lower two male and the terminal female or hermaphrodite
Symplectrodia Gk *syn*, together with. Sharing characters in common with both *Plectrachne* and *Triodia*
sympodica Gk *syn*, together with; *pous*, foot; *-ica*, belonging to. Male spikelets sessile, female spikelets stalked
Synaphe Gk *syn*, together with; *apto*, adhere. The caryopsis adheres to the glumes
Syntherisma Gk *syn*, together with; *therismos*, reaping. Weeds associated with reaping, that is with cereal crops
syreistschikovii, syreistschikowii In honor of Dimitri Petrovich Syreishchikov (1868–1932) Russian botanist
syriacum L. *-acum*, pertaining to. From Syria
syrtic-a, -us L. *syrtis*, sand bank; *-ica*, belonging to. Growing on sandbanks
Syurus Gk *sys*, pig; *oura*, tail. Inflorescence narrow and unbranched
syvaschica L. *-ica*, belonging to. From Sivash on the Black Sea

syzigachne Gk *syzigos*, paired; *achne*, scale. The spikelets fall entire so the pair of conspicuously compressed subtending glumes appear to be fused at their bases

szaboi In honor of Zóltan Szabo (1882–1944) Hungarian botanist

szechuanensis L. *-ensis*, denoting origin. From Sichuan or Szechuan Province, China

szechuensis L. *-ensis*, denoting origin. From Szechuan or Sichuan Province, China

szowitsiana L. *-ana*, indicating connection. In honor of Johann Nepomuk Szovitz (?–1830) Hungarian-born apothecary and botanist who collected in the Ukraine

T

T

tabacaria L. *tabacum*, derived from *tabacco*, the Spanish word for the pipe used to inhale the smoke of burning leaves or *Nicotiana tabacum* or cigars made from leaves of the same species; *-aria*, pertaining to. Stems used for making pipes for smoking

tabascoense L. *-ense*, denoting origin. From Tabasco State, Mexico

taborense L. *-ense*, denoting origin. From Tabora, Tanzania

tabulatum L. *tabula*, table; *-ata*, possessing. Palea oblong

tacanae From Mt Tacana, Guatemala

tacazensis L. *-ensis*, denoting origin. From Tacaza River, Ethiopia

tacuara From Department of Tacuarembó, Uruguay

tacubayensis L. *-ensis*, denoting origin. From Tacubaya, Mexico

tadulingamii In honor of C. Tadulinga Mudaliar (1878–?) Indian botanist

Taeniatherum Gk *taenia*, tape; *ather*, barb or spine. The awns of the lemma are flattened at the base

Taeniorhachis Gk *tainia*, ribbon; *rhachis*, backbone. Rhachis winged

taffzagra From Taffzagra, Ethiopia

taganrocense L. *-ense*, denoting origin. From Taganrog, Ukraine

tagoara One of the spellings for the Brazilian vernacular name for a number of woody grasses

taguara See *tagoara*

tahitensis L. *-ensis*, denoting origin. From Tahiti

taigae Russian, evergreen coniferous forests of sub-arctic regions

taimyrensis L. *-ensis*, denoting origin. From Taimyr (Taymyr) Peninsula, Siberia

taimyrica L. *-ica*, belonging to. See *taimyrensis*

tainanensis L. *-ensis*, denoting origin. From Tainan, Japan

taitensis L. *-ensis*, denoting origin. From Tait, more generally known as Tahiti

taiwanensis L. *-ensis*, denoting origin. From Taiwan

taiwaniana L. *-ana*, indicating connection. From Taiwan

taiwanicola L. *-cola*, dweller. Growing in Taiwan

taiwanicus L. *-icus*, belonging to. From Taiwan

tajimana L. *-ana*, indicating connection. From Tajima Province, now northern Hyogo Prefecture, Japan

takaensis L. *-ensis*, denoting origin. From Taka, Zaire

takasagoensis L. *-ensis*, denoting origin. From Takasago, Honshu Island, Japan

takasagomontana L. *mons*, mountain; *-ana*, indicating connection. From Takasago, Honshu Island, Japan

takedana L. *-ana*, indicating connection. In honor of Hisayohi Takeda (1883–1972) Japanese botanist

takeoi In honor of Takeo Ito (1911–) Japanese botanist

takeshimana L. *-ana*, indicating connection. From Takeshima on Utsuryoto Island, Korea

takizawana L. *-ana*, indicating connection. From Takizuna, Rikuchu Province, now part of Iwate and Akita Prefectures, Japan

talamancae From Cordillera de Talamanca, Costa Rica

talamancensis L. *-ensis*, denoting origin. See *talamancae*
talariata L. *talaria*, robe reaching to the ankles; *-ata*, possessing. Lemma base encircled by long callus hairs
Talasium Gk *talasia*, wool spinning; *-ium*, characteristic of. Spikelets spindle-shaped
talbotii In honor of William Alexander Talbot (1847–1917) Irish-born Indian forester
taldyksuensis L. *-ensis*, denoting origin. From Taldyksu (Taldyk-Su) River, Kyrgyzstan
taliensis L. *-ensis*, denoting origin. From Tali Mountains, Yunnan
talievii In honor of Walery Ivanovich Taliev (1872–1932) Russian botanist
tallanum L. *-anum*, indicating connection. From Talla, Sierra Leone
tallonii In honor of Gabriel Tallon, French botanist
talpensis L. *-ensis*, denoting origin. From Municipio de Talpa de Allende, Mexico
taltalensis L. *-ensis*, denoting origin. From Taltal, Antofagasta Department, Chile
taluh From the vernacular name for the species in Bali, Indonesia
tamanquareana L. *-ana*, indicating connection. From Ilha Tamanquare, Amazonas, Brazil
tamatavense L. *-ense*, denoting origin. From Tamatave, Madagascar
tamaulipense L. *-ense*, denoting origin. From Tamaulipas State, Mexico
tamayonis In honor of Francisco Tamayo (1902–1985) who collected in Venezuela
tamba Vernacular name of the species in Ethiopia
tambacoundense L. *-ense*, denoting origin. From Tambacounda, Senegambia
tambaensis L. *-ensis*, denoting origin. From Tamba Province, now part of Kyoto and Hyogo Prefectures, Japan
tanahashiana L. *-ana*, indicating connection. In honor of K. Tanashashi (fl. 1935) Japanese botanist
tanaiticum L. *-icum*, belonging to. From river Tanais, now Don, Russian Federation
tanakae In honor of Takesi Tanaka (1907–1997) Japanese botanist
tanatrich-a, -um Gk *tanos*, long; *thrix*, hair. Upper glume long-awned
tancitaroensis L. *-ensis*, denoting origin. From Mount Tancítaro, Michoacán, Mexico
tandilensis L. *-ensis*, denoting origin. From Sierra de Tandil, Argentina
tanegasimensis L. *-ensis*, denoting origin. From Tanegashima, an island in Ohsumi Prefecture, Japan
tanfiljewii In honor of Gavril Ivanovich Tanfiljev (1857–1928) Russian botanist
tangaensis L. *-ensis*, denoting origin. From Tanga, Tanzania
tangii In honor of Tang Tsin (1897–1984) Chinese botanist
tangoensis L. *-ensis*, denoting origin. From Tango Prefecture, Japan
tangoyosaensis L. *-ensis*, denoting origin. From Yosagunn, Kyoto Prefecture, Japan
tangutorum Of the Tangutes, the inhabitants of north-eastern Tibet. From Tibet Autonomous Region, China
tanimbarensis L. *-ensis*, denoting origin. From Tanimbar, one of the Lesser Sunda islands, Indonesia
Tansaniochloa Gk *chloa*, grass. Type species collected in Tanzania
tanzawana L. *-ana*, indicating connection. From Tanzawa, Kanagawa Prefecture, Japan
taolanensis L. *-ensis*, denoting origin. From Taolana, Madagascar
taphrophyllum Gk *taphros*, ditch; *phyllon*, leaf. Leaf-blades with dot-like depressions
taquara See *tagoara*
taquetii In honor of E. J. Taquet (fl. 1907–1912) who collected in Korea
tarapacana L. *-ana*, indicating connection. From Tarapacá Province, Chile
tarapotana L. *-ana*, indicating connection. From Tarapoto, Peru
tararaensis L. *-ensis*, denoting origin. From Tarara, Papua New Guinea
tarbagataicus L. *-icum*, belonging to. From Tarbagatai Mountains, Kazakhstan
tarda L. lingering. Flowering late in the season
Tarigidia An anagram of *Digitaria*
tarijensis L. *-ensis*, denoting origin. From Tarija, Bolivia

tarijianus L. *-anus*, indicating connection. From Tarija Department, Bolivia
tarmensis L. *-ensis*, denoting origin. From Tarma Province, Peru
tarnowskii In honor of Stanislaus Tarnowski (1837–1917) Polish historian and President of the Jagellonian University
taropotana L. *-ana*, indicating connection. From Tarapoto, Peru
tarraconensis L. *-ensis*, denoting origin. From Tarija Department, Bolivia
tashiroi In honor of Zentaro Tashiro (1921–1924) Japanese botanist
tashirozentaroana L. *-ana*, indicating connection. As for *tashiroi*
tasmanica L. *-ica*, belonging to. From Tasmania, Australia
tataric-a, -um L. *-ica*, belonging to. From Tataria now Russian Federation, east of the River Don
tatei In honor of George Henry Hamilton Tate (1894–1953) English-born United States botanist
tatewakiana In honor of Misao Tatewaki (1899–?) Japanese botanist
tatewakii As for *tatewakiana*
tateyamensis L. *-ensis*, denoting origin. From Tateyama (Mt Tate), a mountain in Toyama Prefecture, Japan
tatianae As for *sendulskyae*
Tatianyx As for *sendulskyae*
tatrae From the Tatra Mountains bordering Poland and Slovakia
tatrorum See *tatrae*
tauri As for *tauricola*
taurica L. *-ica*, belonging to. See *tauricola*
tauricola L. *-cola*, dweller. From Tauria, now the Crimea, Ukraine
taurinum L. *-inum*, indicating possession. From Tauria, now the Crimea, Ukraine
tauschii In honor of Ignaz Friedrich Tausch (1793–1848) Bohemian botanist
tavoyana L. *-ana*, indication connection. From Tavoy, Tennaseria, Myanmah
taxodiorum Of *Taxodium* swamps, Louisiana, USA
taygetana L. *-ana*, indicating connection. As for *taygetea*
taygetea From Mt Taygeto, Laconia, Greece
taylorii In honor of – (1) William Ernest Taylor (1856–1927) English cleric who collected in tropical East Africa. *Agrostis taylorii* – (2) George Taylor (1904–1993) British botanist. *Melica taylorii* – (3) Peter Geoffrey Taylor (1926–) English botanist. *Chaetopoa taylorii*
tcheliensis L. *-ensis*, denoting origin. From Chihli, China
teba Vernacular name for the species in Java, Indonesia
teberdens-e, -is L. *-ense*, denoting origin. From the valley of the Teberda River, in the northern Caucasus, Russian Federation
teberdensis L. *-ensis*, denoting origin. From forest drained by the Teberda River, Caucasus, Russian Federation
technicum L. *techne*, craft; *-icum*, belonging to. Species cultivated for broom making
tect-a, -um, -us L. *tego*, cover. Covered, usually with leaf-sheaths, as of culms
tectoneticola L. *-etum*, place of growth; *-cola*, dweller. Growing in *Tectona*, that is in teak forests
tectori-a, -us L. that which serves as a covering. Covered, usually with leaf-sheaths, as of culms
tectorum L. *tectum*, roof. Commonly, but not exclusively, grows on roofs
tect-um, -us See *tecta*
tef Amharic. Origin of the name is uncertain but may derive from the Arabic *tahf* (good), a name applied by the Semites of South Arabia to a similar wild grass, the grain of which is collected at times of food scarcity
tehuacanensis L. *-ensis*, denoting origin. From Tehuacan, Mexico
tehuelcha Collective name for a group of Patagonian tribes
teijiroana L. *-ana*, indicating connection. In honor of Teijiro Suzuki (fl. 1932) Japanese botanist
Teinostachyum Gk *teino*, stretch; *stachys*, spike as of an ear of wheat. The spikelets are long and narrow

tejucense L. *-ense*, denoting origin. From Tejuca, Brazil

tekserah Vernacular name of the species in Assam State, India

telata L. *tela*, web; *-ata*, possessing. Lemmas sparsely hairy

Telepogon See *Thelepogon*

telmatica Gk *telmatos*, pond; *-ica*, belonging to. Growing around pond margins

telmatophila Gk *telmatos*, pond; *phileo*, love. Growing in swamps

telmat-um, -us Gk *telmatos*, pond. Growing in swamps or along swamp margins

Tema Origin not given by Adanson, the author of the name, but possibly a reference to Tema, a town in Ghana near to Senegal where he once resided

Temburongia Known only from the Temburong area of Brunei

Temochloa Gk *chloa*, grass. In honor of Tem Smitinand (1920–1995) Thai botanist

temomairemensis L. *-ensis*, denoting origin. From Mt Temomairem, Territory of Amapá, Brazil

tempisquense L. *-ense*, denoting origin. From Rio Tempisque, Costa Rica

temulent-a, -um L. drunken. – (1) Eating of the diseased grain has been long and widely associated with vomiting, staggering and impaired vision. *Lolium temulentum* – (2) derivation uncertain. *Aristida temulenta*

tenacissim-a, -um, -us L. *tenax*, tenacious; *-issima*, most. Foliage persistent

tenax L. holding firmly together, persistent. Densely tufted

tenell-a, -um, -us L. slender. Culms or inflorescence branches slender

tenell-um, -us See *tenella*

tener, -a, -um, -us L. thin. Culms slender

teneriffae Growing on Teneriffe, one of the Canary Islands, Spanish territory in the Atlantic

tenerrima L. very thin. Culms or inflorescence branches thin

tener-um, -us L. thin. See *tener*

tenerus See *tener*

tennantiana L. *-ana*, indicating connection. In honor of John Smaillie Tennant (1865–1958) New Zealand botanist and educator

tennesseens-e, -is L. *-ense*, denoting origin. From Tennessee, USA

tennokawensis L. *-ensis*, denoting origin. From Tennokawamura, Nara Prefecture, Japan

tenorei In honor of Michele Tenore (1780–1861) Italian botanist

tenorii See *tenorei*

tenryuensis L. *-ensis*, denoting origin. From Tenryu, southern Honshu Island, Japan

tenryuriparia L. *riparia*, river bank. From the banks of the Tenryu River, Japan

tentoensis L. *-ensis*, denoting origin. From Tento, Manchuria, now comprising the Provinces of Lianoning, Jilin and Heilongjiang in north-east China

tenu-e, -is L. thin. Culms, leaf-blades or pedicels, slender

tenuiberbis L. *tenuis*, thin; *berbe*, beard. Awn slender

tenuicul-a, -us L. *tenuis*, thin; *-ula*, diminutive. Culms very delicate

tenuiculm-is, -um, -us L. *tenuis*, thin; *culmis*, stem. Culms thin

tenuiculus See *tenuicula*

tenuiflor-a, -um, -us L. *tenuis*, slender; *flos*, flower. With a delicate inflorescence

tenuifoli-a, -um, -us L. *tenuis*, slender; *folium*, leaf. Leaf-blades narrow

tenuilignea L. *tenuis*, weak; *lignum*, wood; *-ea*, resembling. Culms hollow, flexuous

tenuior L. more slender. More delicate than related species

tenuipedicellatus L. *tenuis*, slender; *pedicellus*, stalk; *-atus*, possessing. Pedicels slender

tenuirachis L. *tenuis*, slender; Gk *rhachis*, backbone. Inflorescence of slender racemes

tenuis See *tenue*

tenuiseta L. *tenuis*, weak; *seta*, bristle. – (1) Bristles of involucre very slender. *Setaria tenuiseta* – (2) awns filiform. *Aristida tenuiseta*

tenuisetulosa L. *tenuis*, slender; *seta*, bristle; *-ula*, diminutive; *-osa*, abundance. Awn well developed

T

tenuispatheus L. *tenuis*, narrow; *spatheus*, spathe. Spathes narrow and tightly inrolled

tenuispica L. *tenuis*, narrow; *spica*, a point; hence, in particular, an ear or spike of grain. Inflorescence a narrow-lanceolate spike or panicle

tenuispiculatum L. *tenuis*, narrow; *spica*, a point; hence, in particular, an ear or spike of grain; *-ula*, diminutive; *-atum*, possessing. Inflorescence slender and spike-like

tenuissim-a, -um, -us L. slender; *-issima*, most. – (1) Inflorescence slender. *Agrostis tenuissima, Atropis tenuissima, Melinis tenuissima, Muhlenbergia tenuissima, Panicum tenuissimum, Podosaemum tenuissimum, Puccinellia tenuissima, Sasa tenuissima, Sporobolus tenuissimus* – (2) lemma awns slender. *Stipa tenuissima*

tenuistriatus L. *tenue*, thin; *stria*, furrow; *-atus*, possessing. Leaf-blades narrowly striate

tephrosanth-os, -um Gk *tephros*, grey; *anthos*, flower. Spikelets grey

tepuianum L. *-anum*, indicating connection. From 'tepui', the local name for sandstone mesas in Venezuela

terecaulis L. *teres*, slender; *caulis*, stalk. Culms slender in comparison with those of related species

teres L. *teres*, cylindrical. Spikelets cylindrical

teretiflorum L. *teres*, narrow; *flos*, flower. Spikelets tending towards cylindrical

teretifoli-a, -um, -us L. *teres*, narrow; *folium*, leaf. Leaf-blades narrow and generally rigid

terminale L. *terminus*, limit; *-ale*, pertaining to. Terminal branches of panicle erect

ternarius L. *terni*, three each; *-arius*, pertaining to. Spikelets occurring to triads

ternat-a, -um, -us L. *terni*, three each; *-ata*, possessing. In clusters of three, especially with reference to inflorescence branches or spikelets

ternipes L. *terni*, three each; *pes*, foot. Lemma terminating in a three-branched awm

Terrella See *Terrellia*

Terrellia Latinized form of Terrell-grass, an English vernacular name for a species of *Elymus*

Terrelymus Hybrids between species of *Terrelia* and *Elymus*

terrestris L. *terra*, land; *-estris*, place of growth. Growing on dry land

teshiwoensis L. *-ensis*, denoting origin. From Teshiwo, Sakhalin Island, Russian Far East

tesioensis L. *-ensis*, denoting origin. As for *teshiwoensis*

teslinense L. *-ense*, denoting origin. From Lake Teslin, Canada

tesquicola L. *tesqua*, wild place; *-cola*, dweller. Uncultivated, growing wild

tesselat-a, -us L. *tessela*, little tile; *-ata*, possessing. – (1) Pattern of veins tile-like on the leaf-sheath. *Gelidocalamus tesselata* – (2) on the leaf-blade. *Arundinaria tesselata, Bambusa tesselata, Indocalamus tesselatus, Nastus tesselata, Planotia tesselata*

tessmannii In honor of Günther Tessmann (fl. 1904–1926) ethnographer in West Tropical Africa and Peru

testudinum L. *testudo*, tortoise; *-inum*, indicating possession. Growing in association with tortoises

Tetrachaete Gk *tetra*, four; *chaete*, bristle. Each pair of spikelets is subtended by four sterile spikelets reduced to bristles

Tetrachne Gk *tetra*, four; *achne*, scale. The lower lemmas are sterile and resemble two extra glumes

Tetragonocalamus Gk *tetra*, four; *gony*, knee; *kalamos*, reed. Reed like plant whose culms are square in transverse section

tetragonus Gk *tetra*, four; *gony*, knee. Caryopsis acutely four-angled

tetrantha Gk *tetra*, four; *anthos*, flower. The spikelets have four florets

Tetrapogon Gk *tetra*, four; *pogon*, beard. Spikelets usually with awns on the subtending glumes and two lemmas

tetraquetra L. having four sides. Rhachis acutely four-sided

Tetrarhena See *Tetrarrhena*

Tetrarrhena Gk *tetra*, four; *arrhen*, male. The flowers have four stamens

tetrastachy-s, -um, -us Gk *tetra*, four; *stachys*, spike as of an ear of wheat. Inflorescence with four panicle arms

tetrastichum Gk *tetra*, four; *stichos*, row. Spikelets arranged in four rows

texan-a, -um, -us L. *-ana*, indicating connection. From Texas, USA

texensis L. *-ensis*, denoting origin. See *texana*

textilis L. *textilis*, intertwined. Used for weaving or thatching

textori-a, -um L. *texo*, weave; *-aria*, indicating function. Leaves used for weaving mats

teyberi In honor of Alais Teyber (1876–1914) Austrian botanist

thailandica L. *-ica*, belonging to. From Thailand

Thalasium See *Talasium*

thalassica Gk *thalassa*, sea; *-ica*, belonging to. Species of sea coasts or salt marshes

thalaw-wa Burmese *thalaw*, "Are you better than me?"; *wa*, bamboo. Local name reflecting the high quality of the culms

Thalysia Gk first fruits of the harvest. An important cereal

Thamnocalamus Gk *thamnos*, shrub; *kalamos*, reed. Plants shrubby or small trees

thaoudar Turkish name for wild wheat

tharpii In honor of Benjamin Carroll Tharp (1885–1964) United States botanist

Thaumastochloa Gk *thaumastos*, to be wondered at; *chloa*, grass. The culms are slender and attractive in appearance

Thedachloa Gk *chloa* grass. From "Theda", a grazing lease near Kalumburu, Western Australia

theinlwinii In honor of U Thein Lwin, Myanmar plant collector

Thelepogon Gk *thele*, wart; *pogon*, beard. The glumes are ornamented with short protrubances

Thellungia In honor of Albert Thellung (1881–1928) Swiss botanist

thellungii As for *Thellungia*

Themeda Arabic *thamada*, depression filled with water after rain. Transliterated by the author as *thaemed*. The reason for the choice of name not given by author

thermal-e, -is L. *thermae*, warm baths; *-ale*, pertaining to. Growing in the immediate vicinity of geysers and hot springs

thermarum L. *thermae*, warm baths. Of warm baths, that is growing near hot springs

thermitaria L. *termes*, white ant or termite; *-aria*, pertaining to. Growing near ant nests

thessala From Thessaly, Greece

thiebautii In honor of Arsenne Thiébaut de Berneaud (1777–1850) French soldier and botanist who collected in Mexico

thimiodorus L. *thymum*, mint plant; *odorus*, sweet smelling. Foliage scented

thinophilum Gk *this*, beach; *phileo*, love. Growing at sea level

Thinopyrum Gk *this*, beach; *pyros*, wheat. Growing on beach dunes

thoi In honor of Yow Pong Tho (1945–1991) Malaysian entomologist

thollonii In honor of François-Romain Thollon (1855–1896) who collected in the Congo

thomae-a, -um L. from Mt St. Thomae near Tranquebar, India

thomasiana In honor of Philippe Thomas (?–1831) who collected in Corsica

thomasii In honor of Arthur Stocker Thomas (1902–?) who collected in Africa

thomassonii In honor of Joseph Raymond Thomasson (1946–) United States agrostologist and palaeobotanist

thominei In honor of Charles Thomine-Desmasures (1799–1824) French lawyer and botanist

thompsoniae In honor of Joy Thompson (1923–) Australian botanist

thompsonii In honor of Edward John Thompson (1949–) Australian ecologist

thomsonianum L. *-anum*, indicating connection. In honor of Thomas Thomson (1817–1878) Scots-born physician in Bengal Army, later Superintendent, Calcutta Botanic Garden

thomsonii (1) See *thomsonianum*. *Glyceria thomsonii*, *Puccinellia thomsonii* – (2) George Malcolm Thomson (1849–1933) New Zealand educator and amateur botanist. *Ehrharta thomsonii*, *Petriella thomsonii*, *Rytidosperma thompsonii*

Thonandia An anagram of *Danthonia*

thonii In honor of Carl Thon, Bohemian zoologist

thonningii In honor of Peter Thonning (1775–1848) Danish physician and natural historian

thorbeckei In honor of Franz H. Thorbecke (1875–1945) German botanist

Thorea See *Thoreochloa*

thorei See *Thoreochloa*

thorelii In honor of Clovis Thorel (1833–1911) French botanist who collected in Laos

Thoreochloa Gk *chloa*, grass. In honor of Jean Thore (1762–1823) French physician and naturalist

thoroldian-a, -um L. *-ana*, indicating connection. As for *thoroldii*

thoroldii In honor of William Grant Thorold (fl. 1890) British surgeon-naturalist who collected in Tibet, China and Ghana

thospiticum L. *-icum*, belonging to. From Thospetis Lake, now Van Gölü, Turkey

Thouarea See *Thuarea*

Thouarsia See *Thuarea*

thouarsian-a, -um, -us L. *-ana*, indicating connection. As for *Thuarea*

thouarsii See *Thuarea*

thouinii In honor of André Thouin (1747–1824) French botanist

thracic-a, -us L. *-ica*, belonging to. From Thrace, in Classical times the mid-Balkan peninsula

Thrasya, thrasya In honor of Thrasyas, Arcadian herbalist of the 5th century B.C.E., said to be able to drink an infusion of hellebore without ill effect

thrasyoides Gk *-oides*, resembling. Resembling *Thrasya* with respect to the inflorescence

Thrasyopsis Gk *opsis*, resemblance. Similar to *Thrasya*

Thrixgyne Gk *thrix*, hair; *gyne*, woman. Pistil densely hairy

Thuarea In honor of Louis Marie Aubert du Petit-Thouars (1758–1831) French botanist

Thuaria See *Thuarea*

thuarii See *Thuarea*

thuillieri In honor of Jean Louis Thuillier (1757–1822) French botanist

thulinii In honor of Mats Thulin (fl. 1992) Swedish botanist who collected in Ethiopia

thunbergii In honor of Carl Pehr Thunberg (1743–1828) Swedish botanist and physician

thurberi As for *Thurberia*

Thurberia In honor of George Thurber (1821–1890) United States botanist

thurberiana L. *-ana*, indicating connection. As for *Thurberia*

thurowii In honor of Friedrich Wilhelm Thurow (1852–1930) German-born United States botanist

thwaitesii In honor of George Henry Kendrick Thwaites (1812–1882) English-born botanist, sometime Botanic Gardens Superintendent, Paradeniya, Sri Lanka

thymiodorus See *thimiodorus*

Thyridachne Gk *thyris*, window; *achne*, scale. The tissue at the base of the lower lemma is translucent

Thyridolepis Gk *thyris*, window; *lepis*, scale. The lower glume has a depressed hyaline patch

Thyridostachyum Gk *thyris*, window; *stachys*, spike as of an ear of wheat. The spikelets are embedded in a thick cylindrical axis

Thyrsia Gk *thyrsos*, an ornamental wand. The racemes resemble a Bacchan wand

thyrsigera Gk *thyrsos*, ornamental wand; L. *gigno*, bear. Inflorescence with many branches

thyrsioides Gk *thyrsos*, an ornamental wand; *-oides*, resembling. Inflorescence plumose

thyrsoid-ea, -es Gk *thyrsos*, ornamental wand; *-oidea*, resembling. Panicle large thyrse-like

Thyrsostachys Gk *thyrsos*, an ornamental wand; *stachys*, spike as of an ear of wheat. The inflorescence is a lax panicle

Thysanachne Gk *thysanos*, fringe; *achne*, scale. Palea of lower floret fimbriate

thysanoglottis Gk *thysanos*, fringe; *glottis*, throat. Leaf-blade with a dense rim of long hairs, immediately behind the ligule

Thysanolaena Gk *thysanos*, fringe; *chlaena*, cloak. The upper lemma is markedly ciliate
Thyssanolaena See *Thysanolaena*
tianschanic-a, -um L. *-ica*, belonging to. From Tienshan, eastern Kazakhstan
tianshanica L. *-ica*, belonging to. From Tianshan Mts, China
Tiarrhena See *Triarrhena*
tiberiadis From Tiberias, Palestine
tibestica L. *-ica*, belonging to. From Tibesti Mountains, on the border of Chad and Libya
tibetic-a,-um,-us L. *-ica*, belonging to. From Tibet Autonomous Region, China
tibeticola L. *-cola*, dweller. Growing in Tibet Autonomous Region, China
tibetic-um, -us See *tibetica*
ticinensis L. *-ensis*, denoting origin. From Ticinense territorium, that is Pavia, Italy
tientaiense L. *-ense*, denoting origin. From Tientaishan, Zhejang Province, China
tiff See *tef*
tiflisiensis L. *-ensis*, denoting origin. From Tiflis, Republic of Georgia
tigrensis L. *-ensis*, denoting origin. From Tigre, north eastern Africa
tigridis From Tigris Valley, Iraq
tigurinus L. *-inus*, indicating possession. From Tigurum, now Zürich, Switzerland
tijucae L. from Pico de Tijuca, near Rio de Janeiro, Brazil
tikusialpina L. *alpes*, mountain; *-ina*, indicating possession. From Tikuri Mountain, Japan
tilcarense L. *-ense*, denoting origin. From Tilcara, Argentina
tileni From serra del Telino, Spain
tilesii In honor of Wilhelm Gottlieb Tilesius von Tilenan (1769–1857) who collected in eastern Siberia
tiliatus Gk *tilos*, fibre; L. *-atum*, possessing. The species is a source of fibre as is the bark of *Tilia* spp. (Linden) from which genus the grass species may directly take its name. The generic name *Tilia* is the same as the vernacular Latin which is probably derived from Greek sources
tillettii In honor of Stephen Szlatenyi Tillett (1930–) United States botanist resident in Venezuela
tillieri In honor of Tillier
timococcum Hybrid between *Triticum timopheevii* and *T. monococcum*
timoleontis In honor of Timoleon Holzmann (1843–?) German Government official
timopheevii In honor of Timopheev
timorens-e, -is L. *-ense*, denoting origin. From Timor, an island in the Indonesian Archipelago, part of which has recently become an independent Republic
Timouria In honor of Timour (1336–1405) otherwise known as Tamerlaine, Tartar king from Uzbekistan
Tinaea In honor of Vincenzo Tineo (1791–1856) Sicilian botanist
tincta L. *tincto*, dye. Inflorescence branches reddish
tinctilimba L. *tingo*, dye; *limbus*, belt. Blades of the culm-sheaths wine-red colored
tinei As for *Tinaea*
tiraquensis L. *-ensis*, denoting origin. From Tiraqui, Bolivia
tiricaense L. *-ense*, denoting origin. From Rio Tirica, Venezuela
tirsa Vernacular name of the species on the southern Russian Steppe
Tisserantiella L. *-ella*, diminutive but here used as a name-forming suffix. In honor of Charles Tisserant (1886–1962) French cleric ethnologist, botanist and traveller
tisserantii See *Tisserantiella*
tiutaroana L. *-ana*, indicating connection. In honor of Tiutaro
tjankorreh The vernacular name of the species, West Java, Indonesia
tjicoyaense L. *-ense*, denoting origin. From Tjikoya, Java, Indonesia
tobaeana L. *-ana*, indicating connection. As for *tobagenzoana*
tobagenzoana L. *-ana*, indicating connection. In honor of Toba Genzo (fl. 1935) Japanese botanist
tobishimensis L. *-ensis*, denoting origin. From Tobishima Island, Yamagata Prefecture, Japan

tobolense L. *-ense*, denoting origin. From Upper Tobol, western Siberia
toca From Toca, Colombia
tocussa See *dagussa*
todari In honor of Agostino Todaro (1818–1892) Sicilian botanist
toetoe Maori, to divide into strips. Maori vernacular name for several grasses
togashiana L. *-ana*, indicating connection. In honor of Kogo Togashi, Japanese agriculturalist and forester
togoens-e, -is L. *-ense*, denoting origin. From Togo
tohoensis L. *-ensis*, denoting origin. From Tohozan, a mountain on Taiwan
tokatiensis L. *-ensis*, denoting origin. From Tokati now Tokachi district of Kokkaido Island, Japan
tokiensis L. *-ensis*, denoting origin. From Tokio, otherwise Tokyo, Japan
tokitana L. *-ana*, indicating connection. In honor of Husae Tokita (1927–) Japanese naturalist
tokugawana L. *-ana*, indicating connection. In honor of Tokugawa
tolange Vernacular name for the species in the Uluguru Mountains of Tanzania
toletanus L. *-anus*, indicating connection. From Toletum, now Toledo, Spain
tolimensis L. *-ensis*, denoting origin. From Tolima Province, Colombia
tolmatchewii In honor of Alexsandr Innokentzevich Tolmachev (1903–1979) Russian botanist
tolmatschewii As for *tolmatchewii*
toluccens-e, -is L. *-ense*, denoting origin. See *tolucensis*
tolucensis L. *-ensis*, denoting origin. From Toluca, Mexico
tomentell-a, -us L. *tomentum*, stuffing material of a pillow; *-ella*, diminutive. Plant invested in part or totally with short hairs
tomentos-a, -um, -us L. *tomentum*, stuffing material of a pillow; *-osa*, abundance. Plant invested in part or totally with short hairs
tomikusensis L. *-ensis*, denoting origin. From Tomikusamura, Nagano Prefecture, Japan
Tomlinsonia In honor of Philip Barry Tomlinson (1932–) English-born United States botanist
tomodensis L. *-ensis*, denoting origin. From Tomoda, Mie Prefecture, Japan
tomookana L. *-ana*, indicating connection. In honor of Hiroshi Tomooka, Japanese botanist
tonamimontana L. *mons*, mountain; *-ana*, indicating connection. From Tonamiyama, a mountain in Yettiu Province, Japan
tonduzii In honor of Adolpheo Tonduz (1862–1921) Swiss-born cleric and Central American botanist
tonensis L. *-ensis*, denoting origin. From Toni, Gunma Prefecture, Japan
tongcalingii From Tongcalinga, Mindanao, Philippines
tonglensis L. *-ensis*, denoting origin. From Mt Tonglo in Himalayas
tongo On account of it having the scent of the tongo bean (*Dipteryx odorata*)
tonkinens-e, -is L. *-ense*, denoting origin. From Tonkin, now Vietnam
tons-a, -um L. shaven. Spikelets in whole or in part glabrous, as if shorn
tootsik Vernacular name of the species in Japan, also written as To-chiku, Kokochiku or Nankin-chiku
toppingii In honor of David LeRoy Topping (1861–1939) United States administrator who collected plants in the Philippines, Borneo, Hawaii and Siberia
Toresia See *Torresia*
Torgesia In honor of Emil Torges (1831–1917) German physician and botanist
torgesian-a, -us L. *-ana*, indicating connection. As for *Torgesia*
torquata L. *torqueo*, twist; *-ata*, possessing. Awn strongly twisted
Torresia In honor of G. de la Torre (fl. 1794) Director of Botanic Gardens at Madrid
torreyanus L. *-anus*, indicating connection. As for *Torreyochloa*
torreyi As for *Torreyochloa*
Torreyochloa Gk *chloa*, grass. In honor of John Torrey (1796–1873) United States botanist, chemist and physician
torridum L. torrid zone. From Hawaii

torta L. *torqueo*, twist. – (1) Awn twisted. *Aristida torta* – (2) leaf-blade twisted. *Setaria torta*

tortilis L. *torqueo*, twist; *-ilis*, indicating ability. Lemma awn hygroscopic

tortuos-a, -us L. *tortus*, twisted; *-osa*, abundance. With conspicuously twisted leaf-blades or awns

tosaensis L. *-ensis*, denoting origin. From Tosa Province, now Kochi Prefecture, Japan

Tosagris An anagram of *Agrostis*

tournefortii In honor of Joseph Pitton de Tournefort (1656–1708) French botanist

tourneuxii In honor of Aristide Horace Le Tourneux (1820–1890) French magistrate and botanist

touzelle French, awnless variety of wheat

tovarensis L. *-ensis*, denoting origin. See *Tovarochloa*

tovarii As for *Tovarochloa*

Tovarochloa Gk *chloa*, grass. In honor of Óscar Tovar (1923–) Peruvian botanist

townsendii In honor of Frederick Townsend (1822–1905) English amateur botanist

Toxeumia Gk *toxeuma*, arrow. The culms may have been used as arrows

toyomurensis L. *-ensis*, denoting origin. From Toyomuramura, Shinano Prov. Japan

Tozzettia In honor of Ottaviano Targioni-Tozzetti (1755–1829) Italian botanist

trabutiana L. *-ana*, indicating connection. In honor of Louis Charles Trabut (1853–1929) French physician and botanist

trabutii As for *trabutiana*

trachyantha Gk *trachys*, rough; *anthos*, flower. Lemma surface rough

trachycarpa Gk *trachys*, rough; *karpos*, fruit. Grain pitted

trachycaul-a, -us Gk *trachys*, rough; *kaulos*, stem. Culms scabrid

trachycaulon Gk *trachys*, rough; *kaulos*, stem. Rhachis scabrid

trachycaulus See *trachycaula*

trachychlaena Gk *trachys*, rough; *chlaena*, cloak. Glumes scabrid

trachycoleon Gk *trachys*, rough; *koleos*, sheath. Base of leaf-sheath scabrid

Trachynia Gk *trachyno*, make rough. The keels of the palea are scabrous

Trachynotia Gk *trachys*, rough; *notos*, back. Glumes scabrid on their backs

Trachyozus, Trechyozus Gk *trachys*, rough; *ozos*, twig. Inflorescence branchlets tipped with recurved spines

trachyphyll-a, -um Gk *trachys*, rough; *phyllon*, leaf. Leaf-blades scabrid

Trachypoa Gk *trachys*, rough; *poa*, grass. Possibly a reference to the rough keels of the lemmas

Trachypogon Gk *trachys*, rough; *pogon*, beard. The lemma-awn is hairy

trachypus Gk *trachys*, rough; *pous*, foot. Peduncle hairs with tuberculate bases

T

trachyrhachis Gk *trachys*, rough; *rhachis*, backbone. Panicle axes scabrous

Trachys Gk *rough*. Inflorescence branches tipped with recurved spines

trachysperm-a, -um Gk *trachys*, rough; *sperma*, seed. Fertile lemma ornamented

trachystachy-a, -um Gk *trachys*, rough; *stachys*, spike as of an ear of wheat. Glumes rough

Trachystachys Gk *trachys*, rough; *stachys*, spike as of an ear of wheat. Inflorescence branches tipped with recurved spines

tracyi In honor of – (1) Samuel Mills Tracy (1847–1920) United States agronomist. *Eragrostis tracyi*, *Erianthus tracyi*, *Poa tracyi* – (2) Joseph Prince Tracy (1879–1953) United States Title Examiner and naturalist. *Festuca tracyi*

Tragus Gk *tragos*, he goat. In honor of Hieronymus Bock (1498–1554) German physician, educator, botanist whose surname translates into Greek as Tragus

traninhensis L. *-ensis*, denoting origin. From Tranink, Laos

tranquillans L. *tranquillo*, make tranquil. Origin uncertain, not given by author

transbaicalens-e, -is L. *-ense*, denoting origin. From the Transbaikal region, Russian Federation

transbaicalica L. *-ica*, belonging to. See *transbaicalense*

transbarbata L. *trans*, across; *barba*, beard; *-ata*, possessing. Lower lemma bearing a transverse fringe of silvery hairs

transcaucasic-a, -um L. *-ica*, belonging to. From the Transcaucasus, that is the region between the Black and Caspian Seas

transhyrcan-a, -um, -us L. *trans*, on the other side. Beyond Hercynia, that is northwest Iran

transiens L. *transeo*, pass by. Pass through a generation within a year

transiliens-e, -is L. *-ense*, denoting origin. From beyond river Ili, Kazakhstan

transilvanica See *transsilvanica*

transmorrisonensis L. *trans*, on the other side of. Similar to but differing from *Agrostis morrisonensis*

T

transnominatum L. *trans*, on the other side of; *nomino*, give a name to. Segregated from another species

transsilvanic-a, -um, transsylvanica L. *-ica*, belonging to. From Transsylvania, an historic Region and Province of central Romania

transvaalensis L. *-ensis*, denoting origin. From the Transvaal, South Africa

transvenulosum L. *trans*, across; *vena*, vein; *-ula*, diminutive; *-osum*, abundance. Lateral nerves of leaf-blades connected with numerous transverse veins

trapezuntina L. *-ina*, indicating possession. From Trapezunta, that is, Trebizond, Turkey

trapnellii In honor of Colin Graham Trapnell (1907–) British ecologist who collected in East Africa

trautvetteri In honor of Ernst Rudolf Trautvetter (1809–1889) Russian botanist

travancorens-e, -is L. *-ense*, denoting origin. See *travancorica*

travancorica L. *-ica*, belonging to. From Travancore, India

Trechyozus See *Trachyozus*

tremul-a, -um, -us L. trembling. Inflorescence subject to movement in a slight breeze

Tremularia L. *tremulus*, trembling; *-aria*, pertaining to. The spikelets tremble in the slightest breeze

tremuloides L. *tremulus*, trembling; Gk *-oides*, resembling. Resembling other species with the epithet *tremula*

tremul-um, -us See *tremula*

trepidari-a, -um, -us L. *trepidus*, restless; *-aria*, pertaining to. The inflorescence is lax and mobile in light winds

trepidula L. *trepidus*, restless; *-ula*, tending to. The inflorescence is lax and mobile in light winds

treutleri In honor of William John Treutler (1841–1915) Indian-born of English parents, physician and plant collector

trevesium L. *-ium*, locality. From Treviso Province, Italy

Triachyrium Gk *treis*, three; *achryon*, chaff. Mature florets with three scales due to the lower glume being fugacious

Triaena, triaena Gk *triaena*, trident. Spikelet with a lower bisexual floret and an upper rudimentary three-partite floret

Triaina See *Triaena*

trianae In honor of José Gerónimo Triana (1828–1890) Columbian botanist

triandra Gk *treis*, three; *aner*, man. – (1) Florets with three anthers. *Ehrharta triandra*, *Leersia triandra* – (2) sessile hermaphrodite spikelet surrounded by three male spikelets. *Themeda triandra*

triangularis L. *tria*, three; *angulus*, angle; *-aris*, pertaining to. Spikelets wedge-shaped in outline

triangulata L. *tria*, three; *angulus*, angle; *-ata*, possessing. – (1) Rhizome buds triangular in outline. *Indosasa triangulata* – (2) rhachis triangular in cross section. *Chloris triangulata*

Trianthera Gk *treis*, three; *antheros*, blooming. The florets have three anthers

Trianthium Gk *treis*, three; *anthos*, flower. The spikelets occur in triads of which one is sessile bisexual and two are stalked and either staminate or sterile

Triarenopsis Gk *opsis*, resemblance. Similar to *Triarrhena*

triaristat-a, -um, -us L. *tria*, three; *arista*, bristle; *-ata*, possessing. – (1) Spikelets with lower glume two-awned and upper glume one-awned. *Lepturopsis triaristata*, *Rhytachne triaristata*, *Rhytidachne triaristata* – (2) with lemma three-awned. *Aegilops triaristata*, *Avena triaristata*, *Bromus triaristatus* – (3) lower glume three-awned. *Schizachyrium triaristatum*

Triarrhena Gk *treis*, three; *arrhen*, male. Florets with three stamens

Triathera, Triatherus Gk *treis*, three; *ather*, barb or spine. Lemma three-awned

Tribolium L. *tria*, three; *bolus*, fiery meteor in the form of an arrow. Spikelets with three florets and coarsely hairy glumes which may be awned or acuminate as are the lemmas, the whole at maturity resembling a comet

tribuloides L. *tribulus*, a four-pronged implement strewn on the ground to impede cavalry and so in general applied to burrs; Gk *-oides*, resembling. The spikelets and attendant bristles bear a fanciful resemblance to the fruits of *Tribulus*

tricarinat-a, -um L. *tria*, three; *carina*, keel; *-ata*, possessing. Lower glume three-nerved

Trichachne Gk *thrix*, hair; *achne*, scale. The glumes and sterile lemma bear long hairs

Trichaeta Gk *treis*, three; *chaete*, bristle. Lemma bifid with a dorsal awn

trichaetum Gk *treis*, three; *chaete*, bristle. Lower glume with two awns, upper with one awn

trichanth-a, -um Gk *thrix*, hair; *anthos*, flower. – (1) Spikelets borne on long pedicels. *Calamagrostis trichantha* – (2) spikelets hairy. *Panicum trichanthum*

trichiata Gk *thrix*, hair; L. *-ata*, possessing. Plant densely hairy

Trichloris L. *tria*, three. Similar to *Chloris* but lemma three-awned

trichocaulos Gk *thrix*, hair; *kaulos*, stem. Culms hairy

Trichochlaena See *Tricholaena*

Trichochloa Gk *thrix*, hair; *chloa*, grass. The lemma bears a hair-like awn

trichoclad-a, -um Gk *thrix*, hair; *klados*, stem. – (1) Culms hairy. *Panicum trichocladum* – (2) inflorescence branches thread-like. *Agrostis trichoclada* – (3) lemma with a knot of hairs at its base. *Arctophila trichoclada*, *Poa trichoclada*

trichocolea Gk *thrix*, hair; *koleos*, sheath. Leaf-sheath invested with white hairs

trichocondylum Gk *thrix*, hair; *kondylos*, knuckle. Sub-cupular tips of pedicels invested with long hairs

trichocordia Gk *thrix*, hair; L. *cordus*, heart. Palea ovate with a hairy margin

trichodes Gk *thrix*, hair; *-odes*, resembling. Leaf-blades slender

Trichodiclida Gk *thrix*, hair; *diklis*, double folding as of doors. Lemma hairy and vertically double-grooved

Trichodium Gk *thrix*, hair; *eidos*, appearance. Panicle branches filiform

trichodon Gk *thrix*, hair; *odous*, tooth. Origin uncertain, not given by author

trichodonta Gk *thrix*, hair; *odous*, tooth. Lemma lobes thread-like

trichoglume Gk *thrix*, hair; L. *gluma*, husk. Glumes pubescent

trichogona Gk *thrix*, hair; *gonia*, angle. Upper surface of leaf-blade hairy towards its base

trichoides Gk *thrix*, hair; *-oides*, resembling. Spikelets sparsely hirsute

Tricholaena Gk *thrix*, hair; *chlaena*, cloak. The glumes and sterile lemma are invested in long hairs

tricholaenoides Gk *-oides*, resembling. Spikelets resemble those of *Tricholaena*

tricholemma Gk *thrix*, hair; *lemma*, scale. Lemma hairy

tricholepis Gk *thrix*, hair; *lepis*, scale. Glumes or lemmas hairy

Trichoneura Gk *thrix*, hair; *neuron*, nerve. The lemma has three ciliate keels

trichonode Gk *thrix*, hair; L. *nodus*, knot. Nodes hairy

Trichoon Gk *thrix*, hair; *oon*, egg. Without meaning unless the writer mistakenly thought the ovary was hairy or interpreted the spikelet as the seed

trichophila Gk *thrix*, hair; *phileo*, love. Plant invested extensively with hairs

trichophor-a, -um Gk *thrix*, hair; *phero*, carry. Bearing hairs, especially with reference to the inflorescence

trichophyll-a, -um Gk *thrix*, hair; *phyllon*, leaf. Leaf-blades hair-like

trichopiptum Gk *thrix*, hair; *pipto*, fall. Hairy leaf-sheaths are ultimately shed thereby exposing glabrous culms

trichopod-a, -on, -us Gk *thrix*, hair; *pous*, foot. – (1) Pedicels bear long hairs towards their apices. *Bromus trichopodus, Digitaria trichopoda, Panicum trichopodon* – (2) Lemmas hairy at the base. *Arctophila trichopoda*

trichopodes Gk *thrix*, hair; *pous*, foot. Lemmas hairy at the base

trichopodia Gk *thrix*, hair; *podion*, small foot. Lemmas hairy at the base

trichopodon See *trichopoda*

trichopodus See *trichopoda*

Trichopteria, Trichopterya See *Trichopteryx*

Trichopterix See *Trichopteryx*

Trichopteryx, trichopteryx Gk *thrix*, hair; *pteron*, wing or feather-like. The margins of the lemma bear tufts of hairs

trichopus Gk *thrix*, hair; *pous*, foot. – (1) Pedicels hairy. *Andropogon trichopus, Eriochloa trichopus, Helopus trichopus, Panicum trichopus, Sorghum trichopus, Urochloa trichopus* – (2) racemes bearded at the base. *Iseilema trichopus*

Trichopyrum Gk *thrix*, hair; *pyros*, wheat. Resembling wheat but excessively hairy in some respect

trichorhachis Gk *thrix*, hair; *rhachis*, backbone. Rhachis hairy

Trichosantha See *Trichosathera*

Trichosathera Gk *thrix*, hair; *ather*, barb or spine. Lemma long-awned

trichospicula Gk *thrix*, hair; L. *spica*, a point; hence, in particular, an ear or spike of grain; *-ula*, diminutive. Awn relatively short compared with the length of the lemma

trichospirus Gk *thrix*, hair; L. *spira*, spiral. Awn shortly ciliate and spirally twisted

trichosticha Gk *thrix*, hair; *stichos*, row. Leaf-blades conspicuously hairy along their veins

trichostomum Gk *thrix*, hair; *stoma*, mouth. Apex of leaf-sheath bearing long hairs

trichotom-a, -um Gk *thrix*, hair; *tome*, end left after cutting off. – (1) Lemma apex crowned with a ring of hairs due to the abscission of the awn. *Piptochaetium trichotomum* – (2) pedicel apex with a few hairs that are conspicuous after the abscission of the spikelet. *Melinis trichotoma*

trichozygus Gk *thrix*, hairs; *zygos*, yoke. Spikelets in pairs subtended by a ring of hairs

tricolor L. *tria*, three; *color*, color. Glumes and sterile lemma green to purple, fertile lemma yellow

tricornis L. *tria*, three; *cornu*, horn. Awn terminally divided into three short segments

tricostata L. *tria*, three; *costa*, rib; *-ata*, possessing. Lemma three-nerved

tricostulata L. *tria*, three; *costus*, rib; *-ula*, dimunitive; *-ata*, possessing. The sterile lemma is conspicuously three-ribbed

tricuspidata L. *tria*, three; *cuspis*, point; *-ata*, possessing. Glumes three-toothed

tricuspidula L. *tria*, three; *cuspis*, point; *-ula*, diminutive. Lemma three-awned

Tricuspis L. *tria*, three; *cuspis*, point. The lemma is three-toothed

Tridens L. *tria*, three; *dens*, tooth. The lemma is shortly three-toothed

tridentat-a, -us L. *tria*, three; *dens*, tooth; *-ata*, possessing. – (1) One or both glumes three-toothed. *Andropogon tridentatus, Apocopis tridentatus, Lophopogon tridentatus, Parahyparrhenia tridentata, Paspalum tridentata, Stereochlaena tridentata* – (2) lemma three-toothed. *Agrostis tridentata*

trifida L. *tria*, three; *findo*, divide. – (1) Lemma conspicuously divided into three awns. *Bouteloua trifida, Eragrostis trifida* – (2) upper glume trifid. *Muhlenbergia trifida*

triflor-a, -um, -us L. *tria*, three; *flos*, flower. – (1) Spikelets with three florets. *Agrostis triflora, Aira triflora, Bromus triflorus, Cenchrus triflorus, Eragrostis triflora, Festuca triflora, Glyceria triflora* – (2) spikelets in clusters of three. *Mnesithea triflora, Panicum triflorum, Rottboellia triflora*

trifolium L. *tria*, three; *folium*, leaf. Culms mostly three-leaved

trifurcatum L. *tria*, three; *furca*, two-pronged fork. Lemma awned with a pair of wing-like outgrowths at its base

trigemina L. *tria*, three; *gemini*, twins. Spikelets with two florets and arranged in groups of three

triglochinoides Gk *-oides*, resembling. Similar to *Triglochin*

Triglossum Gk *treis*, three; *glossa*, tongue. Lodicules three, each longer than the ovary

triglum-e, -is L. *tria*, three; *gluma*, glume. Spikelets with two glumes and a sterile lemma

triglumis L. *tria*, three; *gluma*, husk. Spikelets often have three subtending glumes

trigonum L. *tria*, three; *gonum*, angle. Grain obtusely three-sided

trigyna Gk *treis*, three; *gyne*, woman. The pistil has three styles

Trikeraia Gk *treis*, three; *keras*, horn. Lemma with a long and two short awns

triloba L. *tria*, three; *lobus*, lobe. Fertile lemma three-lobed

Trilobachne L. *tria*, three; *lobus*, lobe; Gk *achne*, scale. The lower glume of the female spikelet is trilobed

trimenii In honor of Henry Trimen (1843–1896) English-born Sri Lankan botanist

trimucronata L. *tria*, three; *mucro*, small projection; *-ata*, possessing. Lemma shortly three-awned

trinervata L. *tria*, three; *nervus*, nerve; *-ata*, possessing. Lemma three-nerved

trinerv-e, -is L. *tria*, three; *nervum*, nerve. – (1) Upper glume three-nerved. *Digitaria trinervis*, *Panicum trinerve* – (2) lemma three-nerved. *Festuca trinervis*, *Poa trinervis* – (3) leaf-blade three-nerved. *Poacites trinervis*

trinervia L. *tria*, three; *nervum*, nerve. Glumes mostly three-nerved

trinervis See *trinerve*

triniana L. *-ana*, indicating connection. As for *Triniochloa*

trinii As for *Triniochloa*

Triniochloa Gk *chloa*, grass. In honor of Carl Bernhard Trinius (1778–1844) German-born Russian physician and agrostologist

trinitensis L. *-ensis*, denoting origin. From Trinidad, one of the two Caribbean islands that constitute the Republic of Trinidad and Tobago

Triniusa As for *Triniochloa*

Triodia Gk *treis*, three; *odous*, tooth. The lemma is three-awned

triodioides Gk *-oides*, resembling. Resembling *Trioida*

Triodon See *Triodia*

triphellon Gk *treis*, three; *phellos*, dark-colored. Subtending glumes and sterile lemma dark-colored

Triphlebia Gk *treis*, three; *phlebos*, vein. Lemma three-nerved

tripinnatum L. *tria*, three; *pinna*, feather; *-atum*, possessing. Inflorescence a tripinnate panicle

Triplachne L. *triploos*, triple; *achne*, scale. The lemma is three-awned

Triplasis Gk *triplasios*, triple. The lemma terminates in a short awn and two subulate lobes

Triplathera Gk *triploos*, triple; *ather*, barb or spine. The lemma is three-awned

triplicifolia L. *triplex*, triple; *folium*, leaf. Basal leaf-blades dimorphic and differing from those of the culm, hence leaves of three kinds

triploideum Gk *triploos*, triple. Species possessing three complete chromosome sets per cell

Triplopogon Gk *triploos*, triple; *pogon*, beard. There are three tufts of hair on the glumes

Tripogon Gk *treis*, three; *pogon*, beard. The lemmas have three apical awns and three basal hair-tufts

tripsacoides Gk *-oides*, resembling. With an inflorescence resembling that of *Tripsacum*

Tripsacum Gk *treis*, three; *psakas*, small pieces. The spikes break up into (at least) three pieces

triquetra L. triangular. Caryopsis trigonous

Triraphis Gk *treis*, three; *rhaphis*, needle. The three main nerves of the lemma project as short awns

Trirhaphis As for *Triraphis*

Triscenia Gk *treis*, three; *skene*, tent. Fertile floret subtended by a lemma and two glumes
Trisecale Hybrids between species of *Triticum* and *Secale*
triset-a, -um L. *tria*, three; *seta*, bristle. – (1) Lemma three-awned. *Avena triseta*, *Rhytachne triseta*, *Rhytidachne triseta* – (2) both glumes and the lemma bearing a simple awn. *Garnotia triseta* – (3) awn trifid. *Aristida trifida* – (4) involucre of three bristles. *Pennisetum trisetum* – (5) palea two-awned and lemma one-awned. *Eriachne triseta*
Trisetaria, Trisetarium L. *tria*, three; *seta*, bristle; *-aria*, pertaining to. The lemma is three-awned
Trisetobromus Resembling *Bromus* but lemma with a geniculate awn as with *Trisetum*
trisetoides Gk *-oides*, resembling. Resembling *Trisetum*
Trisetokoeleria Hybrids between species of *Trisetum* and *Koeleria*
Trisetum L. *tria*, three; *seta*, bristle. The lemma is three-awned
Trisiola L. *tria*, three. Distinguished from *Uniola* by the possession of three anthers
trisperma Gk *treis*, three; *sperma*, seed. The spikelets commonly produce three grains
trispiculata L. *tria*, three; *spica*, a point; hence, in particular, an ear or spike of grain; *-ula*, diminutive; *-ata*, possessing. The upper glume of the sessile spikelet is three-dentate
tristachy-a, -on, -um, -us Gk *treis*, three; *stachys*, spike as of ear of wheat. Arranged in threes as of spikelets or racemes
Tristachya Gk *treis*, three; *stachys*, spike as of ear of wheat. Spikelets borne in threes with their pedicels fused
tristachyoides Gk *-oides*, resembling. As for *tristachya*
tristachyon As for *tristachya*
Tristania In honor of Jules Marie Claude de Tristan (1776–1861) French botanist
Tristegis Gk *treis*, three; *stege*, cover. Fertile floret subtended by an empty lemma and two glumes
tristigmatica L. *tria*, three; Gk *stigma*, mark; L. *-ica*, belonging to. Pistil with three stigmas
tristis L. dull-colored. Leaf-blades greyish, often as a result of being invested with short hairs
Trithordeum, Tritordeum Intergeneric hybrids between species of *Triticum* and *Hordeum*
Triticale Hybrids between species of *Triticum* and *Secale*
tritice-um, -us Resembling *Triticum* with respect to the inflorescence
Triticoides Gk *-oides*, resembling. Fossil grasses similar to *Triticum*
triticoides Gk *-oides*, resembling. Resembling *Triticum* usually with reference to the inflorescence
Triticosecale Hybrids between species of *Triticum* and *Secale*
Triticum The Roman vernacular name for wheat. Based on *tritum*, which in turn comes from *tero*, grind, because the grain is ground into flour
Trititrigia Hybrids between species of *Elytrigia* and *Triticum*
Tritordeum See *Trithordeum*
triuncialis L. *tria*, three; *uncus*, hook; *-alis*, pertaining to. Glumes and/or lemmas three-awned
Triunila Distinguished from *Uniola* by the possession of three stamens
trivalvis L. *tria*, three; *valva*, leaf of a folding door. With spikelets occurring in threes
trivialiformis L. commonplace; *forma*, appearance. Resembling a related species with the epithet *trivialis*
trivialis L. commonplace. Widespread species
Trixostis Gk *trixos*, triple; *osteon*, a bone. The lemma terminates in three rigid awns
trochainii In honor of Jean Trochain (1903–1976) French botanist
Trochera, trochera Gk *trocheros*, round. The culm is swollen at the base
trochlearis L. pully-shaped. The twisted column of the lemma awn resembles a rope wound around a pulley axil

troctolepis Gk *troktos*, gnawed; *lepis*, scale. Lemma irregularly dentate

trogloditarum Of the Troglodytes, an Ethiopian people referred to by Herodotus, but the name later came to mean cave dwellers in general or people that lived before the Biblical flood and hence an epithet applied to fossil species

trollii In honor of Carl Troll (1899–1975) German botanist

tropic-a, -um, -us Gk *tropikos*, regions in which the sun is overhead twice a year. From tropical regions

tropidoblephare Gk *tropis*, ship's keel; *blepharon*, eye-lid. Lower glume strongly keeled with a single ciliate nerve

trotteri In honor of Alessandro Trotter (1874–1967) Italian botanist

truchmenorum Arabic *tourdjouman*, interpreter. A Latinized form of the Arabic, possibly in honor of the interpreters associated with the collector

truncat-a, -um, -us L. *trunco*, shorten by cutting off. Truncate with respect to – (1) apices of lemmas or glumes. *Andropogon truncatus, Avena truncata, Chloris truncata, Enneapogon truncatus, Eremochloa truncata, Heteropogon truncatus, Isachne truncata, Panicum truncatum, Phalaris truncata, Poa truncata, Raddiella truncata, Rottboellia truncata, Trachypogon truncatus, Uranthoecium truncatum* – (2) culm-sheaths. *Dinochloa truncata, Pleioblastus truncatum* – (3) ligules. *Anthephora truncata*

truncatella L. *trunco*, shorten by cutting off; *-ella*, diminutive. Glume apices truncate or with shallow notches

truncatiglume L. *trunco*, shorten by cutting off; *gluma*, scale. The apex of the lower spikelet is truncate

truncatula L. *trunco*, shorten by cutting off; *-ula*, indicating tendency. Ligules much reduced

truncat-um, -us See *truncata*

trypheron Gk *trypheros*, soft or delicate. Loosely tufted annual

tsangii In honor of Wai Tak Tsang (fl. 1927–1938) collector of the type

tsaratananens-e, -is L. *-ense*, denoting origin. From Mt Tsaratanana, Madagascar

tschatkalica L. *-ica*, belonging to. From Chatkal'skiy Khrebet Mountains, Kyrgyzstan

tschegolevii In honor of Tschegolev

tschimganic-a, -um, -us L. *-ica*, belonging to. From the Chimgan Range, Uzbekistan

Tschompskia Presumably in honor of Tschompski (or Tschompsky) but origin not given by author

Tschonoskia In honor of Chônosuke Sugawa (1841–1925) Japanese botanist who collected for Maximowicz

tsiafajavonensis L. *-ensis*, denoting origin. From Mt Tsiafajavona, Madagascar

tsiangii In honor of Tsiang Ying (1898–1982) Chinese botanist

tsitondroinensis L. *-ensis*, denoting origin. From Mt Tsitondroina, Madagascar

tsuboiana L. *-ana*, indicating connection. In honor of Tsuboi, Japanese botanist

tsugetorum L. of *Tsuga* woodlands. Type collected in Hemlock Grove of the New York Botanical Garden

tsukubanantaicola L. *-cola*, dweller. From Tsukubasan, a mountain in Ibaraki Prefecture, Japan

tsukubensis L. *-ensis*, denoting origin. From Tsukubasan, Ibaraki Prefecture, Japan

tsukushiense L. *-ense*, denoting origin. From Chikuzen Province, part of Fukuoka Prefecture, Japan

tsurumachiana L. *-ana*, indicating connection. In honor of Tsurumachi, Japanese botanist

tsurumatiana L. *-ana*, indicating connection. In honor of H. Tsurumati, Japanese botanist

tsushimensis L. *-ensis*, denoting origin. From Tsushima Island, Nagasake Prefecture, Japan

tsutsuiana L. *-ana*, indicating connection. In honor of S. Tsutsui, Japanese botanist

tuaensis L. *-ensis*, denoting origin. From Tua, Zaire

tuberculat-a, -um, -us L. *tuberculus*, wart; *-ata*, possessing. – (1) Lemmas or glumes bearing wart-like projections. *Andropogon tuberculatus, Berriochloa tuberculata, Castellia tuberculata, Danthoniopsis tuberculata, Dichanthium tuberculatus, Eremopogon tuberculatus, Oryzopsis tuberculata, Piptochaetium tuberculatum, Sporobolus tuberculatus, Tristachyum tuberculata* – (2) culms bearing abundant wart-like projections. *Chaetaria tuberculata, Chimonobambusa tuberculata* – (3) hairs of leaf-sheaths with tubercle-bases. *Panicum tuberculatum, Pogonathera tuberculata* – (4) margin of leaf-blades with wart-like projections. *Eriachne tuberculata*

tuberculiflorum L. *tuberculus*, wart; *flos*, flower. Lower glume reduced to a fleshy ring

tuberculos-a, -um L. *tuberculus*, wart; *-osa*, abundance. Lemmas or glumes bearing warty projections

tuberifera L. *tuber*, swelling; *fero*, carry or bear. Basal internodes thickened

tuberos-a, -um, -us L. *tuber*, swelling; *-osa*, abundance. – (1) Culms thickened at the base. *Avena tuberosa, Holcus tuberosus, Micropyropsis tuberosa, Phalaris tuberosa* – (2) culm bases matted with mud and mistakenly regarded as tuberous. *Panicum tuberosum, Paspalum tuberosum*

tubulosa L. *tubus*, tube; *-ulus*, tending to; *-osa*, well developed. The bony involucre subtending the inflorescence is subgloblular

tubus L. tube. Anthoecia cylindrical

tuckeri In honor of Gerard Tucker (1854–1930) Australian farmer

tuckermanii In honor of Edward Tuckerman (1817–1886) United States botanist

Tuctoria An anagram of *Orcuttia*

tucumana From Tucumán, Argentina

tucumanica L. *-ica*, belonging to. See *tucumana*

tuerckheimii In honor of Hans von Tuerckheim (1853–1920) German botanist

tugarinovii In honor of Tugarinov

tuitensis L. *-ensis*, denoting origin. From Municipio El Tuito, Mexico

tulcanensis L. *-ensis*, denoting origin. From Tulcán, Ecuador

tulcumbense L. *-ense*, denoting origin. From Tulcumbah, New South Wales, Australia

tulda The Benghali name for the species

tuldoides Gk *-oides*, resembling. Similar to *Bambusa tulda*

tumbuckianus L. *-anus*, indicating connection. From Tumbuck, South Africa

tumescens L. *tumesco*, swell up. Spikelets subspherical

tumidinoda L. *tumidus*, swollen; *nodus*, knot. Nodes swollen

tumidulus L. *tumidus*, swollen; *-ulus*, diminutive. Rhachis internodes and pedicels stouter than those of related species

tumidum L. swollen. Spikelets inflated

tunetana L. *-ana*, indicating connection. From Tunetum now known as Tunisia

tungnathii From Tungnath, a mountain in Uttar Pradesh, India

tunicata L. *thin*, separable covering; *-ata*, possessing. Leaf-sheaths loose about culm

turbaria Middle English *turbary*, medieval L. *turbaria*, peat-bog or peat-moss. Occurring in peat-bog

turbinat-a, -um, -us L. *turbo*, a top; *-ata*, possessing. Spikelets top-shaped

turcic-a, -um L. *-ica*, belonging to. From Turcia, now Turkey

turcomanic-a, -um, -us L. *-ica*, belonging to. From Turcomania, that is the lands of the Turkmen people of central Asia

turczaninoviana L. *-ana*, indicating connection. In honor of Porphir Kiril Nicolas Stepanovich Turczaninov (1796–1864) Russian botanist

turczaninovii, turczaninowii As for *turczaninoviana*

turfos-a, -um L. from a peat bog. Growing in swamps and grasslands

turgaicus L. *-icus*, belonging to. From Turgai, Kustanai, Kazakhstan

turgid-a, -um L. swollen. With swollen spikelets

turgidovillosum A hybrid between *Triticum turgidum* and *Triticum villosum*

turgidul-a, -um L. *turgidus*, swollen; *-ula*, diminutive. With somewhat swollen spikelets

turgidum See *turgida*
turkestanic-a, -um, -us L. *-ica*, belonging to. From Turkestan region of Kazakhstan
turneri In honor of – (1) J. Turner (fl. 1880s) who collected in New Ireland and New Britain, Bismarck Archipelago, Papua New Guinea. *Ischaemum turneri* – (2) G. E. Turner (fl. 1942) who collected in Alberta, USA. ×*Agroelymus turneri*
turnerian-a, -um L. *-ana*, indicating connection. In honor of Fred Turner (1856–1939) English-born Australian botanist
turonensis L. *-ensis*, denoting origin. From Turon, now Tours, France
Turraya Origin obscure, not given by author
turrialbae From Mt Turrialba, a volcano in Costa Rica
turriforme L. *turris*, tower; *formis*, appearance. A robust species
turuchanens-e, -is L. *-ense*, denoting origin. From Turuchan, that is Turukhansk
tuskaulensis L. *-ensis*, denoting origin. From Tuskaul in Central Asia
tuyamae In honor of Takasi Tuyama (1910–) Japanese botanist
tuzsonii In honor of János Tuzson (1870–1943) Hungarian botanist
tweedyi In honor of Frank Tweedy (1854–1937) United States surveyor and plant collector
tylanthum Gk *tylos*, knot; *anthos*, flower. The solitary subsphaerical spikelets borne on filiform pedicels resemble small clubs
Tylothrasya Gk *tylos*, knot. Similar to *Thrasya* in the form of the inflorescence but with the pedicels short and swollen
typhoid-ea, -es, -eum Gk *-oidea*, resembling. With an inflorescence similar to *Typha*
Typhoides Gk *-oides*, resembling. The dense sparsely branched panicle resembles that of *Typha*
typhoideum See *typhoidea*
typhur-a, -um Gk *typhos*, snake; *oura*, tail. Inflorescence a spicate panicle
typic-a, -us L. typical. Typical of the genus
tysonii In honor of William Tyson (1851–1920) Jamaican-born South African teacher and plant collector
tytthanthus Gk *tytthos*, small; *anthos*, flower. Inflorescence with few spikelets
tyttholepis Gk *tytthos*, small; *lepis*, scale. Glumes smaller than lemmas
tyuhgokensis L. *-ensis*, denoting origin. From Tyuhgoke, Hiroshima Prefecture, Japan
Tzvelevia L. *-ana*, indicating connection. In honor of Nikolai Nikolaievich Tzvelev (also as Tsvelev, Tsvelov and Tsvelyov) (1925–) Russian agrostrologist
tzveleviana L. *-ana*, indicating connection. As for *Tzvelevia*
tzvelevii As for *Tzvelevia*

U

uberior L. more fruitful. The inflorescences have more spikelets than those of related species
ubinica L. *-ica*, belonging to. From the Czernaja Ubi Valley in the western Altai Mountains, Mongolia and extending into China
ubsunurica L. *-ica*, belonging to. From Ubsu-Nur Province, Mongolia
uchidae In honor of Shigetarô Uchida (1885–?) Japanese agriculturalist and forester
uchidana L. *-ana*, indicating connection. As for *uchidae*
uchikawae In honor of T. Uchikawai (fl. 1942) who collected in the Provinces of Lianoning, Jilin and Heilongjiang in northeast China
uclueletensis L. *-ensis*, denoting origin. From Ucluelet, British Columbia, Canada
ucrainica L. *-ica*, belonging to. From the Ukraine
ucranica L. *-ica*, belonging to. From the Ukraine
ud-a, -um L. damp. Growing in swampy areas
udawnensis L. *-ensis*, denoting origin. From Udawn, Thailand
udensis L. *-ensis*, denoting origin. From the *Uda* River, Russian Federation

udum See *uda*
uechtritziana L. *-ana*, indicating connection. In honor of Rudolf von Uechtritz (1838–1886) German botanist
ugamic-a, -um, -us L. *-ica*, belonging to. From Ugam River near Tashkent, Uzbekistan
ugandensis L. *-ensis*, denoting origin. From Uganda
uhligii In honor of Victor Karl Uhlig (1857–1911) German geologist who collected in Tanzania
uii In honor of N. Ui (fl. 1909–1918) Japanese botanist
uinuizoana L. *-ana*, indicating connection. In honor of Ui Nuizo (fl. 1934–1940) Japanese botanist

U

ukishiba Vernacular name for the species in Japan
ulei In honor of Ernst Heinrich Georg Ule (1854–1915) German botanist and plant explorer
uliginos-a, -um L. *uligo*, wetness of the earth; *-osa*, abundance. Growing in swampy places
ullungdoensis L. *-ensis*, denoting origin. From Ullung, Korea
ulochaeta Gk *ulos*, woolly; *chaete*, bristle. The lemma awn shortly hairy
ultramafica Latinized form of ultramafic. Growing on soils derived from ultramafic rocks
ulugurensis L. *-ensis*, denoting origin. From the Uluguru Mountains, Tanzania
umbellat-a, -um, -us L. *umbella*, parasol; *-ata*, possessing. Inflorescence branches whorled
umbonulatum L. *umbo*, beak; *-ulus*, tending toward; *-atum*, possessing. The lemma of the fertile floret is somewhat beaked
umbraphilus L. *umbra*, any shady place; Gk *phileo*, love. Growing in forest shade
umbraticola L. *umbraticus*, belonging to the shade; *-cola*, dweller. Growing in shady places
umbratil-e, -is L. *umbra*, any shady place; *-atile*, place of growth. Growing on wet, shaded river-banks
umbricola L. *umbra*, any shady place; *-cola*, dweller. Growing in shady places
umbros-a, -um, -us L. *umbra*, any shady place; *-osa*, abundance. Growing in shady places
unarede Vernacular name employed for the species by the Maori people at Akaroa, New Zealand
uncinat-a, -um L. *uncinus*, hook; *-ata*, possessing. – (1) Upper glume drawn out into a hook. *Hemarthria uncinata* – (2) sterile lemma and upper glume with hooked hairs. *Panicum uncinatum* – (3) glumes and lemmas with hooked tips. *Australopyrum uncinatum*
uncinioides Gk *-oides*, resembling. Resembling *Uncinia* because of its hooked lemma
uncinulat-a, -um L. *uncinulus*, small hook; *-ata*, possessing. With hooked hairs on the glumes and sterile lemma
unciphyllum L. *uncus*, hook; Gk *phyllon*, leaf. Apices of leaf-blades forming a hook
undat-a, -um, -us L. wavy. Leaf-blade assumes the form of a shallow wave
underwoodii In honor of Lucien Marcus Underwood (1853–1907) United States botanist
unduavensis L. *-ensis*, denoting origin. From Unduavi, Bolivia
undulat-a, -um, -us L. wavy. – (1) Sterile lemmas transversely rugose. *Andropogon undulatus*, *Paspalum undulatum* – (2) margin of leaf-blade undulate. *Ehrharta undulata*
undulatifoli-a, -um, -us L. *undulatus*, wavy; *folium*, leaf. The surface of the leaf-blade assumes the form of shallow waves
undulat-um, -us See *undulata*
ungavens-e, -is L. *-ense*, denoting origin. From Baie de Ungava, Quebec, Canada
unguiculatum L. *unguis*, claw; *-ula*, diminutive; *-atum*, possessing. Fertile lemma mucronate
uniaristata L. *unus*, one; *arista*, bristle; *-ata*, possessing. Lemma one-awned
unica L. *unicus*, singular. Lower glume one-nerved
unifaria L. *unus*, one; *-aria*, in a row. Spikelets borne in a single row on the inflorescence branches

uniflor-a, -um, -us L. *unus*, one; *flos*, flower. – (1) Spikelets with one fertile floret. *Aciachne uniflora, Airochloa uniflora, Avena uniflora, Cenchrus uniflorus, Centotheca uniflora, Chusquea uniflora, Ehrharta uniflora, Eragrostis uniflora, Koeleria uniflora, Leptochloa uniflora, Melica uniflora, Pentameris uniflora, Poa uniflora* – (2) spikes of one spikelet. *Bouteloua uniflora* – (3) spikelets solitary within involucre. *Pennisetum uniflorum*

unifolia L. *unus*, one; *folium*, leaf. Culms with a single leaf

uniglum-e, -is L. *unus*, one; *gluma*, husk. – (1) Lower glume absent or reduced to a minute callus. *Eragrostis uniglumis, Festuca uniglumis, Sporobolus uniglumis, Vulpia uniglumis* – (2) glumes missing but sterile lemma present. *Digitaria uniglumis, Panicum uniglume*

unilateral-e, -is L. *unus*, one; *latus*, side; *-ale*, pertaining to. – (1) Inflorescence with branches directed towards one side. *Andropogon unilateralis, Aristida unilateralis, Calamagrostis unilateralis, Nardurus unilateralis, Poa unilateralis, Triticum unilaterale, Vulpia unilateralis* – (2) rhachis flattened on one side only. *Stenotaphrum unilaterale*

unilineatum L. *unus*, one; *linea*, line, one-twelfth of an inch; *-atum*, possessing. An obsolete unit of measurement but usually employed without qualification. One commonly accepted value is based on the "Paris inch", which yields a length of about 2.25 mm

uninervia L. *unus*, one; *nervus*, nerve. Glumes one-nerved

uninodis L. *unus*, one; *nodus*, knot. Flowering culms with a single node

Uniola Roman name for a species of uncertain identity

uniolae Resembling *Uniola*

unioloides Gk *-oides*, resembling. With spikelets resembling those of *Uniola*

unionis L. *unio*, unity. Of the Union, that is from United States of America

uniplumis L. *unus*, one; *pluma*, feather. Central arm of trifid awn of lemma plumose

uniramosa L. *unus*, one; *ramus*, branch; *-osa*, abundance. Culms with one branch per node

uniseriatum L. *unus*, one; *series*, row; *-atum*, possessing. The spikelets are widely separated seeming to form a single row

uniset-a, -um, -us L. *unus*, one; *seta*, bristle. – (1) Lemma with a single awn. *Aegopogon unisetus, Aristida uniseta, Tripogon unisetus* – (2) spikelet subtended by a single bristle. *Beckera uniseta, Beckeropsis uniseta, Gymnothrix uniseta, Ixophorus unisetus, Panicum unisetum, Pennisetum unisetum, Setaria uniseta, Urochloa uniseta*

unispicat-a, -um L. *unus*, single; *spica*, a point; hence, in particular, an ear or spike of grain; *-ata*, possessing. Inflorescences usually of a single raceme thereby resembling a spike

unispice-a, -us L. *unus*, one; *spica*, a point; hence, in particular, an ear or spike of grain. Inflorescence a single spike

unispiculata L. *unus*, one; *spica*, a point; hence, in particular, an ear or spike of grain; *-ula*, diminutive; *-ata*, possessing. Inflorescence usually a single raceme

unoi In honor of Kakuo Uno (fl. 1920) Japanese botanist

Urachne Gk *oura*, tail; *achne*, scale. The lemma is awned

uralens-e, -is L. *-ense*, denoting origin. From Ural Mountains, Russian Federation

Uralepis Gk *oura*, tail; *lepis*, scale. The lemma apex is drawn out

Uralepsis See *Uralepis*

Uranthoecium Gk *oura*, a tail; *anthoecium*, inflorescence. The lemmas of both upper and lower florets are tailed or have acute apices

urartu Assyrian name for Mt Ararat in Armenia

urbanian-a, -us L. *-ana*, indicating connection. As for *urbanii*

urbanii In honor of Ignatz Urban (1848–1931) German botanist

urceolat-a, -um, -us L. *urceus*, jug; *-ola*, diminutive; *-ata*, possessing. – (1) Lower glume inflated and contracted towards its apex. *Leptaspis urceolata, Pharus urceolatus, Scrotochloa urceolata* – (2) rhachis joints clavate and with a hollow apex. *Andropogon urceolatus, Schizachyrium urceolatum*

Urelytrum Gk *oura*, tail; *elytrum*, cover. The lower glume of the pedicelled spikelet is long-awned
ureneiana L. *-ana*, indicating connection. From Uren, Gifu Prefecture, Japan
urgutina L. *-ina*, indicating possesion. From Urgut Kishlyak, Iran
urjanchaica L. *-ica*, belonging to. From Urjanchai district, Siberia
Urochlaena Gk *oura*, tail; *chlaena*, cloak. The lemma has a curved awn
Urochloa Gk *oura*, tail; *chloa*, grass. The fertile lemma contracts abruptly to a tail-like awn
U
urochloides Gk *-oides*, resembling. Resembling *Urochloa*
Urochondra Gk *oura*, tail; *chondros*, grain. The pericarp has a beak formed by the style base
ursina L. *ursus*, bear; *-ina*, indicating possession. From localities inhabited by bears
ursorum L. *ursus*, bear; of the bears. From pastures in Kamchatka (Russian Far East) frequented by bears
urssulensis L. *-ensis*, denoting origin. From Urssul, Altai Mountains, Russian Federation
ursulae In honor of Ursula Scholtz (fl. 1979) who collected in Togo
ursulus L. *ursa*, bear; *-ulus*, diminutive. Peduncles and sessile spikelets densely covered with reddish-yellow hairs and so resemble the pelts of young bears
urticans L. *urtica*, nettle. The leaf-sheaths bear rigid stinging hairs
uruguayens-e, -is L. *-ense*, denoting origin. From Uruguay
uruguense L. *-ense*, denoting origin. From Uruguay
urvillean-a, -um, -us In honor of Jules Sébastien César Dumort d'Urville (1790–1842) French Naval Officer and botanist
urvillei As for *urvilleana*
usambarensis L. *-ensis*, denoting origin. From Usambara Mts, Tanzania
usambarica L. *-ica*, belonging to. From Usambara Mts, Tanzania
usawae In honor of Usawa, Japanese botanist
ushae In honor of Usha Ganguli Lachungpa, of the Forest Department, Sikkim State, India
usitata L. common. Widespread in the Andes
usorum L. *usus*, use; *habit*, custom. Origin uncertain, not given by author but possibly a reference to being cultivated by Kaffirs
uspallatensis L. *-ensis*, denoting origin. From Paramillo de Uspallata near Mendoza, Argentina
ussuriensis L. *-ensis*, denoting origin. From Ussuri, Russian Far East
usterii In honor of Alfred Usteri (1869–1948) Swiss horticulturalist and agriculturalist
ustilata L. *ustilo*, burn. Involucral bristles darkly colored in the upper part
ustulata L. *ustulo*, crisp the hair. Hairs on lemma apex curled
usuiensis L. *-ensis*, denoting origin. From Usui, Nagano Prefecture, Japan
utilis L. useful. – (1) Grain used as a cereal. *Echinochloa utilis* – (2) culms used for timber. *Fargesia utilis* – (3) peduncles used for manufacture of hats. *Aristida utilis* – (4) culms used for paper making. *Dinochloa utilis*
utowanaeum Commemorating the yacht "Utowana" which served as a base for an expedition of United States scientists visiting the Caribbean
utriculat-a, -us L. *utriculus*, little belly; *-ata*, possessing. – (1) Upper leaf-sheath inflated. *Alopecurus utriculatus* – (2) seed fertile floret enclosed in remains of male florets. *Hierochloe utriculata, Torresia utriculata*
utriculos-a, -um L. *utriculus*, little belly; *-osa*, abundance. The inflorescence is enclosed in the sheath of the flag-leaf
uvida L. moist. Grows along the margins of seasonal watercourses
uvulatum L. *uva*, grape; *-ula*, diminutive; *-atum*, possessing. The spikelets are clustered at the ends of branches and resemble small grapes
uyemurana L. *-ana*, indicating connection. In honor of Katsuji Uyemura, Japanese agriculturalist and forester
uyenoensis L. *-ensis*, denoting origin. From Uyens, Mie Prefecture, Japan
uyetsuensis L. *-ensis*, denoting origin. From Yettsui Province, Japan
uyucensis L. *-ensis*, denoting origin. From Cerro Uyuca, Honduras

V

vaccarian-a, -um L. *-ana*, indicating connection. In honor of Antonio Vaccari (1867–1961) Italian physician and botanist

vachanica L. *-ica*, belonging to. From Vachan in the Western Pamirs, on the border of Kyrgyzstan and Tajikistan

vachellii In honor of George Harvey Vachell (1799–?) who collected in China

vacillans L. *vacillo*, wave to and fro. Inflorescence branches slender so readily waving in the breeze

Vacoparis L. *vaco*, empty; *paris*, equal to another. The paired stalked spikelets are much reduced and similar

vagans L. *vago*, wander. – (1) Culms much branched. *Aegilops vagans, Andropogon vagans, Aristida vagans, Microstegium vagans, Triticum vagans* – (2) rhizomes widely creeping. *Arundinaria vagans, Sasa vagans*

vagiflorum L. *vagus*, wandering; *flos*, flower. Panicle branches very slender

vaginaeflor-a, -um See *vaginiflorum*

vaginalis L. *vagina*, sheath; *-alis*, pertaining to. Leaf-sheath densely hirsute

vaginans L. *vagina*, sheath; *-ans*, assuming the form of. Leaf-sheaths inflated

vaginat-a, -um, -us L. *vagina*, sheath; *-ata*, possessing. – (1) Leaf-sheaths conspicuous. *Agrostis vaginata, Apocopis vaginatus, Aristida vaginata, Arundinaria vaginata, Cenchrus vaginatus, Colobachne vaginata, Digitaria vaginata, Dimeiostemon vaginatus, Festuca vaginata, Pappophorum vaginatum, Perotis vaginata, Puccinellia vaginata, Rottboellia vaginata, Sanguinaria vaginata, Spodiopogon vaginatus, Stipa vaginata* – (2) inflorescences concealed or partially concealed in uppermost leaf-sheaths. *Apocopis vaginata, Hemarthria vaginata, Paspalum vaginatum* – (3) leaf-sheaths subtending inflorescence branches. *Andropogon vaginatus*

vaginiflor-um, -us L. *vagina*, sheath; *flos*, flower. – (1) The inflorescence hardly exceeds the ensheathing upper leaf. *Iseilema vaginiflorum* – (2) inflorescence remaining tightly wrapped in subtending leaf-sheath. *Sporobolus vaginiflorus* – (3) the inflorescence arising on a long peduncle conspicuously invested by the upper leaf-sheath. *Paspalum vaginiflorum*

vaginiviscosum L. *vagina*, sheath; *viscosum*, sticky. Leaf-sheath viscid

vahlian-a, -um L. *-ana*, indicating connection. As for *Vahlodea*

vahlii See *Vahlodea*

Vahlodea In honor of Martin Vahl (1749–1804) Danish botanist

vaillantianum L. *-anum*, indicating connection. In honor of Sébastien Vaillant (1669–1722) French botanist

valdesii In honor of Jesús Valdés Reyna

valdivian-a, -us L. *-ana*, indicating connection. From Valdivia, Chile

valdiviensis L. *-ensis*, denoting origin. See *valdiviana*

valentina From Valentina, now Valencia, Spain

valenzuelanum L. *-anum*, indicating connection. In honor of José Moria Valenzuela (fl. 1833) who collected in Cuba

valesiac-a, -um From Valesiacus, now Valois, France

valesiana See *vallesiana*

valid-a, -um, -us L. robust. Culms erect, stout

Valiha Madagascan name for a musical instrument resembling a tube-zither

vallesiaca See *valesiaca*

vallesiana L. *-ana*, indicating connection. From Valesia, now Canton of Valais, Switzerland

vallicola L. *vallus*, valley; *-cola*, dweller. Growing in mountain valleys

Vallota, Valota In honor of Antoine Vallot (1594–1671) French Garden's Director

vallsiana L. *-ana*, indicating connection. As for *vallsii*

vallsii In honor of José Francisco Montenegro Valls (1945–) Brazilian agrostologist

Valota See *Vallota*

valvata L. *valva*, leaf of a folding door; *-ata*, possessing. Palea and lemma oblong in outline and similar in length
vancouverensis L. *-ensis*, denoting origin. From Vancouver Island, Canada, or mistakenly believed to come from that island
vanderystii As for *Ystia*
vandovii In honor of Vandov
vaneedenii In honor of Willem Frederik van Eeden (1829–1901) Dutch botanist and museum director, Netherlands
vannum L. a winnowing fan. Basal leaves conspicuously distichous, their blades forming a fan-shaped cluster
vansonii In honor of Georges Van Son (1898–1967) Russian-born South African botanist
vargasii In honor of Cesar Vargas, also known as Julio Cesar Vargas-Calderón (1907–1960?) who collected in Peru
vari-a, -us L. variable. Species polymorphic and usually with many varieties
variabil-e, -is L. *varius*, variable; *-abilis*, indicating capacity. See *varia*
variana L. *varius*, variable; *-ana*, indicating connection. Fertile lemma mottled
varians L. *vario*, diversify. Variable in some respect
varicosa L. with swollen veins. Glumes rugose
variegat-a, -um, -us L. *variegatio*, diversify. – (1) Spikelets, leaf-blades or stems variable in color. *Andropterum variegatum*, *Arundinaria variegata*, *Bambusa variegata*, *Bromus variegatus*, *Calamagrostis variegata*, *Catabrosa variegata*, *Colpodium variegatum*, *Eragrostis variegata*, *Paspalum variegatum*, *Pleioblastus variegatus*, *Poa variegata*, *Sehima variegatum*, *Sporobolus variegatus*, *Stipa variegata* – (2) anthoecia variable in shape. *Stipidium variegatum*
variostriatus L. *varius*, variable; *striatus*, striped. The internodes are variously striped
varius See *varia*
varnense L. *-ense*, denoting origin. From Varnam, Bulgaria
vasaria L. *vas*, vessel; *-aria*, pertaining to. Culms used for making vessels
vasconcensis L. *-ensis*, denoting origin. From Vascon, Spain
vasconica L. *-ica*, belonging to. From Vascon, Spain
Vaseya As for *Vaseyochloa*
vaseyan-a, -um L. *-ana*, indicating connection. As for *Vaseyochloa*
vaseyi As for *Vaseyochloa*
Vaseyochloa In honor of George Vasey (1822–1893) United States botanist
vassiljevii In honor of N. Vasil'ev (fl. 1940) Russian botanist
vatkeana L. *-ana*, indicating connection. In honor of Georg Karl Wilhelm Vatky (1849–1889) German botanist
vatovae In honor of Aristocle Vatova (1897–1992) who collected in Eritrea
vatroensis L. *-ensis*, denoting origin. From Vatro, Argentina
vaviloviana L. *-ana*, indicating connection. As for *vavilovii*
vavilovii In honor of Nikolai Ivanovich Vavilov (1887–1942) Russian plant geneticist
vegeta L. vigorous. Plant growing strongly and freely
veitchiana L. *-ana*, indicating connection. In honor of John Gould Veitch (1839–1870) English nurseryman
veitchii As for *veitchiana*
velatus L. *velo*, conceal. Inflorescence base enclosed by sheath of subtending leaf
veldkampii In honor of Jan Frederik Veldkamp (1941–) Dutch botanist
velenovskyi In honor of Josef Velenovsky (1858–1949) Bohemian botanist
vella L. *vellus*, wool. Lemma and palea white-woolly
vellarianus L. *-anus*, indicating connection. From Vellarimala, India
velutin-a, -um, -us L. velvety. Plant in whole or in part covered with dense short hairs
velutinos-a, -um L. *velutina*, velvety; *-osum*, abundance. Plant densely covered with short hairs
velutin-um, -us See *velutina*
velutinus L. *vellutus*, shaggy; *-inus*, indicating resemblance. Auricles densely hairy

veneris From Portovenere, Liguria, Italy
venesuelae See *venezuelae*
veneta L. sea-green. – (1) Foliage glaucous. *Pentaschistis veneta* – (2) from Veneto, Italy. *Stipa veneta*
venezuelae From Venezuela
venezuelana L. *-ana*, indicating connection. As for *venezuelae*
venos-a, -um L. *vena*, vein; *-osa*, abundance. Veins conspicuous or many branched
ventanicola L. *-cola*, dweller. From Sierra de la Ventana, Argentina
Ventenata In honor of Étienne Pierre Ventenat (1757–1808) French botanist
ventenatii As for *Ventenata*
ventosa L. *ventus*, wind; *-osa*, abundance. Native to high peaks of the Pyrenees, a mountain range straddling the border of France and Spain
ventricos-a, -um, -us L. *venter*, belly; *-osa*, abundance. – (1) Spikelets inflated. *Agrostis ventricosa, Alopecurus ventricosus, Chloris ventricosa, Gastridium ventricosum, Isachne ventricosa, Olyra ventricosa, Stipidium ventricosum* – (2) internodes swollen. *Bambusa ventricosa*
ventriosa L. *venter*, belly; *-osa*, abundance. Fertile lemma at maturity conspicuously swollen
venturii In honor of Santiago Venturi (fl. 1910–1923) Argentinian botanist
venulosum L. *vena*, vein; *-ula*, diminutive; *-osum*, abundance. Glumes conspicuously veined
venust-a, -um, -us L. beautiful, graceful. Mostly a reference to habit
venustula L. *venusta*, beautiful; *-ula*, diminutive. The small and pendulous spikelets giving the plant an attractive appearance
venustuloides Gk *-oides*, resembling. Similar to *Aristida venustula*
venust-um, -us See *venusta*
veralensis L. *-ensis*, denoting origin. From El Veral, Cuba
verdcourtii In honor of Bernard Verdcourt (1925–) English botanist
verdickii In honor of Edgard Verdick (fl. 1899–1903) who collected in the Congo
veresczaginii In honor of Victor Ivanovich Vereschagin (1871–1956) Soviet botanist
Verinea L. *verinus*, tendril; *-ea*, indicating resemblance. Central nerve of lower glume prolonged
vernal-e, -is L. *vernus*, spring-like; *-ale*, pertaining to. – (1) Spring flowering. *Agrostis vernalis, Milium vernale* – (2) dying back in the autumn and re-emerging in the spring. *Panicum vernale*
vernicos-um, -us L. varnished. Spikelets glossy
vernix L. varnish. Spikelets glossy
verruciferum L. *verruca*, wart; *fero*, carry or bear. Fertile lemma with a rugose surface
verrucos-a, -um L. *verruca*, wart; *-osa*, abundance. – (1) The glumes and sterile lemmas bear verrucose hairs. *Digitaria verrucosa* – (2) glumes with warty protuberances. *Paspalum verrucosum*
verruculosa L. *verruca*, wart; *-ula*, diminutive; *-osa*, abundance. Lemma surface warty
versicolor L. variously colored. Spikelets variously colored as with glumes being green in the lower third, purple in the middle and brown in the upper third
versuta L. *versuta*, deceitful. A replacement name to eliminate a homonym
verticillat-a, -um, -us L. whorl; *-ata*, possessing. Primary inflorescence branches whorled
verticilliflor-a, -um, -us L. *verticillus*, whorl; *flos*, flower. Inflorescence branches whorled
vesc-a, -um L. containing little nutrition. Spikelets few-flowered or foliage sparse
Veseyochloa Gk *chloa*, grass. In honor of Leslie Desmond Edward Foster Vesey-Fitzgerald (ca. 1910–1974) British-born East African ecologist
vesiculosa L. *vesicula*, vesicle or blister; *-osa*, abundance. Leaf-blades have abundant vesicles
vestit-a, -us L. *vestio*, clothe. Plant in some respect hairy
Vetiveria Tamil *vetti*, khus-khus; *ver*, root. The rhizomes possess an aromatic oil

vettonica L. *-ica*, belonging to. From the region of Spain known as Vettones to the Romans
vetus L. old age. Pedicels of sterile florets bearded with abundant long hairs
vexillare L. *vexillium*, flag; *-are*, pertaining. Racemes few and held at right angles to axis of inflorescence
vexillifera L. *vexillium*, flag; *fero*, carry or bear. The awned spikelets, when flapping in the wind, resemble small flags
viale L. *via*, way; *-ale*, pertaining to. Growing on roadsides
viancinii In honor of Viancin who collected in Oubangui, Central African Republic
viatic-a, -um L. *viaticus*, relating to a journey. Growing along paths
vicarium L. substitute. Closely resembling another species
viciniflorum L. *vicinus*, neighbouring; *flos*, flower. Unlike those of related species the ultimate divisions of the inflorescence are contracted along the primary branches
vicin-um, -us L. near. Applied to species closely resembling others
viciosorum In honor of Benito (1850–1929) and Carlos (1897–1968) Vicioso
vickeryae As for *vickeryana*
vickeryana L. *-ana*, indicating connection. In honor of Joyce Winifred Vickery (1908–1979) Australian botanist
vickeryi As for *vickeryana*
victorialis L. *victoria*, victory; *-alis*, pertaining to. Commemmorating the Chinese victory in the Anti-Japanese War
victoriana L. *-ana*, indicating connection. From Depto. Santa Victoria, Argentina
vicunarum Spanish *vicuña*. Of the vicuñas, that is a constituent of their pastures
vidalii In honor of Francesco Vidal Gormaz, Chilean Army Officer
vierhapperi In honor of Fritz Vierhapper (1876–1902) Austrian botanist
vietbacensis L. *-ensis*, denoting origin. From Vietbac, Vietnam
vietnamense L. *-ense*, denoting origin. From Vietnam
vietnamica L. *-ica*, belonging to. See *vietnamense*
Vietnamocalamus Resembling *Calamus* and from Vietnam
Vietnamochloa Gk *chloa*, grass. From Vietnam
Vietnamosa L. *-osa*, abundance. From Vietnam
Vietnamosasa From Vietnam and resembling *Sasa*
vigens L. *vigeo*, thrive. Abundantly branching from lower culm nodes
vigoratum L. *vigor*, vigor; *-atum*, possessing. Culms coarse, erect from well developed rhizomes
Viguierella L. *-ella*, diminutive but here used as a name-forming suffix. In honor of René Viguier (1880–1931) French botanist who collected on Madagascar
viguieri See *Viguierella*
vihorlatica L. *-ica*, belonging to. From Vihorlát Mountains of Czech Republic
Vilfa Meaning obscure, origin of name not given by author
Vilfagrostis Combination of *Vilfa* and *Agrostis*. Origin uncertain, not given by author
vilfifolia L. *folium*, leaf. Leaf-blades like those of *Vilfa*
vilfoidea Gk *-oidea*, resemblance. Similar to *Vilfa*
vilis L. worthless. The grain is of no value as a cereal
villamontana L. *-ana*, indicating connection. From Villamontes, Bolivia
villanensis L. *-ensis*, denoting origin. From Villa Ana, Santa Fe Province, Argentina
villaricens-e, -is L. *-ense*, denoting origin. From Vaillari, Paraguay
villaroelii In honor of Arthuro Villaroel (fl. 1878) who collected in Chile
villarsii In honor of Dominique Villars (1745–1814) French botanist
villiculmis L. *villi*, long weak hairs; *culmus*, stalk. Except for the spikelets the plant is densely villous
villiferum L. *villi*, long weak hairs; *fero*, carry or bear. Plants densely pubescent
villiflor-a, -us L. *villi*, long weak hairs; *flos*, flower. Lemmas hairy

villifolium L. *villi*, long weak hairs; *folium*, leaf. Leaf-blades densely covered with long hairs
villiglumis L. *villi*, long weak hairs; *gluma*, husk. Glumes and sterile lemma densely hairy
villipalea L. *villi*, long weak hairs; *palea*, chaff. Palea densely hairy
villos-a, -um, -us L. *villi*, long weak hairs; *-osa*, abundance. The plant in whole or in part covered with long hairs
villosipes L. *villi*, long weak hairs; *-osa*, abundance; *pes*, foot. With hairy pedicels, spikelet bases or basal leaf-sheaths
villosissim-a, -um, -us L. most hairy. Plant densely hairy overall or in part
villosul-a, -um, -us L. *villi*, long weak hairs; *-osa*, abundance; *-ula*, diminutive. Sparsely hairy
villos-um, -us See *villosa*
vilmorianum L. *-anum*, indicating connection. In honor of a member of the Vilmorin family, several generations of whom were nurserymen, horticulturalists and writers of botanical memoirs
vilnensis L. *-ensis*, denoting origin. From Vilna, Lithuania
vilvoides Gk *-oides*, resembling. Similar to *Vilfa* also sometimes spelled *Vilva*
vimine-um, -us L. with long flexible shoots as used for wicker work
vincentianum L. *-anum*, indicating connection. From St Vincent, one of the Cape Verde Islands
vindobonensis L. *-ensis*, denoting origin. From Vindobona, now Vienna, Austria
vinealis L. *vinea*, vineyard; *-alis*, pertaining to. Growing in vineyards
vinhphuensis L. *-ensis*, denoting origin. See *vinhphuica*
vinhphuica L. *-ica*, belonging to. From Vinhphu, Vietnam
vinnulum L. delightful. Anthoecium white and shining
vinos-a, -um L. purplish-red. Inflorescence purplish-red
vinzentii In honor of Vinzent (fl. c. 1847) who collected in Texas, USA
violace-a, -um, -us L. violet. – (1) Spikelets, stigmas or anthers violet-colored. *Agrostis violacea, Aira violacea, Andropogon violaceus, Arthraxon violaceus, Bellardiochloa violacea, Calamagrostis violacea, Colpodium violaceum, Danthonia violacea, Erianthus violaceus, Festuca violacea, Hordeum violaceum, Lucaea violacea, Melica violacea, Panicum violaceum, Pennisetum violaceum, Poa violacea, Rytidosperma violacea, Stipa violacea, Triticum violaceum* – (2) culms and foliage violet-colored. *Saccharum violaceum*
violaceapurpurea L. *violacea*, violet; *purpurea*, purple. Spikelets purple-violet
violace-um, -us See *violacea*
violascens L. *violesco*, become violet. Anthers, stigmas, glumes or whole inflorescences blue to purple
virens L. *vireo*, be green. Leaf-blades or culms unusually bright yellow-green
virescens L. *viresco*, become green. Panicle shiny-green
vireta L. *vireo*, become green; *-eta*, place of growth. Growing in grasslands
virgat-a, -um, -us L. *virga*, broom; *-ata*, possessing. Inflorescence arms or culms held erect
virginic-a, -um, -us L. *-ica*, belonging to. From Virginia, USA
virgultorum L. *virgultus*, thicket. Growing amongst shrubs
virid-e, -is L. green. Widely applied but especially to species with green spikelets
viridearistata L. *viridis*, green; *arista*, bristle; *-ata*, possessing. The tips of the awns are pale-green
viridescens L. *viridesco*, become green. The plant in whole or in part bright-green
viridiflor-a, -um L. *viridis*, green; *flos*, flower. Spikelets bright-green
viridiglaucescens L. *viridis*, green; *glaucesco*, become bluish-green. Culms dark olive-green
viridiglumis L. *viridis*, green; *gluma*, husk. Glumes conspicuously green
viridis See *viride*
viridissima L. *viridis*, green; *-issima*, most. Spikelets very green

viridistriat-a, -us L. *viridis*, green; *striatus*, striated. Leaf-blades variegated
viridul-a, -us L. *viridis*, green; *-ula*, diminutive. Plant in whole or in part pale-green, often glaucous
virletii In honor of Pierre-Théodore Virlet d'Aoust (1800–?) who collected in Mexico
virolinens-e, -is L. *-ense*, denoting origin. From Corregimineto Virolín, Colombia
viscid-a, -um L. sticky. Plant sticky to touch
viscidellum L. *viscidus*, sticky; *-ellum*, diminutive. Panicle branches slightly sticky
viscidula L. *viscida*, sticky; *-ula*, diminutive. Invested with small glandular tubercles
viscidum See *viscida*
viscosa L. sticky. Panicle-branches or foliage sticky
vitiense L. *-ense*, denoting origin. Name derived from Viti Levu, the largest island in the Republic of the Fiji Islands, but widely applied to any island of the whole archipelago
vittat-a, -us L. *vitta*, band; *-ata*, possessing. Leaf-blades marked with transverse white stripes
vivax L. long-lived. Culms long-lived
vivipar-a, -um L. *vivus*, living; *pario*, bring forth. – (1) With bulbils replacing spikelets. *Agrostis vivipara*, *Deyeuxia vivipara*, *Poa vivipara* – (2) with culms much branched. *Panicum viviparum*
viviparoidea L. *vivus*, living; *parturo*, bring forth young; *-oidea*, indicating resemblance. Lemmas replaced by bracts that subtend pseudoviviparous shoots
vizzavonae From Col de Vizzavona, Corsica
vlassovii, vlassowii In honor of Osip Fedorovic Vlassov
voeltzkowii In honor of Alfred Voeltzkow (1860–1946) German botanist
vogelian-um, -us L. *-anum*, indicating connection. In honor of Julius Rudolph Theodor Vogel (1812–1841) German botanist
vogulic-a, -us L. *-ica*, belonging to. From Vogul in the Urals, Russian Federation
vohiboryensis L. *-ensis*, denoting origin. From the Voribory Range, Madagascar
vohitrense L. *-ense*, denoting origin. From Vohitra, Madagascar
volcanensis L. *-ensis*, denoting origin. From Volcán, Argentina
volcanicus L. *Vulcanus*, god of fire; *-icus*, belonging to. Collected from a volcanic peak in Costa Rica
volckmannii In honor of Herman Volckmann (fl. 1857–1861) who collected in Chile
volgens-e, -is L. *-ense*, denoting origin. From the Volga River, Russian Federation
volhynensis L. *-ensis*, denoting origin. From Volhynia, Ukraine
volhynicum L. *-icum*, belonging to. See *volhynensis*
volkensii In honor of Georg Ludwig August Volkens (1855–1917) German botanist
vollesenii In honor of Kaj B. Vollesen (1946–) Danish botanist
volutans L. *voluto*, tumble about. The mature panicle breaks off a unit and in windy weather tumbles along the ground
vorobievii In honor of Vorobiev
voroninii In honor of Michael Stepanowitch Woronin (1838–1903) Russian botanist
Vossia In honor of John Heinrich Voss (1751–1826) German poet
vrangelica L. *-ica*, belonging to. From Ostrov Vrangelya (Wrangel Island), an island in the East Siberian Sea
vriesii In honor of Willem Hendrik de Vriese (1806–1862) Dutch botanist
vryburgensis L. *-ensis*, denoting origin. From Vryburgh, Cape Province, South Africa
vuilletii In honor of André Vuillet (1883–1914) French plant pathologist
vulcanalis L. *vulcanus*, volcana; *-alis*, pertaining to. Growing on the slopes of Chiriqui, a volcano in Panamá
vulcanic-a, -um L. *vulcanus*, volcano; *-ica*, belonging to. Growing on the slopes of volcanoes
vulgar-e, -is L. *vulgus*, public; *-are*, pertaining to. Common in the wild or in cultivation
vulgatus L. *vulgo*, make public. Species recognized by its formal publication

vulnerans L. *vulnero*, wound. – (1) The mature spikelet has a sharp callus capable of wounding. *Aristida vulnerans*, *Arundo vulnerans*, *Phragmites vulnerans*, *Stipagrostis vulnerans* – (2) leaf-blade rigid and needle-like. *Triodia vulnerans*

Vulpia In honor of Johann Samuel Vulpius (1760–1846) German apothecary and botanist

vulpiaeformis L. *forma*, appearance. Inflorescence resembling that of *Vulpia*

vulpiastrum L. *-astrum*, resembling imperfectly. Similar to *Vulpia*

Vulpiella L. *-ella*, diminutive here used as a name-forming suffix. Resembling *Vulpia*

vulpin-a, -um, -us L. *vulpus*, fox; *-ina*, indicating resemblance. – (1) Inflorescences with racemes resembling fox tails. *Anthisteria vulpina*, *Hyparrhenia vulpina*, *Penicillaria vulpina* – (2) inflorescences spicate resembling fox tails. *Elymus vulpinus*, *Panicum vulpinum*, *Pennisetum vulpinum*, *Setaria vulpina*

vulpioides Gk *-oides*, resembling. Inflorescences similar to those of *Vulpia*

vulpiset-a, -um L. *vulpus*, fox; *seta*, bristle. With an inflorescence resembling a fox-tail

vurilochensis L. *-ensis*, denoting origin. From Vuriloche Pass, Argentina

vvedenskyi In honor of Aleksei Ivanovich Vvedenskii (1911–1929) Russian botanist

W

wabo Burmese *wa*, bamboo; *bo*, grandfather. Vernacular name in Myanmar for an edible bamboo with very large culms

wacei In honor of Nigel Morritt Wace (1929–2005) English-born Australian botanist and geographer

wachteri In honor of Willem Hendrick Wachter (1882–1946) Dutch botanist

wagenerianum L. *-anum*, indicating connection. In honor of Hermann Wagener (1823–1877) who collected in Colombia

wagneri In honor of János (Joannes) Wagner (1870–1955) Hungarian botanist

wagnerianum See *wagenerianum*

wahowensis L. *-ensis*, denoting origin. From Wahu, now O'ahu, one of the Hawaiian Islands

waibeliana L. *-ana*, indicating connection. In honor of Leo Waibel (fl. 1911–1916) who collected in South Africa

waikoloaense L. *-ense*, denoting origin. From Waikoloa Gulch on Oahu, one of the Hawaiian islands

waimeaense L. *-ense*, denoting origin. From Waimea on Kauai, one of the Hawaiian Islands

waishanensis L. *-ensis*, denoting origin. From Wai Shan, on the Jiangai-Fujian Provincial boundary, China

wakha Burmese *wa*, bamboo; *kha*, bitter. Young shoots inedible

wakoolica L. *-ica*, belonging to. From Wakool Shire, New South Wales, Australia

waldsteinii In honor of Franz de Paula Adam Waldstein (1759–1823) Austrian soldier and botanist

walense L. *-ense*, denoting origin. From Walo in Senegal

walkeri In honor of – (1) George Warren (?–1844) British soldier and his wife A. W. Walker, who collected in Sri Lanka. *Eragrostiella walkeri*, *Eragrostis walkeri*, *Isachne walkeri* – (2) Walker (fl. 1885) a New Zealand farmer. *Atropis walkeri*, *Poa walkeri*

walkeriana In honor of A. W. Walker; see *walkeri*

wallichian-a, -um L. *-ana*, indicating connection. In honor of Nathanial Wallich (1786–1854) Danish-born physician and sometime superintendent of Calcutta Botanic Gardens

wallichii As for *wallichiana*

wallii In honor of Arnold Wall (1869–1966) Sri-Lankan born, New Zealand Professor of English and amateur botanist

wallisii In honor of Gustav Wallis (1830–1878) German botanist

wallowaens-e, -is L. *-ense*, denoting origin. From Wallowa Mountains, Oregon, USA

walpersii In honor of Walpers but origin unclear, not given by author

walteri In honor of – (1) Thomas Walter (1740–1789) United States botanist. *Echinochloa walteri, Oplismenus walteri, Panicum walteri* – (2) either H. or E. Walter (fl. 1937) German botanists, who collected jointly on occasion in south-west Africa. *Aristida walteri, Eragrostis walteri*

walterianum L. *-anum*, indicating connection. As for *walteri* (2)

wamin Burmese *wa*, bamboo; *min*, king. A giant bamboo

wanet Burmese *wa*, bamboo; *net*, black. Culms black

Wangenheimia In honor of Friedrich Adam Julius von Wangenheim (1749–1800) Polish botanist

W

warburgii In honor of Otto Warburg (1859–1938) German botanist

wardiana L. *-ana*, indicating connection. In honor of Francis Kingdon-Ward (1885–1958) English botanist

wardii As for *wardiana*

warmingian-a,-um L. *-ana*, indicating connection. In honor of Johannes Eugenius Bülow Warming (1841–1924) Danish botanist

warmingii As for *warmingiana*

wasaensis L. *-ensis*, denoting origin. From Wasa, Zaire

Wasatchia From Wasatch Range in the Rocky Mountains, USA

washingtonica L. *-ica*, belonging to. From Washington State, USA

watense See *walense*

waterbergensis L. *-ensis*, denoting origin. From Waterberg, Transvaal State, South Africa

watsoniana L. *-ana*, denoting connection. In honor of Sereno Watson (1826–1892) United States botanist

wawawaiensis L. *-ensis*, denoting origin. From Wawawai, Washington State, USA

wayanadense L. *-ense*, denoting origin. From Wayanad District, India

webberi In honor of David Gould Webber (1809–1883?) United States physician, miner and miller

webberianum L. *-anum*, indicating connection. In honor of H. J. Webber (fl. 1894) United States botanist

webbiana L. *-ana*, indicating connection. In honor of Philip Barker Webb (1793–1854) English botanist

weberae In honor of Erna Weber

weberbaueri In honor of August Weberbauer (1871–1948) German botanist resident in Peru

weberi In honor of P. Weber, United States botanical illustrator

websteri In honor of Robert Dale Webster (1950–) United States botanist

weigeltiana L. *-ana*, indicating connection. In honor of Christoph Weigelt (?–1828) German physician and plant collector in Surinam

weilleri In honor of Marc Weiller (1880–1945) French botanist

Weingaertneria In honor of Johann Christoph Weingärtner (1771–1833) mathematician and pharmacist at Erfurt, Germany

weinmannii In honor of Johann Anton Weinmann (1782–1858) German-born Russian botanist

weixiensis L. *-ensis*, denoting origin. From Weixi, Yunnan Province, China

wellwitschii See *welwitschii*

welwitschii In honor of Friedrich Martin Josef Welwitsch (1806–1872) Austrian-born botanist, physician and traveller

wenchouensis L. *-ensis*, denoting origin. From Wenchou, Zhejiang Province, China

wendelboi In honor of Per Erland Berg Wendelbo (1927–1981) Norwegian botanist

werdermannii In honor of Erich Werdermann (1892–1959) German botanist

werneri In honor of – (1) William C. Werner (1851–1935) United States florist and plant collector. *Panicum werneri* – (2) Werner, origin unclear, not given by the author. *Hordeum werneri*

westernwoldicum L. *-icum*, belonging to. A corruption of the German place name to Westernwolth's Rye, a cultivar from New South Wales, Australia

westii In honor of Oliver West (1910–) Zimbabwean agronomist

wettsteinii In honor of Richard Wettstein Westerheim (1863–1931) Austrian botanist

Whalleya In honor of Ralph Derwyn Broughton Whalley (1933–) Australian grassland ecologist

wheeleri In honor of George Montague Wheeler (fl. 1871–1875) United States engineer and explorer

whitean-a,-um L. *-anum*, indicating connection. As for *Whiteochloa*

whitei (1) In honor of Samuel Albert White (1870–1954) South Australian naturalist and plant collector. *Panicum whitei* – (2) as for *Whiteochloa. Amphibromus whitei, Schizostachyum whitei*

Whiteochloa In honor of Cyril Tenison White (1890–1950) Australian botanist

whitneyi In honor of Leo David Whitney (1908–1937)

whytei In honor of Alexander Whyte (1834–1908) Scots-born botanist who collected in Malawi

wibeliana L. *-ana*, indicating connection. In honor of August Wilhelm Eberhard Christoph Wibel (1775–1814) German physician and botanist

wiehei In honor of Paul Octave Wiehe (fl. 1938–1975) who collected in Malawi

wiesneri In honor of Wiesner

Wiestia In honor of Anton Wiest (1801–1835) German plant collector in Egypt

wiestii As for *Wiestia*

wightian-a,-um,-us L. *-ana*, indicating connection. In honor of Robert Wight (1796–1872) Scots-born physician and sometime Superintendent, Botanic Gardens, Madras

wightii As for *wightiana*

wilburii In honor of Robert Wilbur Lynch (1925–) United States botanist

wilcoxianum L. *-anum*, indicating connection. In honor of Timothy Erastus Wilcox (1840–1932) United States botanist

wilczekiana L. *-ana*, indicating connection. As for *wilczekii*

wilczekii In honor of Ernest Wilczek (1867–1948) Swiss botanist, pharmacist and Gardens Director

wildemannii In honor of Émile Auguste Joseph De Wildeman (1866–1947) Belgian botanist

wildtii In honor of Albin Wildt (1845–1927) Czech botanist

wilhelminae From Mount Wilhelmina, Papua, Indonesia

Wilhelmsia In honor of Christian Wilhelms (fl. 1819) German-born Russian apothecary and botanist at Tiflis, Republic of Georgia

wilhelmsii As for *Wilhelmsia*

Wilibalda, Willibaldia See Franz Wilibald Schmidt under entry for *Schmidtia*

Wilibald-Schmidtia See Franz Wilibald Schmidt under entry for *Schmidtia*

wiliwilinuiense L. *-ense*, denoting origin. From Wiliwilinui Ridge on Molokai, one of the Hawaiian islands

wilkesii In honor of Charles Wilkes (1798–1877) United States Naval Officer and explorer

Willbleibia German *bleiben*, remain. The first four letters are all that remain of *Willkommia*, the generic name it was intended to replace

willdenoviana See *willdenowiana*

willdenowian-a, -um L. *-ana*, indicating connection. In honor of Carl Ludwig Willde-now (1765–1812) German botanist

willdenowii As for *willdenowiana*

willemetiana L. *-ana*, indicating connection. In honor of Pierre Remy Willemet (1735–1807) French apothecary and botanist

willemetii In honor of Hubert-Félix Soyer-Willemet (1791–1867) French librarian and amateur botanist

williamsii In honor of – (1) Robert Statham Williams (1859–1946) United States botanist. *Poa williamsii, Trisetum williamsii* – (2) Thomas Albert Williams (1865–1900) United States botanist. *Agrostis williamsii, Stipa williamsii* – (3) Samuel Wills Williams (1812–1884) United States botanist who collected in China. *Panicum williamsii* – (4) Louis Otto Williams (1908–1991) United States botanist. *Bouteloua williamsii* – (5) Leonard Howard John Williams (1915–) British botanist. *Erianthus williamsii, Saccharum williamsii*

Willibaldia See *Wilibalda*

Willkommia In honor of Heinrich Moritz Willkomm (1821–1895) German botanist
wilmaniae In honor of Maria Wilman (1867–1957) South African botanist and geologist
wilmingtonense L. *-ense*, denoting origin. From Wilmington, North Carolina, USA
wilsonii In honor of Ernest Henry Wilson (1876–1930) English-born United States botanist who collected in China
wiluic-a, -um L. *-ica*, belonging to. From the Wilu (Vilui, Viluy) River, Yakutsk Province, central Siberia
windersii In honor of C. W. Winders (fl. 1931) who collected in northern Queensland, Australia
windischii In honor of Paulo Guenter (Guenther) Windisch (1948–) Brazilian botanist
Windsora See *Windsoria*
Windsoria In honor of John Windsor (1787–1868) English physician and amateur botanist
winkleri As for *winklerianus*
winklerianus L. *-anus*, indicating connection. In honor of Moritz Winkler (1812–1889) Austrian botanist
winterianus L. *-anus*, indicating connection. In honor of August Wilhelm Winter who collected in Sri Lanka
wippraensis L. *-ensis*, denoting origin. From thc Chaussee Grillenberg-Wippra, southeast Harz, Germany
Wirtgenia In honor of Philipp Wilhelm Wirtgen (1806–1870) German teacher
wirtgeniana L. *-ana*, indicating connection. As for *Wirtgenia*
wisean-a L. *-ana*, indicating connection. In honor of Frank Joseph Scott Wise (1897–1986) Australian politician
wisselii In honor of F. J. Wissel (1907–) Dutch engineer and plant collector
wittei In honor of de Witte (fl. 1931) who collected in Zaire
woeltzkowii See *voeltzkowii*
wolfii In honor of John Wolf (1820–1897) United States botanist
wolgens-e, -is See *volgense*
wombaliensis L. *-ensis*, denoting origin. From Wombali, Zaire
woodii In honor of – (1) John Medley Wood (1827–1915) English-born South African botanist. *Setaria woodii* – (2) David George Wood (1939–) English botanist. *Chloris woodii*
woodrovii, woodrowii As for *Woodrowia*
Woodrowia In honor of George Marshall Woodrow (1846–1911) Scots-born Indian teacher and plant collector
woodrowii As for *Woodrowia*
woronowii In honor of Georg Jierii Nikolaewitsch Voronov (1874–1931) Russian botanist
wrangelica See *vrangelica*
wrayi In honor of Leonard Wray (1853–1942) British Colonial gardens superintendent
wrightianum L. *-anum*, indicating connection. In honor of Leonard Wright
wrightii In honor of Charles Wright (1811–1885) United States botanical collector
wulfeniana L. *-ana*, indicating connection. In honor of Franz Xaver von Wulfen (1728–1805) Balkan cleric
wuliangshanensis L. *-ensis*, denoting origin. From Wuliangshan, Yunnan Province, China
wulingensis L. *-ensis*, denoting origin. From Mount Wu-Ling, Manchuria, now comprising the Provinces of Lianoning, Jilin and Heilongjiang in north-east China
wullschlaegelii In honor of Heinrich Rudolph Wullschlägel (1805–1864) who collected in Guyana
wunthoensis L. *-ensis*, denoting origin. From Wuntho, Myanmar
wurdackii In honor of John Julius Wurdack (1921–1998) United States botanist
wuyiensis L. *-ensis*, denoting origin. From Wuyi, Fujian Province, China
wuyishanensis L. *-ensis*, denoting origin. From Wuyi Shan, Fujian Province, China
wuyishanicum L. *-icum*, belonging to. From Wuyi Shan, Fujian Province, China
wyomingensis L. *-ensis*, denoting origin. From Wyoming, USA

X

xalapense L. *-ense*, denoting origin. From Xalapa, alternatively Jalapa, Mexico

Xanthanthos Gk *xanthos*, yellow; *anthos*, flower. Spikelets yellow

xanthina Gk *xanthos*, yellow; *-ina*, indicating resemblance. Spikelets yellowish

xanthoblepharis Gk *xanthos*, yellow; *blepharis*, eyelash. Rhachis and glumes with long yellow hairs

xanthodas Gk *xanthos*, yellow; *das*, torch. The panicles are dense and golden

xantholeuc-a, -um Gk *xanthos*, yellow; *leukos*, white. Foliage yellowish-green and rhachis ciliate with white hairs

Xanthonanthos See *Xanthonanthus*

Xanthonanthus Gk *xanthos*, yellow; *anthos*, a flower. Spikelets yellow-green at maturity

xanthophysum Gk *xanthos*, yellow; *physa*, bladder. Spikelets yellow-green when dried

xanthospermum Gk *xanthos*, yellow; *sperma*, seed. Anthoecium yellow

xanthotrich-a, -um Gk *xanthos*, yellow; *thrix*, hair. The upper glume and sterile lemma are invested in yellow hairs

xenica Gk *xenikos*, alien. It was for some time thought that the species was not endemic

Xenochloa Gk *xenos*, stranger; *chloa*, grass. Far from Europe from whence described

xenophontis In honor of Xenophon (431–c. 350 B.C.E.) Greek historian who, as a soldier, passed through the area where the species was collected

xerachne Gk *xeros*, dry; *achne*, scale. Upper glume leathery

xerampelina Gk color of withered vine leaves. Panicle dark-purple and yellow

Xerochloa Gk *xeros*, dry; *chloa*, grass. Growing in arid regions

Xerodanthia Gk *xeros*, dry. Xeromorphic species related to *Danthonia*

xerophil-a, -us Gk *xeros*, dry; *phileo*, love. – (1) Desert species. *Eragrostis xerophila*, *Iseilema xerophila*, *Lepturus xerophilus*, *Neurachne xerophila*, *Thyridolepis xerophila* – (2) growing in exposed rocky situations as in the Marquesas. *Eragrostis xerophila* (not the desert species listed above), *Leptochloa xerophila*

xestophyllus Gk *xestos*, shaved; *phyllon*, leaf. Leaf-blades glabrous

xichangensis L. *-ensis*, denoting origin. From Xichang, Sichuan Province, China

xinanensis L. *-ensis*, denoting origin. From Xinan, Sichuan Provice, China

xinwuense L. *-ense*, denoting origin. From Xinwu, Jiangxi Province, China

Xiphagrostis Gk *xiphos*, sword; *agrostis*, grass. Margins of leaf-blades armed with sharp teeth

xizangensis L. *-ensis*, denoting origin. From Xizang Autonomous Region, China

xylosa Gk *xylon*, wood; L. *-osa*, abundance. Culms woody at the base

Xyochlaena Gk *xyo*, polish; *chlaena*, cloak. The glumes and sterile lemmas are glabrous or nearly so

Xystidium Gk *xystis*, robe with a sweeping train; *-idium*, diminutive but here used as a name-forming suffix. Glumes long awned

xystrophyllus Gk *xyster*, file; *phyllon*, leaf. Margins of leaf-blades with stiff retrorse hairs

Y

yabeana L. *-ana*, indicating connection. In honor of Y. Yabe (fl. 1909) Japanese botanist

Yadakeya Japanese *yadake*, arrow bamboo. Arrow shafts are made from the culms

yadkinense L. *-ense*, denoting origin. From the river Yadkin in North Carolina, USA

yadongensis L. *-ensis*, denoting origin. From Yadong Xian, Tibet Autonomous Region, China

yaganica L. *-ica*, belonging to. From Yagan in the south of South America

yagiana L. *-ana*, indicating connection. In honor of Shigeiti Yagi (fl. 1934) Japanese plant collector
yaguaronense L. *-ense*, denoting origin. From Yaguaron, Paraguay
yahikoensis L. *-ensis*, denoting origin. From Yahiko, Niigata Prefecture, Japan
yajiangensis L. *-ensis*, denoting origin. From Yajiang, China
Yakirra Arandic *yakerre*. Name used by Alyawarre people of Central Australia for the type species, meaning unknown
yakusimensis L. *-ensis*, denoting origin. From Yakushina, Ohsumi Prefecture, Japan
yamadoriana L. *-ana*, indicating connection. In honor of K. Yamadori, Japanese botanist
yamakitensis L. *-ensis*, denoting origin. From Yamakita, Kanagawa Prefecture, Japan
yamatensis L. *-ensis*, denoting origin. From Yamato Province, now Nara Prefecture, Japan
yangambiense L. *-ense*, denoting origin. From Yangambi, Zaire
yangii In honor of Jun-Liang Yang (1930–) Chinese botanist
yanyuanensis L. *-ensis*, denoting origin. From Yanyuan, Sichuan Province, China
yarochenkoi In honor of Yarochenkov
yarrabensis L. *-ensis*, denoting origin. From Yarraba, Queensland, Australia
yasaburoana L. *-ana*, indicating connection. In honor of Yasa Buro, Japanese botanist
yashadake Vernacular name for the species in Japan
yasokichii In honor of Yasokichi Kinoshita (fl. 1940) Japanese plant collector
yasuianus In honor of Yasui
yasuokensis L. *-ensis*, denoting origin. From Yasuoka, Nagano Prefecture, Japan
yatsugatakensis L. *-ensis*, denoting origin. From Yatsuga Dake, Honshu Prov., Japan
yaviensis L. *-ensis*, denoting origin. From Yavi Department, Argentina
yavitaense L. *-ense*, denoting origin. From Yavita, Venezuela
yeizanensis L. *-ensis*, denoting origin. From Hiyeizan, a mountain near Kyoto, Japan
yemenensis L. *-ensis*, denoting origin. From Yemen
yemenic-a, -um L. *-ica*, belonging to. From Yemen
yenaensis L. *-ensis*, denoting origin. From Yenasan, a mountain in Gifu Prefecture, Japan
yentuensis L. *-ensis*, denoting origin. From Yen Tu, Ha Nam Ninh Province, Vietnam
yessaensis L. *-ensis*, denoting origin. From Yetsigo Province, Japan
yessoensis L. *-ensis*, denoting origin. From Yezo, Kitami Province, Hokkaido, Japan
yettiuensis L. *-ensis*, denoting origin. From Yettiu Province, Japan
yezoalpina L. *alpes*, mountains; *-ina*, indicating possession. From Yezo Mountains, Kitami Province, Hokkaido, Japan
yezoens-e, -is L. *-ense*, denoting origin. From Yezo, Kitami Province, Hokkaido, Japan
yezolasioderma L. from Yezo, Kitami Province, Hokkaido, Japan and belonging to *Lasioderma* section of *Sasa*
yezomontana L. *mons*, mountain; *-ana*, indicating connection. From Mt Yezo, Kitami Province, Hokkaido, Japan
yinduensis L. *-ensis*, denoting origin. From Yindu, Zaire
yiuensis See *yiwuensis*
yiwuensis L. *-ensis*, denoting origin. From Yiwu (Yiu), Xinjiang Uyghur Autonomous Region, China
yixingensis L. *-ensis*, denoting origin. From Yixing, Jiangsu Province, China
yoigana L. *-ana*, indicating connection. From Iga Province, now western Mie Prefecture, Japan
yokoensis L. *-ensis*, denoting origin. From Yoko, Belgian Congo
yokotae In honor of Yokota, Japanese botanist
yonaiensis L. *-ensis*, denoting origin. From Yonaimura, Rikuchiu province, Japan
yongshanensis L. *-ensis*, denoting origin. From Yongsang, Yunnan Province, China
yonoskei In honor of Yônoske Tutui (fl. 1935) Japanese botanist
Yorkia From York County, Pennsylvania, USA

yosaensis L. *-ensis*, denoting origin. From Yosa, Japan

yoshikawana L. *-ana*, indicating connection. In honor of J. Yoshikawa (fl. 1935) Japanese botanist

yoshinoi In honor of Zensuke Yoshino (fl. 1901) Japanese botanist

yosiokae In honor of Sigeo Yosioka (fl. 1934) Japanese botanist

youngianum In honor of Stephen M. Young (fl. 1980) collector of bamboos in Ecuador

youngii In honor of William Spearman Young (1842–1912) New Zealand surveyor

Ystia In honor of Hyacinthe Robert Julien Vanderyst (fl. 1907–1925) Belgian cleric and botanist

yuanjiangensis L. *-ensis*, denoting origin. From Yuangiang County, Yunnan Province, China

yuanmouensis L. *-ensis*, denoting origin. From Yuanmou County, Yunnan Province, China

yuanmounensis L. *-ensis*, denoting origin. From Yuanmou County, Yunnan Province, China

yubaridakensis Japanese *take*, bamboo; L. *-ensis*, denoting origin. A bamboo from Yubari, a Japanese mountain

yucatan-a, -us L. *-ana*, indicating connection. From Yucatán, Mexico

yukonens-e, -is L. *-ense*, denoting origin. From the Yukon, Alaska

yulongshanensis L. *-ensis*, denoting origin. From Yulong Shan, Yunnan Province, China

yulungschanica L. *-ica*, belonging to. See *yulongshanensis*

yunguensis L. *-ensis*, denoting origin. From Yungu, Zaire

yunhoensis L. *-ensis*, denoting origin. From Yunghe Xian, Zhejiong Province, China

yunnanens-e, -is L. *-ense*, denoting origin. From Yunnan Province, China

yunzalinensis L. *-ensis*, denoting origin. From Yunzalin (Yunxalin), Myanmar

Yushania From Mount Yu Shan, Taiwan, also known as Mount Morrison

yushuensis L. *-ensis*, denoting origin. From Yushu, Quinghai Province, China

yutajensis L. *-ensis*, denoting origin. From Cerro Yutaje, Venezuela

yutakana L. *-ana*, indicating connection. In honor of Hukuda Yutaka (fl. 1935) Japanese botanist

Yvesia In honor of Alfred Marie Augustine Saint-Yves (1855–1933) French soldier and amateur agrostologist

yvesiana L. *-ana*, indicating connection. As for *Yvesia*

yvesii As for *Yvesia*

Z

zaissanica L. *-ica*, belonging to. From the steppes about Lake Zaissan, Kazakstan

zaleshii See *zalesskii*

zalesskii In honor of Viatscheslav Konstantinovic Zalessky (1871–1936) Russian botanist

zambesiense L. *-ense*, denoting origin. From the Zambesi region of southern tropical Africa, now included in Mozambique and Malawi

zaprjagajevii In honor of F. Zapirjagaev (fl. 1932)

zarubinii In honor of Zarubin

zavadilianum L. *-anum*, indicating connection. In honor of Zavadil

zayuensis L. *-ensis*, denoting origin. From Zaya Xian, Xizang Province, China

Zea Gk *zeia*, one-seeded wheat. Resembling one-seeded wheat (*Triticum monococcum*), or the genus *Zea* to which *Triticum* is unrelated

zea Resembling *Zea*

zeae As for *Zea*

zeelandicum L. *-icum*, belonging to. From New Zealand

zehntneri In honor of Leo Zehntner (fl. 1912) who collected in Brazil

Zeia Gk one-seeded wheat. A superfluous name for *Triticum* and in no way related to *Zea*

Zeites Gk *-ites*, resembling. Fossil leaf-blades resembling those of *Zea*

zejensis L. *-ensis*, denoting origin. From Zeja, Amur Province, Russian Far East
zelanica L. *-ica*, belonging to. From Zelania, now Sri Lanka
zelayens-e, -is L. *-ense*, denoting origin. From Zelaya, Mexico
zenkeri As for *Zenkeria*
Zenkeria In honor of Jonathan Karl Zenker (1799–1837) German botanist
zenkowskii In honor of Leo de Cienkowski (1822–1887) Polish-born Russian botanist
Zeocriton, Zeocritum Gk *zeia*, name of a one-seeded wheat. Similar to spelt wheat
zeocriton See *Zeocriton*. Formerly cultivated in England as Fulham Barley
Zeocritum See *Zeocriton*
zephyrina L. *-ina*, indicating resemblance. Leaf-blades broad as in *Zephyra*, a Chilean member of Tecophilaeaceae
Zerna The Classical Greek name for a plant, possibly a *Cyperus* species, eaten by horses
zerninensis L. *-ensis*, denoting origin. From the shores of Lake Zernin in north-western Germany
zerovii In honor of Dimitri Konstantinovich Zerov (1895–1971) and Konstantin Konstantinovich Zerov (1899–?) Russian botanists
Zeugites The Classical Greek name for an unidentified reed
zeugites Resembling *Zeugites*
zeyheri As for *zeyheriana*
zeyheriana L. *-ana*, indicating connection. In honor of Carl Ludwig Philipp Zeyher (1799–1858) German-born South African plant collector
zeylanic-a, -um, -us L. *-ica*, belonging to. From Zeylona, the name in Classical times for Ceylon, now Sri Lanka
zeylonica See *zeylanica*
zhongbaensis L. *-ensis*, denoting origin. From Chungba, Xizang Province, China
zhukovskyi In honor of Pyotr Mykhailovich Zhukovskii (1888–1975) Russian botanist
zifukuensis L. *-ensis*, denoting origin. From Jifuku-mura, Nagato Prefecture, Japan
ziganensis L. *-ensis*, denoting origin. From Mt Zigana, Turkey
zigzag French, of symbolic formation suggesting alteration of direction and first applied to fortifications. Form of culms zigzag at the base
Zingeria In honor of Nikola Wassiljevicz Zinger (1836–1907) Russian botanist
Zingeriopsis Gk *opsis*, resemblance. Similar to *Zingeria*
zingiberina L. *-ina*, indicating resemblance. Leaf-blades broad like those of *Zingiber*
zinserlingii In honor of Yuri Dmitrievich (George) Zinserling (1894–1938) Russian botanist
zittelii In honor of Karl Alfred von Zittel (1839–1904) German geologist
Zizania Gk *zizanion*, a weed growing amongst wheat, probably darnel (*Lolium temulentum*) but now applied to a quite different genus
zizanioides Gk *-oides*, resembling. Resembling *Zizania*
Zizaniopsis Gk *opsis*, appearance. Some species resemble those of *Zizania*
zizinii In honor of N. V. Zizin (fl. 1955)
zobelii In honor of August Zobel (fl. 1905) school-teacher at Dessau, Germany
Zoisia See *Zoysia*
zollingeri In honor of Heinrich Zollinger (1818–1859) Swiss botanist
zonal-e, -is Gk *zone*, girdle; L. *-ale*, pertaining to. The leaf-blades are marked with dark transverse bars
zongbaensis See *zhongbaensis*
Zonotriche Gk *zone*, girdle; *thrix*, hair. The lemmas bear transverse rows of hair tufts
zopilotense L. *-ense*, denoting origin. From Canada del Zopilote, Mexico
Zotovia In honor of Victor Dmitrievich Zotov (1906–1974) Russian born New Zealand botanist
Zoydia See *Zoysia*
Zoysia In honor of Karl Zois Edelstein (1756–1800) Austrian botanist
zukovskyi In honor of Waldemar Zukovsky (fl. 1934) Russian botanist

zuloagae In honor of Fernando Omar Zuloaga (1951–) Argentinian botanist

zuvantica L. *-ica*, belonging to. From Zuvant, Azerbaijan

zwierleinii In honor of Zwierlein (fl. 1884) who collected in Sicily

Zygochloa Gk *zygos*, pair; *chloa*, grass. The species is dioecious and so the male and female florets occur on different plants

zygomeris Gk *zygos*, yoke; *meros*, part. Origin uncertain, not given by the author

GPSR Compliance

The European Union's (EU) General Product Safety Regulation (GPSR) is a set of rules that requires consumer products to be safe and our obligations to ensure this.

If you have any concerns about our products, you can contact us on ProductSafety@springernature.com

In case Publisher is established outside the EU, the EU authorized representative is:

Springer Nature Customer Service Center GmbH
Europaplatz 3
69115 Heidelberg, Germany

Batch number: 10761599

Printed by Printforce, the Netherlands